QUANTITATIVE FINANCE

QUANTITATIVE FINANCE

Its Development, Mathematical Foundations, and Current Scope

T. W. Epps
University of Virginia

WILEY

A JOHN WILEY & SONS, INC., PUBLICATION

Published by John Wiley & Sons, Inc., Hoboken, New Jersey.
Published simultaneously in Canada.

For general information on our other products and services or for technical support, please contact our Customer Care Department within the United States at (800) 762-2974, outside the United States at (317) 572-3993 or fax (317) 572-4002.

Wiley also publishes its books in a variety of electronic formats. Some content that appears in print may not be available in electronic format. For information about Wiley products, visit our web site at www.wiley.com.

Library of Congress Cataloging-in-Publication Data:

Epps, T. W.
  Quantitative finance : its development, mathematical foundations, and current scope / T.W. Epps.
    p. cm.
  Includes bibliographical references and index.
  ISBN 978-0-470-43199-3 (cloth)
  1. Finance—Mathematical models. 2. Investments—Mathematical models. I. Title.
  HG106.E67 2009
    332.01'5195—dc22   2008041830

Printed in the United States of America

10 9 8 7 6 5 4 3 2 1

In loving memory of my mother and father
Jane Wakefield Epps, 1918-2008
Thomas L. Epps, 1920-1980

CONTENTS

Preface

This work gives an overview of core topics in the "investment" side of finance, stressing the quantitative aspects of the subject. The presentation is at a moderately sophisticated level that would be appropriate for masters or early doctoral students in economics, engineering, finance, and mathematics. It would also be suitable for advanced and well motivated undergraduates—*provided* they are adequately prepared in math, probability, and statistics. Prerequisites include courses in (1) multivariate calculus; (2) probability at the level of, say, Sheldon Ross' *Introduction to Probability Models*; and (3) statistics through regression analysis. Basic familiarity with matrix algebra is also assumed. Some prior exposure to key topics in real analysis would be extremely helpful, although they are presented here as well. The book is based on a series of lectures that I gave to fourth-year economics majors as the capstone course of a concentration in financial economics. Besides having the math preparation, they had already acquired a basic familiarity with financial markets and the securities that are traded there. The book is presented in three parts. Part I, "Perspective and Preparation,"begins with a characterization of assets as "bundles" of contingent claims and of markets as ways of "transporting" those claims from those who value them less to those who value them more. While this characterization will be unfamiliar to most readers, it has the virtue of stripping financial theory down to its essentials and showing that apparently disparate concepts really do fit together. The two remaining chapters in Part I summarize the tools of analysis and probability that will be used in the remainder of the book. I chose to put this material up front rather than in an appendix so that all readers would at least page through it to see what is there. This will bring the necessary concepts back into active memory for those who have already studied at this level. For others, the early perusal will show what tools are there and where to look for them when they are needed. Part II, "Portfolios and Prices,"presents researchers' evolving views on how individuals choose portfolios and how their collective choices determine the prices of primary assets in competitive markets. The treatment, while quantitative, follows roughly the historical development of the subject. Accordingly, the material becomes progressively more challenging as we range from the elementary dividend-discount models of the early years to modern theories based on rational expectations and dynamic optimization. Part III, "Paradigms for Pricing,"deals with relations among prices that rule out opportunities for riskless gains—that is, opportunities for *arbitrage*. After the first chapter on "static" models, the focus is entirely on the pricing of financial derivatives. Again tracking the historical development, we progress from the (now considered elementary) dynamic replication framework of Black–Scholes and Merton to the modern theory of martingale pricing based on changes of measure. Chapters 22 and 23 apply martingale pricing in advanced models of price dynamics and are the most mathematically demanding portion of the book. Each of Chapters 4-23 concludes with exercises of progressive difficulty that are designed both to consolidate and to extend what is covered in the text. Complete solutions to selected problems are collected in the appendix, and solutions to all the exercises are available to instructors who submit requests to the publisher on letterhead stationery. At the ends of Chapters 4, 5, 7, 10, 13, 18, and 23 are empirical projects that would be suitable for students with moderate computational skills and access to standard statistical software. Some components of these require programming in Matlab® or a more basic language. The necessary data for the projects can be obtained via FTP from ftp://ftp.wiley.com/public/sci_tech_med/quantitative_finance. Reviews of a preliminary manuscript and many valuable suggestions were provided by Lloyd Blenman, Jason Fink, Sadayuki Ono, and William Smith. Perhaps my gratitude is best indicated by the fact that so many of the suggestions have been implemented in the present work. As one of the reviewers pointed out, the phrase "its current scope" in the title is something of an exaggeration. Clearly, there is nothing here on the corporate side of finance, which lies almost wholly outside my area of expertise. There is also a significant omission from the investment side. While I have described briefly the classic Vasicek and Cox–Ingersoll–Ross models of the short rate of interest, I have omitted entirely the subject of derivatives on fixed-income products. Accordingly, there is nothing here on the modern Heath-Jarrow-Morton approach to modeling the evolution of the forward-rate structure nor on the LIBOR-market model that seeks to harmonize HJM with the elementary methods that traders use to price caps and floors. There is also nothing here on credit risk. While no one would deny the importance of fixed-income models in finance, perhaps some would agree with me that it is hard to do justice to a subject of such breadth and depth in a single survey course.

I found it reassuring that the reviewer who drew attention to the omission had the same view of things. Having thanked the reviewers, I cannot fail to thank my economist-wife, Mary Lee, for her unfailing encouragement of my efforts and her tolerance of my many selfish hours at the computer. A great debt is owed, as well, to the legions of my former students, many of whom have made substantial contributions to the evolution of quantitative finance.

THOMAS W. EPPS

Charlottesville, Viginia
September 2008

ACRONYMS AND ABBREVIATIONS

a.e.	almost everywhere
APT	arbitrage-pricing theory
AR	autoregressive
ARCH	AR conditional heteroskedasticity
a.s.	almost sure(ly)
BM	Brownian motion
CAPM	capital asset pricing model
CDF	cumulative distribution function
CEV	constant elasticity of variance
CF	characteristic function
CLT	central-limit theorem
CRRA	constant relative risk aversion
EMM	equivalent martingale measure
EU	expected utility
GARCH	generalized ARCH
GBM	geometric BM
i.i.d.	independent and identically distributed
JD	jump diffusion
MA	moving average
MGF	moment-generating function
PDE	partial differential equation
PDF	probability density function
PGF	probability-generating function
PMF	probability mass function
RV	random variable
SDE	stochastic differential equation
SLLN	strong law of large numbers
SML	security market line
SV	stochastic volatility
VG	variance-gamma

A-D	Anderson-Darling
B-G-W	Bienaymé-Galton-Watson
B-S	Black-Scholes
L-S	Lebesgue-Stieltjes
R-N	Radon-Nikodym
R-S	Riemann-Stieltjes

PART I

PERSPECTIVE AND PREPARATION

CHAPTER 1

INTRODUCTION AND OVERVIEW

Our subject in this book is financial assets—how people choose them, how their prices are determined, and how their prices relate to each other and behave over time. To begin, it helps to have a clear and simple conception of what assets *are,* why people desire to hold and trade them, and how the allocation of resources to financial firms and markets can be justified.

1.1 AN ELEMENTAL VIEW OF ASSETS AND MARKETS

Economists usually think of assets as "bundles" of time–state-contingent claims. A metaphor helps to see what they mean by this. When events unfold through time it is as if we are moving along a sequence of time-stamped roulette wheels. At time t nature spins the appropriate wheel and we watch to see in which slot the ball settles. That slot defines the "state of the world" at t. When the state is realized, so is the cash value of each asset at time t, which is thus *contingent* on the state and the time. From our point of view the state itself is just a description of current reality in sufficient detail that we know what each asset is worth at the time.

1.1.1   Assets as Bundles of Claims

The simplest conceivable financial asset is one that entitles the holder to receive one unit of cash when the wheel for some particular date selects one particular state—and nothing otherwise. There are no exact counterparts in the real financial world, but the closest would be an insurance contract that pays a fixed amount under a narrowly defined condition. The next simpler conception is a "safe" asset that yields a specified cash payment at t regardless of where the wheel stops. A government-backed, default-free "discount" bond that matures at t would be the nearest example, since the issuer of the bond promises to pay a fixed number of units of cash regardless of the conditions at t. A default-free bond that matures at t_n and makes periodic payments of interest ("coupons") at $t_1, t_2, ..., t_n$ is like a portfolio of these state-independent discount bonds. A forward contract to exchange a fixed number of units of cash at future date T for a fixed number of units of a commodity is a simple example of an asset whose value at T does depend on the state. One who is to pay the cash and receive the commodity has a state-independent liability (the cash that is owed) and a state-dependent receipt (the value of the commodity). At times before their maturities and expirations, values of marketable bonds and forward contracts alike are state dependent. Unlike either of these instruments, shares of stock have lifetimes without definite limit. A share of stock offers bundles of alternative state-contingent payments at alternative future dates out to some indefinite time at which a state is realized that corresponds to the company's liquidation. Dividends are other time-stamped, state-contingent claims that might be paid along the way. A European-style *call option* on the stock offers claims that are tied to states defined explicitly in terms of the stock's price at a fixed expiration date. One who holds such an option that expires at date T can pay a fixed sum (the "strike" price) and receive the stock on that date, but would choose to do so only in states in which the stock's price exceeds the required cash payment. If the option is so "exercised" at T, the former option holder acquires the same state-contingent rights as the stockholder from that time.

Each day vast numbers of these and other time–state-contingent claims are created and passed back and forth among individuals, financial firms, and nonfinancial businesses. Some of the trades take place in central marketplaces like the New York Stock Exchange (NYSE) and affiliated European exchanges in Euronext, the Chicago Mercantile Exchange (CME), the Chicago Board Options Exchange (CBOE), and exchanges in other financial centers from London to Beijing. Other trades occur over computer-linked networks of dealers and traders such as the NASDAQ market and Instinet. Still other trades are made through agreements and contracts negotiated directly between seller and buyer with no middleman involved. In modern times political boundaries scarcely impede the flow of these transactions, so we now think of there being a "world" financial market. Worldwide, the process involves a staggering expenditure of valuable human labor and physical resources. Yet, when the day's trading is done, not one single intrinsically valued physical commodity has been produced. Is this not remarkable?

1.1.2   Financial Markets as Transportation Agents

What justifies and explains this expenditure of resources? Since the transactions are made freely between consenting parties, each party to a trade must consider that what has been received compensates for what has been given up. Each party, if asked the reason for the trade it has made, would likely give an explanation that was highly circumstantial, depending on the transactor's particular situation and beliefs. Nevertheless, when we view assets through the economist's lens as time–state-contingent claims, a coherent image emerges: *Trading assets amounts to transferring resources across time and across states.* Thus, one who uses cash in a liquid, well managed money-market fund to buy a marketable, default-free, T-maturing discount bond gives up an indefinite time sequence of (almost) state-independent claims for a sequence of alternative state-*dependent* claims terminating with a state-independent receipt of principal value at T. The claims prior to T are those arising from potential sales of the bond before maturity, the amounts received depending on current conditions. Of course, the claims at all dates after any date $t \leq T$ are forfeited if the bond is sold at t. One who commits to hold the bond to T just makes a simple transfer across time. By contrast, one who trades the money-market shares for shares of common stock in XYZ company gives up the (almost) state-independent claims for an *indefinite* time sequence of claims that are highly state dependent. The exchange amounts to transferring or *transporting* claims from states that are unfavorable or merely neutral for XYZ to states that are favorable.

Once we recognize trading as such a transportation process, it is not so hard to understand why individuals would devote resources to the practice, any more than it is difficult to understand why we pay to have goods (and ourselves) moved from one place to another. We regard assets as being valued not for themselves but for the opportunities they afford for consumption of goods and services that do have intrinsic value. Just as goods and services are more valuable to us in one place than in another, opportunities for consumption are more valued at certain times and in certain states. Evidently, we are willing to pay enough to brokers, market makers, and financial firms to attract the resources they need to facilitate such trades. Indeed, we are sufficiently willing to allow governments at various levels to siphon off consumption opportunities that are generated by the transfers.

1.1.3   Why Is Transportation Desirable?

What is it that accounts for the subjective differences in value across times and states? Economists generally regard the different subjective valuations as arising from an inherent desire for "smoothness" in consumption, or, to turn it around, as a distaste for variation. We take out long-term loans to acquire durable goods that yield flows of benefits that last for many years; for example, we "issue" bonds in the form of mortgages to finance the purchases of our dwellings. This provides an alternative to postponing consumption at the desired level until enough is saved to finance it ourselves. We take the other side of the market, lending to banks through saving accounts and certificates of deposit (CDs) and buying bonds, to provide for consumption in later years when other resources may be lacking. While the consumption opportunities that both activities open up are to some extent state dependent, the usual primary motivation is to transfer over time.

Transfers across states are made for two classes of reasons. One may begin to think that certain states are more likely to occur than considered previously, or one may begin to regard consumption in those states as more valuable if they do occur. In both cases it becomes more desirable to place "bets" on the roulette wheel's stopping at those states. One places such bets by buying assets that offer higher cash values in the more valuable states—that is, by trading assets of lesser value in such states for those of higher value. Two individuals with different beliefs about the likelihood of future states, the value of consumption in those states, or a given asset's entitlements to consumption in those states will want to trade the asset. They will do so if the consumption opportunities extracted by the various middlemen and governments are not too large. The "speculator" in assets is one who trades primarily to expand consumption opportunities in certain states. The "hedger" is one who trades mainly to preserve existing state-dependent opportunities. Claims for payoffs in the various states are continually being passed back and forth between and within these two classes of transactors.

1.1.4 What Vehicles Are Available?

The financial instruments that exist for making time–state transfers are almost too numerous to name. Governments at all levels issue bonds to finance current expenditures for public goods or transfers among citizens that are thought to promote social welfare. Some of these are explicitly or implicitly backed by the taxation authority of the issuer; others are tied to revenues generated by government-sponsored or government-aided entities. Corporate debt of medium to long maturity at initiation is traded on exchanges, and short-term corporate "paper" is shuffled around in the institutional "money" market. Such debt instruments of all sorts—short or long, corporate or government—are referred to as "fixed income" securities. Equity shares in corporations consist of "common" and "preferred" stocks, the latter offering prior claim to assets on liquidation and to revenues that fund payments of dividends. Most corporate equity is tradable and traded in markets, but private placements are sometimes made directly to institutions. There are exchange-traded funds that hold portfolios of bonds and of equities of various special classes (e.g., by industry, firm size, and risk class). Shares of these are traded on exchanges just as are listed stocks. Mutual funds offer stakes in other such portfolios of equities and bonds. These are managed by financial firms, with whom individuals must deal directly to purchase and redeem shares. There are physical commodities such as gold—and nowadays even petroleum—that do have intrinsic consumption value but are nevertheless held mainly or in part to facilitate time–state transfers. However, since the production side figures heavily in determining value, we do not consider these to be *financial* assets.

We refer to stocks, bonds, and investment commodities as *primary* assets, because their values in various states are not linked *contractually* to values of other assets. The classes of assets that are so contractually linked are referred to as *derivatives*, as their values are derived from those of "underlying" primary financial assets or commodities. Thus, stock options—puts and calls—yield cash flows that are specified in terms of values of the underlying stocks during a stated period; values of commodity futures and forward contracts are specifically linked to prices of the underlying commodities; options and futures contracts on stock and bond indexes yield payoffs determined by the index levels, which in turn depend on prices of the component assets; values of interest-rate caps and swaps depend directly on the behavior of interest rates and ultimately on the values of debt instruments traded in fixed-income markets, lending terms set by financial firms, and actions of central banks. Terms of contracts for ordinary stock and index options and for commodity futures can be sufficiently standardized as to support the liquidity needed to trade in organized markets, such as the CBOE and CME. This permits one easily both to acquire the obligations and rights conferred by the instruments and to terminate them before the specified expiration dates. Thus, one buys an option either to get the right to exercise or to terminate the obligation arising from a previous net sale. Direct agreements between financial firms and individuals and nonfinancial businesses result in "structured" or "tailor-made" products that suit the individual circumstances. Typically, such specialized agreements must be maintained for the contractually specified terms or else terminated early by subsequent negotiation between the parties.

1.1.5   What Is There to Learn about Assets and Markets?

Viewing assets as time–state claims and markets as transporters of those claims does afford a useful conceptual perspective, but it does not give practical normative guidance to an investor, nor does it lead to specific predictions of how investors react to changing circumstances or of how their actions determine what we observe at market level. Without an objective way to define the various states of the world, their chances of occurring, and their implications for the values of specific assets, we can neither advise someone which assets to choose nor understand the choices they have made. We would like to do both these things. We would also like to have some understanding of how the collective actions of self-interested individuals and the functioning of markets wind up determining the prices of primary assets. We would like to know why there are, on average, systematic differences between the cash flows (per unit cost) that different classes of assets generate. We would like to know what drives the fluctuations in their prices over time. We would like to know whether there are in these fluctuations certain patterns that, if recognized, would enable one with some consistency to achieve higher cash flows; likewise, whether there is other publicly available information that would make this possible. Finally, we would like to see how prices of derivative assets prior to expiration relate to the prices of traded primary assets and current conditions generally. In the chapters that follow we will see some of the approaches that financial economists have taken over the years to address issues such as these. Although the time–state framework is not directly used, thinking in these terms can sometimes help us see the essential features of other approaches.

1.1.6   Why the Need for *Quantitative* Finance?

We want to know not just what typically happens but *why* things happen as they do, and attaining such understanding requires more than merely documenting empirical regularities. Although we concede up front that the full complexity of markets is beyond our comprehension, we still desire that the abstractions and simplifications on which we must rely yield useful predictions. We desire, *in addition*, that our abstract theories make us feel that they capture the essence of what is going on or else we would find them unsatisfying. The development of satisfying, predictive theories about quantifiable things requires building formal models, and the language in which we describe quantifiable things and express models is that of mathematics. Moreover, we need certain specific mathematical tools. If we regard the actors and transactors in financial markets as purposeful individuals, then we must think of them as having some way of ranking different outcomes and of striving to achieve the most preferred of these. Economists regard such endeavor as *optimizing* behavior and model it using the same tools of calculus and analysis that are used to find extrema of mathematical functions—that is, to find the peaks and troughs in the numerical landscape. But in financial markets nothing is certain; the financial landscape heaves and tosses through time in ways that we can by no means fully predict. Thus, the theories and predictions that we encounter in finance inevitably refer to *uncertain* quantities and future events. We must therefore supplement the tools of calculus and analysis with the developed mathematical framework for characterizing uncertainty— probability theory. Through the use of mathematical analysis and probability theory, *quantitative finance* enables us to attain more ambitious goals of understanding and predicting what goes on in financial markets.

1.2   WHERE WE GO FROM HERE

The two remaining chapters of this preliminary part of the book provide the necessary preparation in analysis and probability. For some, much of this will be a review of familiar concepts, and paging through it will refresh the memory. For others much of it will be new, and more thoughtful and deliberate reading will be required. However, no one who has not seen it before should expect to master the material on the first pass. The objective should be to get an overall sense of the concepts and remember where to look when they are needed. The treatment here is necessarily brief, so one will sometimes want to consult other sources.

Part II presents what most would consider the core of the "investment" side of financial theory. Starting with the basic arithmetic of bond prices and interest rates in Chapter 4, it progresses in the course of Chapters 5–10 through single-period portfolio theory and pricing models, theories and experimental evidence on choices under uncertainty, and empirical findings about marginal distributions of assets' returns and about how prices vary over time. Chapter 11, "Stochastic Calculus," is another "tools" chapter, placed here in proximity to the first exposure to models of prices that evolve in continuous time. Chapters 12 and 13 survey dynamic portfolio theory, which recognizes that people need to consider how current decisions affect constraints and opportunities for the future. Chapter 14 looks at the implications of optimal dynamic choices and optimal information processing for the dynamic behavior of prices. Part II concludes with some empirical evidence of how well information is actually processed and how prices actually do vary over time.

The pricing models of Part II are based on a concept of market equilibrium in which prices attain values that make everyone content with their current holdings. Part III introduces an alternative paradigm of pricing by "arbitrage." Within the time–state framework, pricing an asset by arbitrage amounts to assembling and valuing a collection of traded assets that offers (or can offer on subsequent reshuffling) the same time–state-dependent payoffs. If such a replicating package could be bought and sold for a price different from that of the reference asset, then buying the cheaper of the two and selling the other would yield an immediate, riskless profit. This is one type of arbitrage. Another would be a trade that confers for free some positive-valued time–state-contingent claim—that is, a free bet on some slot on the wheel. Presuming that markets of self-interested and reasonably perceptive individuals do not let such opportunities last for long, we infer that the prices of any asset and its replicating portfolio should quickly converge. Chapter 16 takes a first look at arbitrage pricing within a static setting where replication can be accomplished through buy-and-hold portfolios. Chapter 17 introduces the Black–Scholes–Merton theory for pricing by dynamic replication. We will see there that options and other derivatives can be replicated by acquiring and rebalancing portfolios over time, so long as prices of underlying assets are not too erratic. The implications of the model and its empirical relevance in today's markets are considered in the chapter that follows. When underlying prices are discontinuous or are buffeted by extraneous influences that cannot be hedged away, not even dynamic replication will be possible. Nevertheless, through "martingale pricing" it is possible at least to set prices for derivatives and structured products in a way that affords no opportunity for arbitrage. The required techniques are explained and applied in the book's concluding chapters.

CHAPTER 2

TOOLS FROM CALCULUS AND ANALYSIS

This chapter serves to highlight specific topics in analysis that are of particular importance for what we do in later chapters and to add depth in a few areas rarely covered in undergraduate courses. To begin, it is helpful to collect in one place for reference the mathematical notation that will be used throughout.

1. Symbols e and $\ln$ represent the exponential and natural log functions (log to base e), respectively.

2. $\{a, b, c\}$ represents a set containing the discrete elements a, b, c. $\{a\}$ denotes a *singleton* set with just one element. $\{a, b, c, ...\}$ represents an indeterminate (possibly infinite) number of discrete elements. $\{x_j\}_{j=1}^{n}$ and $\{x_j\}_{j=1}^{\infty}$ are alternative representations of sets with finitely (but arbitrarily) and infinitely many elements, respectively.

3. $\aleph = \{1, 2, ...\}$ represents the positive integers (the *natural* numbers), and $\aleph_0 = \{0, 1, 2, ...\}$ represents the nonnegative integers.

4. $\Re$ and $\Re^+$ represent the real numbers and the nonnegative real numbers, respectively.

5. $\Re_k$ represents k-dimensional Euclidean space.

6. Symbol $\times$ represents Cartesian product. Thus, $\Re_k = \Re \times \Re \times \cdots \times \Re$.

7. Symbols (a, b), $[a, b)$, $(a, b]$, $[a, b]$ indicate *intervals* of real number that are, respectively, open (not containing either endpoint), left-closed, right-closed, and closed. In such cases it is understood that $a < b$. Thus, $\Re = (-\infty, \infty)$ and $\Re^+ = [0, \infty)$, while $(a, b) \times (c, d)$, $[a, b] \times [c, d]$ represent open and closed rectangles (subsets of $\Re_2$).

8. Derivatives to third order are indicated with primes, as $f'(x) = df/dx$, $f''(x) = d^2 f(x)/dx^2$, $f'''(x) = d^3 f(x)/dx^3$. Higher-order derivatives are indicated with superscripts, as $f^{(k)}(x)$.

9. Partial derivatives are indicated with subscripts, as

$$f_x(x, y) = \frac{\partial f(x, y)}{\partial x}$$

$$f_{xx}(x, y) = \frac{\partial^2 f(x, y)}{\partial x^2}$$

$$f_{xy}(x, y) = \frac{\partial^2 f(x, y)}{\partial x \, \partial y},$$

and so forth. Subscripts on functions are sometimes used for other purposes also, but the context will make the meaning clear.

10. If Ω is a set, ω is a generic element of that set, $\mathcal{A}$ is a class of subsets of Ω, and A is a member of that class, then $\mathbf{1}_A(\omega)$ denotes a function from $\mathcal{A} \times \Omega$ to $\{0, 1\}$. This *indicator function* takes the value unity when $\omega \in A$ and the value zero otherwise. Thus, with $\Omega = \Re$ and $\mathcal{A}$ containing all the intervals on $\Re$, the function

$$f(x) = \begin{cases} 0, & x < 0 \\ x, & 0 \le x < 1 \\ x^2, & 1 \le x \end{cases}$$

can be represented far more compactly as $f(x) = x\mathbf{1}_{[0,1)}(x) + x^2\mathbf{1}_{[1,\infty)}(x)$.

11. Symbols representing matrices and vectors are in boldface. Vectors are understood to be in column form unless transposition is indicated by a prime, as $\mathbf{a}' = (a_1, a_2, ..., a_n)$. The symbol $\mathbf{1}$ (without subscript or argument) represents a column vector of units. Symbol $\mathbf{I}$ represents the identity matrix.

2.1   SOME BASICS FROM CALCULUS

1. **Order notation**. Often we need to approximate functions of certain variables when those variables take extreme values, and for this it often helps to determine limiting values as the variables approach zero or infinity. *Order notation* facilitates analysis of limits by distinguishing the relevant from the irrelevant

parts of expressions at extreme values of the arguments. For example, given a *sequence* of real numbers $\{A_n\}_{n=1}^{\infty}$—a mapping from positive integers $\aleph$ to real numbers $\Re$—we might want to approximate A_n for large values of n. Alternatively, we could have a real-valued function f, a mapping from $\Re$ to $\Re$ or from $\Re^+$ to $\Re$, say, and may want to approximate $f(x)$ for values of x near zero.

If $\lim_{n\to\infty} A_n = a$ in the first situation, we could write $A_n = a + o\left(n^0\right)$ as a first-order approximation. The expression $o\left(n^0\right) \equiv o(1)$ (lowercase letter *O*) represents the residual part of A_n that approaches zero as $n \to \infty$, without specifying its form exactly. In words, we would say that A_n equals a plus terms of *order less than one*. Thus, an expression that is $o(1)$ goes to zero *faster than* unity as $n \to \infty$, which just means that it goes to zero. For example, $A_n = a + b/n + c/n^2 = a + o\left(n^0\right)$. Alternatively, as a second-order approximation we might write $A_n = a + b/n + o\left(n^{-1}\right)$. This means that $A_n - a - b/n$ goes to zero faster than n^{-1}, or equivalently that $n\left(A_n - a - b/n\right) \to 0$. In general, an expression represented as $o(n^{-k})$ is such that $n^k o\left(n^{-k}\right) \to 0$ as $n \to \infty$. To specify precisely the slowest rate at which an expression vanishes, we would use the "big O" (uppercase) notation $O\left(n^{-k}\right)$. Writing $B_n = O\left(n^{-k}\right)$ signifies that $n^k B_n$ approaches a *nonzero, finite* constant as $n \to \infty$. Thus, $A_n = a + b/n + c/n^2 = a + O\left(n^{-1}\right)$ if $b \neq 0$, and $A_n = a + b/n + O\left(n^{-2}\right)$ if $c \neq 0$.

In the situation in which $f(x)$ is to be approximated for values of x near zero, we simply treat x like n^{-1}. Specifically, expressions $g(x) = o\left(x^k\right)$ and $h(x) = O\left(x^k\right)$ signify that $x^{-k} g(x) \to 0$ and that $x^k h(x) \to c$ (a finite, nonzero constant) as $x \to 0$. Thus, $f(x) = a + bx + cx^2$ can be written as $f(x) = a + o\left(x^0\right)$ as $x \to 0$ or (if $b \neq 0$) as $f(x) = a + O(x)$ or as $f(x) = a + bx + o(x)$ or (if $c \neq 0$) as $f(x) = a + bx + O\left(x^2\right)$.

2. **A continuity theorem.** *If* $\{x_n\}_{n=1}^{\infty}$ is a sequence of real numbers converging to the real number x, and if function f is defined at each x_n and continuous in a neighborhood of x, then $\lim_{n\to\infty} f(x_n) = f\left(\lim_{n\to\infty} x_n\right) = f(x)$. To prove it we must show that for each $\varepsilon > 0$ there is an $N(\varepsilon)$ such that $|f(x_n) - f(x)| \leq \varepsilon$ for all $n \geq N(\varepsilon)$. Since f is continuous near x, there is a $\delta > 0$ (depending, in general, on ε) such that $|f(y) - f(x)| \leq \varepsilon$ for all y such that $|y - x| \leq \delta(\varepsilon)$. But since $x_n \to x$, there is for any such $\delta > 0$ an integer $M(\delta)$ such that $|x_n - x| \leq \delta$ for all $n \geq M(\delta)$. Taking $N(\varepsilon) = M(\delta(\varepsilon))$ completes the proof.

3. **Taylor's theorem with remainder.** It is easy to see how to approximate a polynomial such as $a + bx + cx^2$ when $|x|$ is small, but how about functions such as e^x, $\sin x$, and $\ln(1 + x)$? For a function $f(x)$ that is continuous and has a continuous derivative in a closed interval $[-\varepsilon, \varepsilon]$ about the origin, we know from the mean-value theorem that there is *some* intermediate point x^* between x and 0 such that $f(x) - f(0) = f'(x^*) x$ when $|x| \leq \varepsilon$, or equivalently that $f(x) = f(0) + f'(x^*) x$. Moreover, the continuity of f'

ensures that $f'(x^*) \to f'(0)$ as $|x| \to 0$, and so for small enough x we can approximate $f(x)$ as the linear function $f(0) + f'(0)x$. Taylor's theorem extends the result to give a kth-order polynomial approximation of a function with at least k continuous derivatives. Specifically, if f has k continuous derivatives in a neighborhood of the origin, then for x in that neighborhood there is an x^* between 0 and x such that

$$f(x) = f(0) + f'(0)x + f''(0)\frac{x^2}{2} + \cdots + f^{(k-1)}(0)\frac{x^{k-1}}{(k-1)!} + f^{(k)}(x^*)\frac{x^k}{k!}.$$

Again, the continuity of $f^{(k)}$ implies $f^{(k)}(x^*) \to f^{(k)}(0)$ as $|x| \to 0$. Thus,

$$f^{(k)}(x^*)x^k = f^{(k)}(0)x^k + \left[f^{(k)}(x^*) - f^{(k)}(0)\right]x^k$$
$$= f^{(k)}(0)x^k + o\left(x^k\right)$$

and hence

$$f(x) = f(0) + f'(0)x + f''(0)\frac{x^2}{2} + \cdots + f^{(k)}(0)\frac{x^k}{k!} + o\left(x^k\right).$$

Of course, this implies also that $f(x) = f(0) + o(1)$, that $f(x) = f(0) + f'(0)x + o(x)$, that $f(x) = f(0) + f'(0)x + O\left(x^2\right)$ if $f''(0) \neq 0$, and so on.

More generally, if f is at least k times continuously differentiable in a neighborhood of a, then in that neighborhood

$$f(x) = f(a) + f'(a)(x-a) + \cdots + f^{(k)}(a)\frac{(x-a)^k}{k!} + o(x-a)^k.$$

As examples, for $k \in \{0, 1, 2, \ldots\}$ and $a = 0$ we have

$$e^x = 1 + x + \frac{x^2}{2} + \cdots + \frac{x^k}{k!} + o\left(x^k\right), x \in \Re$$

$$\sin x = x - \frac{x^3}{3!} + \frac{x^5}{5!} + \cdots + \frac{(-1)^k x^{2k+1}}{(2k+1)!} + o\left(x^{2k+1}\right), x \in \Re$$

$$\ln(1+x) = x - \frac{x^2}{2} + \frac{x^3}{3} - \frac{x^4}{4} + \cdots + (-1)^k \frac{x^{k+1}}{k+1} + o\left(x^{k+1}\right), x > -1.$$

Thinking of the polynomial part of each expression as the kth member of a sequence $\{P_k(x)\}_{k=0}^\infty$, the sequences for e^x and $\sin x$ converge to those functions as $k \to \infty$ for each real x; however, the sequence for $\ln(1+x)$ converges only for $x \in (-1, 1)$.

4. **A particular limiting form**. We often use the fact that $\left[1 + a/n + o\left(n^{-1}\right)\right]^n \to e^a$ as $n \to \infty$ for any $a \in \Re$, and the proof applies the three previous results. Function e^x is continuous for all $x \in \Re$, and $\ln x$ is continuous for $x \in (0, \infty)$, so for any positive sequence $\{b_n\}_{n=1}^\infty$ that attains a positive limit we have

$\lim_{n \to \infty} b_n = \lim_{n \to \infty} \exp(\ln b_n) = \exp(\lim_{n \to \infty} \ln b_n)$ by the continuity theorem. Thus, starting with n large enough that $1 + a/n + o(n^{-1}) > 0$,

$$
\begin{aligned}
\lim_{n \to \infty} \left[1 + \frac{a}{n} + o(n^{-1}) \right]^n &= \exp \left\{ \lim_{n \to \infty} n \ln \left[1 + \frac{a}{n} + o(n^{-1}) \right] \right\} \\
&= \exp \left\{ \lim_{n \to \infty} n \left[\frac{a}{n} + o(n^{-1}) \right] \right\} \text{ (Taylor)} \\
&= \exp \left\{ \lim_{n \to \infty} \left[a + o(n^0) \right] \right\} \\
&= e^a.
\end{aligned}
$$

2.2 ELEMENTS OF MEASURE THEORY

2.2.1 Sets and Collections of Sets

The concept of a set is understood intuitively, and one is assumed to be familiar with the basic set operations and relations: intersection ($\cap$), union ($\cup$), subset ($\subset$ and $\subseteq$), difference ($\backslash$). If all sets $A, B, \ldots$ being considered are subsets of some other set Ω, then we refer to Ω as the relevant *space*. With this as the frame of reference the *complement* of a set A, denoted A^c, is $\Omega \backslash A$. For any set A we have $A \backslash A = \emptyset$ (the empty set), and by convention we suppose that $\emptyset \subset A$ and $A \subset A$ for every A. Sets A and B are said to be *equal* if both $A \subset B$ and $B \subset A$ are true, and in that case $A \backslash B = \emptyset$ and $B \backslash A = \emptyset$. The natural numbers (positive integers) $\aleph$, the nonnegative integers $\aleph_0 = \aleph \cup \{0\}$, the real numbers $\Re$, the nonnegative reals $\Re^+$, and the rational numbers Q are sets to which we frequently refer. Each element $q \in Q$ is expressible as the ratio of integers $\pm m/n$, with $m \in \aleph_0$ and $n \in \aleph$. The *irrational* numbers constitute the set $\Re \backslash Q$.

A set A is *countable* if its elements can be put into one-to-one correspondence with elements of $\aleph$; that is, if they can be removed and counted out one by one in such a way that no element would remain if the process were to continue indefinitely. Of course, any finite set is countable, but so are some infinite sets, such as the set Q. Positive elements m/n of Q can all be found among elements of a matrix with columns corresponding to $m \in \aleph$ and rows corresponding to $n \in \aleph$, as

$$
\begin{array}{ccccc}
\frac{1}{1} & \frac{2}{1} & \frac{3}{1} & \frac{4}{1} & \cdots \\
\frac{1}{2} & \frac{2}{2} & \frac{3}{2} & \frac{4}{2} & \cdots \\
\frac{1}{3} & \frac{2}{3} & \frac{3}{3} & \frac{4}{3} & \cdots \\
\frac{1}{4} & \frac{2}{4} & \frac{3}{4} & \frac{4}{4} & \cdots \\
\vdots & \vdots & \vdots & \vdots & \ddots
\end{array}
$$

Elements of this matrix can ultimately be counted out by moving diagonally from each element of the first row to the corresponding element in the first column; as $\frac{1}{1}, \frac{2}{1}, \frac{1}{2}, \frac{3}{1}, \frac{2}{2}, \frac{1}{3}$, and so forth. The negatives $-\frac{m}{n}$ can be counted out in the same way from a like matrix, and the entire set Q can be counted by starting with 0 and then alternately picking from the matrices of positives and negatives. The sets $\Re$ and $\Re \backslash Q$ cannot be counted in this (or any other) way, nor can the reals or irrationals in any open interval (a, b).

Any set A with finitely many elements n contains 2^n subsets, if we include $\emptyset$ and A itself. (Any subset corresponds to a string of n elements consisting of "yes"s and "no"s, where the entry in the jth position of the string indicates whether the jth element of A is to be included. There are 2^n such strings.) An infinite set—even one that is countable—contains uncountably many subsets.

A collection $\mathcal{F}$ of subsets of a space Ω is called a *field* if (1) $A^c \in \mathcal{F}$ whenever $A \in \mathcal{F}$ and (2) $A \cup B \in \mathcal{F}$ whenever A and B belong to $\mathcal{F}$. (Note that A is a *subset* of the space Ω but is an *element* of the collection of subsets $\mathcal{F}$; i.e., $A \subset \Omega$ but $A \in \mathcal{F}$.) Together these imply that $\Omega = A \cup A^c \in \mathcal{F}$, that $\emptyset = \Omega^c \in \mathcal{F}$, and that $\cup_{j=1}^n A_j = A_1 \cup A_2 \cup \cdots \cup A_n \in \mathcal{F}$ for each finite n whenever all the $\{A_j\}$ are in $\mathcal{F}$. Thus, fields are "closed" under complementation and finite unions. de Morgan's laws—$\left(\cup_{j=1}^n A_j\right)^c = \cap_{j=1}^n A_j^c$ and $\left(\cap_{j=1}^n A_j\right)^c = \cup_{j=1}^n A_j^c$—imply that closure extends to finite intersections as well. For example, with $\Omega = \Re$ as our space, suppose that we start with the collection of all (the uncountably many) right-closed intervals $(a, b]$ with $a < b$, together with the infinite intervals of the form (b, ∞). Since $(a, b]^c = (-\infty, a] \cup (b, \infty)$ is not an interval at all, the collection is not a field; however, it becomes one if we add in all finite unions of its elements. The smallest field that contains the space Ω consists of just Ω and the empty set, $\emptyset$. This is called the *trivial* field and denoted $\mathcal{F}_0$.

A field $\mathcal{F}$ of subsets of a space Ω is called a *sigma field* (σ field) or *sigma algebra* if it is also closed under *countably many* set operations. Thus, if $\{A_j\}_{j=1}^\infty \in \mathcal{F}$ then $\cup_{j=1}^\infty A_j \in \mathcal{F}$ and $\cap_{j=1}^\infty A_j \in \mathcal{F}$. The field comprising finite unions of the intervals $(a, b]$ and (b, ∞) in $\Re$ is *not* a σ field, since it does not contain finite open intervals (a, b). It becomes a σ field if we add in the countable unions, since $(a, b) = \cup_{n=1}^\infty (a + b - n^{-1}]$.

Of course, the collection of *all* subsets of a space is automatically a σ field (if the space is finite, we can take countably many unions of any set with itself), but in complicated spaces like $\Re$ we typically want to deal with simpler collections. The reason is that some subsets of $\Re$ cannot be *measured* in the sense to be described. The *measurable* sets of $\Re$ include what are called the *Borel* subsets, the σ field $\mathcal{B}$ that contains countable unions and complements of the intervals $(a, b]$. Because $\mathcal{B}$ is the smallest such σ field that contains all such intervals, it is said to be the σ field *generated* by the intervals $(a, b]$. It is also the σ field generated by open intervals (a, b) and by closed intervals $[a, b]$ and by left-closed intervals $[a, b)$, since any set in $\mathcal{B}$ can be constructed through countably many operations on members of each of these classes; for example, $\cap_{n=1}^\infty (a, b + n^{-1}) = (a, b]$.

2.2.2 Set Functions and Measures

Given any set in some collection, we can set up a principle for assigning a unique real number to that set. Such a rule constitutes a function f, which in this case is called a *set function* since its argument is a set rather than just a number. Thus, if $\mathcal{C}$ is our collection of sets—a class of subsets of some Ω—then $f : \mathcal{C} \to \Re$ is a real-valued set function on the domain $\mathcal{C}$. For example, if Ω is itself countable and A is any subset, the function $\mathcal{N}(A) = $ "# of elements of A" is a kind of set function with which we are already perfectly familiar: the *counting* function. Of course, $\mathcal{N}(A) = +\infty$ if A is countably infinite, and so in our definition we would expand the range of $\mathcal{N}$ to include $+\infty$, as $\mathcal{N} : \mathcal{C} \to \aleph \cup \{+\infty\}$, where $\mathcal{C}$ is the σ field of *all* subsets of the countable space Ω. Another very important example is the set function $\lambda : \mathcal{B} \to \Re \cup \{+\infty\}$ that assigns to each member of the Borel sets its *length*. This function has the property that $\lambda((a, b]) = b - a$ for intervals. Set functions $\mathcal{N}$ and λ are of the special variety known as *measures*; namely, set functions that are (1) nonnegative and (2) countably additive. That a measure $\mu : \mathcal{C} \to \Re$ is nonnegative has the obvious meaning that $\mu(A) \geq 0$ for each $A \in \mathcal{C}$. That μ is countably additive means that if $\{A_j\}_{j=1}^{\infty}$ are disjoint sets in $\mathcal{C}$ (i.e., $A_j \cap A_k = \emptyset$ for $j \neq k$), then $\mu\left(\cup_{j=1}^{\infty} A_j\right) = \sum_{j=1}^{n} \mu(A_j)$; in words, the measure of any set is the sum of the measures of its (disjoint) parts. Likewise, if $A \subset B$, then $B = A \cup (B \backslash A)$ implies that $\mu(B \backslash A) = \mu(B) - \mu(A)$. We would insist that measures of things we deal with in ordinary experience (mass, distance, volume, time, etc.) have these properties.

Consider now a *monotone* sequence of measurable sets $\{A_n\}_{n=1}^{\infty}$; that is, either $A_n \subseteq A_{n+1}$ for each n (an *increasing* sequence, denoted $\{A_n\} \uparrow$) or else $A_{n+1} \subseteq A_n$ for each n (a *decreasing* sequence, denoted $\{A_n\} \downarrow$). In the former case we define $\lim_{n \to \infty} A_n$ as $\cup_{n=1}^{\infty} A_n$; in the latter, as $\cap_{n=1}^{\infty} A_n$. Then the countable additivity of measures also confers the following *monotone property*: (1) $\lim_{n \to \infty} \mu(A_n) = \mu(\lim_{n \to \infty} A_n)$ if $\{A_n\} \uparrow$ and (2) $\lim_{n \to \infty} \mu(A_n) = \mu(\lim_{n \to \infty} A_n)$ if $\{A_n\} \downarrow$ and $\mu(A_1) < \infty$. Thus, for *finite* measures it is always true that $\lim_{n \to \infty} \mu(A_n) = \mu(\lim_{n \to \infty} A_n)$. A measure on $(\Omega, \mathcal{F})$ that is not necessarily finite is nevertheless said to be σ-*finite* if there is a countable sequence $\{A_n\}$ such that $\cup_{n=1}^{\infty} A_n = \Omega$ with $\mu(A_n) < \infty$ for each n.

The length measure λ is known as *Lebesgue* measure. Its natural domain is the Borel sets, since any set in $\mathcal{B}$ can be constructed by countably many operations on intervals and can thus be measured by adding and subtracting the measures of the pieces. In fact, there are sets in $\Re$ that cannot be measured, given the properties that we want length measure to have. Of course, λ is not a finite measure since $\lambda(\Re) = +\infty$, but it is nevertheless σ-finite, since $\Re = \cup_{n=1}^{\infty} [-n, n]$ is the union of a countable sequence of sets of finite length. The countable additivity of λ has the following implication. We know from its basic property that $\lambda((0, 1]) = 1 - 0 = 1$, but suppose we want to measure the *open* interval $(0, 1)$. Since

$$(0, 1) = \left(0, \frac{1}{2}\right] \cup \left(\frac{1}{2}, \frac{3}{4}\right] \cup \left(\frac{3}{4}, \frac{7}{8}\right] \cup \cdots = \cup_{n=0}^{\infty} \left(\frac{2^n - 1}{2^n}, \frac{2^{n+1} - 1}{2^{n+1}}\right]$$

(a countable union of disjoint sets), countable additivity implies

$$\lambda((0, 1)) = \sum_{n=0}^{\infty} \lambda\left(\left(\frac{2^n - 1}{2^n}, \frac{2^{n+1} - 1}{2^{n+1}}\right]\right)$$

$$= \sum_{n=0}^{\infty} \frac{1}{2^{n+1}}$$

$$= 1.$$

Alternatively, we could reach the same conclusion by applying the monotonicity property of λ:

$$\lambda\left((0,1)\right) = \lambda\left(\lim_{n\to\infty}\left(0,1-\frac{1}{n}\right]\right) = \lim_{n\to\infty}\lambda\left(\left(0,1-\frac{1}{n}\right]\right) = \lim_{n\to\infty}\left(1-\frac{1}{n}\right) = 1.$$

Thus, $\lambda((0,1]) = \lambda((0,1))$, and so $\lambda(\{1\}) = \lambda((0,1]\setminus(0,1)) = \lambda((0,1]) - \lambda((0,1)) = 0$. In words, the length measure associated with the singleton set $\{1\}$ is zero. Likewise, $\lambda(\{x\}) = 0$ for each $x \in \Re$, and so $\lambda([a,b]) = \lambda([a,b)) = \lambda((a,b]) = \lambda((a,b))$ for all $a,b \in \Re$. Moreover, $\lambda(\cup_{n=1}^{\infty}\{x_n\}) = \sum_{n=1}^{\infty}\lambda(\{x_j\}) = 0$ for any countable number of points in $\Re$. In particular, $\lambda(\mathcal{Q}) = 0$, $\lambda(\Re\setminus\mathcal{Q}) = +\infty$, and $\lambda([0,1]\setminus\mathcal{Q}) = \lambda([0,1]) = 1$.

Let $\mathcal{S} \in \mathcal{B}$ be a measurable set in $\Re$ such that $\lambda(\mathcal{S}) = 0$. If C is some condition or statement that holds for each real x except for points in $\mathcal{S}$, then C is said to hold *almost everywhere with respect to Lebesgue measure*. This is often abbreviated as "C a.e. λ" or as "C a.e.," Lebesgue measure being understood if no other is specified. Thus, the statement "x is an irrational number" is true a.e.

Let F be a nondecreasing, finite-valued function on $\Re$; that is, F may be constant over any interval, but nowhere does it decrease. Such a function is said to be *monotone increasing*. Such a monotone function *may* be everywhere continuous. If it is not, then it has at most countably many discontinuities. Looking toward applications, let us consider only monotone functions F that are right continuous. Thus, $F(x+) \equiv \lim_{n\to\infty} F(x+1/n) = F(x)$ at each $x \in \Re$. (Of course, if F is continuous, then it is both left and right continuous.) Now construct a set function μ_F on Borel sets $\mathcal{B}$ such that $\mu_F((a,b]) = F(b) - F(a)$ for intervals $(a,b]$. If $F(x) = x$ for $x \in [0,\infty)$, then $\mu_F = \lambda$ and we are back to Lebesgue measure on the Borel sets of $\Re^+ = [0,\infty)$; otherwise, set function μ_F can be shown to be countably additive on $\mathcal{B}$ and therefore a new measure in its own right: $\mu_F : \mathcal{B} \to \Re^+$. By countable additivity, the measure associated with a single point b is $\mu_F(\{b\}) = \lim_{n\to\infty} \mu_F((b-1/n,b]) = \lim_{n\to\infty}[F(b) - F(b-1/n)] = F(b) - F(b-)$. If F is continuous at b (and therefore left continuous), then $\mu_F(\{b\}) = 0$; otherwise, $\mu_F(\{b\}) > 0$. Thus, $\mu_F((a,b]) > \mu_F((a,b))$ if F is discontinuous at b.

2.3   INTEGRATION

The ordinary Riemann integral $\int_a^b g(x) \cdot dx$ of a real-valued function g is familiar from a first course in calculus. Of course, we think of the definite integral as the net area between the curve g and that part of the x axis that extends from a to b. It is constructed as the limit of approximating sums of inscribed or of superscribed rectangles, the limits of these being the same when g is Riemann integrable, as it is when g is bounded on $[a,b]$ with at most finitely many discontinuities. The *Riemann–Stieltjes* integral is a straightforward extension that is extremely useful in applications. Some familiarity with a more radical departure from the Riemann construction, the *Lebesgue–Stieltjes* integral, will be helpful also.

2.3.1   Riemann–Stieltjes

Introduce a monotone increasing, right-continuous function F and its associated measure μ_F. As in the construction of the Riemann integral, let $a = x_0 < x_1 < \cdots < x_{n-1} < x_n = b$ be a partition of $[a, b]$, and form the sum

$$S_n\left((a, b]\right) = \sum_{j=1}^{n} g\left(x_j^*\right)\left[F\left(x_j\right) - F\left(x_{j-1}\right)\right],$$

where x_j^* is an arbitrary point on $[x_{j-1}, x_j]$ for each j. If $\lim_{n\to\infty} S_n\left((a, b]\right)$ exists and has the same value for each partition $(a, b]$ for which

$$\lim_{n\to\infty} \max_{j\in\{1,2,\dots,n\}} (x_j - x_{j-1}) = 0$$

and each choice of intermediate point x_j^*, then g is said to be Riemann–Stieltjes integrable on $(a, b]$ with respect to F, and we write $\int_{(a,b]} g\left(x\right)\cdot dF\left(x\right)$ or just $\int_{(a,b]} g\cdot dF$ for the common value of the limit. The integral does exist if g is bounded on $(a, b]$, has at most finitely many discontinuities there, and is continuous at every point on $(a, b]$ at which F is discontinuous. There are three prominent cases to consider.

1. When F is continuous and has derivative $F' > 0$ except possibly at finitely many points on $[a, b]$, then it is possible to set up partitions so that for each n and each $j \in \{1, 2, \dots, n\}$ there is an $x_j^{**} \in (x_{j-1}, x_j)$ such that $F\left(x_j\right) - F\left(x_{j-1}\right) = F'(x_j^{**})\left(x_j - x_{j-1}\right)$. In that case

$$\lim_{n\to\infty} S_n\left((a, b]\right) = \int_a^b g\left(x\right) F'\left(x\right)\cdot dx,$$

   an ordinary Riemann integral.

2. When F is a monotone-increasing, right-continuous *step function*—one for which $F'\left(x\right) = 0$ a.e. but $F\left(x_j\right) - F\left(x_j-\right) > 0$ at finitely many $\{x_j\}$ on $(a, b]$—then

$$\lim_{n\to\infty} S_n\left((a, b]\right) = \sum_{j:x_j\in(a,b]} g\left(x_j\right)\left[F\left(x_j\right) - F\left(x_j-\right)\right].$$

   Note that when F has discontinuities at either a or b the limits of $S_n\left((a, b]\right)$, $S_n\left([a, b)\right)$, $S_n\left((a, b)\right)$, and $S_n\left([a, b]\right)$ will not all be the same. Thus, the notation $\int_a^b g\cdot dF$ is ambiguous when F has discontinuities, and to be explicit we write $\int_{(a,b]} g\cdot dF$ or $\int_{[a,b)} g\cdot dF$ or $\int_{(a,b)} g\cdot dF$ or $\int_{[a,b]} g\cdot dF$ as the specific case requires.

3. When F can be decomposed as $F = G + H$, where (1) G is continuous and has derivative $G'\left(x\right) > 0$ at all but finitely many points on $(a, b]$ and (2) H is

a monotone-increasing, right-continuous step function with discontinuities at finitely many points $\{x_j\}$ on $(a, b]$, then

$$\lim_{n \to \infty} S_n\left((a, b]\right) = \int_a^b g\left(x\right) G'\left(x\right) \cdot dx + \sum_{j: x_j \in (a, b]} g\left(x_j\right) \left[H\left(x_j\right) - H\left(x_j-\right)\right].$$

Of course, this case subsumes cases 1 and 2.

In all three cases above we would write the Riemann–Stieltjes (R-S) integral as $\int_{(a, b]} g \cdot dF$. More generally, for any Borel set S for which a decomposition of case 3 applies, we would write $\int_S g \cdot dF$ as the integral over that set.

2.3.2   Lebesgue/Lebesgue–Stieltjes

The Riemann construction of the integral of a function g requires that its domain $\mathcal{D}$ be totally *ordered*, since we must work with partitions in which $x_{j-1} < x_j$. Of course, since the real numbers are totally ordered, this presents no inherent problem for functions $g : \Re \to \Re$. However, the Lebesgue construction allows us to consider integrals of functions $g : \Omega \to \Re$ whose domains are not so ordered, and it turns out that the construction also generalizes the class of functions on $\Re$ that can be considered integrable.

The expressions $\int g \cdot d\mu$, $\int g\left(\omega\right) \cdot d\mu\left(\omega\right)$, $\int g\left(\omega\right) \mu\left(d\omega\right)$ are alternative but equivalent ways to denote the Lebesgue–Stieltjes (L-S) integral of a function g with respect to some measure μ that maps a σ field $\mathcal{F}$ of subsets of Ω to $\Re^+$. The definition proceeds in stages, beginning with nonnegative *simple* functions $g : \Omega \to \Re^+$ such that $g\left(\omega\right) = \sum_{j=1}^n \gamma_j \mathbf{1}_{S_j}\left(\omega\right)$. Here (1) $\{\gamma_j\}_{j=1}^n$ are nonnegative constants; (2) $\mathbf{1}_{S_j}$ is an indicator function that takes the value unity when argument ω is in the set S_j and the value zero otherwise; and (3) $\{S_j\}_{j=1}^n$ are sets that partition Ω, meaning that $S_j \cap S_k = \emptyset$ for $j \neq k$ and that $\cup_{j=1}^n S_j = \Omega$. The L-S integral of such a simple function g with respect to measure μ is defined as $\int g \cdot d\mu = \sum_{j=1}^n \gamma_j \mu\left(S_j\right)$. Notice that the construction amounts to partitioning Ω according to the values that function g assigns to its elements, since g takes the constant value γ_j on each set S_j. In this way we get around the problem that Ω itself may not be ordered. The interpretation of the integral at this stage is entirely natural; we are merely weighting the value of the function on each set by the set's measure, and then adding up the results. This is just what is done in approximating the Riemann integral as a finite sum of heights of rectangles times their Lebesgue (length) measures.

The second stage of the construction of $\int g \cdot d\mu$ extends to nonnegative measurable functions g, using the fact that any such function can be represented as the limit of an increasing sequence of nonnegative simple functions; i.e., $g\left(\omega\right) = \lim_{n \to \infty} \sum_{j=1}^n \gamma_j \mathbf{1}_{S_j}\left(\omega\right)$. (Function g is measurable—more precisely, $\mathcal{F}$-measurable—if to any Borel set B there corresponds a set $S \in \mathcal{F}$ such that $\omega \in S$ whenever $g\left(\omega\right) \in B$. We express this in abbreviated form as $g^{-1}\left(B\right) \in \mathcal{F}$.) For such nonnegative measurable functions we set $\int g \cdot d\mu = \lim_{n \to \infty} \sum_{j=1}^n \gamma_j \mu\left(S_j\right)$. Such a limit always exists, although the limit may well be $+\infty$, and it can be shown that the limit is the same for *any* sequence $\{g_n\}$ of simple functions that converges up to g.

The third and last stage extends to measurable functions g that are not necessarily nonnegative. The functions

$$g^+(\omega) \equiv \max[g(\omega), 0]$$
$$g^-(\omega) \equiv \max[-g(\omega), 0]$$

are nonnegative, measurable, and therefore integrable as previously defined, and $g \equiv g^+ - g^-$. If both $\int g^+ \cdot d\mu$ and $\int g^- \cdot d\mu$ are finite, then we say that g is *integrable* (with respect to μ), and we put $\int g \cdot d\mu = \int g^+ \cdot d\mu - \int g^- \cdot d\mu$. Since $g^+ + g^- = |g|$, the integrability condition amounts to the requirement that $\int |g| \cdot d\mu < \infty$.[1] The integral $\int_S g \cdot d\mu$ over an arbitrary $\mathcal{F}$-set S is defined simply as $\int g \mathbf{1}_S \cdot d\mu$.

In the special case $\Omega = \Re, \mathcal{F} = \mathcal{B}, \mu = \lambda$ the integral $\int g \cdot d\lambda \equiv \int g(x) \cdot d\lambda(x) \equiv \int g(x)\lambda(dx)$ is referred to simply as the *Lebesgue integral* (as opposed to Lebesgue–Stieltjes). The intuitive (but really meaningless) expression $dx = \lambda((a, x + dx]) - \lambda((a, x])$ for $a < x$ supports the convention of writing the integral in the familiar Riemann form $\int g(x) \cdot dx$. In the same way, if μ_F is the measure based on a monotone increasing, right-continuous function F, then we can write $\int g \cdot d\mu_F$ in the R-S form $\int g(x) \cdot dF(x)$. In practice, it is best just to think of the latter expression as defined by the former.

2.3.3 Properties of the Integral

If g is a function on $\Re$ and if $|g|$ is Riemann integrable, then it is not difficult to see that the Lebesgue and Riemann constructions give the same answers for definite integrals. This means that the usual integration formulas and techniques (change of variables, integration by parts, etc.) apply to Lebesgue integrals of ordinary Riemann-integrable functions. Thus, if interested just in functions of real variables (but we are not, as the next chapter will indicate!), one may ask what things the Lebesgue construction buys us that the Riemann construction lacks. They are as follows:

1. **Greater generality.** There are Lebesgue-integrable functions that are not Riemann integrable. A standard example is the function $g(x) = \mathbf{1}_{\mathcal{Q}}(x)$ that takes the value unity on the rationals and zero elsewhere. On any partition $a = x_0 < x_1 < \cdots < x_n = b$, the sum $\sum_{j=1}^n g(x_j^*)(x_j - x_{j-1})$ equals $b - a$ for all n whenever each $x_j^* \in [x_{j-1}x_j]$ is chosen to be a rational number, but the sum is zero for all n when each x_j^* is chosen to be irrational. Thus, the limits differ, and g is not Riemann integrable. However, $g = \mathbf{1}_{\mathcal{Q}}$ is a simple function, and the Lebesgue integral is simply

$$\int_a^b \mathbf{1}_{\mathcal{Q}}(x) \cdot dx = 1 \cdot \lambda(\mathcal{Q} \cap [a, b]) + 0 \cdot \lambda([a, b] \setminus \mathcal{Q}) = 1 \cdot 0 + 0 \cdot (b - a) = 0.$$

---

[1] There are three ways in which g could fail to be integrable: (1) $\int g^+ \cdot d\mu = +\infty$ and $\int g^- \cdot d\mu < \infty$; (2) $\int g^+ \cdot d\mu < \infty$ and $\int g^- \cdot d\mu = +\infty$; and (3) $\int g^+ \cdot d\mu = +\infty$ and $\int g^- \cdot d\mu = +\infty$. In cases 1 and 2, respectively, we put $\int g \cdot d\mu = +\infty$ and $\int g \cdot d\mu = -\infty$. In case 3 we say that $\int g \cdot d\mu$ *does not exist*, since the operation $(+\infty) + (-\infty)$ is not defined.

2. **Lack of ambiguity.** In the Riemann sense the integral $\int_a^\infty g(x) \cdot dx$ is defined as $\lim_{n \to \infty} \int_a^{b_n} g(x) \cdot dx$ for any sequence $\{b_n\} \uparrow +\infty$, provided that the integral exists for each b_n and that the sequence has a limit. Likewise $\int_{-\infty}^b g(x) \cdot dx = \lim_{n \to \infty} \int_{a_n}^b g(x) \cdot dx$ for any sequence $\{a_n\} \downarrow -\infty$. The doubly infinite integral $\int_{-\infty}^\infty g(x) \cdot dx$ could be calculated as $\lim_{n \to \infty} \int_{a_n}^\infty g(x) \cdot dx$ for $\{a_n\} \downarrow -\infty$ or as $\lim_{n \to \infty} \int_{-\infty}^{b_n} g(x) \cdot dx$ for $\{b_n\} \uparrow +\infty$ or as $\lim_{n \to \infty} \int_{-b_n}^{b_n} g(x) \cdot dx$. However, these limits might well differ. For example, if $g(x) = x$ for all real x, then $\int_{a_n}^\infty g(x) \cdot dx = +\infty$ for each n and so $\lim_{n \to \infty} \int_{a_n}^\infty g(x) \cdot dx = +\infty$, whereas $\lim_{n \to \infty} \int_{-\infty}^{b_n} g(x) \cdot dx = \lim_{n \to \infty}(-\infty) = -\infty$ and $\lim_{n \to \infty} \int_{-b_n}^{b_n} g(x) \cdot dx = 0$. The requirement that $\int |g(x)| \cdot dx < \infty$ for Lebesgue integrability eliminates this ambiguity.

3. **Simpler conditions for finding limits of sequences of integrals.** Suppose that we have a sequence $\{g_n\}_{n=1}^\infty$ of functions $\Omega \to \Re$, each of which is integrable and such that $\lim_{n \to \infty} g_n(\omega) = g(\omega)$ for all $\omega \in \Omega$, and that we wish to evaluate $\lim_{n \to \infty} \int g_n \cdot d\mu$. For L-S integrals we have two crucial theorems for this purpose, which do not necessarily apply under the R-S construction:[2]

   - **Monotone convergence.** If $\{g_n\}$ is an increasing sequence of nonnegative, measurable functions converging up to g (i.e., if for each ω we have $0 \leq g_n(\omega) \leq g_{n+1}(\omega)$ for each n and $g_n(\omega) \to g(\omega)$), then $\lim_{n \to \infty} \int g_n \cdot d\mu = \int g \cdot d\mu$. Here the integral on the right may well be infinite, but the sequence of integrals on the left converges to whatever value it has.

   - **Dominated convergence.** If h is a nonnegative function such that $\int h \cdot d\mu < \infty$, and if $\{g_n\}$ is a sequence of functions (not necessarily nonnegative) such that (a) $|g_n(\omega)| \leq h(\omega)$ for each ω and each n and (b) $g_n(\omega) \to g(\omega)$, then g itself is integrable, and $\lim_{n \to \infty} \int g_n \cdot d\mu = \int g \cdot d\mu$.

Note that the conclusions of both theorems remain valid even if the conditions on the $\{g_n\}$ fail on sets of zero μ measure; that is, so long as the conditions hold a.e. μ. Clearly, what the theorems do is authorize an interchange of the operations of taking limits and integration. As an example, consider the sequence of integrals $\{\int g_n(x) \cdot dx\}_{n=1}^\infty$ with $g_n(x) = (1-1/n)\mathbf{1}_{[0,1]\setminus \mathcal{Q}}(x)$, where $\mathbf{1}_{[0,1]\setminus \mathcal{Q}}(x)$ takes the value unity on the irrationals of $[0,1]$. Functions $\{g_n\}$ are nonnegative and bounded by the integrable function $\mathbf{1}_{[0,1]}(x)$, and $g_n(x) \uparrow g(x) = \mathbf{1}_{[0,1]\setminus \mathcal{Q}}(x)$ for all x. Thus, both theorems apply under the Lebesgue construction, and so $\lim_{n \to \infty} \int g_n(x) \cdot dx = \int \mathbf{1}_{[0,1]\setminus \mathcal{Q}}(x) \cdot dx = 1$; however, neither g nor any of the $\{g_n\}$ is Riemann integrable.

---

[2]Proofs of these can be found in standard texts on real analysis and measure theory, such as Billingsley (1995).

The dominated convergence theorem leads to the following extremely useful result.

- **Differentiation of integrals with respect to parameters**. Let $g\left(\cdot,\cdot\right) : \Omega \times \Re \to \Re$ be a measurable function with derivative $g_t\left(w,t\right) \equiv \partial g\left(w,t\right)/\partial t$ for t in some neighborhood of a point t_0, and for such t suppose that $\left|g_t\left(w,t\right)\right| \le h\left(w\right)$ a.e. μ, where $\int h\cdot d\mu < \infty$. Then the derivative of the integral, when evaluated at t_0, is the integral of the derivative:

$$\frac{\partial}{\partial t}\int g\left(w,t\right)\mu\left(dw\right)\bigg|_{t_0} = \int g_t\left(w,t_0\right)\mu\left(dw\right).$$

Thus, to differentiate "under the integral sign" we need only verify that g itself is integrable, that it is differentiable a.e. μ, and that its derivative is bounded by an integrable function.

Here are some other essential properties of the integral that follow from the definition:

- **Linearity**: $\int\left(a_0 + a_1 g_1 + \cdots + a_k g_k\right)\cdot d\mu = a_0 + a_1\int g_1\cdot d\mu + \cdots + a_k\int g_k\cdot d\mu$.

- **Domination**: $h \ge g$ a.e. μ implies $\int h\cdot d\mu \ge \int g\cdot d\mu$. In particular, $\int\left|g\right|\cdot d\mu \ge \int g\cdot d\mu$.

- **Integrals over sets**: If $\left\{A_j\right\}_{j=1}^{\infty}$ are disjoint $\mathcal{F}$ sets, μ is a measure on $\left(\Omega,\mathcal{F}\right)$, and $g : \Omega \to \Re$ is integrable, then

$$\int_{\cup_{j=1}^{\infty}A_j} g\cdot d\mu = \sum_{j=1}^{\infty}\int_{A_j} g\cdot d\mu = \sum_{j=1}^{\infty}\int \mathbf{1}_{A_j}\left(w\right)g\left(w\right)\mu\left(dw\right). \quad (2.1)$$

2.4   CHANGES OF MEASURE

If μ is a measure on $\left(\Omega,\mathcal{F}\right)$, then $\mu\left(A\right)$ can be represented as the Lebesgue–Stieltjes integral $\int_A d\mu = \int \mathbf{1}_A\left(w\right)\mu\left(dw\right)$ for every $A \in \mathcal{F}$. Now introduce a nonnegative, $\mathcal{F}$-measurable $g : \Omega \to \Re^+$ and define $\nu\left(A\right) = \int_A g\cdot d\mu$. The mapping $v : \mathcal{F} \to \Re^+$ is a nonnegative, real-valued set function on $\left(\Omega,\mathcal{F}\right)$, and from the property of integrals stated in (2.1) it follows that ν is also countably additive; thus, ν is a measure. If g is such that $\nu\left(A\right) < \infty$ whenever $\mu\left(A\right) < \infty$, and if μ is σ-finite, then ν is σ-finite as well. (Of course, if g is integrable with respect to μ, then ν is a finite measure.) If both μ and ν are σ-finite, then ν has a certain continuity property. Specifically, if $\left\{A_j\right\}_{j=1}^{\infty}$ is a monotone decreasing sequence of $\mathcal{F}$ sets with $\lim_{n\to\infty} A_n = \cap_{n=1}^{\infty} A_n = \emptyset$, then $\lim_{n\to\infty} \mathbf{1}_{A_n}\left(w\right) = 0$ a.e. μ, so dominated convergence implies that

$$\lim_{n\to\infty}\nu\left(A_n\right) = \lim_{n\to\infty}\int \mathbf{1}_{A_n}\left(w\right)g\left(w\right)\mu\left(dw\right) = \int \lim_{n\to\infty}\mathbf{1}_{A_n}\left(w\right)\cdot g\left(w\right)\mu\left(dw\right) = 0.$$

Clearly, it also true that $\nu(A) = 0$ whenever $\mu(A) = 0$. In general, if $\mu(A) = 0$ implies $\nu(A) = 0$ for measures on a space $(\Omega, \mathcal{F})$, then ν is said to be *absolutely continuous* with respect to μ.

From what we have just seen, it is possible to create a new σ-finite measure ν from another such measure μ given the introduction of a suitable function g. The *Radon–Nikodym theorem* tells us that there is a converse; specifically, if ν and μ are σ-finite measures on $(\Omega, \mathcal{F})$, and if ν is absolutely continuous with respect to μ, then there exists an $\mathcal{F}$-measurable $g : \Omega \to \Re^+$ such that $\nu(A) = \int_A g \cdot d\mu$ for each $A \in \mathcal{F}$. In fact, there are infinitely many *versions* of such a function, each agreeing with the others except possibly on sets of μ measure zero. Function g is called the *Radon–Nikodym* (R-N) *derivative* of ν with respect to μ. It is often written stylistically as $d\nu/d\mu$ in view of the formal expression $\nu(A) = \int_A (d\nu/d\mu) \cdot d\mu$.

Two measures μ and ν each of which is absolutely continuous with respect to the other are said to be *equivalent*. Thus, μ and ν are equivalent if $\mu(A) = 0$ whenever $\nu(A) = 0$ and conversely; in other words, equivalent measures have the same "null" sets. In this case there exist both R-N derivatives $d\nu/d\mu$ and $d\mu/d\nu$.

To summarize, a finite or σ-finite measure μ can always be changed to another measure ν by integrating with respect to μ a suitable real-valued function; and if two measures μ and ν are equivalent, then each can be represented as an integral with respect to the other.

CHAPTER 3

PROBABILITY

3.1 PROBABILITY SPACES

There are two common intuitive conceptions of probability. In the *frequentist* view
the probability of an event represents the relative frequency with which it would occur in infinitely many repeated trials of a chance experiment. In the *subjectivist* view
probability is just a "degree of belief" that the event will occur. Either way, probabilities are just numbers assigned to events—sets of one or more distinct "outcomes."
The mathematical description of probability formalizes this functional representation
without taking a stand on the underlying intuition. Formally, given a set Ω of outcomes ω of which all events of interest are composed and a σ field $\mathcal{F}$ of subsets of Ω
(events), probability is a countably additive set function that maps $\mathcal{F}$ onto $[0, 1]$—in
other words, it is a measure whose range is limited to the unit interval in $\Re$. The con-
textual setting for probability in any given application is fully described by specifying
the outcome set or sample space Ω, the class of sets $\mathcal{F}$ that are considered measurable,
and the particular measure $\mathbb{P}$ that maps $\mathcal{F}$ onto $[0, 1]$. The pair $(\Omega, \mathcal{F})$ that sets the
stage for $\mathbb{P}$ is called a *measurable space*, and the triple $(\Omega, \mathcal{F}, \mathbb{P})$ that completes the
description is called a *probability space*.

Quantitative Finance. By T.W. Epps

For example, if Ω is any finite and nonempty set, if $\mathcal{F}$ is the collection of all its subsets, and if $\mathcal{N}$ is counting measure on $\mathcal{F}$, then $\mathbb{P}(A) = \mathcal{N}(A)/\mathcal{N}(\Omega)$ assigns to any A in $\mathcal{F}$ (i.e., any $A \subset \Omega$) a number equal to the proportion of all possible outcomes that lie in A. This is the typical model for ordinary games of chance in which outcomes are defined in such a way as to be considered "equally likely." Alternatively, if $\Omega = \Re$, $\mathcal{F} = \mathcal{B}$ (the Borel subsets), and λ represents Lebesgue measure, then $\mathbb{P}(A) = \lambda(A \cap [0,1))$ assigns to any $A \in \mathcal{B}$ the length measure of that portion of A that lies on $[0,1)$. This would be a reasonable model for the clockwise distance from "0" of a pointer that is spun and that comes to rest on a circular scale of unit circumference. (Imagine a loosely suspended hand of a clock that could be spun around with the finger.) Note that for any $x \in \Re$ we have $\mathbb{P}(\{x\}) = \lambda(\{x\} \cap [0,1))$. If the point x lies outside the unit interval, then $\{x\} \cap [0,1) = \emptyset$ and event $\{x\}$ is said to be *impossible*. However, if $x \in [0,1)$ then $\{x\}$ still has zero $\mathbb{P}$ measure even though it is *possible*. Events of zero $\mathbb{P}$ measure, whether possible or not, are said to be "null" or "$\mathbb{P}$-null." From the properties of Lebesgue measure it follows that $\mathbb{P}(\mathcal{Q}) = \lambda(\mathcal{Q} \cap [0,1)) = 0$, so that the entire set $\mathcal{Q}$ of rational numbers is itself $\mathbb{P}$-null. Accordingly, that a pointer on a circular unit scale will wind up pointing to an *irrational* number is an event of probability one, even though a rational outcome is possible. In general, if $A \in \mathcal{F}$ is such that $A^c \neq \emptyset$ but $\mathbb{P}(A) = 1$, we say that the occurrence of event A is *almost certain,* or that it will happen *almost surely,* under measure $\mathbb{P}$. This is the same as saying that A holds a.e. $\mathbb{P}$.

Probability measures, like finite measures generally, possess the *monotone* property that $\lim_{n \to \infty} \mathbb{P}(A_n) = \mathbb{P}(\lim_{n \to \infty} A_n)$ if either $\{A_n\} \uparrow$ or $\{A_n\} \downarrow$. More generally, if $\{A_n\}_{n=1}^{\infty}$ is any sequence of sets (not necessarily monotone), then we define $\lim_n \sup A_n$ as $\cap_{n=1}^{\infty} \cup_{m=n}^{\infty} A_m$ and $\lim_n \inf A_n$ as $\cup_{n=1}^{\infty} \cap_{m=n}^{\infty} A_m$. $\lim_n \sup A_n$ represents an event that will occur at *some* point after *any* stage of the sequence. Since we can never reach a stage beyond which such an event *cannot* occur, we interpret $\lim_n \sup A_n$ as the set of outcomes that will occur *infinitely often*. $\lim_n \inf A_n$ represents an event that will occur at *every* stage from *some* point forward. Since such an event occurs infinitely often, it follows that $\lim_n \inf A_n \subset \lim_n \sup A_n$. (Note that event $(\lim_n \sup A_n)^c$ will never occur from some point forward, so $(\lim_n \sup A_n)^c = \lim_n \inf A_n^c$.) Of course, that an event occurs at infinitely many stages does not imply that it will from some point occur at every stage, so it is not generally true that $\lim_n \sup A_n \subset \lim_n \inf A_n$. When this does happen to be the case, then $\lim_n \sup A_n = \lim_n \inf A_n$. This is always true for monotone sequences but is sometimes true otherwise. When it is, we define $\lim_{n \to \infty} A_n$ as the common value. We then have the general result that $\lim_{n \to \infty} \mathbb{P}(A_n) = \mathbb{P}(\lim_{n \to \infty} A_n)$ for any sequence of sets that approaches a limit.

Given a particular probability space $(\Omega, \mathcal{F}, \mathbb{P})$ and an $A \in \mathcal{F}$ with $\mathbb{P}(A) > 0$, the set function $\mathbb{P}_{\cdot | A}$ with value $\mathbb{P}_{\cdot | A}(B) \equiv \mathbb{P}(B \cap A)/\mathbb{P}(A)$ for any $B \in \mathcal{F}$ defines a new measure on $(\Omega, \mathcal{F})$—the *conditional* probability measure *given* event A. It is customary to write $\mathbb{P}_{\cdot | A}(B)$ more simply as $\mathbb{P}(B \mid A)$. If $\mathbb{P}(A) = 0$ (A being then $\mathbb{P}$-null) $\mathbb{P}_{\cdot | A}(B)$ is simply not defined. Now when we condition on A in this way we are effectively using partial information about the experiment, namely, that the outcome ω was in A. Typically, but not always, this information leads to a new probability assessment, in that $\mathbb{P}(B \mid A)$ typically differs from *unconditional* probability $\mathbb{P}(B)$ that would apply in the absence of the information. For example, if all 52 cards from a deck are equally likely to be drawn, then the probability of getting the ace of spades on one draw is 1 in 52, whereas the probability of getting that card *given* that the card was an ace is 1 in 4. On the other hand, the probability of getting a spade (without regard to denomination) is the same regardless of whether one has partial knowledge that the card was an ace. In this case events $A =$ "ace" and $B =$ "spade" are said to be *independent*, and $\mathbb{P}(B \mid A) = \mathbb{P}(B)$.

Regardless of whether A and B are independent, the definition $\mathbb{P}(B \mid A) = \mathbb{P}(B \cap A) / \mathbb{P}(A)$ implies the *multiplication rule* for determining the probability that A and B *both* occur: $\mathbb{P}(B \cap A) = \mathbb{P}(B \mid A)\mathbb{P}(A)$. To handle the case $\mathbb{P}(A) = 0$ in which $\mathbb{P}(B \mid A)$ is not defined, we can simply *define* independence through the condition $\mathbb{P}(B \cap A) = \mathbb{P}(B)\mathbb{P}(A)$. This makes sense in that $B \cap A \subset A$ and $\mathbb{P}(A) = 0$ imply $\mathbb{P}(B \cap A) = 0$. With this convention any event B is independent of any null event A, and B and A^c are also independent.[3]

Any set $A \in \mathcal{F}$ by itself can be said to *generate* a field or σ field. That is, we take the field to consist of A itself, its complement A^c, and the union and intersection of these, $\Omega = A \cup A^c$ and $\emptyset = A \cap A^c$. Representing the generated field as $\mathcal{F}_A$, we can think of $\mathcal{F}_A$ as the partial information about whether A occurred. For any $B \in \mathcal{F}$ we can then regard $\mathbb{P}(B \mid \mathcal{F}_A)$ as the representing the probability that would be assigned to B given such knowledge about A. The *smallest* field (or σ field) containing Ω is the trivial field $\mathcal{F}_0 = (\Omega, \emptyset)$, an information set that just tells us what outcomes are possible. Thus, $\mathbb{P}(B \mid \mathcal{F}_0)$ and "unconditional" probability $\mathbb{P}(B)$ are just the same.

Any countable collection of events $\{A_j\}_{j=1}^{\infty} \in \mathcal{F}$ that are *exclusive* (i.e., disjoint) and such that $\cup_{j=1}^{\infty} A_j = \Omega$ are said to *partition* Ω. Just as do A and A^c any such partition generates a σ field $\mathcal{F}_{\{A_j\}}$ consisting of the $\{A_j\}$ themselves, their unions, intersections, and complements. The symbol $\mathbb{P}(B \mid \mathcal{F}_{\{A_j\}})$ for any $B \in \mathcal{F}$ would represent the probability assigned to B once we know which specific event in $\{A_j\}_{j=1}^{\infty}$ occurred. Since $B = B \cap \Omega = B \cap (\cup_{j=1}^{\infty} A_j) = \cup_{j=1}^{\infty}(B \cap A_j)$, the events $\{A_j\}$ partition B as well as Ω. Because the components $\{B \cap A_j\}_{j=1}^{\infty}$ are exclusive, we have $\mathbb{P}(B) = \sum_{j=1}^{\infty} \mathbb{P}(B \cap A_j)$. Restricting consideration now to partitions for which $\mathbb{P}(A_j) > 0$ for all j, we have the *law of total probability*:

$$\mathbb{P}(B) = \sum_{j=1}^{\infty} \mathbb{P}(B \mid A_j)\mathbb{P}(A_j). \tag{3.1}$$

This is often helpful in simplifying probability calculations. For example, suppose that two cards are *dealt* (drawn without replacement) from a deck of 52. Letting A_j be the event of ace on draw $j \in \{1, 2\}$, we have $A_2 = (A_2 \cap A_1) \cup (A_2 \cap A_1^c)$ and,

---

[3] $\mathbb{P}(A) = 0$ implies $\mathbb{P}(A^c) = 1$ and so $\mathbb{P}(B)\mathbb{P}(A^c) = \mathbb{P}(B) = \mathbb{P}(B \cap (A \cup A^c)) = \mathbb{P}(B \cap A) + \mathbb{P}(B \cap A^c) = \mathbb{P}(B \cap A^c)$.

assuming that all $\binom{52}{2}$ pairs are equally likely to be drawn, we obtain

$$\begin{aligned} \mathbb{P}\left(A_2\right) &= \mathbb{P}\left(A_2 \mid A_1\right) \mathbb{P}\left(A_1\right) + \mathbb{P}\left(A_2 \mid A_1^c\right) \mathbb{P}\left(A_1^c\right) \\ &= \frac{3}{51} \cdot \frac{4}{52} + \frac{4}{51} \frac{48}{52} = \frac{4}{52}. \end{aligned}$$

3.2   RANDOM VARIABLES AND THEIR DISTRIBUTIONS

In some experiments, as in spinning the pointer on the $[0, 1)$ scale, the outcomes ω have a natural interpretation as real numbers. When they do not, we can nevertheless form a rule for assigning numbers to outcomes. If a single number is assigned to each outcome, such a rule amounts to a function $X : \Omega \to \Re$, with $X\left(\omega\right)$ representing the (one and only) number assigned to $\omega \in \Omega$. The function itself is called a *random variable* (RV) and, as above, is usually denoted by a capital letter, such as X, Y, Z, with generic values represented in lowercase, such as $X\left(\omega\right) = x, Y\left(\omega\right) = y$. Since values of X reside in $\Re$, and since all the Borel sets $\mathcal{B}$ of $\Re$ can be measured, it is natural to require of X that it link $\mathcal{B}$ and $\mathcal{F}$ in such a way that we can calculate probabilities of events like $\{\omega : X\left(\omega\right) \in B\}$ for any Borel set B. Such an event would be described in words as the event that X takes a value in B. It can be represented more compactly as $X^{-1}\left(B\right)$. If $(\Omega, \mathcal{F}, \mathbb{P})$ is our probability space, then we "connect" $\mathcal{B}$ and $\mathcal{F}$ by insisting of a RV X that $X^{-1}\left(B\right) \in \mathcal{F}$ for each $B \in \mathcal{B}$; in other words, we insist that X be $\mathcal{F}$-*measurable*. In this way $\mathbb{P}\left[X^{-1}\left(B\right)\right]$ gives the *probability that X takes a value in B*, which statement we express in convenient shorthand notation as $\Pr\left(X \in B\right)$. With this setup any RV X *induces* a new probability space $(\Re, \mathcal{B}, \mathbb{P}_X)$, where $\mathbb{P}_X\left(B\right) = \mathbb{P}\left[X^{-1}\left(B\right)\right]$ for each $B \in \mathcal{B}$.

It is helpful to develop a deeper understanding of the condition that X be $\mathcal{F}$-measurable. Recall that $\mathcal{F}$ is the σ field of subsets of Ω to which it is possible to assign numbers that have the properties that we want probabilities to have. If we know the outcome of the experiment ω, then we clearly know in which of the sets of $\mathcal{F}$ it resides. Conversely, if we know which sets of $\mathcal{F}$ contain ω, then we know the outcome itself, since "individual" outcomes are themselves $\mathcal{F}$ sets. In this sense, "knowing" $\mathcal{F}$ amounts to knowing what happened in the experiment, so we can think of $\mathcal{F}$ as full information about the experiment. Then to require that RV X be $\mathcal{F}$-measurable is to require that we know the value of X itself when we have information $\mathcal{F}$. In fact, though, a smaller information set than all of $\mathcal{F}$ may suffice to determine X. For example, suppose that our experiment is to flip a coin and that a description of the outcome ω includes both the face that turns up and the number of times that the coin flipped over before landing. In that case the value of a RV X that takes the value unity if heads and zero if tails is known as soon as we see the face on the settled coin, even though we did not see how many times it turned over. The *smallest* information set that determines the value of X—the smallest σ field with respect to which X is measurable—is called the σ field *generated* by X and is denoted $\sigma\left(X\right)$. Thus, if X is a RV on measurable space $(\Omega, \mathcal{F})$, then it is always true that $\sigma\left(X\right) \subset \mathcal{F}$, but it is *not* necessarily also true that $\mathcal{F} \subset \sigma\left(X\right)$ (i.e., that $\mathcal{F} = \sigma\left(X\right)$).

A set $\mathbb{X} \in \mathcal{B}$ such that $\mathbb{P}_X(\mathbb{X}) = 1$ is called a *support* of the RV X. Just as $\mathbb{P}$ shows how the unit "mass" of probability is distributed over subsets of Ω, $\mathbb{P}_X$ shows how the unit mass is distributed over any support $\mathbb{X}$, and hence over all of $\Re$. A more convenient such representation for applied work is the *cumulative distribution function* (CDF) of X. This is the function $F : \Re \to [0, 1]$ defined as $F(x) = \mathbb{P}_X((-\infty, x])$ for each $x \in \Re$. Thus, in our shorthand $F(x)$ represents $\Pr(X \in (-\infty, x])$ or simply $\Pr(X \leq x)$—the probability that X takes a value no greater than x. The term "cumulative" distribution function is appropriate because $F(x)$ represent the *accumulated* probability mass on that part of $\Re$ up to and including x. Clearly $F(x + \varepsilon) - F(x) = \mathbb{P}_X((-\infty, x + \varepsilon] \setminus (-\infty, x]) = \Pr(x < X \leq x + \varepsilon) \geq 0$ for all x and ε; thus, CDFs are *nondecreasing* functions. Note that

$$\Pr(X = x) = \mathbb{P}_X(\{x\}) = \mathbb{P}_X((-\infty, x] \setminus (-\infty, x))$$
$$= \mathbb{P}_X((-\infty, x]) - \mathbb{P}_X((-\infty, x))$$
$$= F(x) - F(x-),$$

where $F(x-) \equiv \lim_{n\to\infty} F(x - 1/n)$ is the *left-hand* limit. Since $\Pr(X = x) = \mathbb{P}_X(\{x\})$ may be positive, F may be discontinuous from the left. On the other hand

$$\lim_{n\to\infty} F(x + 1/n) - F(x) = \lim_{n\to\infty} \mathbb{P}_X\left(\left(-\infty, x + \frac{1}{n}\right] \setminus (-\infty, x]\right)$$
$$= \lim_{n\to\infty} \mathbb{P}_X\left(\left(x, x + \frac{1}{n}\right]\right)$$
$$= \mathbb{P}_X\left(\lim_{n\to\infty}\left(x, x + \frac{1}{n}\right]\right) \quad \text{(monotone property)}$$
$$= \mathbb{P}_X(\emptyset) = 0.$$

Thus, a CDF F is always continuous from the right. The fact that

$$\lim_{n\to\infty} F(-n) = \lim_{n\to\infty} \mathbb{P}_X((-\infty, -n])$$
$$= \mathbb{P}_X(\cap_{n=1}^{\infty}(-\infty, -n])$$
$$= \mathbb{P}_X(\emptyset)$$

implies that $\lim_{x\to-\infty} F(x) = 0$ for any CDF. Similarly, $\lim_{x\to+\infty} F(x) = 1$.

Random variables whose CDFs are continuous (i.e., from the left as well as the right) are called *continuous* RVs. For any such continuous RV X we have $\Pr(X = x) = \mathbb{P}_X(\{x\}) = 0$ for each real x, yet $\Pr(a < X < b) = \mathbb{P}_X((a, b)) > 0$ for *some* interval (a, b). Since $\Pr(X = a) = \Pr(X = b) = 0$, it follows that $\Pr(a < X < b) = \Pr(a \leq X \leq b)$ when X is continuous. RVs whose CDFs increase only at countably many points $\{x_j\}$ (at which the CDF is discontinuous) are called *discrete* RVs. In other words, a discrete RV X is supported on some *countable* set $\mathbb{X}$. RVs whose CDFs are increasing and continuous on some interval and yet have discontinuities at one or more points are of the *mixed* variety—neither purely continuous nor purely discrete.

Among the continuous RVs, the ones of usual interest in applied work are those whose CDFs have a stronger form of continuity, known as *absolute* continuity. These have (nonnegative) derivatives that are strictly positive on some set of unit $\mathbb{P}_X$ measure.[4] For an absolutely continuous RV there exists a function $f : \Re \rightarrow \Re^+$ called a *probability density function* (PDF) (more formally, a density with respect to Lebesgue measure), such that

$$\Pr(X \in B) = \mathbb{P}_X(B) = \int_B f(x) \cdot dx \tag{3.2}$$

for any Borel set B. For example, $\Pr(0 \leq X \leq 1) = \Pr(0 < X < 1) = \int_0^1 f(x) \cdot dx$ and $\int_{-\infty}^{\infty} f(x) \cdot dx = \Pr(X \in \Re) = 1$. In particular $F(x) = \Pr(X \leq x) = \int_{-\infty}^{x} f(t) \cdot dt$. From the discussion of change of measure, we recognize the density f as the Radon–Nikodym derivative of $\mathbb{P}_X$ with respect to Lebesgue measure: $f = d\mathbb{P}_X/d\lambda$.

At any point x at which CDF F is differentiable, we can represent $f(x)$ as $F'(x)$, and since F is nondecreasing, we must have $f(x) \geq 0$. At other points, and indeed at all points in any countable set, f can be defined arbitrarily, since doing so does not alter the value of $\int_B f(x) \cdot dx$. Thus, there are infinitely many *versions* of f that represent the distribution of X, but in applications we simply work with some particular one that is convenient. For example, if X represents the clockwise distance from 0 of a pointer that comes to rest after being spun on a circular scale of unit circumference, a reasonable model for the CDF is $F(x) = x\mathbf{1}_{[0,1]}(x) + \mathbf{1}_{(1,\infty)}(x)$. This has derivative $F'(x) = 0$ for $x < 0$ and $x > 1$ and $F'(x) = 1$ on the open interval $(0,1)$, but F' does not exist on $\{0,1\}$. Either $f(x) = \mathbf{1}_{(0,1)}(x)$ (unity on the open interval) or $f(x) = \mathbf{1}_{[0,1]}(x)$ serves perfectly well as a convenient version of f. While we *could* make some arbitrary change at one or more points, as $f(x) = \mathbf{1}_{[0,.5) \cup (.5,1]}(x) + 10^{23}\mathbf{1}_{\{.5\}}(x)$, this would be pointless and certainly not convenient.

For a discrete RV X there is another function f, called a *probability mass function* (PMF), such that

$$\Pr(X \in B) = \sum_{x \in \mathbb{X} \cap B} f(x) = \sum_{x \in \mathbb{X} \cap B} \mathbb{P}_X(\{x\}), \tag{3.3}$$

where the summation is over the countably many points of support $\mathbb{X}$ that are in B. Thus, $f(x) = \Pr(X = x) = \mathbb{P}_X(\{x\}) \in [0,1]$ for any discrete RV. To link the concepts of PMF and PDF, let $\mathcal{N}$ be counting measure on the collection of all subsets of $\mathbb{X}$. Thus, $\mathcal{N}$ assigns to any finite set $A \subset \mathbb{X}$ an integer equal to the number of elements of A, and it assigns to an infinite set the symbol $+\infty$. Next, let $\mathcal{N}_{\mathbb{X}}$ be the

---

[4]Random variable X is absolutely continuous if its induced measure $\mathbb{P}_X$ is absolutely continuous with respect to Lebesgue measure, λ. Although rarely of interest in applications, there do exist continuous *singular* RVs whose CDFs are continuous but have $F' = 0$ almost everywhere with respect to Lebesgue measure. In this case $\mathbb{P}_X$ is singular with respect to λ, meaning that there is a set $\mathcal{S}$ with $\mathbb{P}_X(\mathcal{S}) = 1$ and $\lambda(\mathcal{S}) = 0$.

measure on $(\Re, \mathcal{B})$ such that $\mathcal{N}_{\mathbb{X}}(B) = \mathcal{N}(\mathbb{X} \cap B)$. In words, $\mathcal{N}_{\mathbb{X}}(B)$ just counts the elements of $\mathbb{X}$ that are also in any *Borel* set B. Now it is clear that $\mathcal{N}_{\mathbb{X}}(B) = 0$ implies $\mathbb{P}_X(B) = 0$, so that $\mathbb{P}_X$ is absolutely continuous with respect to $\mathcal{N}_{\mathbb{X}}$. We can thus interpret PMF f as the Radon–Nikodym derivative of $\mathbb{P}_X$ with respect to counting measure:

$$\mathbb{P}_X(B) = \int_B \frac{d\mathbb{P}_X}{d\mathcal{N}_{\mathbb{X}}} \cdot d\mathcal{N}_{\mathbb{X}} = \sum_{x \in \mathbb{X} \cap B} \frac{d\mathbb{P}_X}{d\mathcal{N}_{\mathbb{X}}}(x) \, \mathcal{N}_{\mathbb{X}}(\{x\}) = \sum_{x \in \mathbb{X} \cap B} \frac{d\mathbb{P}_X}{d\mathcal{N}_{\mathbb{X}}}(x).$$

Any two different *versions* of $f = d\mathbb{P}_X / d\mathcal{N}_{\mathbb{X}}$ would agree on $\mathbb{X}$ but could take different values elsewhere on $\Re$. Of course, it is most convenient—and much safer—just to put $f(x) = 0$ for $x \notin \mathbb{X}$, and this is what we always do.

Whether X is absolutely continuous, discrete, or mixed we can represent probabilities in one consistent and convenient way as Stieltjes integrals:

$$\Pr(X \in B) = \int_B dF(x) = \int \mathbf{1}_B(x) \cdot dF(x).$$

When X is continuous with PDF f this is evaluated as in (3.2), and when X is discrete with PMF f it is evaluated as in (3.3). In theoretical work, when the specific nature of X does not matter, using the Stieltjes representation accommodates this generality.

If X is a RV with support $\mathbb{X}$ and if $g : \mathbb{X} \to \mathbb{Y}$ is a function such that $g^{-1}(B) \equiv \{x : g(x) \in B\} \in \mathcal{B}$ whenever $B \in \mathcal{B}$ (i.e., such that the inverse image of any Borel set is itself Borel), then $Y = g(X)$ is a RV on a new induced space $(\Re, \mathcal{B}, \mathbb{P}_Y)$, where $\mathbb{P}_Y(B) = \mathbb{P}_X \left[g^{-1}(B) \right]$. If X is a discrete RV with PMF f_X, then Y is necessarily discrete as well, and its PMF at any $y \in \Re$ is

$$f_Y(y) = \mathbb{P}_Y(\{y\}) = \mathbb{P}_X \left[g^{-1}(\{y\}) \right] = \sum_{x : g(x) = y} f_X(x).$$

If g is a one-to-one function, then there is an inverse *function* $g^{-1} : \mathbb{Y} \to \mathbb{X}$ in the usual sense,[5] and so we just have $f_Y(y) = f_X \left[g^{-1}(y) \right]$. If X is a continuous RV with PDF f_X and CDF F_X, then Y may be continuous, discrete, or mixed depending on the nature of g. (For example, $\mathbb{P}_Y(\{3\}) = 1$ if $g(x) = 3$ for all $x \in \mathbb{X}$.) In general, the value of the CDF of Y must be determined at each $y \in \mathbb{Y}$ as

$$F_Y(y) = \mathbb{P}_X \left[g^{-1}((-\infty, y]) \right] = \int_{x : g(x) \leq y} f_X(x) \cdot dx.$$

Things are simpler if g happens to be one-to-one, since then Y is also continuous. Its CDF is given by

$$F_Y(y) = \Pr(Y \leq y) = \Pr\left(X \leq g^{-1}(y)\right) = F_X \left[g^{-1}(y) \right]$$

if g is strictly increasing and by

$$F_Y(y) = \Pr\left(X \geq g^{-1}(y)\right) = 1 - F_X \left[g^{-1}(y) \right]$$

[5] In this case $g^{-1}(\{y\})$ is the singleton *set* $\{g^{-1}(y)\}$.

if g is strictly decreasing. In either of these one-to-one cases the PDF of Y can be found via the *change-of-variable* formula

$$f_Y(y) = f_X\left[g^{-1}(y)\right]\left|\frac{dg^{-1}(y)}{dy}\right|.$$

More than one RV can be defined on a given outcome set Ω in other ways than by simple transformations of an original variable. For example, if the experiment is to draw one individual at random from a group, then $(X(\omega), Y(\omega), Z(\omega)) \equiv (X, Y, Z)(\omega)$ could represent the height, weight, and age of outcome (person) ω. More generally, a function $\mathbf{X} = (X_1, X_2, ..., X_k)' : \Omega \to \Re_k$ would be a k-vector-valued RV if to each Borel set $B_k \in \mathcal{B}_k$ (the σ field generated by the "rectangular" sets $(a_1, b_1] \times \cdots \times (a_k, b_k]$) there is a corresponding $\mathbf{X}^{-1}(B_k) \in \mathcal{F}$. In this case, as in the univariate case $k = 1$, there is an induced measure $\mathbb{P}_{\mathbf{X}}$ on $(\Re_k, \mathcal{B}_k)$ such that $\mathbb{P}_{\mathbf{X}}(B_k) = \mathbb{P}\left[\mathbf{X}^{-1}(B_k)\right]$ for each $B_k \in \mathcal{B}_k$, and there is a *joint* CDF $F_{\mathbf{X}}(\mathbf{x}) = \Pr(X_1 \leq x_1, ..., X_k \leq x_k) = \mathbb{P}_{\mathbf{X}}((-\infty, x_1] \times \cdots \times (-\infty, x_k])$ for each $\mathbf{x} = (x_1, x_2, ..., x_k)' \in \Re_k$. When $\mathbf{X}$ is (absolutely) continuous, then $F_{\mathbf{X}}$ is continuous with derivative $\partial^k F_{\mathbf{X}}(\mathbf{x})/\partial x_1 \cdots \partial x_k$ a.e., and there is a joint PDF $f_{\mathbf{X}}$ such that

$$\Pr(\mathbf{X} \in B_k) = \int\int\cdots\int f_{\mathbf{X}}(x_1, ..., x_k) \cdot dx_1 \cdots dx_k,$$

where the integral is over the region $B_k \subset \Re_k$. An operable version of $f_{\mathbf{X}}$ can be defined so as to equal $\partial^k F_{\mathbf{X}}(\mathbf{x})/\partial x_1 \cdots \partial x_k$ almost everywhere. Now "integrating out" all but x_k, say, in the expression for $f_{\mathbf{X}}$ gives the *marginal* CDF of X_k as

$$\begin{aligned} F_k(x_k) &= \Pr(X_1 \in \Re, ..., X_{k-1} \in \Re, X_k \leq x_k) \\ &= F_{\mathbf{X}}(+\infty, ..., +\infty, x_k) \\ &= \int_{-\infty}^{x_k}\int_{-\infty}^{\infty}\cdots\int_{-\infty}^{\infty} f_{\mathbf{X}}(t_1, ..., t_{k-1}, t_k) \cdot dt_1 \cdots dt_{k-1}dt_k \end{aligned}$$

and the marginal PDF $f_k(x_k) = \int_{-\infty}^{\infty}\cdots\int_{-\infty}^{\infty} f_{\mathbf{X}}(t_1, t_2, ..., t_{k-1}, x_k)\cdot dt_1 \cdots dt_{k-1}$. Likewise, the *joint marginal* PDF of any subset of $X_1, ..., X_k$ is found by integrating out the remaining variables. Focusing on the bivariate case with $\mathbf{Z} = (X, Y)'$ for simplicity, we have $f_X(x) = \int_{-\infty}^{\infty} f_{\mathbf{Z}}(x, y) \cdot dy$ and $f_Y(y) = \int_{-\infty}^{\infty} f_{\mathbf{Z}}(x, y) \cdot dx$ as the two marginal PDFs. If (X, Y) are discrete RVs supported on a measurable, countable set $\mathbb{Z} \subseteq \mathbb{X} \times \mathbb{Y} \subset \Re_2$,[6] then the conditional CDF of Y given $X = x$ is $F_{Y|X}(x, y) = F_{\mathbf{Z}}(x, y)/f_X(x)$ for any $y \in \mathbb{Y}$ and any x such that $f_X(x) > 0$. Likewise, the conditional PMF is $f_{Y|X}(x, y) = f_{\mathbf{Z}}(x, y)/f_X(x)$ when $f_X(x) > 0$. Hence, we would calculate $\Pr(Y \in B \mid X = x)$ as $\sum_{y \in \mathbb{Y} \cap B} f_{Y|X}(x, y)$.

---

[6]If $\mathbb{X}$ and $\mathbb{Y}$ are respective *marginal* supports for X and Y, then $\mathbb{X} \times \mathbb{Y}$ will always support $\mathbf{Z} = (X, Y)'$; however, if X and Y are not *independent* (see below), there may be a proper subset $\mathbb{Z} \subset \mathbb{X} \times \mathbb{Y}$ with $\Pr(\mathbf{Z} \in \mathbb{Z}) = 1$.

When X is continuous, the event $X = x$ is $\mathbb{P}_X$-null for each $x \in \Re$, and so the conditional probability $\Pr(Y \in B \mid X = x)$ has no obvious meaning. However, *some* such null event regarding X necessarily occurs, and so it does make sense to think of conditioning on it. For example, if X is the clockwise distance from origin on the spin of the $[0, 1)$ pointer and Y is the time it takes for the pointer to come to rest, then event $X = .3$ is certainly *possible*—indeed, we might have been told that it *happened*. Thus, we might well want to know the probability that $Y \leq 0.5$ conditional on $X = .3$. The solution to the difficulty is to think of $\Pr(Y \in B \mid X = x)$ as the limit of $\Pr(Y \in B \mid X \in (x - \varepsilon, x + \varepsilon))$ as $\varepsilon \to 0$. This works and yields a meaningful probability because the ratio $\Pr[Y \in B \cap (X \in (x - \varepsilon, x + \varepsilon))] / \Pr[X \in (x - \varepsilon, x + \varepsilon)]$ is always between zero and unity when $f(x) > 0$. Generally, if continuous RVs X, Y have joint CDF $F_{\mathbf{Z}}$ and PDF $f_{\mathbf{Z}}$, and if marginal PDF f_X is positive at x, then the conditional CDF and PDF of Y given $X = x$ are defined as $F_{Y|X}(y \mid x) = F_{\mathbf{Z}}(x, y) / f_X(x)$ and $f_{Y|X}(y \mid x) = f_{\mathbf{Z}}(x, y) / f_X(x)$. We would then calculate $\Pr(Y \leq .5 \mid X = .3)$ as $F_{Y|X}(.5 \mid .3) = \int_{-\infty}^{.5} f_{Y|X}(y \mid .3) \cdot dy$.

Let $\mathbf{X}$ be a continuous k-vector-valued RV with joint PDF $f_{\mathbf{X}}$ and support $\mathbb{X} \subset \Re_k$. With $\mathbf{g} : \mathbb{X} \to \mathbb{Y} \subset \Re_k$ as a one-to-one function, the transformation $\mathbf{Y} = \mathbf{g}(\mathbf{X})$ yields a new k-vector-valued RV with joint PDF $f_{\mathbf{Y}}$ and support $\mathbb{Y}$. Letting $\mathbf{h} = \mathbf{g}^{-1}$ represent the unique inverse function such that $\mathbf{h}[\mathbf{g}(\mathbf{X})] = \mathbf{X}$, we have the multivariate change-of-variable formula

$$f_{\mathbf{Y}}(\mathbf{y}) = f_{\mathbf{X}}[\mathbf{h}(\mathbf{y})] \left\| \frac{d\mathbf{h}}{d\mathbf{y}} \right\|, \qquad (3.4)$$

where $\|d\mathbf{h}/d\mathbf{y}\|$ is the absolute value of the determinant ($|d\mathbf{h}/d\mathbf{y}|$) of (Jacobian) matrix

$$\frac{d\mathbf{h}}{d\mathbf{y}} = \begin{pmatrix} \frac{\partial h_1}{\partial y_1} & \frac{\partial h_1}{\partial y_2} & \cdots & \frac{\partial h_1}{\partial y_k} \\ \frac{\partial h_2}{\partial y_1} & \frac{\partial h_2}{\partial y_2} & \cdots & \frac{\partial h_2}{\partial y_k} \\ \cdots & \cdots & \ddots & \cdots \\ \frac{\partial h_k}{\partial y_1} & \frac{\partial h_k}{\partial y_2} & \cdots & \frac{\partial h_k}{\partial y_k} \end{pmatrix}.$$

3.3   INDEPENDENCE OF RANDOM VARIABLES

We have described the sense in which σ field $\mathcal{F}$ in a probability space $(\Omega, \mathcal{F}, \mathbb{P})$ represents an information set. We have also seen that the definition of a RV X requires it to be an $\mathcal{F}$-measurable function and explained that X might be measurable with respect to a smaller σ field $\sigma(X) \subset \mathcal{F}$. Now consider a second RV Y and its generated σ field $\sigma(Y) \subset \mathcal{F}$. We interpret $\sigma(X)$ as the minimal set of information about the experiment that pins down the value of X; likewise, for $\sigma(Y)$ and Y. Suppose that for *each* $A \in \sigma(X)$ and *each* $B \in \sigma(Y)$ we have $\mathbb{P}(A \cap B) = \mathbb{P}(A)\mathbb{P}(B)$, so that events A and B are independent. In this case we say that the σ fields $\sigma(X)$ and $\sigma(Y)$ themselves are independent. Intuitively, this means that knowledge of the value taken on by X tells us nothing about the value that might be assumed by Y, and vice versa. We then say that the RVs X and Y are independent.

Taking $A = X^{-1}\left((-\infty, x]\right) = \{\omega : X(\omega) \leq x\}$ and $B = Y^{-1}\left((-\infty, y]\right)$ for arbitrary real x, y shows that

$$F_{\mathbf{Z}}(x, y) = \mathbb{P}(A \cap B) = F_X(x) F_Y(y)$$

when $\mathbf{Z} = (X, Y)'$ and X and Y are independent. Conversely, since the rectangles $(-\infty, x] \times (-\infty, y]$ generate the Borel sets $\mathcal{B}_2$ of $\Re_2$, just as the intervals $(-\infty, x]$ generate the Borel sets $\mathcal{B}$ of $\Re$, the above equality *implies* that X and Y are independent if it holds for *all* x, y. If $\mathbf{Z} = (X, Y)'$ is continuous, then $f_{\mathbf{Z}}(x, y) = \partial^2 F_{\mathbf{Z}}(x, y) / \partial x \partial y$ almost everywhere, and so the condition $f_{\mathbf{Z}}(x, y) = f_X(x) f_Y(y)$ for (almost) all x, y is necessary and sufficient for continuous RVs to be independent.[7] The same relation holds for discrete RVs in terms of their PMFs. When X and Y are independent, the conditional and marginal distributions of each variable given the other are the same; thus, $f_{Y|X}(y \mid x) = f_Y(y)$ for any x such that $f_X(x) > 0$, and $f_{X|Y}(x \mid y) = f_X(x)$ for any y such that $f_Y(y) > 0$.

Note that independent RVs are always supported on a *product space*; that is, if X and Y are supported marginally on $\mathbb{X}$ and $\mathbb{Y}$, respectively, and if X and Y are independent, then $\mathbf{Z} = (X, Y)'$ is necessarily supported on $\mathbb{Z} = \mathbb{X} \times \mathbb{Y}$.

The concept of and conditions for independence of two RVs extend to countable collections $\{X_j\}_{j=1}^{\infty}$. These are independent if and only if $F_{\mathbf{X}_n}(\mathbf{x}_n) = F_{j_1}(x_{j_1}) \cdot \cdots \cdot F_{j_n}(x_{j_n})$ for each $\mathbf{x}_n = (x_{j_1}, ..., x_{j_n})' \in \Re_n$ and each *finite* collection $\mathbf{X}_n = (X_{j_1}, ..., X_{j_n})'$.

3.4   EXPECTATION

Suppose that we have a probability space $(\Omega, \mathcal{F}, \mathbb{P})$ for which sample space Ω is a finite set. If $X : \Omega \to \Re$ is a random variable on this space, then, since $\sum_{\omega \in \Omega} \mathbb{P}(\omega) = 1$, the summation $\sum_{\omega \in \Omega} X(\omega) \mathbb{P}(\omega)$ represents a *weighted average* of the values of X. Thinking of the weight $\mathbb{P}(\omega)$ as the relative frequency with which outcome ω would occur in infinitely many trials of the random experiment, it is clear that the sum represents an average of the values that X would assume in those trials. We call such an hypothetical "long-run average" value the *expected value* or *mathematical expectation* of X and use the shorthand notation EX to represent it. Now the finite collection $\{\omega\}$ of individual outcomes is itself a finite partition of Ω, and RV X is clearly constant on each set (each ω) in the partition, so X is a "simple" function as that was defined in the development of the Lebesgue integral. From that development we see at once that the expected value of X corresponds to the Lebesgue–Stieltjes integral of X:

$$EX = \sum_{\omega \in \Omega} X^+(\omega) \mathbb{P}(\omega) - \sum_{\omega \in \Omega} X^-(\omega) \mathbb{P}(\omega) = \int X \cdot d\mathbb{P},$$

---

[7] Strictly, the condition has to hold for some *version* of the PDFs.

where $X^+ \equiv \max(X, 0)$ and $X^- \equiv \max(-X, 0)$. Note that since elements ω of Ω need have no natural order, this is an instance in which the Lebesgue construction is essential.

Now we can generalize to an arbitrary space $(\Omega, \mathcal{F}, \mathbb{P})$ in which Ω is not necessarily finite. If $\int |X| \cdot d\mathbb{P} = \int (X^+ + X^-) \cdot d\mathbb{P} < \infty$, then $X : \Omega \to \Re$ is integrable with respect to $\mathbb{P}$, and in that case we define the expected value of X as

$$EX = \int X \cdot d\mathbb{P} \equiv \int X(\omega) \mathbb{P}(d\omega) = \int X^+(\omega) \mathbb{P}(d\omega) - \int X^-(\omega) \mathbb{P}(d\omega).$$
(3.5)

If precisely one of $\int X^+ \cdot d\mathbb{P}$ and $\int X^- \cdot d\mathbb{P}$ is finite, then we define EX as $+\infty$ or as $-\infty$ (the choice of sign being obvious). If both integrals are infinite, then EX is said not to "exist."

Although this is the most insightful definition of EX, there are more practical ways to calculate it in applications. Since X induces a new measure $\mathbb{P}_X$ on $(\Re, \mathcal{B})$, as $\mathbb{P}_X(B) = \mathbb{P}[X^{-1}(B)] = \mathbb{P}(\{\omega : X(\omega) \in B\})$, the Lebesgue–Stieltjes integral $\int x \cdot d\mathbb{P}_X(x) \equiv \int x \mathbb{P}_X(dx)$ is another way of averaging the values that X would take on in infinitely many trials. Here we weight the distinct *values* that X can assume, whereas (3.5) weights the values associated with distinct *outcomes*. (Recall that function X assigns just one value to each ω but that the correspondence $\omega \leftrightarrow X(\omega)$ it is not necessarily one to one.) Since the measure $\mathbb{P}_X((a, b])$ associated with interval $(a, b]$ is given by $F(b) - F(a)$, we can also write the expectation as

$$EX = \int x \cdot dF(x).$$

When X is continuous with PDF f this is calculated as $\int x f(x) \cdot dx$, and when X is discrete with countable support $\mathbb{X}$ and PMF f it is calculated as $\sum_{x \in \mathbb{X}} x f(x)$.

If we make the transformation $Y(\omega) = g[X(\omega)]$, then EY is represented most directly in terms of X as

$$EY = \int g[X(\omega)] \mathbb{P}(d\omega) = \int g(x) \mathbb{P}_X(dx) = \int g(x) \cdot dF(x),$$

assuming that the integral of at least one of g^+ and g^- (the positive and negative parts) is finite. Alternatively, we can work with the induced distribution $\mathbb{P}_Y$ of Y and its associated representations through F_Y or f_Y.

If $\mathbf{X}$ is a k-dimensional RV with joint CDF F, then the expected value of a function $g : \Re_k \to \Re$ is $Eg(\mathbf{X}) = \int_{\Re_k} g(\mathbf{x}) \cdot dF(\mathbf{x})$. In the continuous case this is $Eg(\mathbf{X}) = \int_{\Re_k} g(\mathbf{x}) f(\mathbf{x}) \cdot d\mathbf{x}$. In the Lebesgue construction $Eg(\mathbf{X})$ would be evaluated as $\int g^+ \cdot d\mathbb{P}_{\mathbf{X}} - \int g^- \cdot d\mathbb{P}_{\mathbf{X}}$, where $\mathbb{P}_{\mathbf{X}}$ is the measure on $(\Re_k, \mathcal{B}_k)$ induced by $\mathbf{X}$. Each of these integrals would be evaluated in turn as limits of integrals $\sum_{j=1}^n \gamma_j \mathbb{P}_{\mathbf{X}}(S_j)$ of increasing sequences of simple functions that take the constant values $\{\gamma_j\}$ on k-dimensional Borel sets $\{S_j\}$ having measures $\{\mathbb{P}_{\mathbf{X}}(S_j)\}$. In *practice*, to avoid this tedious process one would try to evaluate $\int_{\Re_k} g(\mathbf{x}) f(\mathbf{x}) \cdot d\mathbf{x}$ as a *repeated* integral, integrating out the variables $x_1, x_2, ..., x_k$ one at a time and treating the other remaining ones as constants. Fubini's theorem from analysis justifies this serial process—conducted in any order—if either (1) g is nonnegative or (2) the repeated integral of $|g|$ is finite when evaluated in any particular order.

The linearity property of the integral implies the linearity property of expectation; i.e., that $E\left[g_1\left(X_1\right)+\cdots+g_k\left(X_k\right)\right]=Eg_1\left(X_1\right)+\cdots+Eg_k\left(X_k\right)$. In particular, $E\left(a_0+a_1X_1+\cdots+a_kX_k\right)=a_0+a_1EX_1+\cdots+a_kEX_k$. In vector form, with $\mathbf{a}=\left(a_1,...,a_k\right)'$ this is $E\left(a_0+\mathbf{a}'\mathbf{X}\right)=a_0+\mathbf{a}'E\mathbf{X}$, where $E\mathbf{X}$ denotes the vector of expectations. On the other hand, it is not generally true that $Eg\left(\mathbf{X}\right)$ and $g\left(E\mathbf{X}\right)$ are equal *unless* g is of this affine form. In particular, $E\left[g_1\left(X_1\right)g_2\left(X_2\right)\cdot\cdots\cdot g_k\left(X_k\right)\right]$ does not in general equal the product of the expectations. However, when RVs $X_1,X_2,...,X_k$ are independent, it *is* true that $E\left[g_1\left(X_1\right)g_2\left(X_2\right)\cdot\cdots\cdot g_k\left(X_k\right)\right]=Eg_1\left(X_1\right)Eg_2\left(X_2\right)\cdot\cdots\cdot Eg_k\left(X_k\right)$. Under the conditions for the conclusion of Fubini's theorem, this is easily verified by writing the expectation on the left as a repeated integral with respect to $dF\left(\mathbf{x}\right)=dF_1\left(x_1\right)\cdot\cdots\cdot dF_k\left(x_k\right)$.

3.4.1 Moments

Returning to the univariate case $X:\Omega\to\Re$, the expectation EX^m for $m\in\aleph$ is called the mth *moment* of X about the origin and is represented conventionally as μ'_m. Following the Lebesgue construction, we say that μ'_m *exists* if at least one of $E\left(X^+\right)^m$ and $E\left(X^-\right)^m$ is finite. Taking $m=1$ gives the *mean* of X, which gets the special, simpler symbol μ (or μ_X if more than one RV is being considered). When it exists and is finite, the mean can be interpreted as the center of mass of the distribution of X. Visually, EX is a point on the real line at which a plot of the PMF or PDF would appear to balance if supported there. The expectation $E\left(X-\mu\right)^m$ is called the mth *central* moment. This is represented as μ_m, without the prime. Obviously, μ_m and μ'_m coincide when $\mu=0$. The *first* central moment, $\mu_1=E\left(X-\mu\right)=EX-\mu$, is identically zero and is never considered. When $m=2$, another special symbol and term are used: $\sigma^2=\mu_2=E\left(X-\mu\right)^2$ is the *variance* of X, while $\sigma\equiv+\sqrt{\sigma^2}$ is the *standard deviation*. We often just write EX for the mean and VX for the variance. Expanding $\left(X-\mu\right)^2$ and applying the linearity property of E show that $VX=EX^2-\mu^2=\mu'_2-\mu^2$. The quantities μ and σ and their interpretations as measures of central value and dispersion of the distribution of X will be familiar from an elementary course in probability or statistics. When $\mu\neq0$ the ratio σ/μ is often referred to as the *coefficient of variation* of RV X. The dimensionless quantities $\kappa_3\equiv\mu_3/\sigma^3$ and $\kappa_4=\mu_4/\sigma^4$, called the *coefficients of skewness* and *kurtosis*, are common (if highly imperfect) scalar indicators of the asymmetry and tail thickness of the distribution of X.

There is a useful alternative way to represent and calculate the mean of a nonnegative RV. Assuming that $\Pr\left(X\geq0\right)=1$ and that $EX<\infty$, we have

$$EX=\int_{[0,\infty)}x\cdot dF(x)=-\int_{[0,\infty)}x\cdot d\left[1-F(x)\right]$$
$$=x\left[1-F(x)\right]\big|_0^\infty+\int_0^\infty\left[1-F(x)\right]\cdot dx.$$

In the first term it is clear that, $\lim_{x\downarrow 0} x \left[1 - F(x)\right]| = 0$ since $F(0) \in [0, 1]$. For the upper limit of the first term, observe that

$$EX - \int_{[0,n)} x \cdot dF(x) = \int_{[n,\infty)} x \cdot dF(x)$$

$$\geq \int_{[n,\infty)} n \cdot dF(x) = n \left[1 - F(n)\right].$$

Letting $n \to \infty$, the left side goes to zero since X is integrable (EX is finite), and so the right side (being nonnegative) must also go to zero. Thus, for a nonnegative X with $EX < \infty$ we have

$$EX = \int_0^\infty \left[1 - F(x)\right] \cdot dx. \tag{3.6}$$

A function g is *convex* if for any x_1 and x_2 in its domain and all $\alpha \in [0, 1]$ we have $g\left[\alpha x_1 + (1 - \alpha) x_2\right] \geq \alpha g(x_1) + (1 - \alpha) g(x_2)$. If g is convex on a support $\mathbb{X}$ of RV X, then *Jensen's inequality* states that $Eg(X) \geq g(EX)$.[8] For example $EX^2 = VX + (EX)^2 \geq (EX)^2$. A function h is *concave* if $-h$ is convex, and then $Eh(X) \leq h(EX)$. For positive integers $m < n$ the function $g = x^{m/n} : \Re^+ \to \Re^+$ is concave, and so $E|X|^k \equiv E\left[\left(|X|^m\right)^{k/m}\right] \leq (E|X|^m)^{k/m}$. Thus, if RV X possesses a finite mth moment, then it has finite moments of all lower positive integer orders.

In the bivariate case $\mathbf{Z} = (X, Y)' : \Omega \to \Re_2$ with $EX = \mu_X$, $VX = \sigma_X^2$, $EY = \mu_Y$, and $VY = \sigma_Y^2$ the *product moment* $E(X - \mu_X)(Y - \mu_Y)$ is called the *covariance* of X and Y. This is represented by the symbols σ_{XY} and $C(X, Y)$. Since $E(X - \mu_X)(Y - \mu_Y) = E(X - \mu_X)Y - E(X - \mu_X)\mu_Y = E(X - \mu_X)Y$, we have the further relations $\sigma_{XY} = E(X - \mu_X)Y = EX(Y - \mu_Y) = EXY - \mu_X\mu_Y$. In particular, $\sigma_{XY} = EXY$ if either X or Y has zero mean. The dimensionless quantity $\rho_{XY} \equiv \sigma_{XY}/(\sigma_X\sigma_Y)$ is called the *coefficient of correlation*. The Schwarz (or *Cauchy–Schwarz*) inequality

$$\left[Eg(X)h(Y)\right]^2 \leq Eg(X)^2 Eh(Y)^2 \tag{3.7}$$

implies (1) that $|\sigma_{XY}| < \infty$ if both σ_X^2 and σ_Y^2 are finite and (2) that $|\rho_{XY}| \leq 1$. When $|\rho_{XY}| = 1$ (the case of *perfect* correlation) there are finite real numbers (constants) a and $b \neq 0$ such that $Y = a + bX$, where the signs of b and ρ_{XY} are the same. Thus, when $|\rho_{XY}| = 1$ the realizations $(X, Y)(\omega)$ lie almost surely on a straight line that intersects each axis at one and only one point. We can therefore think of the value of $|\rho_{XY}|$ as representing the extent to which realizations of X and Y are *linearly* related. When X, Y are independent we can simply *define* both σ_{XY} and ρ_{XY} to be zero (regardless of whether the expectations exist and are finite). However, that X and Y are *uncorrelated* ($\rho_{XY} = 0$) does *not* imply that they are independent, since there can be highly nonlinear relations.

---

[8]For a proof of Jensen's inequality see Fig. 7.1 and the final text paragraph in Section 7.3.3.

Covariance and correlation are inherently *bivariate* concepts, but (when they exist) they can be determined for all k^2 pairs of members of any collection $\{X_j\}_{j=1}^k$, as $\sigma_{jm} = E(X_j - \mu_j)(X_m - \mu_m)$ and $\rho_{jm} = \sigma_{jm}/(\sigma_j\sigma_m)$. It is then often useful to array the covariances as a $k \times k$ matrix Σ with σ_{jm} in row j and column m. Such a *covariance matrix* is symmetric since $\sigma_{jm} = \sigma_{mj}$, and its principal diagonal contains the variances, $\sigma_{jj} = VX_j = \sigma_j^2$. When dealing with a vector-valued RV $\mathbf{X} = (X_1, X_2, ..., X_k)'$ we sometimes write $V\mathbf{X}$ for the covariance matrix, just as we write $E\mathbf{X}$ for the vector of means. A *correlations matrix* $\mathbf{R}$ just replaces each σ_{jm} in Σ by correlation coefficient ρ_{jm}, with $\rho_{jj} = 1$.

For *affine* functions $g(\mathbf{X}) = a_0 + a_1X_1 + \cdots + a_kX_k = a_0 + \mathbf{a}'\mathbf{X}$ we have seen that $Eg(\mathbf{X}) = a_0 + \mathbf{a}'\mu = \sum_{j=1}^k a_j\mu_j$, when means $\mu = (\mu_1, \mu_2, ..., \mu_k)'$ exist. The variance of such an affine function can be represented in terms of the covariance matrix as $V(a_0 + \mathbf{a}'\mathbf{X}) = \mathbf{a}'\Sigma\mathbf{a} = \sum_{j=1}^k \sum_{m=1}^k a_j a_m \sigma_{jm}$. This does not depend on a_0 since the dispersion of a RV's distribution is not influenced by an additive constant.

3.4.2 Conditional Expectations and Moments

If $\mathbf{Z} = (X, Y)' : \Omega \to \Re_2$ is a bivariate RV with support $\mathbb{Z} \subseteq \mathbb{X} \times \mathbb{Y}$, joint CDF $F_{\mathbf{Z}}$, and conditional CDF $F_{Y|X}(y|x)$, then the conditional expectation of Y given $X = x$ is $E(Y \mid x) = \int y \cdot dF_{Y|X}(y|x)$. We refer to this as the *conditional mean* of Y given $X = x$. In the discrete and continuous cases, this becomes $\sum_{y:(x,y)\in\mathbb{Z}} y f_{Y|X}(y \mid x)$ and $\int_{-\infty}^{\infty} y f_{Y|X}(y \mid x) \cdot dy$, respectively. These indicate the average of the realizations of Y among those (infinitely many) replications of the chance experiment that produce the result $X = x$. It helps to think of $E(Y \mid x)$ as the value we "expect" Y to have, given the partial information about the experiment's outcome that $X(\omega) = x$. Regarded as a function $E(Y \mid x) : \mathbb{X} \to \Re$, the conditional expectation is known as the *regression function* of Y on X. The case of linear regression, $E(Y \mid x) = \alpha + \beta x$, and the statistical estimation of intercept α and slope β by ordinary (or general) least squares should be familiar from a first course in statistics.

If we put $\mathcal{R}(x) \equiv E(Y \mid x)$ to emphasize any such functional relation (linear or otherwise), we see that evaluating $\mathcal{R}$ as function of the *random variable* X just produces a new RV $\mathcal{R}(X)$, as with any transformation of one RV to another. Now we can think of $\mathcal{R}(X)$ as the (as yet unknown) value we *will* expect Y to have once we acquire the information about the value of X. Given our interpretation of $\sigma(X)$ (the σ field generated by X) as the minimal information that tells us the value of X, we could as well write $\mathcal{R}(X) = E(Y \mid X)$ as $E(Y \mid \sigma(X))$. So much for background; now for some formal manipulation. Treating just the continuous case for brevity and recalling that $f_{\mathbf{Z}}(x, y) = f_{Y|X}(y \mid x) f_X(x)$ and $f_Y(y) = \int_{-\infty}^{\infty} f_{\mathbf{Z}}(x, y) \cdot dx$, we have

$$E[E(Y \mid \sigma(X))] = \int_{-\infty}^{\infty} E(Y \mid x) f_X(x) \cdot dx$$
$$= \int_{-\infty}^{\infty} \left[\int_{-\infty}^{\infty} y f_{Y|X}(y \mid x) \cdot dy\right] f_X(x) \cdot dx$$
$$= \int_{-\infty}^{\infty} \int_{-\infty}^{\infty} y f_{\mathbf{Z}}(x, y) \cdot dy\, dx$$

$$= \int_{-\infty}^{\infty} y \left[\int_{-\infty}^{\infty} f_{\mathbf{Z}}(x, y) \cdot dx \right] \cdot dy$$

$$= \int_{-\infty}^{\infty} y f_Y \cdot dy = EY.$$

A like result holds in the discrete case and even when $\mathbf{Z}$ is neither purely discrete nor purely continuous. In words, this says that our expectation of the value we *will* expect Y to have once we learn the value of X is just our expectation of Y based on the trivial information $\mathcal{F}_0 = (\emptyset, \Omega)$ that we have at the outset. The result $E[E(Y \mid \sigma(X))] = EY$ is known to probabilists as the *tower property* of conditional expectation. In economics and finance it is usually referred to as the *principle of iterated expectation*. With $\mu_Y \equiv EY$ and $\mu_{Y \mid X} \equiv E[Y \mid \sigma(X)]$ as the unconditional and conditional means, we have the simple relation

$$\mu_Y = E\mu_{Y \mid X}. \tag{3.8}$$

The expectation $E\left[Y^2 \mid \sigma(X)\right] - \mu_{Y\mid X}^2 \equiv V[Y \mid \sigma(X)] \equiv \sigma_{Y\mid X}^2$ represents the *conditional variance* of Y given X. The relation between the unconditional and conditional variances, σ_Y^2 and $\sigma_{Y\mid X}^2$, is a bit more complicated than that between the unconditional and conditional means. To see it, first note that the conditional mean itself has variance

$$V\mu_{Y\mid X} = E\mu_{Y\mid X}^2 - \left(E\mu_{Y\mid X}\right)^2 = E\mu_{Y\mid X}^2 - \mu_Y^2.$$

Next, the expectation of the conditional variance is

$$E\sigma_{Y\mid X}^2 = E\left\{E\left[Y^2 \mid \sigma(X)\right]\right\} - E\mu_{Y\mid X}^2 = EY^2 - E\mu_{Y\mid X}^2.$$

Adding the two together, we have

$$E\sigma_{Y\mid X}^2 + V\mu_{Y\mid X} = EY^2 - \mu_Y^2 = \sigma_Y^2. \tag{3.9}$$

Thus, the variance of Y is the *expectation* of its conditional variance plus the *variance* of its conditional mean.

The concept of conditioning on a σ field is sufficiently abstract and yet sufficiently important in finance that a deeper understanding is required. Let us start by considering more carefully the meaning of $E(Y \mid \sigma(X))$ in the case that X is a discrete RV with a finite support, $\mathbb{X} = \{x_1, x_2, ..., x_n\}$. The subset of sample space Ω that corresponds to any event $X = x_j$ is $X^{-1}(\{x_j\}) = \{\omega : X(\omega) = x_j\}$. For brevity, let S_j represent this set. Since the $\{x_j\}$ are distinct and X is single-valued, the sets $\{S_j\}_{j=1}^n$ are disjoint. Moreover, their union is Ω itself, since X must take on one of the values $\{x_j\}_{j=1}^n$. Thus, $\{S_j\}_{j=1}^n$ form a partition, and $\sigma(X)$ is simply the σ field generated by this partition; that is, $\sigma(X)$ contains the $\{S_j\}_{j=1}^n$ themselves, their unions, intersections, and complements. Now, suppose that we have partial information about the experiment that $X = x_j$. In effect, this tells us that the outcome ω, although not fully known, lies in S_j. The conditional expectation $E(Y \mid x_j)$ can then be considered the *realization* of $E(Y \mid \sigma(X))$ for this value of ω; that is, $E(Y \mid \sigma(X))$ is a random variable such that $E(Y \mid \sigma(X))(\omega) = E(Y \mid x_j)$ when $\omega \in S_j$. The tower property of conditional expectation simply affirms that this random variable has the same expectation as does Y itself. (It should go without saying that $E(Y \mid \sigma(X))$ is itself $\sigma(X)$-measurable.) More generally, the RV $E(Y \mid \sigma(X))$ has the property that $Eg(X)Y = E\{E[g(X)Y \mid \sigma(X)]\} = E[g(X)E(Y \mid \sigma(X))]$ for any bounded $\sigma(X)$-measurable function g. Thus, in particular, the conditional expectation *given* $\sigma(X)$ of any integrable Y times a function $g(X)$ equals $g(X)$ times the conditional expectation of Y. All this holds as well in the general case that X is not restricted to be a discrete RV with finite support.

When X is discrete with PMF f_X and support $\mathbb{X}$, $E(Y \mid x)$ could be defined arbitrarily for any $x \notin \mathbb{X}$ without altering $E[E(Y \mid X)] = \sum_{x \in \mathbb{X}} E(Y \mid x) f_X(x)$; that is, $E(Y \mid x)$ could be defined arbitrarily on any set of zero $\mathbb{P}_X$ measure. Likewise, when X is continuous, $E(Y \mid x)$ could be assigned arbitrary values on any countable set of points $\{x_j\}$ without altering the value of $EY = \int E(Y \mid x) f_X(x) \cdot dx$. Thus, we always work just with some "version" of the conditional expectation. Fortunately, the analysis of a specific problem will ordinarily produce a particular version that applies for all $x \in \mathbb{X}$ and has some definite interpretation, and the fact that other versions exist can just be ignored. The issue corresponds precisely to the fact that there are infinitely many versions of a PDF f_X, each of which yields the same calculation of $\mathbb{P}_X(B) = \int_B f_X(x) \cdot dx$ and the same value of $Eg(X) = \int g(x) f_X(x) \cdot dx$ for measurable functions g.

3.4.3 Generating Functions

The convergence of Taylor expansion $e^{\zeta x} = 1 + \zeta x + \zeta^2 x^2/2 + \zeta^3 x^3/3! + \cdots$ for any real numbers ζ and x implies a relation between $Ee^{\zeta X}$ and the moments of X and suggests that $Ee^{\zeta X}$ "explodes" faster as $|\zeta|$ increases than does any polynomial in the moments. In fact, even if *all* positive integer moments of X exist and are finite (i.e., if $E|X|^m < \infty$ for all $m \in \aleph$), it may still turn out that $Ee^{\zeta X}$ is infinite for all $\zeta \neq 0$. However, when the distribution of X is such that $Ee^{\zeta X}$ is finite for all $\zeta \in [-\varepsilon, \varepsilon]$ and some $\varepsilon > 0$, then the function $\mathcal{M}(\zeta) \equiv Ee^{\zeta X} : [-\varepsilon, \varepsilon] \to \Re^+$ is called the *moment-generating function* (MGF) of X. The existence of $\mathcal{M}$ implies the finiteness of *all* positive integer moments and the expansion $\mathcal{M}(\zeta) = 1 + \zeta \mu + \zeta^2 \mu_2'/2 + \zeta^3 \mu_3'/3! + \cdots$ for ζ near 0. Thus, if the function $\mathcal{M}(\zeta)$ has been deduced by once working out the integral $\int e^{\zeta x} \cdot dF(x)$, it can be used to "generate" all the moments just by differentiation, as $\mathcal{M}'(0) \equiv d\mathcal{M}(\zeta)/d\zeta|_{\zeta=0} = \mu$, $\mathcal{M}''(0) = \mu_2'$, and so on, with $\mathcal{M}^{(m)}(0) = \mu_m'$ for $m = 0, 1, \ldots$.

Two other properties of $\mathcal{M}$ are of importance for our purposes. First, if RVs $\{X_j\}_{j=1}^k$ are independent with MGFs $\{\mathcal{M}_j(\zeta)\}_{j=1}^k$, then the MGF of the sum $S_k \equiv \sum_{j=1}^k X_j$ also exists and equals the *product* of the MGFs:

$$\mathcal{M}_{S_k}(\zeta) = Ee^{\zeta(X_1 + \cdots + X_k)} = E\left(e^{\zeta X_1} \cdots e^{\zeta X_k}\right)$$
$$= Ee^{\zeta X_1} \cdots Ee^{\zeta X_k} = \mathcal{M}_1(\zeta) \cdots \mathcal{M}_k(\zeta).$$

Second, if RV X has an MGF, then the function $\mathcal{M}$ uniquely identifies its distribution. In other words, RVs with distinct MGFs have distinct distributions, and RVs with distinct distributions have distinct MGFs (if they have MGFs at all). Thus, if we can identify the MGF of some RV S_k (as above) as corresponding to a certain family of distributions, then we can be assured that S_k has a distribution in that family.

If $\mathbf{X} = (X_1, X_2, ..., X_k)'$ is a k-vector-valued RV the function $\mathcal{M}_{\mathbf{X}}(\zeta) = E e^{\zeta'\mathbf{X}}$, if it exists for all real vectors ζ in some neighborhood of the origin (e.g., for all ζ such that $\zeta'\zeta \leq \varepsilon$), is the *joint* MGF of $\mathbf{X}$. Partial derivatives of $\mathcal{M}_{\mathbf{X}}(\zeta)$ generate the moments about the origin, as

$$\frac{\partial \mathcal{M}_{\mathbf{X}}(\mathbf{0})}{\partial \zeta_j} \equiv \left. \frac{\partial \mathcal{M}_{\mathbf{X}}(\zeta)}{\partial \zeta_j} \right|_{\zeta=0} = EX_j$$

$$\frac{\partial^2 \mathcal{M}_{\mathbf{X}}(\mathbf{0})}{\partial \zeta_j^2} = EX_j^2$$

$$\frac{\partial^2 \mathcal{M}_{\mathbf{X}}(\mathbf{0})}{\partial \zeta_j \, \partial \zeta_k} = EX_j X_k,$$

and so on. Evaluating $\mathcal{M}_{\mathbf{X}}$ at $\zeta = (0, ..., 0, \zeta, 0, ..., 0)'$ (with scalar ζ in the jth position) gives the *marginal* MGF of X_j: $\mathcal{M}_j(\zeta) = E e^{\zeta X_j}$.

A related concept, the *probability-generating function* (PGF), is particularly useful for RVs supported on nonnegative integers $\aleph_0 = \{0, 1, 2, ...\}$. If X is such a RV with PMF f, then the function $\Pi(\zeta) \equiv E\zeta^X = \sum_{x=0}^{\infty} \zeta^x f(x)$ necessarily exists and is finite for $|\zeta| \leq 1$. It is easy to see that $\Pi(0) = f(0)$, that $\Pi'(0) \equiv d\Pi(\zeta)/d\zeta|_{\zeta=0} = f(1)$, and, in general, that $\Pi^{(m)}(0)/m! = f(m)$. When $E\zeta^X$ is finite also for $\zeta \in [1, 1 + \varepsilon)$ and some $\varepsilon > 0$, then the identity $E\zeta^X = E e^{X \ln \zeta} = \mathcal{M}(\ln \zeta)$ shows that X also possesses an MGF—and therefore has finite positive integer moments. The derivatives of Π evaluated at $\zeta = 1$ then generate what are called the *descending factorial moments*: $\Pi'(1) = \mu, \Pi''(1) = \mu_{[2]} = EX(X-1) = \mu_2' - \mu, ..., \Pi^{(m)}(1) = \mu_{[m]} = EX(X-1) \cdots (X - m + 1)$.

Another related concept, the *characteristic function* of the RV X, is of great importance in modern financial applications. It is discussed at length in Chapter 22.

3.5 CHANGES OF PROBABILITY MEASURE

The general considerations regarding changes of measure that were discussed in Section 2.4 naturally apply to probability measures in particular, but there are a few special points of interest. Moreover, as we shall see in Chapter 19, changing from one probability measure to another is a crucial step in the arbitrage-free pricing of financial derivatives, so putting special emphasis on this special case is not unwarranted.

Given a probability space $(\Omega, \mathcal{F}, \mathbb{P})$, introduce a nonnegative, $\mathcal{F}$-measurable $G : \Omega \to \Re$ (a nonnegative RV) such that $EG = \int G(\omega) \mathbb{P}(d\omega) = 1$, and put $\hat{\mathbb{P}}(A) = E(G \mathbf{1}_A) = \int_A G(\omega) \mathbb{P}(d\omega)$ for any measurable set A. Then $\hat{\mathbb{P}}$ is a nonnegative, countably additive set function on $(\Omega, \mathcal{F})$ with the property that $\hat{\mathbb{P}}(\Omega) = EG = 1$, and so $\hat{\mathbb{P}}$ is a probability measure. Moreover, since $\hat{\mathbb{P}}(A) = 0$ whenever $\mathbb{P}(A) = 0$, $\hat{\mathbb{P}}$ is absolutely continuous with respect to $\mathbb{P}$. Now to go back the other way, let $\hat{\mathbb{P}}$ and $\mathbb{P}$ be probability measures on $(\Omega, \mathcal{F})$, and suppose that $\hat{\mathbb{P}}$ is absolutely continuous with respect to $\mathbb{P}$. Then, since probability measures are finite, the Radon–Nikodym (R-N) theorem applies and tells us that there is *some* measurable G—some RV—such that $\hat{\mathbb{P}}(A) = EG\mathbf{1}_A$ for any $A \in \mathcal{F}$. Thus, we can recognize G as a version of the R-N derivative $d\hat{\mathbb{P}}/d\mathbb{P}$. Moreover, if $\hat{\mathbb{P}}$ and $\mathbb{P}$ are *equivalent* measures—having the same null sets—then there is also a $G(\omega)^{-1} = 1/G(\omega) \equiv (d\mathbb{P}/d\hat{\mathbb{P}})(\omega)$ such that $\mathbb{P}(A) = \int_A G(\omega)^{-1}\hat{\mathbb{P}}(dw) = \hat{E}G^{-1}\mathbf{1}_A$ for any measurable A.

Changes of measure can be effected also by taking expectations with respect to measures on $(\Re, \mathcal{B})$ that are induced by random variables. Thus, with X as a RV on $(\Omega, \mathcal{F})$, $\mathbb{P}_X$ as the induced measure on $(\Re, \mathcal{B})$, and $g : \Re \to \Re$ a nonnegative, $\mathcal{B}$-measurable function with $Eg(X) = 1$, the mapping $\mathbb{P}_Y : \mathcal{B} \to [0, 1]$ with $\mathbb{P}_Y(B) = Eg(X)\mathbf{1}_B(X) = \int_B g(x)\mathbb{P}_X(dx) = \int_B g(x) \cdot dF(x)$ for $B \in \mathcal{B}$ is a new measure, absolutely continuous with respect to $\mathbb{P}_X$. A simple view of this comes by considering the case of a discrete RV X with PMF f_X such that $f_X(x) > 0$ for $x \in \mathbb{X} \subset \Re$ and $f_X(x) = 0$ for $x \in \mathbb{X}^c$. The corresponding measure is then $\mathbb{P}_X(B) = \sum_{x \in \mathbb{X} \cap B} f_X(x)$, $B \in \mathcal{F}$. Now, if f_Y is another PMF such that $f_Y(x) > 0$ for $x \in \mathbb{Y} \subset \mathbb{X}$ and $f_Y(x) = 0$ elsewhere, then $\mathbb{P}_X(B) = 0$ for some B implies that $\mathbb{P}_Y(B) = \sum_{x \in \mathbb{X} \cap B} f_Y(x) = 0$ also. Thus, $\mathbb{P}_Y$ is absolutely continuous with respect to $\mathbb{P}_X$. We can then relate the measures as $\mathbb{P}_Y(B) = \sum_{x \in \mathbb{X} \cap B} g(x) f_X(x)$ just by taking $g(x) \equiv f_Y(x)/f_X(x)$. If the measures are actually equivalent—that is, if $\mathbb{Y} = \mathbb{X}$—then we can take $g(x)^{-1} = 1/g(x) = f_X(x)/f_Y(x)$ and also write $\mathbb{P}_X(B) = \sum_{x \in \mathbb{X} \cap B} g(x)^{-1} f_Y(x)$. Likewise, if X and Y are continuous RVs with PDFs f_X and f_Y having *versions* that are positive on the same set and zero elsewhere, then putting $g(x) \equiv f_Y(x)/f_X(x)$ and $g(x)^{-1} \equiv 1/g(x)$ gives

$$\mathbb{P}_Y(B) = \int_B g(x) f_X(x) \cdot dx$$

$$\mathbb{P}_X(B) = \int_B g(x)^{-1} f_Y(x) \cdot dx.$$

3.6 CONVERGENCE CONCEPTS

We know that a sequence of *numbers* $\{a_n\}_{n=1}^{\infty}$ converges to some finite number a if all values of the sequence beyond some point N can be made arbitrarily close to a by taking N sufficiently large. Specifically, this means that for each $\varepsilon > 0$ there is an integer $N(\varepsilon)$ such that $|a_n - a| \le \varepsilon$ for all $n \ge N(\varepsilon)$. In such a case we write $\lim_{n \to \infty} a_n = a$ or just $a_n \to a$. Of course, some sequences, such as $\{a_n = (-1)^n\}$, do not have limits. In other cases, such as $\{a_n = n\}$, the sequence has no *finite* limit, in that for any arbitrary positive number b there is an $N(b)$ such that $|a_n| \ge b$ for $n \ge N(b)$. In such a case we would say that $\{a_n\}$ *diverges*.

Next, suppose that we have a sequence $\{g_n\}_{n=1}^{\infty}$ of *functions* $g_n : \Re \to \Re$. There are two prominent senses in which we say that the sequence converges to some function $g : \Re \to \Re$. The first is *pointwise* convergence, in which for *each* $x \in \Re$ the sequence $\{g_n(x)\}_{n=1}^{\infty}$ converges to $g(x)$. The precise meaning is the same as for a sequence of numbers $\{a_n\}_{n=1}^{\infty}$; specifically, there is an integer $N(\varepsilon, x)$ such that $|g_n(x) - g(x)| \leq \varepsilon$ for all $n \geq N(\varepsilon, x)$. Here, as the notation indicates, the threshold value of N may well depend on the point x. If, on the other hand, there is a single $N(\varepsilon)$ that controls the discrepancy $|g_n(x) - g(x)|$ for all x at once, then we say that $\{g_n\}_{n=1}^{\infty}$ converges to g *uniformly*. This is the second usual sense in which we speak of convergence of functions.

Now suppose we have not a sequence of numbers or of functions of a real variable but a sequence of RVs $\{X_n\}_{n=1}^{\infty}$; that is, a sequence of functions $X_n : \Omega \to \Re$ on some probability space $(\Omega, \mathcal{F}, \mathbb{P})$. There are several senses in which one can say that such a sequence converges. We shall need to understand the following ones:

1. **Convergence in distribution** (also called *weak* convergence and convergence *in law*). Each member X_n of our sequence $\{X_n\}_{n=1}^{\infty}$ has a distribution that can be represented by a CDF $F_n(x) = \Pr(X_n \leq x), x \in \Re$. Accordingly, there is a sequence $\{F_n\}_{n=1}^{\infty}$ of CDFs. If these are all evaluated at some fixed $x \in \Re$, then we have merely a sequence of numbers (probabilities), $\{F_n(x)\}_{n=1}^{\infty}$. Such a sequence can never diverge since its members all lie on $[0, 1]$, but it may well not have a limiting value. Suppose, though, that a limit does exist for *each* real x, and represent the limiting function as $F_\infty(x), x \in \Re$; that is, we suppose that $\{F_n\}_{n=1}^{\infty}$ converges *pointwise* to F_∞. We would be tempted to think of F_∞ as the "limiting distribution" of $\{X_n\}$, but there is a catch: F_∞ may not have the properties (right continuity, etc.) that qualify it to be a distribution function. For example, let $F_n(x) = nx\mathbf{1}_{(0,1/n]}(x) + \mathbf{1}_{[1/n,\infty)}(x)$ for each $n \in \aleph$. Then $F_n(0) = 0$ for all n and so $F_\infty(0) = \lim_{n\to\infty} 0 = 0$, whereas $F_\infty(x) = 1$ for any $x > 0$. Thus F_∞ is not right continuous. Still, since the probability mass associated with X_n all lies on $(0, 1/n]$ and since this interval shrinks ever closer to the origin, it seems obvious that in some sense the $\{X_n\}$ are just converging to zero. How can we reconcile fact with intuition? To do so, let F (without a subscript) be the CDF of some other RV X, and suppose that F_∞ and F coincide at each point at which F is continuous. We could then say that $\{X_n\}$ *converges in distribution* to F. (We sometimes say instead—somewhat sloppily—that $\{X_n\}_{n=1}^{\infty}$ converges in distribution to X.) This is written symbolically in various ways: $X_n \to^d F$, $X_n \Longrightarrow F$, and $X_n \rightsquigarrow F$. In the example, F would be the CDF $F(x) = \mathbf{1}_{[0,\infty)}(x)$, a step function that puts all the probability mass at 0, just as our intuition demands. With this as a definition, we have the nice implication that $Eg(X_n) = \int g(x) \cdot dF_n(x) \to \int g(x) \cdot dF(x) = Eg(X)$ for all bounded, (uniformly) continuous g whenever $\{X_n\}_{n=1}^{\infty}$ converges in distribution to F.

2. **Convergence in probability**. To generalize slightly the example above, suppose $F(x) = \mathbf{1}_{[c,\infty)}(x)$ is the limiting distribution of $\{X_n\}_{n=1}^{\infty}$, where c is some arbitrary constant. It is then clear that the probability mass associated with the $\{X_n\}_{n=1}^{\infty}$ is piling up around c, so that for any $\delta > 0$ the sequence of probabilities $\{\Pr(|X_n - c| \geq \delta)\}_{n=1}^{\infty}$ is approaching zero. That is, given any

$\delta > 0$ and any $\varepsilon > 0$ there is an $N(\delta, \varepsilon)$ such that $\Pr(|X_n - c| \geq \delta) \leq \varepsilon$ for all $n \geq N(\delta, \varepsilon)$. When $\Pr(|X_n - c| \geq \delta) \to 0$ for some constant c, then we say either that $\{X_n\}_{n=1}^{\infty}$ *converges in probability* to c or that c is the *probability limit* of $\{X_n\}_{n=1}^{\infty}$. We express these equivalent statements symbolically as $X_n \to^{\mathrm{P}} c$ and as $P\lim_{n\to\infty} X_n = c$. Likewise, if X is some RV, and if $\Pr(|X_n - X| \geq \delta) \to 0$ for each $\delta > 0$, then we say that $X_n \to^{\mathrm{P}} X$. In effect, this means that probability mass is piling up on a set of outcomes ω for which $X_n(\omega)$ and $X(\omega)$ are "close." As one would suppose, if $X_n \to^{\mathrm{P}} X$ and F is the distribution of X, then $\{X_n\}_{n=1}^{\infty}$ converges in distribution to F (or to X). However, that the $\{X_n\}$ converge in distribution to X just means that the distributions are becoming close, not that the realizations themselves are so. For example, the numbers that turn up on a roll of two dice of the same physical composition have the same distribution, but the numbers themselves are not necessarily close.

3. **Almost-sure (a.s.) convergence**. That $\Pr(|X_n - c| \geq \delta) \to 0$ for each $\delta > 0$ just ensures that when n is large there is small probability *at each separate stage* that X_n will take on a value much different from c. This does not necessarily mean that there is small probability of a large deviation *somewhere* down the line. In other words, the fact that $\Pr(|X_n - c| \geq \delta) \leq \varepsilon$ for each separate $n \geq N$ does not necessarily make small the probability that *any* of $X_N, X_{N+1}, \ldots$ are more than δ away from c. On the other hand, when the probability of *ever* seeing a large deviation *can* be made arbitrarily small by taking N sufficiently large, then we say that $\{X_n\}_{n=1}^{\infty}$ converges to c *almost surely*. In symbols we write this as $X_n \to^{\mathrm{a.s.}} c$. Of course, $X_n \to^{\mathrm{a.s.}} c$ clearly implies that $X_n \to^{\mathrm{P}} c$, since the probability of a large deviation at any stage is necessarily small if the probability is small for a deviation from that stage on.

The *Borel–Cantelli lemma* gives a sufficient condition for a.s. convergence and a sufficient condition for $\{X_n\}$ *not* to converge. To explain the result, let $B_n(\delta) \equiv \{\omega : |X_n(\omega) - c| \geq \delta\}$ for brevity. Think of this as the event of a "big" deviation from c at stage n. The event $\lim_n \sup B_n(\delta)$ is the event that infinitely many such big deviations occur. The *convergence* part of Borel–Cantelli tells us that if $\sum_{n=1}^{\infty} \mathbb{P}[B_n(\delta)] < \infty$ then $\mathbb{P}[\lim_n \sup B_n(\delta)] = 0$. Thus, if the sum of probabilities of δ discrepancies converges for all $\delta > 0$, then $X_n \to^{\mathrm{a.s.}} c$. The *divergence* part of Borel–Cantelli tells that if $\sum_{n=1}^{\infty} \mathbb{P}[B_n(\delta)] = \infty$ *and* if the events $\{B_n(\delta)\}_{m=1}^{\infty}$ are independent—which is true if the $\{X_n\}_{n=1}^{\infty}$ are independent—then $\mathbb{P}[\lim_n \sup B_n(\delta)] = 1$. Thus, if the sum diverges and if the $\{X_n\}$ are independent, then a δ discrepancy is almost sure to occur at a future stage, no matter how large is n. (Of course, the proviso "almost" just means that the set of outcomes on which discrepancies do *not* occur infinitely often has measure zero.)

3.7   LAWS OF LARGE NUMBERS AND CENTRAL-LIMIT THEOREMS

Laws of large numbers are theorems that state conditions under which *averages* of RVs converge (in some sense) to a constant as the number of variables being averaged increases. Specifically, let $\{X_j\}_{j=1}^{\infty}$ be a sequence of RVs each having mean $EX_j = \mu$, and let $\bar{X}_n \equiv n^{-1}\sum_{j=1}^{n}X_j$ be the arithmetic average of the first n of these. Then laws of large number state conditions under which the sequence $\{\bar{X}_n\}_{n=1}^{\infty}$ converges in some sense to μ. Our interest will be in averages of independent and identically distributed (i.i.d.) RVs. For this situation the strongest result is supplied by *Kolmogorov's strong law of large numbers* (SLLN):[9] If $\{X_j\}_{j=1}^{\infty}$ are i.i.d. with expected value μ, then $\bar{X}_n \to^{\text{a.s.}} \mu$ as $n \to \infty$. Of course, this implies that $\bar{X}_n \to^{\text{P}} \mu$, a result known as the *weak* law of large numbers. (The weak law had been proved before the strong law was known to hold.)

The SLLN is actually a broader result than it may appear to be. If $\{X_j\}_{j=1}^{\infty}$ are i.i.d. RVs (so that each has the same marginal distribution as some generic X) and if g is such that $Eg(X)$ exists, then $n^{-1}\sum_{j=1}^{n}g(X_j) \to^{\text{a.s.}} Eg(X)$. Thus: (1) if X has mth moment μ'_m, then $n^{-1}\sum_{j=1}^{n}X_j^m \to^{\text{a.s.}} \mu'_m$; (2) if $\mathcal{M}(\zeta) = Ee^{\zeta X}$ exists for all ζ in some neighborhood of the origin, then $n^{-1}\sum_{j=1}^{n}e^{\zeta X_j} \to^{\text{a.s.}} \mathcal{M}(\zeta)$ for each such ζ; (3) $E\mathbf{1}_B(X) = \Pr(X \in B)$ always exists for any Borel set B, so $n^{-1}\sum_{j=1}^{n}\mathbf{1}_B(X_j) \to^{\text{a.s.}} \mathbb{P}_X(B)$. Moreover, by the continuity theorem presented in Section 2.1, if $Eg(X) = \gamma$, say, and if h is continuous in a neighborhood of γ, then $h\left[n^{-1}\sum_{j=1}^{n}g(X_j)\right] \to^{\text{a.s.}} h(\gamma)$.

We have seen that a sequence of random variables converges in *probability* to a constant c if and only the sequence converges in *distribution* to the step function $\mathbf{1}_{[c,\infty)}(x)$. We often refer to this as a *degenerate* distribution, since all the probability mass resides at a single point. Of course, sequences of RVs sometimes converge to distributions that are not degenerate. *Central-limit theorems* give conditions under which sequences of sums or averages of RVs converge in distribution to the normal, when properly centered and scaled. For the case of i.i.d. RVs the best result is provided by the *Lindeberg–Lévy theorem*: If $\{X_j\}_{j=1}^{\infty}$ are i.i.d. with expected value μ and finite variance σ^2, then the sequence $\left\{Z_n \equiv \sqrt{n}\left(\bar{X}_n - \mu\right)/\sigma\right\}_{n=1}^{\infty}$ converges in distribution to standard normal—that is, a normal distribution with mean zero and unit variance. In symbols we would write $Z_n \rightsquigarrow N(0,1)$. Since $E\bar{X}_n = \mu$ and $V\bar{X}_n = \sigma^2/n$, we can write Z_n as $\left(\bar{X}_n - E\bar{X}_n\right)/\sqrt{V\bar{X}_n}$—that is, as the *standardized* arithmetic mean. Now putting $S_n \equiv \sum_{j=1}^{n}X_j = n\bar{X}_n$, we have $ES_n = n\mu$ and $VS_n = n\sigma^2$, so Z_n also equals the standardized *sum*, $(S_n - ES_n)/\sqrt{VS_n}$. Accordingly, standardized sums of i.i.d. RVs with finite variance converge in distribution to standard normal, just as do standardized means.

---

[9]After the brilliant Russian probabilist A. N. Kolmogorov.

3.8 IMPORTANT MODELS FOR DISTRIBUTIONS

Here we describe some features of the models for distributions that are commonly encountered in finance.

3.8.1 Continuous Models

1. **Normal family**. The univariate normal family of distributions is represented by the collection of PDFs on $\mathbb{X} = \Re$ given by

$$f(x; \alpha, \beta) = \left(2\pi\beta^2\right)^{-1/2} \exp\left[-\frac{(x - \alpha)^2}{2\beta^2}\right], \alpha \in \Re, \beta > 0.$$

Since $\int x f(x; \alpha, \beta) \cdot dx = \alpha$ and $\int (x - \alpha)^2 f(x; \alpha, \beta) \cdot dx = \beta^2$ for any such α and β, the PDF is usually written as

$$f(x; \mu, \sigma^2) = \frac{1}{\sqrt{2\pi\sigma^2}} \exp\left[-\frac{(x - \mu)^2}{2\sigma^2}\right].$$

Thus, if X has a distribution in the normal family, its mean and variance suffice to identify it, and the symbols $X \sim N(\mu, \sigma^2)$ completely specify its distribution. The rapid decay of $f(x; \mu, \sigma^2)$ as $|x - \mu| \to \infty$ ensures the finiteness of all moments and the existence of the MGF:

$$\mathcal{M}(\zeta) = \exp\left(\zeta\mu + \frac{\zeta^2\sigma^2}{2}\right), \zeta \in \Re.$$

Thus, if the logarithm of a RV's MGF, $\mathcal{L}(\zeta) \equiv \ln\mathcal{M}(\zeta)$, (which is referred to as the *cumulant-generating function*) is a quadratic function, then the RV necessarily has a distribution in the normal family, and the coefficients on ζ and ζ^2 correspond to the mean and half the variance, respectively. If $S_n = a_0 + \sum_{j=1}^{n} a_j X_j$, where the $\{a_j\}$ are constants and the $\{X_j\}_{j=1}^{n}$ are independent with $X_j \sim N(\mu_j, \sigma_j^2)$, then S_n has MGF

$$\mathcal{M}_{S_n}(\zeta) = e^{\zeta a_0}\mathcal{M}_{X_1}(\zeta a_1) \cdots \mathcal{M}_{X_n}(\zeta a_n)$$

$$= \exp\left[\zeta\left(a_0 + \sum_{j=1}^{n} a_j\mu_j\right), \frac{\zeta^2\left(\sum_{j=1}^{n} a_j^2\sigma_j^2\right)}{2}\right]$$

$$= \exp\left(\zeta ES_n, \frac{\zeta^2 VS_n}{2}\right),$$

which shows that affine functions of independent normals have normal marginal distributions. Putting $Z \equiv (X_1 - \mu_1)/\sigma_1$ (the special case $n = 1$, $a_0 = -\mu_1/\sigma_1, a_1 = 1/\sigma_1$) gives $M_Z(\zeta) = \exp(\zeta^2/2)$, showing directly that $EZ = 0$ and $VZ = 1$. We write $Z \sim N(0, 1)$ to denote such a *standard normal*

RV. Its PDF is $f(z) = (2\pi)^{-1/2} \exp(-z^2/2)$. All odd moments of Z are zero. In particular, the coefficient of skewness is $\kappa_3 = \mu_3/\sigma^3 = \mu_3 = 0$. The coefficient of kurtosis is $\kappa_4 = \mu_4/\sigma^4 = \mu_4 = 3$. Since κ_3 and κ_4 are invariant under location-scale changes, these same values apply to all normal RVs. The fact that $(X - \mu)/\sigma \sim N(0, 1)$ when $X \sim N(\mu, \sigma^2)$ is crucial for determining probabilities such as $\Pr(a < X < b)$, since there is no simple expression for definite integral $\int_a^b f(x; \mu, \sigma^2) \cdot dx$. By *standardizing* X, we can evaluate such expressions as integrals of the standard normal PDF, as

$$\Pr(a < X < b) = \Pr\left(\frac{a - \mu}{\sigma} < \frac{X - \mu}{\sigma} < \frac{b - \mu}{\sigma}\right)$$

$$= \Pr\left(\frac{a - \mu}{\sigma} < Z < \frac{b - \mu}{\sigma}\right)$$

$$= \Phi\left(\frac{b - \mu}{\sigma}\right) - \Phi\left(\frac{a - \mu}{\sigma}\right).$$

Here $\Phi(z) = \Pr(Z \leq z) = \int_{-\infty}^z (2\pi)^{-1/2} e^{-t^2/2} \cdot dt$ is the usual special symbol for the standard normal CDF. This function is tabulated in all elementary statistics texts and generated in standard statistical software. The symmetry of the distribution of Z about the origin implies that $\Phi(z) = 1 - \Phi(-z)$, which is why $\Phi(z)$ is usually tabulated only for $z \geq 0$.

2. **Multivariate normal family.** The univariate normal model is easily extended to the multivariate case. Let $\mathbf{Z} = (Z_1, Z_2, ..., Z_n)'$ be a vector of independent standard normals. The joint PDF $f_{\mathbf{Z}}(\mathbf{z})$ is just the product of the marginals, or

$$f_{\mathbf{Z}}(\mathbf{z}) = (2\pi)^{-n/2} \exp\left(-\frac{1}{2}\sum_{j=1}^n z_j^2\right) = (2\pi)^{-n/2} e^{-\mathbf{z}'\mathbf{z}/2},$$

and the joint MGF is likewise the product of the marginal MGFs:

$$\mathcal{M}_{\mathbf{Z}}(\zeta) = Ee^{\zeta'\mathbf{Z}} = Ee^{\zeta_1 Z_1} \cdot Ee^{\zeta_2 Z_2} \cdot \ldots \cdot Ee^{\zeta_n Z_n}$$

$$= \exp\left(\frac{1}{2}\sum_{j=1}^n \zeta_j^2\right) = e^{\zeta'\zeta/2}.$$

Introducing a vector of constants $\mathbf{a} = (a_1, a_2, ..., a_n)'$ and a nonsingular $n \times n$ matrix of constants $\mathbf{B}$, the transformation $\mathbf{X} = \mathbf{a} + \mathbf{BZ}$ yields a new collection of n RVs. Letting $\mathbf{b}_j$ be the jth *row* of $\mathbf{B}$, the jth element of $\mathbf{X}$ is

$$X_j = a_j + \sum_{i=1}^n b_{ji} Z_i = a_j + \mathbf{b}_j \mathbf{Z} \sim N(a_j, \mathbf{b}_j \mathbf{b}_j').$$

The covariance between X_j and X_k is $E(\mathbf{b}_j \mathbf{Z})(\mathbf{b}_k \mathbf{Z}) = E(\mathbf{b}_j \mathbf{Z} \mathbf{Z}' \mathbf{b}_k') = \mathbf{b}_j \mathbf{I} \mathbf{b}_k' = \mathbf{b}_j \mathbf{b}_k'$, and the covariance matrix of $\mathbf{X}$ is $V\mathbf{X} = \mathbf{BB}'$, with element

$\mathbf{b}_j \mathbf{b}_k'$ in the jth row and kth column. Putting $\mu \equiv \mathbf{a}$ and $\boldsymbol{\Sigma} \equiv \mathbf{BB}'$, we thus have $E\mathbf{X} = \mu$ and $V\mathbf{X} = \boldsymbol{\Sigma}$. The joint MGF of $\mathbf{X}$ is

$$\mathcal{M}_\mathbf{X}(\zeta) = E \exp[\zeta'(\mathbf{a} + \mathbf{BZ})] = e^{\zeta' \mathbf{a}} \mathcal{M}_\mathbf{Z}(\mathbf{B}'\zeta)$$
$$= e^{\zeta' \mathbf{a}} \exp\left(\frac{1}{2}\zeta'\mathbf{BB}'\zeta\right)$$
$$= \exp\left(\zeta'\mu + \frac{1}{2}\zeta'\boldsymbol{\Sigma}\zeta\right). \qquad (3.10)$$

The joint PDF is found from multivariate change-of-variable formula (3.4) with $\mathbf{g}(\mathbf{Z}) = \mathbf{a} + \mathbf{BZ}$ and $\mathbf{h}(\mathbf{X}) = \mathbf{B}^{-1}(\mathbf{X} - \mathbf{a})$. With $\mathbf{a} = \mu$ and $\mathbf{B}^{-1} = \boldsymbol{\Sigma}^{-1/2}$ this gives

$$f_\mathbf{X}(\mathbf{x}) = f_\mathbf{Z}\left[\boldsymbol{\Sigma}^{-1/2}(\mathbf{x} - \mu)\right]\left|\boldsymbol{\Sigma}^{-1/2}\right|$$
$$= (2\pi)^{-n/2}\left|\boldsymbol{\Sigma}^{-1/2}\right|\exp\left[-\frac{1}{2}(\mathbf{x} - \mu)'\boldsymbol{\Sigma}^{-1}(\mathbf{x} - \mu)\right]. \qquad (3.11)$$

(Covariance matrix $\boldsymbol{\Sigma}$ is positive definite, and so $\left\|\boldsymbol{\Sigma}^{-1/2}\right\| = \left|\boldsymbol{\Sigma}^{-1/2}\right|$.) A vector-valued RV with PDF (3.11) and MGF (3.10) is said to have a *multivariate normal distribution*. Evaluating $\mathcal{M}_\mathbf{X}(\zeta)$ at $\zeta = (0, 0, ..., \zeta, 0, ..., 0)'$ (with ζ in the jth position) gives $\exp\left(\zeta\mu_j + \zeta^2\sigma_j^2/2\right)$ as the marginal MGF of X_j, where $\mu_j = a_j$ and $\sigma_j^2 = \sigma_{jj} = \sum_{i=1}^n b_{ji}^2 = \mathbf{b}_j\mathbf{b}_j'$ is the jth diagonal element of $\boldsymbol{\Sigma}$. We thus see again that $X_j \sim N\left(\mu_j, \sigma_j^2\right)$. Now X_j corresponds to the product $\mathbf{c}'\mathbf{X}$ with $\mathbf{c}' = (0, 0, ..., 1, 0, ..., 0)$. It is easy to see that the transformation $Y = \mathbf{c}'\mathbf{X}$ for any $\mathbf{c} = (c_1, c_2, ..., c_n)'$ such that $\mathbf{c}'\mathbf{c} > 0$ yields a univariate normal RV, $Y \sim N(\mathbf{c}'\mu, \mathbf{c}'\boldsymbol{\Sigma}\mathbf{c})$. The fact that the normal family is "closed" under such affine transformations accounts for much of the model's usefulness in financial (and other) applications. One caution: If we know that $\{X_j\}_{j=1}^k$ are *independent* with normal marginal distributions, then it does follow that $\mathbf{X} = (X_1, X_2, ..., X_k)'$ is multivariate normal. This is easily seen on observing that the product of the marginal PDFs has the form (3.11) with a diagonal matrix $\boldsymbol{\Sigma}$. However, there are *dependent* RVs with normal marginals that are not multivariate normal.[10] A necessary and sufficient condition that $\mathbf{X}$ be multivariate normal is that $Y_\mathbf{a} \equiv \mathbf{a}'\mathbf{X}$ have a normal marginal distribution for *all* constants $\mathbf{a}$ such that $\mathbf{a}'\mathbf{a} > 0$.

3. **Lognormal family**. With $X \sim N(\mu, \sigma^2)$, the transformation $Y = g(X) = e^X$ is a one-to-one mapping from $\mathfrak{R}$ to $(0, \infty)$. The inverse function being $\ln Y$,

---

[10]An example is the pair Z_1, Z_2 with joint PDF $f_\mathbf{Z}(z_1, z_2) = \pi^{-1}\exp\left[-\left(z_1^2 + z_2^2\right)/2\right]$ when $z_1 z_2 > 0$ and $f_\mathbf{Z}(z_1, z_2) = 0$ otherwise. These are dependent since support $\mathbb{Z} = \{z_1, z_2 : z_1 z_2 > 0\}$ is not a product space, yet each of Z_1, Z_2 is distributed as $N(0, 1)$.

the change-of-variable formula gives as the PDF

$$f_Y(y) = f_X(\ln y) \cdot \frac{d\ln y}{dy}$$

$$= \frac{1}{y\sqrt{2\pi\sigma^2}} e^{-(\ln y - \mu)^2/(2\sigma^2)} \mathbf{1}_{(0,\infty)}(y).$$

A RV Y with this PDF is said to have a *lognormal distribution* with parameters μ, σ^2. We write $Y \sim LN(\mu, \sigma^2)$ for short. (Just remember that the logarithm of a lognormal is a normal.) The moments of Y are easily found from the MGF of X. Specifically, for $t \in \Re$ we have

$$EY^t = Ee^{tX} = \mathcal{M}_X(t) = e^{t\mu + t^2\sigma^2/2}.$$

Although all moments exist (even negative and fractional ones), Y itself has no MGF, since $\int_0^\infty e^{\zeta y} f_Y(y) \cdot dy$ does not converge for any $\zeta > 0$. If $\{Y_j\}_{j=1}^k$ are independent lognormals with parameters $\{\mu_j, \sigma_j^2\}$, and if $\{a_j\}_{j=0}^k$ are real constants with $a_0 > 0$ and at least one of $\{a_j\}_{j=1}^n$ not equal to zero, then the RV $P_k = a_0 Y_1^{a_1} Y_2^{a_2} \cdots Y_k^{a_k}$ also has a lognormal distribution. This is true because $\ln P_k = \ln a_0 + a_1 \ln Y_1 + \cdots + a_k \ln Y_k$ is an affine function of independent (and therefore multivariate) normals. Thus, while the normal family is closed under affine transformations, the lognormal family is closed under power transformations and multiplication by positive numbers. Because the compounding effect of investment returns is multiplicative, this feature gives lognormal distributions important application in finance.

4. **Gamma family**. With $Z \sim N(0,1)$ put $W \equiv Z^2$. Then

$$F_W(w) = \Pr(Z^2 \le w) = \Phi(\sqrt{w}) - \Phi(-\sqrt{w}) = 2\Phi(\sqrt{w}) - 1$$

for $w \ge 0$. Differentiating at any $w > 0$ gives for the PDF

$$f_W(w) = 2f_Z(\sqrt{w}) \cdot \frac{w^{-1/2}}{2} = \frac{1}{\sqrt{2\pi}} w^{-1/2} e^{-w/2} \mathbf{1}_{(0,\infty)}(w).$$

Familiar from statistics, this PDF represents the *chi-squared* distribution with one degree of freedom; in symbols, $W \sim \chi^2(1)$. The MGF is

$$\mathcal{M}_W(\zeta) = \int_0^\infty e^{\zeta w} \frac{1}{\sqrt{2\pi}} w^{-1/2} e^{-w/2} \cdot dw.$$

Changing variables as $w(1 - 2\zeta) = y^2$ for $\zeta < \frac{1}{2}$ gives

$$\mathcal{M}_W(\zeta) = (1 - 2\zeta)^{-1/2} 2 \int_0^\infty \frac{1}{\sqrt{2\pi}} e^{-y^2} \cdot dy = (1 - 2\zeta)^{-1/2}, \zeta < \frac{1}{2}.$$

Now letting $\{W_j\}_{j=1}^{k}$ be independent $\chi^2(1)$ RVs and putting $X_k = \sum_{j=1}^{k} W_j$, we see at once that $\mathcal{M}_{X_k}(\zeta) = (1 - 2\zeta)^{-k/2}$, and an integration along the lines of that for $\mathcal{M}_W$ shows that this MGF corresponds to a RV with PDF

$$f_{X_k}(x) = \frac{1}{(k-1)!2^{k/2}} x^{k/2-1} e^{-x/2} \mathbf{1}_{(0,\infty)}(x). \tag{3.12}$$

This represents the chi-squared distribution with k degrees of freedom, $\chi^2(k)$. The χ^2 family can be generalized further. The *gamma* function, $\Gamma : (0, \infty) \to (0, \infty)$, is defined through the integral expression

$$\Gamma(\alpha) = \int_0^{\infty} y^{\alpha-1} e^{-y} \cdot dy.$$

This has the properties (1) $\Gamma\left(\frac{1}{2}\right) = \sqrt{\pi}$, (2) $\Gamma(1) = 1$, and (3) $\Gamma(\alpha) = (\alpha-1)\Gamma(\alpha-1)$ for $\alpha > 1$. When $\alpha = k$ (a positive integer) the last property implies that $\Gamma(k) = (k-1)!$, and this with property 2 justifies the usual convention $0! = 1$. Now, putting $F(y) = \Gamma(\alpha)^{-1} \int_0^y t^{\alpha-1} e^{-t} \cdot dt$ for $y > 0$ and $F(y) = 0$ for $y \leq 0$ gives a nondecreasing, right-continuous function with $F(-\infty) = F(0) = 0$ and $F(+\infty) = 1$; that is, F is a CDF, and $f(y) = F'(y)\mathbf{1}_{(0,\infty)}(y) = \Gamma(\alpha)^{-1} y^{\alpha-1} e^{-y}\mathbf{1}_{(0,\infty)}(y)$ is the PDF of some RV Y. Putting $X = \beta Y$ for some $\beta > 0$ and changing variables, we have

$$f_X(x; \alpha, \beta) = \frac{1}{\Gamma(a)\beta^{\alpha}} x^{\alpha-1} e^{-x/\beta} \mathbf{1}_{(0,\infty)}(x).$$

A RV with this PDF is said to have a *gamma* distribution with shape parameter α and scale parameter β; in symbols, $X \sim \Gamma(\alpha, \beta)$. Comparison with (3.12) shows that $\chi^2(k)$ corresponds to $\Gamma(k/2, 2)$. It is easy to see that $\mathcal{M}_X(\zeta) = (1 - \beta\zeta)^{-\alpha}$ for $\zeta < \beta^{-1}$. Thus, gamma RVs possess all their positive integer moments; indeed, it is not difficult to see that $EX^t = \Gamma(\alpha + t)\beta^t$ for all real $t > -\alpha$. In particular, $EX = \alpha\beta$ and $VX = \alpha\beta^2$. In the special case $\alpha = 1$ we have an *exponential distribution* with scale parameter (and mean) β and CDF $F_X(x) = \left(1 - e^{-x/\beta}\right)\mathbf{1}_{(0,\infty)}(x)$.

5. **Uniform distributions.** Let F be the CDF of *any* continuous RV X with support $\mathbb{X}$, and consider the transformation $U = g(X) = F(X)$. Clearly, this is a one-to-one mapping from $\mathbb{X}$ to $[0, 1]$. The CDF of U is

$$F_U(u) = \Pr[F(X) \leq u] = \Pr[X \leq F^{-1}(u)] = F[F^{-1}(u)] = u$$

for any $u \in (0, 1)$, with $F_U(u) = 0$ for $u \leq 0$ and $F_U(u) = 1$ for $u \geq 1$. Any of $\mathbf{1}_{(0,1)}(u)$, $\mathbf{1}_{[0,1)}(u)$, $\mathbf{1}_{(0,1]}(u)$, and $\mathbf{1}_{[0,1]}(u)$ is a convenient version of the PDF, and from these functions of uniform density comes the name *uniform distribution*. To generalize, put $X = \alpha + (\beta - \alpha)U$ for $\alpha \in \Re$ and $\beta > \alpha$. Then the change-of-variable formula gives $f_X(x) = (\beta - \alpha)^{-1} \mathbf{1}_{(\alpha,\beta)}(x)$. We then say that X is distributed *uniformly* on (α, β) or, in symbols, $X \sim U(\alpha, \beta)$.

6. **Cauchy family**. Let $U \sim U(-\pi/2, \pi/2)$ be uniformly distributed on an interval of length π centered on the origin, and put $Y = g(U) = \tan U$, a one-to-one mapping from $(-\pi/2, \pi/2)$ to $(-\infty, \infty)$. Then the change-of-variable formula gives

$$f_Y(y) = f_U\left(\tan^{-1} y\right) \cdot \left| \frac{d\tan^{-1} y}{dy} \right| = \frac{1}{\pi(1+y^2)}.$$

A RV with this PDF is said to have a *standard Cauchy distribution*. Like the standard normal, this is symmetric about the origin and rather bell-shaped, but the function decays much more slowly as $|y| \to \infty$—so slowly in fact that

$$E|Y| = \int_{-\infty}^{\infty} \frac{|y|}{\pi(1+y^2)} \cdot dy = +\infty.$$

Since both $EY^+ = E\max(Y, 0) = +\infty$ and $EY^- = E\max(-Y, 0) = +\infty$, the Cauchy distribution simply has no mean. (Despite the symmetry, one cannot balance $+\infty$ and $-\infty$, which are the moments of mass on the positive and negative parts of the line—and indeed to left and right of *any* point on the line.) Putting $X = \alpha + \beta Y$ with $\beta > 0$ generalizes to a Cauchy with center (median) α and scale parameter β:

$$f_X(x) = \frac{1}{\beta\pi\left[1 + (x-\alpha)^2/\beta^2\right]}.$$

3.8.2 Discrete Models

1. **Bernoulli**. This is a building block for other discrete models. The outcome set of an experiment is partitioned into two sets, typically characterized as "success" $\mathcal{S}$ or "failure" $\mathcal{S}^c$, with $\mathbb{P}(\mathcal{S}) = \theta \in (0,1)$. Putting $X(\omega) = \mathbf{1}_\mathcal{S}(\omega)$—that is, $X(\omega) = 1$ if $\omega \in \mathcal{S}$ and $X(\omega) = 0$ otherwise—we have $\mathbb{P}_X(\{1\}) = f_X(1;\theta) = \theta = 1 - \mathbb{P}_X(\{0\}) = 1 - f_X(0;\theta)$. Clearly, $\mu'_k = EX^k = \theta$ for all $k > 0$, and the MGF is $\mathcal{M}_X(\zeta) = \theta e^\zeta + 1 - \theta = 1 + \theta(e^\zeta - 1)$.

2. **Binomial**. An experiment consists of n independent Bernoulli "trials," with independent Bernoulli RVs $\{X_j\}_{j=1}^n$ indicating whether the respective trials produce successes or failures. Putting $Y_n \equiv \sum_{j=1}^n X_j$ gives a new RV supported on $\mathbb{Y} = \{0, 1, 2, ..., n\}$, and for each $y \in \mathbb{Y}$ we have $f_{Y_n}(y; n, \theta) = \binom{n}{y}\theta^y(1-\theta)^{n-y}$. Here $\binom{n}{y} = n!/[y!(n-y)!]$ counts the number of arrangements of the y successes among n outcomes, and $\theta^y(1-\theta)^{n-y}$ is the probability of any one such arrangement. A RV with this PMF is said to have a *binomial distribution* with parameters $n \in \mathbb{N}$ and $\theta \in (0,1)$; in symbols, $Y_n \sim B(n, \theta)$. The MGF is most easily found from that of the Bernoulli as $\mathcal{M}_{Y_n}(\zeta) = \mathcal{M}_X(\zeta)^n = \left[1 + \theta(e^\zeta - 1)\right]^n$, and one finds either from this or directly that $\mu = n\theta$, $\sigma^2 = n\theta(1-\theta)$.

3. **Poisson**. If a sequence of Bernoulli trials with fixed $\theta \in (0, 1)$ is extended indefinitely by sending $n \to \infty$, then the number of sucesses also diverges almost surely, in that $\Pr(Y_n > b) \to 1$ as $n \to \infty$ for any $b \in \Re$. The divergent behavior can be seen in the MGF, since $\lim_{n \to \infty} \mathcal{M}_{Y_n}(\zeta) = \infty$ for all $\zeta \neq 1$. Of course, this is neither surprising nor very interesting. However, if we make the probability of success decline inversely with n as $\theta = \lambda/n$ for some $\lambda > 0$ and all $n > \lambda$, then we have

$$\lim_{n \to \infty} \mathcal{M}_{Y_n}(\zeta) = \lim_{n \to \infty} \left[1 + \frac{\lambda}{n} \left(e^\zeta - 1 \right) \right]^n = \exp\left[\lambda \left(e^\zeta - 1 \right) \right],$$

a function that is finite for all ζ. Using Stirling's formula to approximate $n!$ as $\sqrt{2\pi n}(n/e)^n$, it is not difficult to see that

$$\lim_{n \to \infty} f_{Y_n}\left(y; n, \frac{\lambda}{n} \right) = \frac{\lambda^y e^{-\lambda}}{y!} \equiv f_Y(y; \lambda)$$

for $y \in \aleph_0 = \{0, 1, 2, ...\}$. The Taylor expansion $e^\lambda = \sum_{n=0}^{\infty} \lambda^y/n!$ shows that $f_Y(y; \lambda)$ is a legitimate PMF. This represents the limiting distribution of Y_n when $\theta = \lambda/n$, and the calculation $\sum_{y=0}^{\infty} e^{\zeta y} f_Y(y; \lambda) = \exp\left[\lambda \left(e^\zeta - 1 \right) \right]$ reveals the comforting fact that the MGF of the limiting distribution is the limit of the sequence of MGFs. A RV with the PMF

$$f_Y(y; \lambda) = \frac{\lambda^y e^{-\lambda}}{n!} \mathbf{1}_{\aleph_0}(y)$$

is said to have a *Poisson distribution* with parameter λ; in symbols, $Y \sim$ Poisson (λ). Both the mean and the variance of Y are equal to λ.

PART II

PORTFOLIOS AND PRICES

CHAPTER 4

INTEREST AND BOND PRICES

One of the most fundamental concepts in finance is the connection between interest rates and prices of bonds. Accordingly, gaining an understanding of the arithmetic of bond prices and rates is a logical first step in a study of quantitative finance. We begin by developing the idea of a *continuously compounded* interest rate, then introduce *discount* bonds and the associated average and continuous *spot* rates of interest, and conclude by explaining the concepts of *forward* bond prices and interest rates.

4.1 INTEREST RATES AND COMPOUNDING

First, let us review some basic concepts regarding compound interest and introduce the notation and time and currency conventions that are followed henceforth. Throughout, we take the unit of currency to be one *dollar*. (Of course, one is free to substitute whatever currency unit one prefers—euro, franc, peso, pound, real, ruble, yen, yuan, etc.) Unless stated otherwise, we take the unit of time to be one year, and we express interest rates in *per annum* terms. Thus, for example, $r = .10$ indicates a rate of 10% per year, so that one dollar invested for one year at the simple rate r would be worth $V_1 \equiv 1 + r = \$1.10$ at the end, while a dollar invested for 1.5 years at the simple rate r would be worth $1 + 3r/2 = 1.15$. The term "simple" indicates that there is no compounding during the period of investment; that is, the interest is paid just at the end and does not itself earn interest during the period. For an N-year investment at constant rate r with annual compounding, the value per dollar invested initially would be $V_N \equiv (1+r)^N$. If there is semiannual compounding, so that an amount $r/2$ per dollar invested at the beginning of the 6-month period is received and reinvested at the end, then the value per dollar after N years would be $V_N = (1+r/2)^{2N}$. Similarly, we would have $V_N = (1+r/m)^{mN}$ if there were compounding m times per year at equal intervals for N years. Of course, if V_0 dollars were invested initially, the terminal value would be $V_N = V_0(1+r/m)^{mN}$.

Three further extensions of this last result are needed. First, we can allow the quoted per annum rates to differ from one period to the next, starting with r_0 for the first period from initial time $t = 0$ out to $t = 1/m$ years, $r_{1/m}$ for the period from $t = 1/m$ to $t = 2/m$, and so on out to $r_{N-1/m}$ for the final period. In this case the value after N years of an initial investment of V_0 is

$$V_N = V_0 \left(1 + \frac{r_0}{m}\right)\left(1 + \frac{r_{1/m}}{m}\right) \cdots \cdots \left(1 + \frac{r_{N-1/m}}{m}\right)$$

$$= V_0 \prod_{j=1}^{Nm} \left(1 + \frac{r_{(j-1)/m}}{m}\right). \tag{4.1}$$

Next, we can allow the lengths of the intervals between payments to differ. Taking $t_0 = 0$ as the time of the initial investment and $0 < t_1 < t_2 < \cdots < t_n \equiv T$ as the dates at which interest payments are received and reinvested, the terminal value would be

$$V_T = V_0 \left[1 + r_0 t_1\right]\left[1 + r_{t_1}(t_2 - t_1)\right] \cdots \cdots \left[1 + r_{t_{n-1}}(T - t_{n-1})\right]$$

$$= V_0 \prod_{j=1}^{n} \left[1 + r_{t_{j-1}}(t_j - t_{j-1})\right]. \tag{4.2}$$

This reduces to (4.1) on taking $t_j - t_{j-1} = 1/m$, $n = Nm$, and $T = N$. Of course, T in (4.2) need not be an integer.

Finally, we can extend to allow *continuous* compounding, the limiting case in which interest is earned and credited to the investment account at each instant. For this let us write

$$V_T = V_0 \exp\left[\ln\left(\frac{V_T}{V_0}\right)\right] = V_0 \exp\left\{\sum_{j=1}^{n} \ln\left[1 + r_{t_{j-1}}(t_j - t_{j-1})\right]\right\},$$

where "ln" denotes the natural logarithm and $\exp(x) \equiv e^x$ for any real number x. A Taylor expansion of $\ln(1+x)$ gives $\ln(1+x) = x - x^2/2 + x^3/3 - + \cdots = x + O(x^2)$ for $|x| < 1$, where $O(x^2)$ has the property that $\lim_{x \to 0} O(x^2)/x^2 = c$ for some nonzero constant c. (Here $c = \frac{1}{2}$, but the precise value does not matter.) Now let $\tau_n \equiv \max_{j \in \{1,2,\ldots,n\}} |t_j - t_{j-1}|$ be the largest of the n time intervals and require this to approach zero as $n \to \infty$ at a rate proportional to $1/n$. We can then represent V_T as

$$V_T = V_0 \exp \left\{ \sum_{j=1}^{n} \left[r_{t_{j-1}} \left(t_j - t_{j-1} \right) + O \left(n^{-2} \right) \right] \right\}$$

$$= V_0 \exp \left[\sum_{j=1}^{n} r_{t_{j-1}} \left(t_j - t_{j-1} \right) + O \left(n^{-1} \right) \right],$$

where we use the obvious fact that $\sum_{j=1}^{n} O \left(n^{-2} \right) = O \left(n^{-1} \right)$. Taking limits now and assuming that the *process* $\{r_t\}_{0 \le t \le T}$ is continuous (or, more generally, just that it is integrable), we have

$$V_T = V_0 \exp \left(\int_0^T r_t \cdot dt \right) \tag{4.3}$$

as the value after T years of an initial investment of V_0 with continuous compounding at rates $\{r_t\}_{0 \le t \le T}$. Of course, this simplifies to $V_T = V_0 e^{rT}$ if $r_t = r$ for all t.

4.2 BOND PRICES, YIELDS, AND SPOT RATES

We all know generally of the intimate (inverse) relation between interest rates and bond prices, but we need to see specifically how rates of various types are related to and can be deduced from the market prices of certain types of debt instruments. First, the basics. From the standpoint of the issuer (the original seller), we think of a bond as an obligation to make payments of stated amounts at one or more future dates. From the standpoint of the holder of the bond, it represents a contingent stream of receipts over time—the contingency being that the issuer does not default on the obligation. We distinguish between "discount" bonds and "coupon" bonds, the former making a contingent payment at just one date, and the latter making a stream of payments over a fixed period. Discount bonds are often called "zero-coupon bonds." For both types of bonds the terminal payment represents the bond's "principal" value—also called "face" value or "maturity" value—and the terminal date itself is said to be the date at which the bond "matures." For example, the issuer of a one-year discount bond of principal value $1000 is obligated to pay that sum to the holder one year after the bond is issued. It is sold initially at a discount—that is, for less than principal value—so as to make the obligation attractive to investors. The issuer of a 10-year $1000 bond with 5% coupon must pay the holder 5% of the principal value ($50) each year in one or more installments, with a terminal payment of $1000 plus the coupon installment at the end of year 10. For example, if coupons were paid semiannually, the holder of the 10-year bond would receive $25 every 6 months for 9.5 years and then $1025 at maturity. The coupon payment is set so as to make the bond sell at near its principal value at the initial offering and does not change over the life of the bond. In resale markets the bond's price will subsequently rise and fall with general credit conditions in a manner to be described.

Note: Finance practitioners generally reserve the term "bond" for long-term, coupon-paying debt instruments, referring to intermediate-term securities as "notes" and to those of short maturity as "bills." For simplicity we shall refer to all of these as "bonds" unless there is specific need to distinguish them.

One of the most useful concepts in financial modeling is an idealization known as the "default-free discount unit bond." Because the bond is default-free, its contractual payments are guaranteed. As a "discount" bond, there is only the single payment at maturity. As a "unit" bond, its principal value is simply one currency unit—one dollar. In the United States, Treasury bills and Treasury strips ("T-bills" and "strips") are the nearest approximations to default-free discount bonds. They are indeed sold at a discount; T-bills are available currently with maturities of 4, 13, and 26 weeks and strips with maturities out to 30 years. (Payments to holders of strips are financed by pooling the principal or interest payments of government coupon-paying bonds.) Also, as they are backed by the taxing power of the federal government, they are generally considered free of any *nominal* risk; that is, the dollar payments are guaranteed, although their "real" value—their purchasing power—depends on the uncertain future rate of inflation. On the other hand, T-bills and strips are certainly not "unit" bonds, since the minimum principal value is $1000. However, we can simply think of the price of a unit bond as the price of a bill or strip *per dollar* of principal value. Table 4.1 shows bid and asked prices of a sample of Treasury strips as of December 7, 2007, all maturing on August 15 of the given year.[11] Prices are quoted as percentages of principal value, so we simply divide by 100 to view them as prices of unit bonds.

The utility of the concept of a default-free unit discount bond is that it represents effectively a *discount factor* that shows the present market value of a *certain* (i.e., *sure*) future one-dollar receipt. We represent the value of such an instrument with the symbol $B(t, T)$, the first and second arguments standing for the current time and the time of maturity, respectively. Thus, the asked price for the first strip shown in Table 4.1 would be presented as $B(t,T) = .97695$, with t corresponding to December 7, 2007 and T corresponding to August 15, 2008. This tells us that a sure claim to receive a dollar on August 15 of the next year was worth roughly 97.7 cents on December 7, 2007. Note that the values $B(t,T_1), B(t,T_2), ..., B(t,T_n)$ of claims maturing at $T_1, T_2, ..., T_n$ are known as of t, since these are the quoted market prices; however, these prices fluctuate with market conditions, and we do not generally know at time t what they will be at later times. Thus, $B(t', T)$ is not observable as of any $t < t' < T$. On the other hand, given that the claim is default-free, we do know in advance that $B(T, T) = 1$.

Since they represent the appropriate discount factors, market prices of unit bonds of various maturities (the modifiers "default-free" and "discount" will be understood from here on) are precisely what are needed in order to value sure future cash streams.

---

[11] Source: *Wall Street Journal*, December 7, 2007. Strips are currently available with maturity dates February 15, May 15, August 15, and November 15. These coincide with the dates on which various outstanding Treasury notes and bonds pay their semiannual coupons.

Table 4.1 Prices and Yields, Selected U.S. Treasury Strips

Maturity	Bid	Asked	Yield
2008	97.675	97.695	3.45
2009	94.853	94.873	3.15
2015	73.230	73.250	4.09
2017	65.823	65.843	4.36
2019	58.869	58.889	4.59
2021	52.717	52.737	4.73
2023	47.894	47.914	4.75
2025	43.165	43.185	4.81
2027	39.388	39.408	4.79
2029	36.214	36.234	4.74

For example, the value at t of sure cash receipts $c_1, c_2, ..., c_n$ at $T_1, T_2, ..., T_n$ would be

$$V_t = \sum_{j=1}^{n} B(t, T_j) c_j.$$

In particular, even without looking at its market price, we should be able to infer the value of a default-free *coupon-paying* bond by observing the prices of unit bonds of various maturities. Thus, a bond that pays $\$c$ per unit principal value at $T_1, T_2, ..., T_n$ and returns the unit principal at T_n should be worth

$$B(t, c; T_1, T_2, ..., T_n) = c \sum_{j=1}^{n} B(t, T_j) + B(t, T_n). \qquad (4.4)$$

Of course, this value would fluctuate through time as the discount factors change with market conditions and as the number of remaining coupons declines on the way to maturity.

Valuing a coupon bond in this way is an example of pricing by *replication*, which is also called "arbitrage pricing." This means that we are determining what a financial instrument *should* sell for by valuing a portfolio that replicates (reproduces) its payments exactly in amounts and in time. Given the prices $\{B(t, T_j)\}_{j=1}^{n}$ per unit principal value of the T-bills and/or strips that mature at a default-free coupon bond's various payment dates, we would have an *arbitrage opportunity* if the bond could be bought or sold at a price other than that given by (4.4). For example, if a coupon bond of unit principal value sold for more, an institution or investor that held it could sell the bond and reproduce its cash flow by buying c units of T-bills or strips maturing at each of $T_1, T_2, ..., T_{n-1}$ plus $c + 1$ units of bills or strips maturing at T_n. The higher price of the coupon bond would more than finance these purchases, and the difference would be spendable cash, obtained at zero risk. If the coupon bond sold for less than the value of such a portfolio, one could make the same sort of gain by reversing the transaction. Of course, all this assumes that trading itself is costless, that bonds can be bought and sold in arbitrary amounts, and that the bond being "priced" in this way has no special features that affect its value;[12] but at least the calculation (4.4) gives a benchmark for what the bond should be worth. We will give a precise definition of "arbitrage" and begin to use techniques of arbitrage pricing in Part III.

The last column of Table 4.1 presents the quoted "yields" of the various Treasury strips. We can now see how these are defined and calculated from the prices of the bonds. To be more general, we consider the possibility of coupon payments. Consider the price $B(0, c; T_1, T_2, ..., T_n)$ at $t = 0$ of a bond that pays coupons of c per unit principal value at $T_1, T_2, ..., T_n$ and matures, paying unit principal, at T_n. The bond's yield to maturity is defined as the (positive, real-valued) annualized rate y that solves the equation

$$B(0, c; T_1, T_2, ..., T_n) = \frac{c}{(1+y)^{T_1}} + \frac{c}{(1+y)^{T_2}} + \cdots + \frac{c}{(1+y)^{T_n}} + \frac{1}{(1+y)^{T_n}}$$

$$= c \sum_{j=1}^{n} (1+y)^{-T_j} + (1+y)^{-T_n}. \tag{4.5}$$

For example, if the coupons are paid annually beginning in one year, we would have

$$B(0, c; 1, 2, ..., n) = c \sum_{j=1}^{n} (1+y)^{-j} + (1+y)^{-n}. \tag{4.6}$$

Thus, the yield to maturity is the single rate y that equates the present value of the bond's cash flow to its market price. Typically, solving for the yield requires a numerical procedure, but an easy way of approximating the solution is simply to graph the right side of (4.5) for a fine grid of y values and find the one that comes near equating the two sides. For example, Fig. 4.1 plots against y the present value of 10 annual payments of $c = .05$ plus a principal payment of one unit in year $n = 10$. As the figure shows, a bond with market price of $B(0, .05; 1, 2, ..., 10) = .98$ would have a yield to maturity of just under .053.

It is easy to find an explicit solution for the yield of a *discount* bond that simply returns principal value at maturity. In that case (4.5) simplifies to $B(0, T) = (1 + y)^{-T}$, and so $y = B(0, T)^{-1/T} - 1$. For example, a discount bond maturing in 251 days or $T = \frac{251}{365} \doteq .68767$ years and priced at $B(0, .68767) = .97695$ would have a yield of $y \doteq .03449$, corresponding to the first line of Table 4.1. More useful for our purposes will be to calculate a continuously compounded version of the yield to maturity, as the solution to $B(0, T) = e^{-yT}$. More generally, the rate

$$r(t, T) = -(T - t)^{-1} \ln B(t, T)$$

that satisfies the equality $B(t, T) = e^{-r(t,T)(T-t)}$ will be referred to as the *average continuously compounded spot rate* at time t for claims due at T. For example, on

---

[12]Examples are "call" provisions that give issuers the option to redeem bonds at stated prices and times before maturity and "sinking fund" provisions that authorize scheduled partial redemptions.

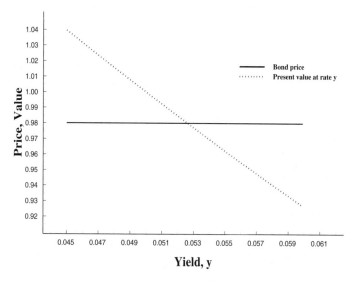

Figure 4.1   Yield to maturity as rate y that equates present value to price.

the basis of the asked price in the first row of Table 4.1 the average spot rate over the period $t =$ December 7, 2007 to $T =$ August 15, 2008 was

$$r(t, T) = -\left(\frac{251}{365}\right)^{-1} \ln(0.97695) \doteq .03391.$$

This represents the continuously compounded rate that would grow an investment of $B(t, T)$ at time t to one dollar at T. It is called a "spot" rate because it is available "on the spot" as of time t for investments out to T, simply by purchasing the bond. ("Forward" rates, discussed below, are *not* available on the spot.) In general, the average spot rate at any given time t depends on the time to maturity over which the rate is averaged. For example, based on the asked price of .36234 in the last row of Table 4.1 the average continuously compounded spot rate from $t = 7$ December 2007 to $T = 15$ August 2029 was

$$r(t, T) = -\left(21 + \frac{251}{365}\right)^{-1} \ln(.36234) \doteq .04681.$$

We shall find another idealization to be useful as we proceed—this time an idealization of the average spot rate. Note first that since $\ln B(T, T) = \ln(1) = 0$ we could represent average rate $r(t, T)$ as

$$r(t, T) = \frac{\ln B(T, T) - \ln B(t, T)}{T - t}.$$

Now consider the average rate as of $t - \Delta t$ on a unit bond that matures at t. This would be

$$r\left(t - \Delta t, t\right) = \frac{\ln B\left(t, t\right) - \ln B\left(t - \Delta t, t\right)}{\Delta t}.$$

Shrinking the time to maturity by letting Δt drop toward zero, and presuming that the limit of the expression on the right does exist, we have what is referred to as the *instantaneous spot rate* or *short rate* as of time t:

$$r_t \equiv \lim_{\Delta t \downarrow 0} \frac{\ln B\left(t, t\right) - \ln B\left(t - \Delta t, t\right)}{\Delta t} = \left.\frac{\partial \ln B\left(t, T\right)}{\partial t}\right|_{T=t}.$$

In practice, we can think of the short rate as the average rate implicit in the price of a discount bond that is just on the verge of maturing. Note that since

$$d\ln B\left(t, t\right) = \left.\frac{\partial \ln B\left(t, T\right)}{\partial t} \cdot dt\right|_{T=t} + \left.\frac{\partial \ln B\left(t, T\right)}{\partial T} \cdot dT\right|_{T=t}$$

and $d\ln B\left(t, t\right) = d\ln\left(1\right) = 0$, we also have

$$r_t = -\left.\frac{\partial \ln B\left(t, T\right)}{\partial T}\right|_{T=t}.$$

Now consider a mutual fund whose investment policy is to hold only such very short-maturity default-free bonds, continually rolling over the principal payments as the bonds mature into new bonds that are also about to mature. A money-market fund that invests in very short term Treasury securities or high-grade commercial paper would be the nearest actual manifestation of this ideal concept. If an ownership share of this fund were priced initially at M_0, and if there were no expense charges or other frictions, then its value at $t > 0$ would be

$$M_t = M_0 \exp\left(\int_0^t r_s \cdot ds\right).$$

This corresponds to expression (4.3) for the value of an investment that grows at continuously compounded rates.

The really nice feature of working with continuously compounded short rates is that they apply additively. For example, the value of our "money fund" at $T > t$ can be expressed as

$$M_T = M_t \exp\left(\int_t^T r_u \cdot du\right)$$

$$= M_0 \exp\left(\int_0^t r_s \cdot ds + \int_t^T r_u \cdot du\right)$$

$$= M_0 \exp\left(\int_0^T r_s \cdot ds\right).$$

We shall make frequent use of this idealized money fund in the treatment of continuous-time finance, beginning in Section 10.3.

4.3 FORWARD BOND PRICES AND RATES

The bond prices that we have considered up to now are "spot" prices, since they represent real trade prices at which one could buy and sell for immediate delivery. Likewise, the rates implicit in holding these bonds to maturity are the rates one could immediately lock in by buying the bonds. By contrast, as we have already seen, no one at time t could say with assurance what would be the spot price of a T-maturing unit bond at some later time t'; likewise, the implicit average spot rate $r(t', T)$ would also be unknown at $t < t'$. Nevertheless, the spectrum of current spot prices of bonds of various maturities contains *implicitly* certain projections—one might call them market forecasts—of future prices and rates. These are known as *forward prices* and *forward rates*.

To understand the concept of a forward bond price, consider an investor's attempt at time t to secure a sure claim on one dollar at future date T. The obvious, direct way to do this is to buy a T-maturing unit bond at price $B(t, T)$. But in fact, if markets allow secure contracting among individuals, there are many other, *indirect* ways. Pick any intermediate date t' between t and T. The investor could enter into a contract to buy a T-maturing unit bond at t' from someone else at a *known* negotiated price, say x. The contract requires no money to change hands at t; it is merely an agreement to transact in the future. Such an agreement represents a *forward contract*, and x represents the t'-forward price for a T-maturing bond as of date t. What would x have to be in a well-functioning market? To ensure the ability to complete the forward transaction at t', the investor could buy x units of t'-maturing bonds now for an amount $xB(t, t')$. At t' their value $xB(t', t') = x$ would provide the cash needed for the contracted purchase of the T-maturing bond. Thus, the direct and indirect ways of securing a sure dollar at T would cost $B(t, T)$ and $xB(t, t')$, respectively. But these must be equal, or else one could work an arbitrage—get a riskless gain—by buying the cheaper package and selling the dearer one. For example, if $B(t, T) > xB(t, t')$, an investor could sell the T-maturing (long) bond and contract ahead to buy it back later for x. This arrangement still provides the same sure claim on a unit of cash at T, and it lets the investor pocket $B(t, T) - xB(t, t')$ immediately. Alternatively, if $B(t, T) < xB(t, t')$ one could sell x units of the short bonds, buy the long bond, and contract ahead to sell it at price x. Again, there would be an immediate gain at t, and at t' the x dollars obtained from the forward transaction would repay the principal due on the short bonds. Thus, in a market without opportunities for arbitrage the t'-forward price as of t is determined by the relative prices of T-maturing and t'-maturing bonds. Representing the forward price now as $B_t(t', T)$ instead of as x, we must have

$$B_t(t', T) = \frac{B(t, T)}{B(t, t')} \tag{4.7}$$

in order to eliminate opportunities for arbitrage.

Of course, the precise equality[13] here holds only if transacting and contracting are costless and if contracts are fully enforceable, but the ratio on the right at least provides a benchmark for contracts to borrow or lend ahead in real markets.

---

[13]To bring this relation quickly to mind, just reflect that $B_t(t', T)$ must be less than unity, which requires that the price of the *long* bond be in the numerator.

Referring to Table 4.1, we can infer from the quoted prices of strips the implicit forward prices of bonds of various intermediate dates t' and maturities T, as of $t = $ December 7, 2007. For example, reckoning from the averages of bid and asked prices, the time t forward price as of $t' = $ August 15, 2019 for a 10-year unit bond was

$$B_t(t',T) = \frac{.36224}{.58879} \doteq .61523.$$

This can be compared with the spot price of a (roughly) 10-year unit bond from the fourth row of the table: .65833. Of course, there was no way of knowing as of t what the *spot* price of a 10-year bond would be in 2019. That would depend on market conditions in 2019 that could be, at best, dimly foreseen in 2007. On the other hand, the forward price as of 2007 does reflect (among other things) some sort of current market consensus about what those future conditions would be.

Just as average and instantaneous spot rates can be deduced from spot prices of bonds of various maturities, their forward counterparts are implicit in bonds' forward prices. Specifically, the average continuously compounded forward rate as of t for riskless loans from t' to T would be

$$r_t(t',T) = -(T-t')^{-1} \ln B_t(t',T) \tag{4.8}$$

for $t < t' < T$, and the (idealized) *instantaneous* forward rate would be

$$r_t(t') = \frac{\partial \ln B_t(t',T)}{\partial t'}\bigg|_{T=t'} = -\frac{\partial \ln B_t(t',T)}{\partial T}\bigg|_{T=t'}$$

for $t < t'$. Of course, $r_t(t)$ would just be the short rate, r_t.

A reasonable way to estimate the course of short rates over periods of a few months is to assume that they will track the instantaneous forward rates $\{r_t(u)\}_{t \leq u \leq T}$. These forward rates would be explicitly observable at t if discount bonds with a continuous range of maturities were traded, but of course they are not. Still, US Treasury bills are normally issued weekly, and there is an active resale market, so that dealers' quotes are obtainable on bills with maturities spaced just one week apart and extending to about 6 months. A reasonable approximation is to set $r_t(u) = r_t(t',t'')$ for values of u falling within any such weekly interval (t',t'') between maturity dates. Of course, the average forward rate $r_t(t',t'')$ can be calculated via (4.8) as

$$\begin{aligned} r_t(t',t'') &= -(t''-t)^{-1} \ln B_t(t',t'') \\ &= (t''-t)^{-1}\left[\ln B(t,t') - \ln B(t,t'')\right]. \end{aligned}$$

As an example, the first two columns in each half of Table 4.2 show a dealer's asked prices as of January 16, 2008 for T-bills (as percentages of maturity value) at weekly maturity intervals $\{[t_{j-1},t_j]\}_{j=1}^{27}$, and column 3 shows the calculated average forward rates $r_t(t_{j-1},t_j)$, based on a 365-day year. With $t_0 = t$ corresponding to date 1/16 and $T = t_{27}$ corresponding to 7/17, the approximation $\int_t^T r_t(u) \cdot du \doteq \sum_{j=1}^{27} r_t(t_{j-1},t_j)(t_j - t_{j-1})$ yields $r(t,T) \doteq 0.0292$ as estimate

Table 4.2

t_j	$B\left(t,t_j\right)$	$r_t\left(t_{j-1},t_j\right)$	t_j	$B\left(t,t_j\right)$	$r_t\left(t_{j-1},t_j\right)$
1/17	99.991	0.0329	4/17	99.242	0.0305
1/24	99.940	0.0266	4/24	99.200	0.0221
1/31	99.982	0.0303	5/01	99.137	0.0331
2/07	99.824	0.0303	5/08	99.069	0.0358
2/14	99.766	0.0303	5/15	99.020	0.0258
2/21	99.708	0.0303	5/22	98.971	0.0258
2/28	99.649	0.0309	5/29	98.907	0.0337
3/06	99.591	0.0304	6/05	98.865	0.0221
3/13	99.535	0.0293	6/12	98.806	0.0311
3/20	99.478	0.0299	6/19	98.745	0.0322
3/27	99.418	0.0315	6/26	98.697	0.0254
4/03	99.361	0.0299	7/03	98.626	0.0375
4/10	99.300	0.0320	7/10	98.600	0.0137
4/17	99.242	0.0305	7/17	98.554	0.0243

of the average short rate out to T, whereas the direct calculation gives $r(t,T) = -\,(365/183)\ln(.98554) \doteq .0291$ to the nearest basis point (.01%).[14]

Just as forward rates can be deduced from bonds' forward prices, we can also go the other way and get the prices from the rates. From (4.8) we see at once that $B_t\left(t',T\right) = e^{-r_t(t',T)(T-t')}$ relates the forward bond price to the average forward rate, just as spot price and average spot rate are linked as $B\left(t,T\right) = e^{-r(t,T)(T-t)}$. Now by subdividing $[t',T]$ as $t' \equiv t_0 < t_1 < t_2 < \cdots < t_{n-1} < t_n \equiv T$ we can express the forward price of lending over $[t',T]$ as the product of forward prices over the subintervals, since the forward loan over the entire period can be replicated by a sequence of shorter loans. Thus

$$B_t\left(t',T\right) = B_t\left(t',t_1\right)B_t\left(t_1,t_2\right)\cdots\cdot B_t\left(t_{n-1},T\right).$$

This, in turn, leads to an expression in terms of average forward rates over the subintervals, as

$$B_t\left(t',T\right) = e^{-r_t(t_0,t_1)(t_1-t_0)}\cdot e^{-r_t(t_1,t_2)(t_2-t_1)}\cdots\cdot e^{-r_t(t_{n-1},t_n)(t_n-t_{n-1})}$$

$$= \exp\left[-\sum_{j=1}^{n} r_t\left(t_{j-1},t_j\right)\left(t_j - t_{j-1}\right)\right].$$

Now, by sending $n \to \infty$ and $\max_{j\in\{1,2,\ldots,n\}}\left(t_j - t_{j-1}\right) \to 0$, we get the time t forward price for loans over $[t',T]$ by exponentiating the integral of instantaneous

[14]2008 was a leap year!

forward rates:

$$B_t\left(t', T\right) = \exp\left[-\int_{t'}^{T} r_t\left(s\right) \cdot ds\right].$$

Finally, by taking $t' = t$ we relate the spot price itself to instantaneous forward rates, as

$$B\left(t, T\right) = \exp\left[-\int_{t}^{T} r_t\left(s\right) \cdot ds\right]. \tag{4.9}$$

What makes relations of this sort possible is that all the forward rates on which these relations depend are deducible just from the current price structure of discount bonds.

Note, however, that it is not possible to express $B\left(t, T\right)$ or any of the forward prices $B_t\left(t', T\right)$ in terms of instantaneous *spot* rates. This is because future spot rates depend on future bond prices, which are *not* known at t. However, as we progress in our study of financial markets, we will often find it useful to abstract from the uncertainty about future bond prices. The reasons are (1) that these tend to be much more predictable over moderate periods than are prices of other assets that interest us (common stocks, stock indexes, currencies, commodities, etc.), and (2) that the assumption of known short rates often greatly simplifies the analysis. Under this helpful (but counterfactual) assumption, short rates and instantaneous forward rates would necessarily coincide, and in that case we would indeed have $B(t, T) = \exp\left(-\int_{t}^{T} r_u \cdot du\right).$

EXERCISES

4.1 Create a table showing the value after 10 years of a $1 investment at annual rate .05 if there is compounding (**a**) once per year, (**b**) semiannually, (**c**) quarterly, and (**d**) continuously.

4.2 Find an expression for the value after t years of a $1 investment that earns continuously compounded interest at each of the following annualized rates; and plot the value functions for $0 \le t \le 10$:

$$r_{1t} = .05$$

$$r_{2t} = \begin{cases} .025, & 0 \le t \le 5 \\ .075, & 5 < t \le 10 \end{cases}$$

$$r_{3t} = .05\left[1 + \sin\left(\frac{\pi t}{5}\right)\right].$$

4.3 Treasury strips of $1000 principal value maturing in 0.5, 1.0, 1.5, and 2.0 years are priced at $975.31, $951.23, $927.74, and $904.84, respectively. Two bonds paying semiannual coupons of $15 and $30 (starting in 0.5 years) and maturing in 2.0 years are being offered for sale at $961.23 and $1012.61. Determine whether these prices offer an opportunity for arbitrage.

4.4 A discount bond maturing in 0.8 years can be bought for $B(t, t + .8) = .94$ dollars per unit principal value. What is the implied average continuously compounded rate of interest over this term, $r(t, t + .8)$?

4.5 A discount bond expiring in .002 year sells for .999900005 dollar per unit principal value. Find the approximate value of the instantaneous spot rate.

4.6 Show that formula (4.6) can be expressed more simply as

$$B(0, c; 1, 2, ..., n) = \frac{c}{y} + \left(1 - \frac{c}{y}\right)(1 + y)^{-n}.$$

4.7 A coupon-paying bond of maturity 2.0 years and $1000 principal value pays semiannual coupons beginning 0.5 year from now. Its yield to maturity is $y = .05$, and it is selling for $1,019.97. What is the amount of its semiannual coupon?

4.8 Based on the average of bid and asked prices for 8/15-maturing strips (i.e., strips that mature on August 15) in Table 4.1, find the implicit continuously compounded average forward rate for 10-year and 20-year loans commencing on August 15, 2009.

EMPIRICAL PROJECT 1

The file BondQuotes.xls in folder Project 1 from the FTP site contains a table of bid and asked quotes for 81 US Treasury securities as of February 18, 2008. The securities consist of Treasury bills, Treasury strips, Treasury notes, and Treasury bonds that mature at 50 distinct dates, all within 10 years of the quote date. The notes and bonds provide semiannual coupon payments at the indicated annual rates; for example, the 3% notes maturing on 2/15/09 were to pay 1.5% of principal value on 8/15/08 and 101.5% of principal value at maturity. Ordinarily, when one buys or sells a coupon bond, the interest accrued from the last payment up to the settlement date is added to the price and is not reflected in the bid and asked quotes. However, all the coupon bonds included in the table mature in integer multiples of 6 months from the quote date (on 8/15/08, 2/15/09, etc.), and so the accrued interest is negligible. The bid and asked quotes can therefore be considered realistic prices at which one could have sold and bought these securities on February 18, 2008.

The purpose of the project is to infer from the quoted prices the present values of sure one-dollar cash receipts on the 50 distinct dates in the table. These values represent the implied market prices of discount (zero-coupon), default-free, *unit* bonds maturing on those dates. Indeed, the Treasury bills and strips in the table *are* discount, default-free bonds, but additional information about market valuations of payments on their maturity dates is also embedded in the prices of bonds that pay coupons on those dates. In a world free of transactions costs and monitoring costs the coupon-paying bonds could be replicated perfectly by a portfolio of zero-coupon bonds. However, since such costs do exist in real markets, replicability is not quite perfect, and so the prices of coupon-paying bonds do indeed provide additional information about the appropriate discount factors for future payments. Here is a way to extract and use that information.

1. Find the discount unit bond prices $\{B\,(0,T_j)\}_{j=1}^{50}$ that are *implied* by the bond quotations using least-squares regression. As dependent or response variable take the average of the bid and asked quotes. There will thus be a vector of 81 prices, calculated as $\{P_i = (B_i + A_i)\,/2\}_{i=1}^{81}$. There will be 50 independent or predictor variables, one for each distinct maturity date in the collection. The ith observation on the jth predictor variable represents the cash payment $c_i\,(T_j)$ received at date T_j by one who holds the ith security. If the ith security matures at T_j, then the cash payment equals the principal value plus one-half the annual coupon rate for the security, as applied to the principal value; otherwise, it equals just half the annual coupon. For simplicity, take all principal values to be 100. We then have an equation of the following sort:

$$P_i = B\,(0,T_1)\,c_i\,(T_1) + B\,(0,T_2)\,c_i\,(T_2) + \cdots + B\,(0,T_{50})\,c_i\,(T_{50}) + e_i,$$

where e_i is an error term. The error picks up the effects of transacting and monitoring costs that could keep the relation from being exact. Notice that there is no intercept or constant term in the regression, since P_i should be zero if all of $\{c_i\,(T_j)\}_{j=1}^{50}$ are zero. However, include an intercept in a first trial and test whether its difference from zero is statistically significant. The R^2 value from this regression will indicate the degree to which the variation in the prices is "explained" by the variation among the payoffs. Use the estimated coefficients $\left\{\hat{B}\,(0,T_j)\right\}_{j=1}^{50}$ from a second trial without the intercept to calculate implied average spot rates, $\{\hat{r}\,(0,T_j)\}_{j=1}^{50}$.

2. Plot the coefficients and the implied average spot rates against time to maturity. Figure 4.2 gives an idea of what to expect.

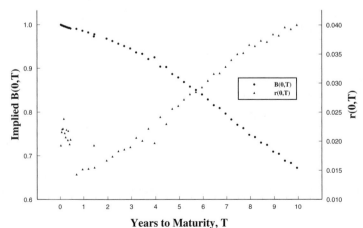

Figure 4.2   Discount factors and average spot rates implied by price quotes for 81 Treasury securities.

3. Compare actual averages of bid and asked prices with the fitted values from the regression

$$\hat{P}_i = \hat{B}\left(0, T_1\right) c_i\left(T_1\right) + \hat{B}\left(0, T_2\right) c_i\left(T_2\right) + \cdots + \hat{B}\left(0, T_{50}\right) c_i\left(T_{50}\right),$$

to get an indication of which bonds would be especially attractive to buy or sell, abstracting from differences in liquidity and special preferences for maturity dates.

4. Look for any apparent opportunities for arbitrage. Such an opportunity exists if one can buy one or more bonds at the asked price and sell one or more at the bid price so as to come up with a cash surplus at some period while having no cash deficit at any other. (We will see in Section 16.3 that there is a systematic way of doing this. If you can't wait, peek ahead.)

CHAPTER 5

MODELS OF PORTFOLIO CHOICE

Financial models are either *normative*, meaning *pre*scriptive, or *positive*, meaning *de*scriptive. Normative models are supposed to guide the decision maker by showing how rational objectives can be met. For example, normative *portfolio* models are supposed to prescribe ways to choose portfolios of assets so as to achieve certain goals. By contrast, positive models are supposed to describe reality (e.g., how economic actors actually *do* choose portfolios). The two approaches to modeling often converge. Economists typically *assume* that individuals will try to do what is best for them and that they have the information and skills for doing so. Thus, normative models often serve as the basis for descriptions of how the economy functions. One builds a model using such rationality assumptions, then tests the implications by comparing the predictions of the model with data. The positivist methodology accepted by most economists requires that models be evaluated on the basis of congruence between prediction and observation, rather than on the descriptive accuracy of assumptions. If the predictions are bad, the usual recourses are to reexamine the empirical methods and reconsider the logic of the normative model and the conception of what constitutes rational behavior. If those seem unassailable, economists as a last resort may begin to question the rationality postulate. We shall see instances of this in Chapter 8 in connection with decisions under uncertainty, where compelling experimental evidence has led many economists to think more broadly about how humans behave. Still, there is much to learn by comparing observations with predictions based on (what we think is) rational behavior. Nothing better illustrates this learning process than the evolution of models of portfolio choice over the past half century. We begin the story in this chapter.

5.1   MODELS THAT IGNORE RISK

One of the earliest prescriptive models for portfolio choice was the *dividend–discount* model. In the simplest form, the advice to an investor was to choose the asset(s) whose future cash flows are of greatest present value relative to current price. Thus, standing at time t and looking ahead, one is supposed to calculate

$$V_t = \sum_{i=1}^{\infty} B\left(t, T_i\right) D_i$$

for each asset and to choose the one(s) for which V_t/P_t is greatest, where P_t is the current price and the $\{D_i\}$ are the dividends received at the future dates $\{T_i\}$. Here "dividend" is construed broadly to encompass all payments received by the holder of the asset. Thus, for a bond of unit principal value that pays periodic coupons and matures at T_n, we have $V_t = c\sum_{i=1}^{n} B\left(t, T_i\right) + B\left(t, T_n\right) = B\left(t, c; T_1, ..., T_n\right).$

There are some other simple special cases. If the asset is a stock that pays dividends at equally spaced dates $\{T_i = t + \tau i\}$, and if the investor perceives them as growing at a constant continuously compounded (annualized) rate ρ from some initial value D_0, then $D_i = D_0 e^{\rho \tau i}$ and

$$V_t = D_0 \sum_{i=1}^{\infty} B\left(t, t + \tau i\right) e^{\rho \tau i}.$$

If, in addition, the average continuously compounded spot rate happens to have the same (annualized) value r from t out to each date $t + \tau i$, then $B\left(t, t + \tau i\right) = e^{-r\tau i}$. In this case, provided $r > \rho$, the valuation formula simplifies to

$$V_t = D_0 \sum_{i=1}^{\infty} e^{-(r-\rho)\tau i} = \frac{D_0}{e^{(r-\rho)\tau} - 1}. \tag{5.1}$$

Of course, if $\rho \geq r$, the value would be infinite, but one presumes that such a rate of growth would be unsustainable.

The obvious limitation of this prescription for choosing assets is that one rarely knows for sure what the future stream of dividends will be. Of course, they are predictable enough for Treasury securities, but with corporate or municipal debt there is always the possibility of default. As for stock dividends, while they are usually fairly stable in the short run, one really cannot be confident of their values even in the near future. A more sophisticated theory recognizes the uncertainty involved in making these predictions and represents that uncertainty by a probability distribution; e.g., by a cumulative distribution function $F_t(x) = \Pr(D_i \le x \mid \mathcal{F}_t)$ or a probability density function $f_t(x) = dF_t(x)/dx$. Here $\mathcal{F}_t$ represents the information available at t—specifically, the σ field of events whose occurrence or nonoccurrence we can verify at t—and the symbol "$\mid$" and the subscripts on F and f indicate that probabilities and densities are conditional on this information. The easiest fix is now to replace D_i with

$$E_t D_i \equiv E(D_i \mid \mathcal{F}_t) = \int x \cdot dF_t(x) = \int x f_t(x) \cdot dx$$

and write

$$V_t = \sum_{i=1}^{\infty} B(t, T_i) E_t D_i.$$

If dividends are modeled as growing *on average* at constant rate ρ, as $D_i = D_0 e^{\rho \tau i} + u_i$ with $E_t u_i = 0$, and if the average spot rate is still treated as a constant, then expression (5.1) still applies for the subjective value of one unit of the asset. With this model the investor "simply" has to know the expected growth rate of dividends in perpetuity and can compare different assets on the basis of their specific values of D_0 and ρ.

Rather than have the investor look into the indefinite future, there is an alternative approach that seems to require less foresight. This supposes that the rational investor has in mind a "holding period" for assets and can forecast their values at the end of that period. To simplify notation, we can just measure time in holding-period units. Thus, an individual who contemplates buying one share of an asset at $t = 0$ would need to consider only its value at $t = 1$ and the value of any cash flow (dividend) received in the interim, or $P_1 + D_1$.[15] This is the amount of cash (per share) that would available to the investor at $t = 1$ if the asset were sold (costlessly) at that time. Expressing this gross cash value relative to the initial cost per unit produces the *total return per dollar*,

$$1 + R \equiv \frac{P_1 + D_1}{P_0}.$$

It is this quantity, or simply the one-period *rate* of return R itself, that is presumed to be of interest to the investor. (Since this is a one-period model, there is no need to give R a subscript.) Of course, as of $t = 0$ this rate of return is a random variable for anyone who lacks perfect foresight. The expectations version of the prescriptive model now has the investor choose the asset or assets for which the *expected* rate of return, $ER = E(P_1 + D_1)/P_0 - 1$, is greatest. (Here and throughout, "$E(\cdot)$" without a subscript stands for $E_0(\cdot) \equiv E(\cdot \mid \mathcal{F}_0)$, the conditioning on initial information being understood.)

As normative theories, all of these expectations models suffer from an obvious limitation. For example, considering the one-period version of the model, suppose

---

[15]Dividends received before the end of the holding period would presumably be reinvested rather than held in cash. For example, they might be used to purchase discount bonds or to buy more of the underlying asset itself. In this case D_1 represents the value of such investments at $t = 1$.

that there are two assets, labeled p and q, with one-period rates of return (random variables) having the same expected value $ER_p = ER_q$. (If no other asset has the same expected rate of return as q, take asset j with $ER_j < ER_q$ and asset ℓ with $ER_\ell > ER_q$, set $R_p \equiv pR_j + (1-p)R_\ell$, and choose p such that $ER_p = ER_q$.) Should rational individuals be indifferent between assets p and q? Is one irrational who thinks that some other aspect of their distributions is relevant? We often say that one who cares about expected value alone is *risk neutral*, because such a person takes no account of the chance that return will differ from its expected value. In other words, such a person just doesn't care about risk, even when large amounts of capital are at stake. There may be such fearless people, but one could scarcely propose to an average person an investment policy that did not take some account of risk.

As positive theories of how individuals actually *do* choose portfolios, the expectations models are miserable failures. To see why, suppose that people did simply try to maximize expected future wealth. Assuming that there are n assets and letting a_j be the number of units of asset j that one holds at $t = 0$, the expected value of wealth at $t = 1$ is

$$EW_1 = \sum_{j=1}^{n} a_j E\left(P_{j1} + D_{j1}\right) = \sum_{j=1}^{n} a_j P_{j0} E\left(\frac{P_{j1} + D_{j1}}{P_{j0}}\right)$$

$$= W_0 \sum_{j=1}^{n} \left(\frac{a_j P_{j0}}{W_0}\right)(1 + ER_j) = W_0 \left(1 + \sum_{j=1}^{n} p_j ER_j\right),$$

where $p_j \equiv a_j P_{j0}/W_0$ is the fraction of current wealth W_0 that is invested in asset j. For an arbitrary, given value of W_0 the optimization problem is then

$$\max_{\{p_1, p_2, \ldots, p_n\}} \sum_{j=1}^{n} p_j ER_j$$

subject to the restriction that $\sum_{j=1}^{n} p_j = 1$. To be more realistic, we could also restrict each p_j to be nonnegative, since individual investors cannot actually reinvest the proceeds of short sales until the shorts are covered. Clearly, if asset i is the one with highest expected rate of return, the solution is then to set $p_i = 1$ and $p_j = 0$ for all $j \neq i$. (We could have $p_i > 1$ for an institutional investor that is able to reinvest the proceeds of short sales of other assets, but there would be some upper limit that depends on whatever credit limitations apply.) Thus, except in the implausible circumstance that the same largest expected return is attained by two or more assets, an individual should and—applying the rationality paradigm—*does* put all wealth into a single asset. We observe, to the contrary, that not many people are so foolish.

Now think about extending to the market level, and consider the implications of such portfolio specialization for prices of assets in a market equilibrium. In a world in which everyone had access to identical information, the models would imply that all assets have exactly the same expected return in any given period. Why? Because if expected returns were initially different, no one would *willingly* hold an asset whose expected return was less than the largest expected return. Efforts to sell losers and buy winners would cause current prices (the denominators) to adjust until

$$E(1 + R_j) = E\left(\frac{P_{j1} + D_{j1}}{P_{j0}}\right) = E\left(\frac{P_{k1} + D_{k1}}{P_{k0}}\right) = E(1 + R_k)$$

for all j, k. In other words, people would have to expect returns of tech stocks, money-market funds, junk bonds, puts and calls, etc. to be precisely the same. (In a world in which people had different information, equilibrium would require that there always be *some* people who expect money-market funds to return more than any other asset—stocks, long-term bonds, junk bonds, etc.) Dissatisfaction with these absurdities led to the next development in portfolio theory, which for the first time took risk and risk aversion into account in a systematic and objective way.

5.2   MEAN–VARIANCE PORTFOLIO THEORY

Harry Markowitz (1952, 1959) made the fundamental observation that investors should (and do) care not just about assets' *expected* returns but about their riskiness as well. He proposed to quantify a portfolio's risk in terms of the variance (or, equivalently, in terms of the standard deviation, $\sigma = +\sqrt{\sigma^2}$) of its one-period rate of return. Markowitz showed that it is possible in principle to identify those portfolios that have the highest level of expected return for each attainable level of risk. Armed with this information, investors could choose risk–return combinations that best suited their tastes. We will now see just how this works. What Markowitz' theory implies about assets' equilibrium prices is discussed in the next chapter.

5.2.1   Mean–Variance "Efficient" Portfolios

We need some notation and definitions. Regard $t = 0$ as the current time, which is when a portfolio is to be chosen, and $t = 1$ as the first date at which the portfolio is to be reevaluated, rebalanced, or cashed out.[16] The unit of time is unspecified; it could be a day, a month, a year, etc., depending on the investor. Initially, we consider there to be n assets with one-period rates of return $\{R_j = (P_{j1} + D_{j1})/P_{j0} - 1\}_{j=1}^{n}$ having mean values $\{\mu_j = ER_j\}_{j=1}^{n}$, positive (but finite) variances $\{\sigma_j^2 \equiv \sigma_{jj} = VR_j\}_{j=1}^{n}$, and covariances $\{\sigma_{jk} = C(R_j, R_k)\}_{j \neq k}$. Let $\mathbf{R} = (R_1, R_2, ..., R_n)'$ be the vector of rates of return, $\mu = (\mu_1, ..., \mu_n)'$ their means, and Σ the $n \times n$ covariance matrix, with σ_{jk} as the element in row j and column k.[17] Vector $\mathbf{a} = (a_1, ..., a_n)'$ represents the number of units (shares) of each asset in a portfolio; and $\mathbf{p} = (p_1, ..., p_n)'$, the *proportions* of current wealth invested, with $p_j = a_j P_{j0}/W_0$ and $\sum_{j=1}^{n} p_j = 1$. We use the terms "wealth" and "portfolio value" interchangeably. Clearly, the vector $\mathbf{p}$

---

[16]If it is merely to be evaluated rather than cashed out, the implicit assumption in the *one-period* framework is that the investor takes no account of how present choices affect future decisions.

[17]Recall that vectors are in column form unless transposition to row form is indicated by a prime. Thus, $\mu = (\mu_1, \mu_2, ..., \mu_n)'$ and $\mu' = (\mu_1, \mu_2, ..., \mu_n)$.

and the level of initial wealth uniquely determine the composition of the portfolio. Its rate of return is the scalar $R_\mathbf{p} \equiv \sum_{j=1}^{n} p_j R_j = \mathbf{p}'\mathbf{R}$, having mean $ER_\mathbf{p} \equiv \mu_\mathbf{p} = \sum_{j=1}^{n} p_j \mu_j = \mathbf{p}'\mu$ and variance $VR_\mathbf{p} \equiv \sigma_\mathbf{p}^2 = \sum_{j=1}^{n} \sum_{k=1}^{n} p_j p_k \sigma_{jk} = \mathbf{p}'\Sigma\mathbf{p}$. Wealth at $t = 1$ can be represented either in terms of the a's and the prices, as $W_1 = \sum_{j=1}^{n} a_j (P_{j1} + D_{j1})$ where $\sum_{j=1}^{n} a_j P_{j0} = W_0$, or in terms of the p's and the returns, as $W_1 = W_0 (1 + R_\mathbf{p}) = W_0 \left(1 + \sum_{j=1}^{n} p_j R_j\right)$. The mean and variance of wealth are related to the mean and variance of rate of return as

$$EW_1 = W_0 (1 + \mu_\mathbf{p}) = W_0 \left(1 + \sum_{j=1}^{n} p_j \mu_j\right) = W_0 (1 + \mathbf{p}'\mu) \qquad (5.2)$$

$$VW_1 = W_0^2 \sigma_\mathbf{p}^2 = W_0^2 \sum_{j=1}^{n} \sum_{k=1}^{n} p_j p_k \sigma_{jk} = W_0^2 \mathbf{p}'\Sigma\mathbf{p}. \qquad (5.3)$$

Of course, the standard deviation (positive square root of the variance) of W_1 is just $W_0 \sigma_\mathbf{p}$. Since W_0 is given, it is only $\sigma_\mathbf{p} = +\sqrt{\mathbf{p}'\Sigma\mathbf{p}}$ and $\mu_\mathbf{p} = \mathbf{p}'\mu$ that matter in choosing the portfolio.

Basic to Markowitz' theory are the following primitive assumptions, referred to collectively as the assumption of *mean–variance dominance*:

1. Of two portfolios having the same mean rate of return, *everyone* will prefer the one with smaller standard deviation (and therefore smaller variance).

2. Of two portfolios whose returns have the same standard deviation (hence, the same variance), *everyone* will prefer the one with larger mean.

Under some further assumptions about completeness and transitivity of the preference ordering, this implies for each individual a set of positively sloped indifference curves in the $(\sigma_\mathbf{p}, \mu_\mathbf{p})$ plane, such as in Fig. 5.1, with higher curves corresponding to more favored combinations.

Markowitz' main contribution was in figuring out the best *attainable* trade-off of mean and standard deviation from a given collection of securities having known means, variances, and covariances. Specifically, given μ and Σ, he showed how to find the *set* of feasible portfolios $\{\mathbf{p}\}$ that have the lowest variance (hence, the lowest standard deviation) for each attainable level of mean return. The problem is represented formally as

$$\min_{\{p_1, \ldots, p_n\}} \sigma_\mathbf{p}^2 = \sum_{j=1}^{n} \sum_{k=1}^{n} p_j p_k \sigma_{jk} = \mathbf{p}'\Sigma\mathbf{p}$$

subject to the constraints $\sum_{j=1}^{n} p_j \mu_j = \mu^*$, $\sum_{j=1}^{n} p_j = 1$, and $p_1 \geq 0, \ldots, p_n \geq 0$. Here μ^* represents any given value between the lowest and highest attainable mean rates of return—that is, one fixes μ at some feasible level μ^* and finds among all sets of investment proportions with this expected rate of return the one having the least variance, repeating the process for *all* feasible values μ^*. Generally, such

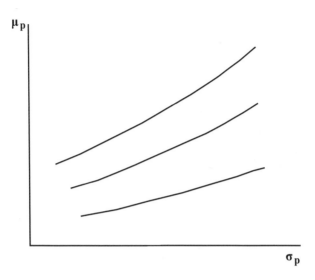

Figure 5.1 Indifference cuves in $(\sigma_{\mathbf{p}}, \mu_{\mathbf{p}})$ plane.

quadratic programming problems (so called because $\mathbf{p}'\mathbf{\Sigma p}$ is a quadratic expression in the $\{p_j\}$) have to be solved numerically, as there is no simple explicit formula for the optimal portfolio weights. However, if the inequality constraints $\{p_j \geq 0\}$ are dropped, allowing assets to be sold short and the proceeds reinvested, then this becomes a simple Lagrange-multiplier problem from multivariate calculus:

$$
\min_{\mathbf{p},\lambda,\phi} \sum_{j=1}^{n} \sum_{k=1}^{n} p_j p_k \sigma_{jk} - \lambda \left(1 - \sum_{j=1}^{n} p_j \right) - \phi \left(\mu^* - \sum_{j=1}^{n} p_j \mu_j \right)
$$
$$
= \min_{\mathbf{p},\lambda,\phi} \left[\mathbf{p}'\mathbf{\Sigma p} - \lambda \left(1 - \mathbf{p}'\mathbf{1} \right) - \phi \left(\mu^* - \mathbf{p}'\mu \right) \right],
$$

where λ and ϕ are Lagrange multipliers and $\mathbf{1}$ is a units vector. An explicit solution for p can in fact be obtained and represented in matrix form.[18]

Since the form of the solution is rather complicated, it is more instructive to look at a simple case and just state a *property* of the general solution. First, consider the simple case in which there are $n = 2$ primitive assets, neither of which dominates the other in the mean–variance sense. Thus, supposing $\mu_1 > \mu_2$ and $\sigma_1 > \sigma_2$ and setting $p_1 = p$ and $p_2 = 1 - p$ to simplify, the portfolio's rate of return $R_p = pR_1 + (1 - p) R_2$ has mean and variance

$$
\mu_p = p\mu_1 + (1 - p)\mu_2 = \mu_2 + p\left(\mu_1 - \mu_2\right)
$$
$$
\sigma_p^2 = p^2 \sigma_1^2 + 2p\left(1 - p\right)\sigma_{12} + \left(1 - p\right)^2 \sigma_2^2.
$$

---

[18]One who wants to see the details can consult Chapter 4 of Ingersoll (1987).

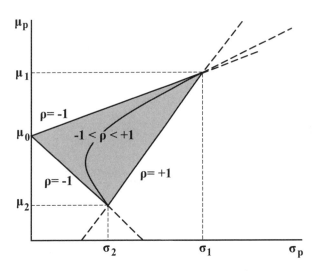

Figure 5.2   Feasible mean, standard deviation loci with two assets.

There are two polar cases, corresponding to perfect positive and perfect negative correlation, $\rho_{12} \equiv \sigma_{12}/(\sigma_1\sigma_2) = \pm 1$. In either case the variance is a perfect square, and $\sigma_p = |p\sigma_1 \pm (1-p)\sigma_2|$. In the case $\rho_{12} = +1$, we can solve for p in terms of σ_p and write

$$p = \frac{\sigma_p - \sigma_2}{\sigma_1 - \sigma_2}$$

$$\mu_p = \mu_2 + \frac{\sigma_p - \sigma_2}{\sigma_1 - \sigma_2}(\mu_1 - \mu_2).$$

The locus of points $\{(\sigma_p, \mu_p)\}_{0 \le p \le 1}$ is a straight line with positive slope in the (σ_p, μ_p) plane.   It represents the points attainable by forming portfolios of these two perfectly correlated assets. Alternatively, when $\rho_{12} = -1$, we get $\sigma_p = |p\sigma_1 - (1-p)\sigma_2|$, which yields one of two expressions, depending on the relation of p to $p_0 \equiv \sigma_2/(\sigma_1 + \sigma_2)$:

$$\sigma_p = \begin{cases} -\sigma_2 + p(\sigma_1 + \sigma_2), & p \ge p_0 \\ \sigma_2 - p(\sigma_1 + \sigma_2), & p \le p_0 \end{cases}.$$

Solving for p in terms of σ_p and plugging into the expression for μ_p in the two cases give a pair of lines intersecting at the point $\mu_0 \equiv (\sigma_1\mu_2 + \sigma_2\mu_1)/(\sigma_1 + \sigma_2)$ on the μ_p axis, as in Fig. 5.2. Thus, when the assets are perfectly negatively correlated there exists a "hedge" portfolio that has deterministic value $R_p = \mu_0$.

The shaded triangular area bounded by the single line for $\rho_{12} = +1$ and the two lines for $\rho_{12} = -1$ contains all the possible loci of (σ_p, μ_p) combinations corresponding to $0 \leq p \leq 1$. When $-1 < \rho_{12} < 1$ the locus is a hyperbola bowed toward the μ_p axis, as in the figure. This shows that if the correlation is small enough, one can attain portfolios with lower standard deviation than that of either asset. The dotted lines show extensions of the loci when we allow $p < 0$ or $p > 1$. These cases correspond to short sales of one asset with proceeds invested in the other.

In the general case of n assets, no two of which are perfectly correlated and all of which have positive, finite variances, it remains true that the minimum-standard-deviation frontier—the locus corresponding to the portfolios with lowest σ_p for a given μ_p—is a hyperbola that faces the μ_p axis. The upper branch of this, which is the locus of portfolios having highest μ_p for given σ_p, is called the "efficient frontier." It is *efficient* in the sense that any portfolio *not* on the frontier would be *dominated*, meaning that everyone would prefer one or more of the portfolios on the frontier. Of course, this relies crucially on the validity of Markowitz' mean–variance dominance *assumption*. We shall consider later how well this really holds up. Figure 5.3 shows the n-asset feasible set as a shaded region and the efficient frontier as the unbroken edge at the top. The figure also shows what happens when we add to the mix a riskless asset with sure simple rate of return $R_0 = r$. Placing fraction p of wealth in some arbitrary asset or portfolio at a point (σ_1, μ_1) and $1 - p$ in the riskless asset yields a portfolio rate of return with mean $\mu_p = r + p(\mu_1 - r)$ and standard deviation $\sigma_p = p\sigma_1$ when $p \geq 0$. The feasible locus is then the straight line $\mu_p = r + \sigma_p[(\mu_1 - r)/\sigma_1]$, which is the thin line in the figure. Portfolios along this line are clearly dominated. Only the "tangency" portfolio on the risky-asset efficient set (designated "T") is now relevant, and the investor would choose some combination of that portfolio and riskless lending or borrowing. Such combinations are represented in the figure by the more thickly drawn line extending from $(0, r)$ to (σ_T, μ_T) and beyond. In the next chapter we shall see what the existence of this tangency portfolio implies for equilibrium market prices under the further assumptions of the famous "capital asset pricing model" (CAPM).

Empirical Project 2 at the end of the chapter offers the opportunity to estimate mean–variance efficient portfolios of stocks of five technology companies on the basis of time series of historical monthly returns.

5.2.2 The Single–Index Model

While it is indeed feasible to calculate the efficient frontier for portfolios comprising a handful of assets, the general procedure cannot be carried out on a large scale because it requires estimates of too many parameters. If there are but two risky assets, one needs just the mean returns, μ_1 and μ_2, the variances, σ_1^2 and σ_2^2, and a single covariance, σ_{12}—a total of five parameters. For five risky assets, there are five means, five variances, and 20 covariances, for a total of 30 parameters; and for $n = 10$ risky assets, the number of parameters increases to 65. Even this is easily managed; however, for portfolios of $n = 100$ assets we require 5150 parameters, and the number grows to about 12.5×10^6 if one considers just the 5000 or so US stocks traded on the NYSE (New York Stock Exchange) and NASDAQ. The problem is that the number of covariances, $n(n-1)/2$, increases with the square of n. Any statistical estimate

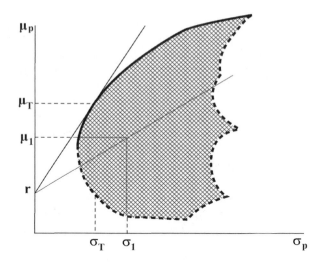

Figure 5.3   Feasible (σ_p, μ_p) points and efficient frontier with and without riskless asset.

of the covariance matrix based on fewer than $n(n-1)/2$ observations of returns would be *singular*, and this would make the necessary computations infeasible.

An approximation that *is* feasible was suggested by William Sharpe (1963). He called it a "single-index" model, but we can present it in slightly different form as a "factor" model. The assumption is that rates of return have the following one-factor structure:

$$R_{jt} = \mu_j + \beta_j \mathfrak{f}_t + u_{jt}, t = 1, 2, \dots . \tag{5.4}$$

Here t counts discrete periods of time, μ_j and β_j are constants specific to asset j, $\mathfrak{f}_t$ (no j subscript!) is a (possibly unobservable) economy-wide variable or *common factor* that influences the return of any asset j whose factor "loading" β_j is not zero, and u_{jt} is a random "idiosyncratic" shock. This shock variable is supposed to have mean zero and to affect just asset j. The last requirement means specifically that shock u_{jt} is uncorrelated with $\mathfrak{f}_t$ and that shocks u_{jt}, u_{kt} for $k \neq j$ are uncorrelated. Factor loading β_j determines the sensitivity of the rate of return of asset j to fluctuations in the common factor. By standardizing (centering and scaling) the common factor so that $E\mathfrak{f}_t = 0$ and $V\mathfrak{f}_t = E\mathfrak{f}_t^2 = 1$, and adjusting the $\{\mu_j, \beta_j\}$ accordingly, one can calculate the covariance between assets j and k as

$$\begin{aligned}
\sigma_{jk} &= E(R_{jt} - \mu_j)(R_{kt} - \mu_k) \\
&= E(\beta_j \mathfrak{f}_t + u_{jt})(\beta_k \mathfrak{f}_t + u_{kt}) \\
&= \beta_j \beta_k.
\end{aligned}$$

With this factor setup, calculating efficient portfolios for n assets requires just $3n$ parameters instead of $2n + n(n-1)/2$; namely, the n sets of means, variances, and factor loadings, $\{\mu_j, \sigma_j^2, \beta_j\}_{j=1}^n$. A proxy that is often used for the common factor is just the standardized rate of return on some broad-based stock index, such as the S&P 500 (Standard & Poor's 500-Stock Price Index) or Wilshire 5000. Letting R_{mt} represent the index rate of return, we set $\mathfrak{f}_t = (R_{mt} - \hat{\mu}_m)/\hat{\sigma}_m$, where $\hat{\mu}_m, \hat{\sigma}_m$ are estimates of the mean and standard deviation; e.g., $\hat{\mu}_m = T^{-1}\sum_{t=1}^{T} R_{mt}$, $\hat{\sigma}_m^2 = (T-1)^{-1}\sum_{t=1}^{T}(R_{mt} - \hat{\mu}_m)^2$. Estimates of μ_j, σ_j^2 for each asset can be found in the same way. The critical β_j parameter can be estimated by ordinary least squares regression, as

$$\hat{\beta}_j = \frac{\sum_{t=1}^{T} R_{jt}\mathfrak{f}_t}{\sum_{t=1}^{T} \mathfrak{f}_t^2}.$$

However, there are many statistical and conceptual issues here. Do returns really have the necessary i.i.d. structure for these to be good estimates? Are historical returns really that good a guide to the future? How long a time series should one use, allowing for the fact that firms seem to go through cycles of growth, maturity, decline, and (sometimes) rebirth? In principle, one could supply one's own subjective estimates on the basis of fundamental analysis of the assets involved, or just seat-of-the-pants judgment. Of course, the usual GIGO (garbage in–garbage out) caveat applies.

One must recognize also that the one-factor structure proposed by Sharpe is just a convenient simplification. In Chapter 16 we will encounter multifactor models that better represent the behavior of assets' returns.

EXERCISES

5.1 The following estimates were obtained for moments of daily rates of return of MSFT (Microsoft) and XOM (Exxon-Mobil) from daily (dividend-adjusted) closing prices during 2006: $\mu_M = .00057$, $\mu_X = .00123$, $\sigma_M^2 = .00017$, $\sigma_X^2 = .00014$, $\sigma_{MX} \doteq .00000$. A portfolio with proportions of value p and $1 - p$ in MSFT and XOM, respectively, would have daily rate of return $pR_M + (1-p) R_X$. Find, subject to the constraint $0 \le p \le 1$,

    (a) The value p_* that would have produced the *minimum-variance* portfolio.

    (b) The mean and standard deviation of the return of the minimum-variance portfolio, μ_* and σ_*.

    (c) The mean and standard deviation of the return of the portfolio with highest mean return.

5.2 For the data in Exercise 5.1 plot the feasible (i.e., attainable) set of points $(\sigma_p, \mu_p)_{0 \le p \le 1}$. Identify on the plot the minimum-variance portfolio and the *efficient* set of portfolios.

5.3 As in Exercises 5.1 and 5.2 there are two assets whose rates of return R_M, R_X are uncorrelated, and we consider portfolios with rate of return $R_p \equiv pR_M +$

$(1 - p) R_X$ having mean and variance μ_p, σ_p^2. Our preferences are expressed by the function

$$U (EW_1, VW_1) = EW_1 - \frac{\rho}{2} \frac{VW_1}{W_0} = W_0 \left[(1 + \mu_p) - \frac{\rho}{2}\sigma_p^2\right],$$

where positive constant ρ governs risk aversion. Since W_0 is given, maximizing this is equivalent to maximizing the following function of the moments of R_p:

$$U^* \left(\mu_p, \sigma_p^2\right) = \mu_p - \frac{\rho}{2}\sigma_p^2.$$

Note that this specification produces *linear* indifference curves $\mu_p = U^* + \rho\sigma_p^2/2$ in the $\left(\sigma_p^2, \mu_p\right)$ plane but *convex* curves in (σ_p, μ_p). Determine the optimal feasible investment proportion p_0 and from this the optimal attainable point (σ_0, μ_0). For what values of ρ is the constraint $0 \leq p_0 \leq 1$ satisfied? Taking $\rho = 10$, calculate the specific values of μ_0 and σ_0 for the data in Exercise 5.1, and indicate this point in the figure created in Exercise 5.2. Finally, plot in the figure over a range of σ values the indifference curve that is tangent to the feasible set at (σ_0, p_0).

5.4 There are n assets with one-period rates of return $R_1, R_2, ..., R_n$. All have the same mean μ and the same variance σ^2, and they are all uncorrelated, so that $\sigma_{jk} = 0$ for $j \neq k$. A portfolio $\mathbf{p} = (p_1, p_2, ..., p_n)'$ (with $\sum_{j=1}^{n} p_j = 1$) has rate of return $R_{\mathbf{p}}$ with mean $\mu_{\mathbf{p}}$ and variance $\sigma_{\mathbf{p}}^2$. What would be the minimum-variance portfolio? What would the feasible set look like if plotted in the $\left(\sqrt{V} = \sigma_{\mathbf{p}}, E = \mu_{\mathbf{p}}\right)$ plane? Which of the feasible portfolios would the Markowitz investor choose?

EMPIRICAL PROJECT 2

The purpose of this project is to give further and more realistic experience in computing mean–variance efficient frontiers.

1. Obtain from the Project 2 folder at the FTP site the file 5TechsS&P.xls. This records beginning-of-month prices of the five tech stocks AAPL, CSCO, DELL, INTC, and MSFT from December 1990 through January 2005, along with values of the S&P 500 index for the same period. The individual stocks' prices have been "corrected" for dividends and splits, meaning that dividends paid during a month have been added into the price to offset the decline in price that occurs when the stock begins to trade ex dividend. From these data calculate the simple monthly rates of return for each stock j in each month t as $R_{jt} = P_{jt}/P_{j,t-1} - 1$, where month $t = 1$ pertains to the month of December 1990 and month $t = 169$ pertains to the month of December 2004 and where P_{jt} is the dividend/split-adjusted price at the beginning of month t. (The assumption implicit in this calculation is that dividends paid during the month were reinvested in the stock.)

2. Calculate and display the vector $\hat{\mu} = (\hat{\mu}_1, \hat{\mu}_2, ..., \hat{\mu}_5)'$ of sample means of the five monthly rates of return and the 5×5 matrix $\hat{\Sigma}$ of sample variances and covariances of the rates of return. (The estimated covariance $\hat{\sigma}_{jk}$ between rates $\{R_j, R_k\}_{t=1}^{169}$ is in row j and column k of $\hat{\Sigma}$, while diagonal element $\hat{\sigma}_{jj} \equiv \hat{\sigma}_j^2$ is the estimated variance of R_j.) Plot the five points $\{(\hat{\sigma}_j, \hat{\mu}_j)\}_{j=1}^{5}$ in a graph whose horizontal and vertical axes are standard deviation and mean monthly rate of return, respectively. Other features will be added to the graph in the steps below.

3. Let $\mathbf{p} = (p_1, p_2, ..., p_5)'$ represent a (column) vector comprising proportions of wealth invested in the five securities, with $\mathbf{p}'\mathbf{1} = \sum_{j=1}^{5} p_j = 1$ (where $\mathbf{1}$ is a vector of units), and let $\mathbf{R}_t = (R_{1t}, ..., R_{5t})'$ be the vector of five rates of return in month t. Then $R_{\mathbf{p}t} = \mathbf{p}'\mathbf{R}_t = \sum_{j=1}^{5} p_j R_{jt}$ is the rate of return in month t on portfolio $\mathbf{p}$. Write out the formula that expresses the estimated mean and variance of $R_{\mathbf{p}t}$, $\hat{\mu}_{\mathbf{p}}$ and $\hat{\sigma}_{\mathbf{p}}^2$, in terms of $\hat{\mu}$ and $\hat{\Sigma}$.

4. Express in symbols the value of $\mathbf{p}$ that corresponds to the global minimum-variance portfolio; i.e., the value $\mathbf{p}_{\min V}$ that solves the problem $\min_{\mathbf{p}} \hat{\sigma}_{\mathbf{p}}^2$ subject to $\sum_{j=1}^{5} p_j = 1$. (Do not impose nonnegativity constraints on the p_j.) Now, using the data, calculate explicitly the minimum-variance portfolio of the five stocks. Present the vector $\mathbf{p}_{\min V}$ and the estimated portfolio mean and standard deviation, $\hat{\mu}_{\min V}$ and $\hat{\sigma}_{\min V}$. Plot these in your graph.

5. Let $\hat{\mu}_{\max\mu}$ be the largest of the estimated mean returns of the five stocks. Divide the interval from $\hat{\mu}_{\min V}$ up through $2\hat{\mu}_{\max\mu}$ into 50 equal parts, and for each $q \in \{0, 1, 2, ..., 50\}$ set

$$\hat{\mu}_q = \hat{\mu}_{\min V} + \frac{q}{50}\left(\hat{\mu}_{\max\mu} - \hat{\mu}_{\min V}\right).$$

For each of these values of the mean, find the portfolio $\mathbf{p}_q$ that solves $\min_{\mathbf{p}} \hat{\sigma}_{\mathbf{p}}^2$ subject to $\sum_{j=1}^{5} p_j = 1$ and $\sum_{j=1}^{n} p_j \hat{\mu}_j = \hat{\mu}_q$. (The portfolio for $q = 0$ should be the same as that in step 4 above.) Tabulate the 51 portfolios and the corresponding means $\hat{\mu}_q$ and standard deviations $\hat{\sigma}_q$ of the portfolios' returns. The 51 $(\hat{\sigma}_q, \hat{\mu}_q)$ points will fall on and illustrate the outline of a portion of the mean–standard-deviation efficient frontier that can be achieved with these five stocks. Plot the points on your graph and connect with a smooth curve.

6. At the end of December 2004 one-month US Treasury bills were quoted at annualized yields of about 2.1%, corresponding to the monthly simple rate of return $r = \sqrt[12]{1.021} - 1 \doteq .0017$. Plot this point on your graph and label appropriately. Now calculate the *tangency* portfolio $\mathbf{p}_T$ corresponding to tangency of the ray extending from the point $(0, r)$ and tangent to the efficient frontier. Determine the estimated mean and standard deviation of the tangency portfolio, μ_T and σ_T, and draw on your graph a line connecting $(0, r)$ with (σ_T, μ_T) and extending to the level of $2\hat{\mu}_{\max\mu}$. This represents a portion of the

mean–standard-deviation efficient frontier that can be achieved with the five stocks plus the riskless asset.

7. From the beginning-of-month values of the S&P index on 5TechsS&P.xls, calculate the monthly simple rates of return in the same way as for the individual stocks. These will serve as proxies for the rates of return on the market portfolio, R_{mt}. Present estimates of the mean and standard deviation of the market rates, $\hat{\mu}_m$ and $\hat{\sigma}_m$.

8. Regress the monthly rate of return of each of the five companies $j \in \{1, 2, ..., 5\}$, one company at a time, on the S&P rate of return, thereby obtaining estimates of each of $\{\beta_j\}_{j=1}^{5}$ in the model

$$R_{jt} = \mu_j + \beta_j \mathfrak{f}_t + u_{jt}, t = 1, 2, ..., 170,$$

where $\mathfrak{f}_t = (R_{mt} - \hat{\mu}_m)/\hat{\sigma}_m$ is the realization of the standardized market "factor" in month t. Now use these estimates of the betas to calculate alternative "single-index" estimates of the covariances of the rates of the five stocks. Form a new covariance matrix $\tilde{\Sigma}$ containing these new covariances and the *original* variance estimates.

9. Repeat steps 4–7 above with $\tilde{\Sigma}$. Construct a new graph on the same scale as the old one, plotting all the same things but using $\tilde{\Sigma}$ instead of $\hat{\Sigma}$. By comparing the graphs, one can get a visual impression of the extent to which the single-index approximation affects the results. Figures 5.4 and 5.5 show what to expect.

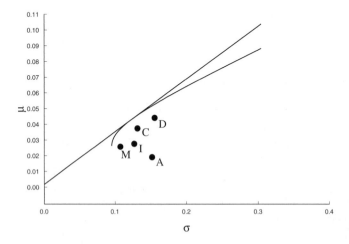

Figure 5.4   Efficient frontier with and without riskless asset based on actual covariances.

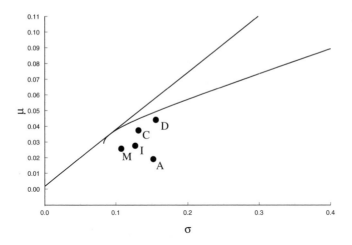

Figure 5.5   Efficient frontier with and without riskless asset based on single-index covariances.

CHAPTER 6

PRICES IN A MEAN–VARIANCE WORLD

In some fleeting instant in which a market might be supposed to be in equilibrium, prices of assets are such as to make all investors content with their current portfolios. This, of course, is simply the defining *condition* for equilibrium. Suppose that all investors followed the Markowitz strategy of choosing portfolios that they considered to be mean–variance efficient. What would this imply about the equilibrium prices of assets? Answer: Not much unless one imposes some further and very strict assumptions. By doing so we get the famous CAPM, which yields some not-so-obvious but simple, interesting, and (more or less) testable predictions that are also intuitively appealing—once one reflects a bit.

6.1 THE ASSUMPTIONS

Here are the sufficient conditions that generate these interesting predictions:

1. All investors choose portfolios that are mean–variance efficient over a common holding period.

2. There is a riskless asset, such as a T-bill, that offers sure rate of return r over the holding period.

3. Markets are "frictionless," meaning that (a) everyone can borrow and lend any amounts at the riskless rate, (b) assets can be sold short and the proceeds invested in other assets, (c) there are no transaction costs, and (d) units of assets (shares, bonds, contracts, etc.) are perfectly divisible.

4. Everyone perceives the same mean vector and covariance matrix of assets' one-period returns.

Clearly, all of these except condition 2 are grossly counterfactual. There are some very astute people who follow stocks actively and perceptively but could not even define the terms "mean" and "variance." Parts of assumption 3 are not too great a stretch for the TIAA-CREFs, Fidelitys, Vanguards, Morgans, Goldmans, etc., who run huge portfolios and have easy access to credit, but it is clear that none of these conditions applies to the individual investor. Assumption 4 is perhaps the hardest to swallow; were 1 and 4 both true, there would be very little trading. But let us move forward and see whether this leads to something worthwhile. There are many derivations of the CAPM, such as in William Sharpe (1964), John Lintner (1965), Eugene Fama (1976), Jonathan Ingersoll (1987), John Cochrane (2001), and in the articles by André Perold (2004) and Fama and Kenneth French (2004). Here is another one.

6.2 THE DERIVATION

Start by formalizing assumption 1 as follows. In equilibrium at any instant each investor $i \in \{1, 2, ..., I\}$ holds a portfolio that solves the problem $\max U^i(EW_{i1}, VW_{i1})$ with respect to investment proportions $p_{i0}, p_{i1}, p_{i2}, ..., p_{in}$ subject to the condition $\sum_{j=0}^{n} p_{ij} = 1$, where (1) W_{i1} is person i's wealth at $t = 1$ (the end of the holding period); (2) W_{i0} is the current wealth; (3) p_{i0} is the fraction of current wealth loaned at riskless rate r (if $p_{i0} > 0$) or borrowed (if $p_{i0} < 0$); (4) p_{ij} is the fraction invested in the jth of n risky assets; and (5) U^i is a function such that the equation $U^i(E, V) = U_0$ for some constant U_0 defines a positively sloped, convex indifference curve in the plane with coordinates $\sqrt{V}, E$. The last means that the person cares only about the mean and variance (standard deviation) of next period's wealth and that increasing levels of $\sqrt{V}$ must be compensated by increasingly large increments of E. The curves in Fig. 5.1 illustrate.

Imposing the constraint $\sum_{j=0}^{n} p_{ij} = 1$, we can set $p_{i0} = 1 - \sum_{j=1}^{n} p_{ij}$ and write for individual i

$$EW_{i1} = W_{i0}\left(1 + p_{i0}r + \sum_{j=1}^{n} p_{ij}\mu_j\right) = W_{i0}\left[1 + r + \sum_{j=1}^{n} p_{ij}(\mu_j - r)\right]$$

$$VW_{i1} = W_{i0}^2 \sum_{j=1}^{n}\sum_{k=1}^{n} p_{ij}p_{ik}\sigma_{jk}.$$

Recalling again the development of Fig. 5.3, we see that each individual will allocate wealth between the riskless asset and a *particular* portfolio of the n risky assets—the "tangency" portfolio in the figure. Assumption 3 implies that it is feasible to do this, since there are no impediments to trading. Now impose assumption 4. If everyone agrees about the first two moments of assets' returns, then the *same* tangency portfolio applies to each person. Everyone who holds risky assets at all will want to hold them in just the same *relative* proportions. It is as if there were some single mutual fund that everyone desired to hold, in amounts that depend on the person's wealth and aversion to risk. That particular "fund" of assets is called the *market portfolio*. It is simply a value-weighted index fund comprising all the risky assets that can be held. The proportions are $\{p_{mj}\}_{j=1}^{n}$, say, with $p_{mj} = (A_j P_{j0})/W_{m0}$, where A_j is the total number of outstanding units of asset j, P_{j0} is the asset's current price per unit, and $W_{m0} = \sum_{j=1}^{n} A_j P_{j0}$ is the total value of all risky assets. In fact, W_{m0} is also the total wealth of all investors, since the borrowing and lending cancel out in the aggregate (and since we ignore nontradeable components such as human capital). Note that if investor i holds a fraction $1 - p_{i0}$ of wealth in this market portfolio, then i's own personal fractional investment in asset j is $p_{ij} = (1 - p_{i0}) p_{mj}$. Summing $(1 - p_{i0}) W_{i0}$ over all I investors gives total market wealth W_{m0}.

The ith investor's optimization problem can now be represented as

$$\max_{\{p_{i1},\ldots,p_{in}\}} U^i \left\{ W_{i0} \left[1 + r + \sum_{j=1}^{n} p_{ij} (\mu_j - r) \right], W_{i0}^2 \sum_{j=1}^{n} \sum_{k=1}^{n} p_{ij} p_{ik} \sigma_{jk} \right\}.$$

Differentiating with respect to p_{ij} for each j and setting equal to zero give n first-order conditions for the maximum, of which the jth is

$$U_E^i W_{i0} (\mu_j - r) + 2U_V^i W_{i0}^2 \sum_{k=1}^{n} p_{ik} \sigma_{jk} = 0.$$

Here, subscripts on U^i indicate partial derivatives; e.g., $U_E^i = \partial U^i / \partial E$. Letting $\gamma_i \equiv -U_E^i/(2U_V^i) > 0$ and simplifying, this equation yields for each asset j the condition

$$\gamma_i (\mu_j - r) = W_{i0} \sum_{k=1}^{n} p_{ik} \sigma_{jk} = W_{i0} (1 - p_{i0}) \sum_{k=1}^{n} p_{mk} \sigma_{jk}.$$

This is simply an optimality condition for the ith investor. Although such a condition must hold for each person when the portfolio weights are optimally chosen, by itself it tells us nothing about how prices are set, since the investor takes market prices as given. However, by summing over all I investors and letting $\sum_{i=1}^{I} \gamma_i \equiv \gamma_m$ and $\sum_{i=1}^{I} (1 - a_{i0}) W_{i0} = W_{m0}$, we get the following *equilibrium* condition for each asset $j \in \{1, 2, \ldots, n\}$:

$$\mu_j - r = \frac{W_{m0}}{\gamma_m} \sum_{k=1}^{n} p_{mk} \sigma_{jk}. \tag{6.1}$$

In the aggregate, the efforts by individual investors to choose optimal portfolios will determine market-clearing prices such that all assets are willingly held at those prices.

Given the expectations of per-share *value* at $t = 1$, $E(P_{j1} + D_{j1})$, the equilibrium current price P_{j0} is implicitly determined from the equilibrium expected rate of return, $\mu_j = E(P_{j1} + D_{j1})/P_{j0} - 1$. What remains is just to put the expression shown above into a form that is more easily interpreted.

Note that the summation $\sum_{k=1}^{n} p_{mk}\sigma_{jk}$ is simply the covariance between the rates of return of asset j and the market, since

$$C\left(R_j, \sum_{k=1}^{n} p_{mk}R_k\right) = \sum_{k=1}^{n} p_{mk}E\left(R_j - \mu_j\right)\left(R_k - \mu_k\right)$$

$$= \sum_{k=1}^{n} p_{mk}\sigma_{jk}.$$

We can denote this covariance simply as σ_{jm}. To interpret the ratio W_{m0}/γ_m, multiply both sides of (6.1) by p_{mj} and sum over j:

$$\sum_{j=1}^{n} p_{mj}\left(\mu_j - r\right) = \frac{W_{m0}}{\gamma_m}\sum_{j=1}^{n}\sum_{k=1}^{n} p_{mj}p_{mk}\sigma_{jk}.$$

The sum on the left equals $\mu_m - r$, where μ_m is the mean rate of return on the market portfolio. The double sum on the right is just the variance of the rate of return on the market portfolio, σ_m^2. We can therefore write

$$\frac{W_{m0}}{\gamma_m} = \frac{\mu_m - r}{\sigma_m^2} \tag{6.2}$$

and so for each asset j

$$\mu_j = r + (\mu_m - r)\frac{\sigma_{jm}}{\sigma_m^2}. \tag{6.3}$$

To get one final, insightful expression, suppose that we were to relate rates of return over time to the rate on the market portfolio, as

$$R_{jt} = \mu_j + \beta_j\left(R_{mt} - \mu_m\right) + u_{jt}, \tag{6.4}$$

where "noise" component u_{jt} has mean zero and is independent of R_{mt}. Here, as in Sharpe's single-index model (5.4), we have attached subscript t to the rates of return because the relation is supposed to apply in each period $t \in \{1, 2, ...\}$. Since this relation implies that $E\left(R_{jt} \mid R_{mt}\right) = \mu_j + \beta_j\left(R_{mt} - \mu_m\right)$, we can interpret $\beta_j = dE\left(R_{jt} \mid R_{mt}\right)/dR_{mt}$ as the sensitivity of the rate for asset j to fluctuations in the market factor. Moreover, it shows that

$$\sigma_{jm} \equiv E\left(R_{jt} - \mu_j\right)\left(R_{mt} - \mu_m\right)$$

$$= \beta_j E\left(R_{mt} - \mu_m\right)^2$$

$$\equiv \beta_j\sigma_m^2,$$

so that $\beta_j = \sigma_{jm}/\sigma_m^2$. Making this substitution in (6.3) gives the usual version of what is called the *security market line* (SML):

$$\mu_j = r + (\mu_m - r)\beta_j, j \in \{1, 2, ..., n\}. \tag{6.5}$$

6.3 INTERPRETATION

Assuming that $\mu_m - r > 0$, as is consistent with a general predisposition to risk aversion, expression (6.5) implies that in equilibrium the expected rate of return of each asset increases linearly with its "beta" coefficient—its sensitivity to fluctuations in the market return. Beta is then the natural and proper measure of an asset's particular risk. If two assets j and k happen to have the same expected future values, $E(P_{j1} + D_{j1}) = E(P_{k1} + D_{k1})$, but if the beta of asset k is higher than that of asset j, then asset k must have higher expected rate of return, $\mu_k = E(P_{k1} + D_{k1})/P_{k0} - 1$, which means that asset k must have lower current price, $P_{k0} < P_{j0}$. Therefore, given the expected end-of-period value of any asset j, its current price depends on its beta and on...*no other feature of the asset itself*—not the variance of its rate of return, σ_j^2 (except as that influences β_j), or (if the asset is a common stock) on the issuing firm's size, its price/dividend ratio, its book-to-market value, etc. Only beta matters.

Does this make any sense at all? It does, on a little reflection; and the fact that it is not totally obvious makes the idea more compelling once it is understood. Recall that beta is proportional to the covariance between the asset's rate of return and that of the market portfolio: $\beta_j = \sigma_{jm}/\sigma_m^2$. Now notice that the covariance σ_{jm} is itself proportional to the rate of change of the variance of the market portfolio with respect to the fraction of total wealth accounted for by asset j:

$$\sigma_{jm} = \sum_{k=1}^{n} p_{mk}\sigma_{jk} = \frac{1}{2}\frac{\partial}{\partial p_{mj}}\left(\sum_{j=1}^{n}\sum_{k=1}^{n} p_{mj}p_{mk}\sigma_{jk}\right) = \frac{1}{2}\frac{\partial \sigma_m^2}{\partial p_{mj}}.$$

In other words, the SML tells us that the increment to an asset's expected return that is required to compensate for its risk is proportional to its own marginal contribution to the variance of the market portfolio—and, therefore, of total wealth. The more an asset contributes to the undesirable riskiness of wealth, the lower will be its equilibrium price, all else equal. The variance of the asset's own rate of return, per se, matters only insofar as its beta depends on it. In fact, expressing β_j in terms of the correlation between R_j and R_m, as

$$\beta_j = \frac{\sigma_{jm}}{\sigma_m^2} = \frac{\rho_{jm}\sigma_j\sigma_m}{\sigma_m^2} = \frac{\rho_{jm}\sigma_j}{\sigma_m}, \tag{6.6}$$

shows that an asset whose rate of return is *uncorrelated* with the market rate ($\rho_{jm} = 0$) is actually priced as if it were riskless, no matter how great its own volatility. In fact, any asset whose return is negatively correlated (if there are any) would actually be priced at a premium over, say, a default-free bond that matures at $t = 1$.

6.4 EMPIRICAL EVIDENCE

The SML relation $\mu_j = r + (\mu_m - r)\beta_j$ for $j \in \{1, 2, ..., n\}$ appears to have several testable implications.

1. Comparing the expected rates of return of all n assets, these are *linear* functions of their beta coefficients. That is, a plot of the points $(\beta_j, \mu_j)_{j=1}^n$ should connect to form a straight line.

2. In the plot of expected rates of return against betas the intercept should equal the riskless rate, and the slope should equal the (positive) average excess rate of return of the market portfolio.

3. No other characteristic of asset j besides its beta affects its expected rate of return.

Unfortunately, determining whether these implications really hold up is complicated by three unpleasant facts of life.

1. Except for r, which is plausibly represented in the United States by rates of return on Treasury bills, nothing in the equation is directly observable, but must be estimated by some statistical procedure.

2. To estimate the unobservable parameters μ_m and β_j statistically requires tracking the returns on a "market" portfolio, but those returns are themselves *unobservable*. Why? In principle, the market portfolio includes *all* risky financial claims in positive net supply: domestic and foreign common and preferred stocks, corporate bonds, government bonds, plus those assets not represented by financial claims, such as gold and other commodities that are held mainly for investment, privately owned real estate (in excess of consumption value), and even potentially marketable collections of art, stamps, coins, baseball cards, and ... you name it. The inability to get accurate price data for all these components means that there is no way to track the value of the market portfolio over time, no way to observe its historical rates of return, $\{R_{mt}\}$, and thus *no* statistical means of estimating μ_m and β_j.

3. As if these problems were not enough, all of the quantities in the SML are actually *time varying*, since μ_j, μ_m, and the components of $\beta_j = \sigma_{jm}/\sigma_m^2$ are really *conditional* moments. That is, the expectations depend on the current state and past history of the economy, and so they all evolve as people's perceptions of the future adjust to new information.

Over the years there have been many ingenious attempts to deal with these problems. Since there seems no way fully to surmount problem 2, there is justification for UCLA economist Richard Roll's famous (1977) critique that the CAPM is really untestable. However, one may hope that some indication of the theory's success can be judged by using some observable *proxy* for R_m. A common choice has been a value-weighted index of prices of the several thousand US common stocks in the CRSP database (Center for Research in Security Prices, Univ. of Chicago). As it happens, results with broader indexes have turned out to be much the same.

Some of the most influential empirical work on the CAPM has been done by Eugene Fama (Chicago Business School) and coauthors. The paper by Fama and

James MacBeth (1973) was the first to deal seriously with the econometric problems. The idea was to perform a series of cross-sectional regressions of monthly excess returns, $\{R_{jt} - r_t\}_{j=1}^{n}$, on time-series estimates of the betas, as

$$R_{jt} - r_t = \gamma_{0t} + \gamma_{1t}\hat{\beta}_{jt} + u_{jt}. \tag{6.7}$$

That is, for each of months $t = 1, 2, ..., T$ we take as dependent (response) variable the excess rates of return of all assets for that month and as independent (explanatory) variable the estimates of their betas, then estimate intercept and slope, γ_{0t} and γ_{1t}, by least squares. This produces a time series of cross-sectional estimates, $\{\hat{\gamma}_{0t}, \hat{\gamma}_{1t}\}_{t=1}^{T}$. If the CAPM is correct, one expects the time average of the intercepts, $\bar{\gamma}_0 = T^{-1} \sum_{t=1}^{T} \hat{\gamma}_{0t}$, to be close to zero and the average slope, $\bar{\gamma}_1$, to be close to the market portfolio's average excess rate of return. The estimate of the beta for each t and each j comes from running a separate time-series regression

$$R_{j\tau} - r_\tau = \delta_j + \beta_{j\tau}(R_{m\tau} - r_\tau) + u_{j\tau}, \tag{6.8}$$

where τ indexes dates prior to t. To make some allowance for the changing of betas over time, Fama and MacBeth run this regression using a rolling window of 60 months' returns prior to t; i.e., with data for $\tau = t - 60$ through $\tau = t - 1$. Unfortunately, these time-series estimates of betas are very noisy, because assets' monthly rates of return themselves have high variances and thus often take values far from their expected values. The resulting large sampling errors in the $\hat{\beta}_{jt}$ lead to a downward bias in the cross-sectional regression estimates of the γ_{1t}'s and an upward bias in estimates of the γ_{0t}'s. To mitigate this, Fama and MacBeth group individual securities into equally weighted portfolios, whose returns have volatilities lower than those of the individual assets. Thus, in the cross section the j's now refer not to individual stocks but to portfolios. Since randomly chosen portfolios would likely have very similar betas, making it difficult to identify the gammas in (6.7) statistically, individual firms are grouped into portfolios at each t on the basis of estimates of betas from data *prior* to $t - 60$.

Using data from the 1930s to mid-1960s, the Fama–MacBeth tests gave general support to the CAPM. The $\hat{\gamma}_{1t}$'s were found to be, on average, positive and statistically significant, and other variables added to the regressions (6.7) ($\hat{\beta}_{jt}^2$ and estimates of the assets' individual volatilities, $\hat{\sigma}_{jt}^2$) had no explanatory power. On the other hand, $\bar{\gamma}_1$ was smaller than the market's average excess return; and $\bar{\gamma}_0$, the average intercept, was positive and significant. Both findings are inconsistent with the basic Sharpe–Lintner version of the CAPM that we have described, but they are not inconsistent with certain later and more general versions, such as that of Fischer Black (1972). In any case, it was thought that the adverse findings might well be due in part to the unobservability of the β_{jt}'s and the market portfolio. The general view through the 1970s and much of the 1980s was that the CAPM does a reasonable job of explaining the cross-sectional variation in assets' average returns.

Contrary evidence began to accumulate during the 1980s. Simple scatter plots of estimates of mean returns on estimates of betas for individual firms showed that small firms tended to plot above the best-fitting regression line (the estimated SML) and large firms tended to plot below. Thus, the *size effect* was born, whereby smaller firms were thought to earn higher average returns than is consistent with their sensitivities to the overall market. Similar results were observed for firms with high earnings/price ratios and firms with high ratios of book value to market value (accounting versus market measures of the value of shareholders' equity). Later a *January effect* was discovered, whereby many firms—especially small ones—tended to earn excessive returns in the first month of the year. Fama and Kenneth French (1997) find in post-1960s data that beta has no bearing on expected return once a firm's size and book-to-market level are taken into account. They conclude that these various covariates must somehow represent risk factors that the market takes into account in setting prices. Lately, however, it looks as if all of these indicators of value (January, E/P, size, and book/market) have lost explanatory power. One conjecture is that the relations between average returns and these characteristics simply had been overlooked by the market until dredged out of the data by academics. Once they were recognized, people's efforts to exploit them—shifting from bonds to stocks in late December, skewing portfolios toward smaller firms and firms with high book/market—eliminated the "bargains" and brought expected returns back to normal levels.

6.5 SOME REFLECTIONS

This current muddle in asset pricing theory arises from a perverse difficulty faced by researchers in finance, and one that is not confronted on the same scale in any other scientific field. It is that the objects of study—human beings, individually and collectively through their participation in markets—can learn and react to the findings of those who study them. While this applies to some extent in all fields that study contemporary human physiology, psychology, and social and economic behavior, the effect is much more dramatic in finance. This is because individuals are intensely motivated to achieve financial gain and because the behavior of financial markets can be substantially influenced by just a few well-informed and well-endowed participants. The problem is somewhat like that faced by researchers in quantum mechanics, where on tiny scales objects seem to "know" that they are being observed and react differently than they otherwise would. However, the financial problem is more severe, in that people alter their behavior individually and purposefully so as to profit from new insights about how they behave collectively. It is as if they actively try to overturn the findings of empirical researchers and thereby vitiate the theories that had been proposed to explain them.[19] Thus, progress in financial research tends to be of the two-steps-forward, one-step-back variety. We shall encounter other instances of this perversity as we continue our survey of quantitative finance. However, the first

[19]Ironically, once-successful theories are sometimes invalidated by reactions on the basis of naive and erroneous interpretations. For example, seeking to increase mean return by loading a portfolio with high-beta stocks is completely inconsistent with the theory that links expected return to beta. In the CAPM the only rational way to increase mean return is to swap riskless bonds for more of the market portfolio, any other portfolio being dominated by some such combination.

task awaiting us is to see if there are better ways to characterize people's actions in the presence of risk than the simple mean–variance approach advocated by Markowitz.

EXERCISES

6.1 If we multiply both sides of Eq. (6.5) by p_{mj} and sum over $j \in \{1, 2, ..., n\}$, then the equality should still hold. What condition does this imply for the $\{\beta_j\}_{j=1}^n$? Show from the definition of β_j in (6.6) that the condition is in fact satisfied.

6.2 What does the condition on the $\{\beta_j\}_{j=1}^n$ from Exercise 6.1 imply about the weighted sum $\sum_{j=1}^n p_{jm} u_{jt}$, where $\{u_{jt}\}_{j=1}^n$ are the noise terms in Eq. (6.4)? Given this, can (6.4) be considered a *factor* model?

6.3 How should one interpret the quantity $\gamma_i \equiv -U_E^i/(2U_V^i) > 0$ that appeared in the development of the CAPM? Suppose that some mass psychological phenomenon caused each individual's γ value suddenly to change by the same factor $\alpha > 0$. How would this be reflected in the findings of empirical tests of the CAPM, assuming that the CAPM is in fact a useful characterization of market equilibrium?

6.4 The following table shows the joint probability distribution of R_A and R_M, representing the rates of return of asset A and market portfolio M. (To interpret the table, both rates will take the value $-.5$ with probability $.3$, R_A will take the value $-.5$ and R_M the value 1.0 with probability $.2$, and so on.)

$R_M \backslash R_A$	$-.5$	1.0
$-.5$	.3	.2
1.0	.2	.3

(**a**) Find the expected rates of return of A and M.

(**b**) Find the beta coefficient of asset A.

(**c**) There is also a riskless asset with sure rate of return $r = .1$ in each period. What expected rate of return for A would be implied by the SML relation of the CAPM?

(**d**) From your answers to parts (a)-(c), determine whether the market portfolio is mean–variance efficient.

CHAPTER 7

RATIONAL DECISIONS UNDER RISK

Choosing a portfolio of assets, deciding whether to buy a lottery ticket, to wait for an oncoming car before making a turn, or to study an extra hour for a particular exam—all of these are examples of decisions under uncertainty. In all these examples, one must choose among actions whose consequences are to some extent unpredictable. A rigorous theory of *rational* decisions under uncertainty was put forth by John von Neumann and Oskar Morgenstern (1944) in the appendix to their *Theory of Games and Economic Behavior.* Although this *expected-utility theory* was intended merely as a tool for analyzing strategic behavior, it has since played a much larger role in efforts to model human decisions.

The approach of von Neumann and Morgenstern was to *axiomatize* the decision making process; that is, to lay down some assumptions about things people would do if they behaved rationally, then show that those assumptions lead to the conclusion of a *theorem* that gives us a way of characterizing decisions. Over the years there have been many refinements and extensions of the theory, in some cases to weaken the primitive assumptions and in others to generalize the setting in which the theory applies. Generally speaking, the weaker the assumptions and the more general the setting, the more difficult is the analysis and the harder it is to prove the final result. We shall focus here on a version that trades generality for simplicity, so that one can get the general idea of the theorem's content without treading through tedious technicalities.

First, some clarification about terminology. Some writers have drawn a sharp line between the terms "uncertainty" and "risk." They apply the former term when the probabilities of different consequences of an action are considered to be unknown; the latter, when probabilities are objectively given. With this distinction, money bet in a game of roulette would be considered at *risk*, while an investment in a portfolio would be considered to have *uncertain* future value. Most economists nowadays proceed as if people form *subjective* probabilities in the uncertain situations and then act on them as if they were objectively given, thus eliminating the need to distinguish the two concepts.[20] We will follow this convention for the most part, but later on will look at a famous experiment whose results suggest that the distinction does indeed have practical relevance in describing human behavior.

7.1 THE SETTING AND THE AXIOMS

We consider decision makers' choices among what will be called risky *prospects,* which we now define. Assume that there is a finite set of *prizes*, $\mathbb{X} = \{x_1, ..., x_n\}$, which could consist of positive or negative amounts of cash, of goods and services, good or bad experiences, and so on. A prospect $\mathbb{P}$ is merely a probability distribution over the prizes; that is, thinking of $\mathbb{X}$ as the outcome set in a chance experiment, $\mathbb{P}$ is a probability measure on the measurable space consisting of $\mathbb{X}$ and the field of all subsets of $\mathbb{X}$. It is convenient to represent $\mathbb{P}$ as a vector $\{p(x_1), ..., p(x_n)\}$ with $p(x_j) \geq 0$ for each j and $\sum_{j=1}^{n} p(x_j) = 1$. In any given situation there is *set* of prospects $\mathcal{P} = \{\mathbb{P}_A, \mathbb{P}_B, ...\}$ over the same measurable space, and the decision maker will have to choose one and only one of these. That "game" will be played once and only once, and the result of the play is that the decision maker will receive one and only one of the prizes. For example, suppose that one must decide whether to buy a lottery ticket that will pay either $10, $1, or $0 with probabilities $\{.05, .20, .75\}$. The lottery ticket costs $1. Counting in the cost of the ticket, the relevant prize space is $\mathbb{X} = \{9, 0, -1\}$ (omitting the $ signs before each), and there are just two relevant prospects: $\mathbb{P}_A = \{.05, .20, .75\}$ if the ticket is bought and $\mathbb{P}_B = \{0, 1, 0\}$ if it is not. Thus, $\mathcal{P} = \{\mathbb{P}_A, \mathbb{P}_B\}$ in this case. In general, for the version of the theorem that we shall consider, there can be any *finite* number n of prizes and *infinitely* many prospects over those prizes from which to choose. The theorem pertains to the decision made by a single decision maker at a single moment in time.

Here, now, are axioms or primitive assumptions that are supposed to characterize a *rational* decision process.

1. **Completeness**. Given a prize set $\mathbb{X}$, the decision maker is able to state for any two prospects $\mathbb{P}_A$ and $\mathbb{P}_B$ either (a) a strict preference for $\mathbb{P}_A$, (b) a strict preference for $\mathbb{P}_B$, or (c) indifference. In symbols, these preferences are represented,

---

[20]Indeed, as Leonard Savage (1954) has shown, it is possible to axiomatize subjective probability along with the theory of choice.

respectively, as $\mathbb{P}_A \succ \mathbb{P}_B$, $\mathbb{P}_B \succ \mathbb{P}_A$, and $\mathbb{P}_A \sim \mathbb{P}_B$. *Weak* preference (either strict preference or indifference) is indicated by "$\succeq$."

2. **Transitivity**. If $\mathbb{P}_A \succ \mathbb{P}_B$ and $\mathbb{P}_B \succ \mathbb{P}_C$, then $\mathbb{P}_A \succ \mathbb{P}_C$. If $\mathbb{P}_A \sim \mathbb{P}_B$ and $\mathbb{P}_B \sim \mathbb{P}_C$, then $\mathbb{P}_A \sim \mathbb{P}_C$.

Together, these first two axioms imply that the elementary prizes can themselves be ranked in order of preference—that is, since prize x_1 is the same as the prospect $\{1, 0, ..., 0\}$ that delivers x_1 for sure, and x_2 is the same as the prospect $\{0, 1, 0, ..., 0\}$ that delivers x_2, pairwise transitive choices among the prizes suffice to put them in rank order by preference. We now number them so that they are ranked in order from most to least preferred, as $x_1 \succeq x_2 \succeq \cdots \succeq x_n$. To keep things from being trivial, we insist that there be strong preference between at least one adjacent pair, so that not all the prizes are considered equivalent.

We now define a *simple* prospect as one in which $p(x_1) + p(x_n) = 1$; that is, a prospect is simple if it offers no chance of getting any except the best and/or worst of the prizes. For example, prospects $\mathbb{P}_A = \{.5, 0, ..., 0, .5\}$, $\mathbb{P}_B = \{.99, 0, ..., 0, .01\}$, $\mathbb{P}_C = \{1, 0, ..., 0, 0\}$, $\mathbb{P}_D = \{0, 0, ..., 0, 1\}$ are all simple.

3. **Dominance**. If prospects $\mathbb{P}_A$ and $\mathbb{P}_B$ are both simple, then $\mathbb{P}_A \succ \mathbb{P}_B$ if and only if $p_A(x_1) > p_B(x_1)$ and $\mathbb{P}_A \sim \mathbb{P}_B$ if and only if $p_A(x_1) = p_B(x_1)$. In other words, if both prospects involve just the best and worst prizes, then the one that offers the greater chance of the better prize is preferred. For example, with the prize space $\mathbb{X}$ consisting of cash amounts $\{9, 0, -1\}$ one would be expected to choose $\mathbb{P}_C = \{.10, 0, .90\}$ over $\mathbb{P}_D = \{.05, 0, .95\}$.

4. **Continuity**. To each prize $x_j \in \mathbb{X}$ there exists an equivalent *simple* prospect; that is, a person would be indifferent between any given prize x_j and *some* simple prospect that gives positive chances of just the best and worst prizes. This simple prospect is defined by just those two probabilities, which we denote $u(x_j)$ and $1 - u(x_j)$. (Why we choose the symbol "u" for a probability will be seen directly.) Thus, $x_j \sim \{u(x_j), 0, ..., 0, 1 - u(x_j)\}$. For example, with $\mathbb{X} = \{9, 0, -1\}$ one might be indifferent between having $x_2 = 0$ for sure and having the simple prospect $\mathbb{P}_A = \{.2, 0, .8\}$ that gives 20% chance of a net gain of \$9. If so, we would write $u(0) = .2$. If, on the other hand, the person says that $x_2 = 0$ is really preferred to $\mathbb{P}_A$, then we could increase incrementally the chance of winning \$9 until the required value $u(0)$ is found. The term "continuity" signifies that such a precise number can be found by incremental adjustment. Note that for a prospect to be equivalent to x_1 (the best prize), it must offer that prize with certainty, so $u(x_1) = 1$. Similarly, $u(x_n) = 0$ since any simple prospect that afforded any positive chance of winning x_1 would be preferred to having x_n for sure. Note that $x_1 \succeq x_2 \succeq \cdots \succeq x_n$ implies (by the dominance axiom) that $1 = u(x_1) \geq u(x_2) \geq \cdots \geq u(x_n) = 0$. Thus, since the function u ranks the prizes themselves in order of preference, we can think of it as the person's *utility function* of the prizes. (That explains the choice of symbol.)

To prepare for the last axiom, suppose for an arbitrary prospect $\mathbb{P} \in \mathcal{P}$ that we replace the jth prize by the simple prospect $\{u(x_j), 0, ..., 0, 1 - u(x_j)\}$ that is considered equivalent to x_j; that is, we remove any chance of winning x_j and, in effect, give the person additional chances to win x_1 and x_n. The new prospect thus created, denoted $\mathbb{P}^j$, delivers x_1 with probability $p(x_1) + p(x_j)u(x_j)$. The first term is the original probability of winning x_1 directly. The second term is the probability of the exclusive event of winning x_1 *indirectly*; that is, it is the product of (1) the probability $u(x_j)$ of winning x_1 *conditional on* winning the x_j-equivalent simple prospect and (2) the probability $p(x_j)$ of winning that equivalent simple prospect. For example, with $\mathbb{X} = \{9, 0, -1\}$, $u(0) = .20$, and $\mathbb{P}_A = \{.05, .20, .75\}$ we would have $\mathbb{P}_A^2 = \{.05 + .20(.20), 0, .75 + .20(1 - .20)\} = \{.09, 0, .91\}$.

5. **Substitution**. The decision maker is indifferent between $\mathbb{P}$ and $\mathbb{P}^j$. This is also called the *independence axiom*, since it holds that decisions are independent of *irrelevant* alternatives. Specifically, if x_j and $\{u(x_j), 0, .., 0, 1 - u(x_j)\}$ are really equivalent, it should be irrelevant whether a prospect delivers one or the other.

7.2   THE EXPECTED-UTILITY (EU) THEOREM

Theorem: If choices satisfy axioms 1–5, then there *exists* a real-valued function u of the prizes, $u : \mathbb{X} \rightarrow \Re$, *unique* up to an increasing, affine transformation, such that for any $\mathbb{P}_A, \mathbb{P}_B \in \mathcal{P}$ we have

$$\mathbb{P}_A \succ \mathbb{P}_B \Longleftrightarrow \sum_{j=1}^{n} p_A(x_j)u(x_j) > \sum_{j=1}^{n} p_B(x_j)u(x_j)$$

and

$$\mathbb{P}_A \sim \mathbb{P}_B \Longleftrightarrow \sum_{j=1}^{n} p_A(x_j)u(x_j) = \sum_{j=1}^{n} p_B(x_j)u(x_j).$$

Before proving the theorem, let us be sure we understand exactly what it says. The assertion about *existence* means that for any individual who satisfies the rationality postulates it should be possible to construct a "utility" function u of the prizes such that the choice between any two prospects can be determined by comparing the respective mathematical expectations of u. When the prizes are all quantifiable and measured in the same units (currency, weight, distance, time, etc.), we can regard the uncertain prize under each prospect as a random variable X that takes values $x_1, x_2, ..., x_n$ with probabilities defined by the prospect itself. For example, with $\mathbb{X} = \{9, 0, -1\}$, the uncertain payoff under prospect $\mathbb{P}_A = \{.05, .20, .75\}$ is a random variable with the distribution

x	$\mathbb{P}_A(x)$
9	.05
0	.20
-1	.75

Thus, when the prizes are quantifiable we have that $\mathbb{P}_A \succ \mathbb{P}_B$ if and only if $E_A u\left(X\right) > E_B u\left(X\right)$ and $\mathbb{P}_A \sim \mathbb{P}_B$ if and only if $E_A u\left(X\right) = E_B u\left(X\right)$. Notice that the theorem does not state that a person actually makes such computations, or even that the decision maker is aware of the existence of the function u. It merely states that such a function does exist—and could in principle be discovered by someone who analyzes the person's decisions—if the decision maker is truly "rational," in the sense of obeying the axioms. Thus, although the theorem allows the actual decision making process to remain a black box, it gives us a consistent, rigorous way of representing and predicting the results.

The assertion that u is *unique* up to an increasing, affine transformation means (1) that another function $u^* = a + bu$ with $b > 0$ would also correctly rank prospects if u does and (2) that any other function u^* that correctly ranks prospects must necessarily be expressible as an affine function of u.

We can now turn to the proof, which is in two parts.

1. **Existence**. By axiom 1 an individual should be able to choose between any two prospects

$$\mathbb{P}_A = \left\{p_A\left(x_1\right), p_A\left(x_2\right), p_A\left(x_3\right), ..., p_A\left(x_n\right)\right\},$$
$$\mathbb{P}_B = \left\{p_B\left(x_1\right), p_B\left(x_2\right), p_B\left(x_3\right), ..., p_B\left(x_n\right)\right\}.$$

By axioms 4 and 5 there exist $u\left(x_2\right)$ and $\bar{u}\left(x_2\right) \equiv 1 - u\left(x_2\right)$ such that $\mathbb{P}_A$ is equivalent to

$$\mathbb{P}_A^2 = \left\{p_A\left(x_1\right) + p_A\left(x_2\right)u\left(x_2\right), 0, p_A\left(x_3\right), ..., p_A\left(x_n\right) + p_A\left(x_2\right)\bar{u}\left(x_2\right)\right\},$$

which substitutes for x_2 its equivalent simple prospect. In turn, $\mathbb{P}_A^2$ is equivalent to the prospect $\mathbb{P}_A^{2,3}$ in which x_3 also is replaced by its equivalent simple counterpart. $\mathbb{P}_A^{2,3}$ has probability $p_A\left(x_1\right) + p_A\left(x_2\right)u\left(x_2\right) + p_A\left(x_3\right)u\left(x_3\right)$ of winning x_1 and zero probability of winning x_2 and x_3. Axiom 2 (transitivity) then implies that $\mathbb{P}_A \sim \mathbb{P}_A^{2,3}$. Continuing in this way to remove intermediate prizes, we can obtain for $\mathbb{P}_A$ an equivalent *simple* prospect $\mathbb{P}_A^{2,3,...,n-1}$, with probability $\mathbb{P}_A^{2,3,...,n-1}\left(x_1\right)$ of winning x_1 equal to

$$p_A\left(x_1\right) + p_A\left(x_2\right)u\left(x_2\right) + \cdots + p_A\left(x_{n-1}\right)u\left(x_{n-1}\right) = \sum_{j=1}^{n} p_A\left(x_j\right)u\left(x_j\right).$$

(The equality holds because $u\left(x_1\right) = 1$ and $u\left(x_n\right) = 0$.) For $\mathbb{P}_B$ there is also an equivalent simple prospect $\mathbb{P}_B^{2,3,...,n-1}$ with $\mathbb{P}_B^{2,3,...,n-1}\left(x_1\right) = \sum_{j=1}^{n} p_B\left(x_j\right)u\left(x_j\right)$. Axiom 3 (dominance) implies that $\mathbb{P}_A^{2,3,...,n-1} \succeq \mathbb{P}_B^{2,3,...,n-1}$ if and only if

$$\sum_{j=1}^{n} p_A\left(x_j\right)u\left(x_j\right) - \sum_{j=1}^{n} p_B\left(x_j\right)u\left(x_j\right) \geq 0, \tag{7.1}$$

with strict preference corresponding to strict inequality. Finally, axiom 2 (transitivity) now implies that (7.1) is necessary and sufficient for $\mathbb{P}_A \succeq \mathbb{P}_B$.

2. **Uniqueness.**

(a) **Sufficiency.** Let $u^*(x) = a + bu(x)$ for each $x \in \mathbb{X}$, with $b > 0$ and a arbitrary. Then

$$
\begin{aligned}
E_A u^*(X) - E_B u^*(X) &= \sum_{j=1}^{n} p_A(x_j) u^*(x_j) - \sum_{j=1}^{n} p_B(x_j) u^*(x_j) \\
&= \sum_{j=1}^{n} [p_A(x_j) - p_B(x_j)] u^*(x_j) \\
&= \sum_{j=1}^{n} [p_A(x_j) - p_B(x_j)] [a + bu(x_j)] \\
&= b \sum_{j=1}^{n} [p_A(x_j) - p_B(x_j)] u(x_j) \\
&= b [E_A u(X) - E_B u(X)] .
\end{aligned}
$$

Since b is positive, u^* and u rank $\mathbb{P}_A$ and $\mathbb{P}_B$ in the same way.

(b) **Necessity.** An example suffices to show that the signs of $E_A u - E_B u$ and $E_A u^* - E_B u^*$ are not always the same if $u^*(x)$ is not of the form $a + bu(x)$. Consider prize space $\mathbb{X} = \{9, 0, -1\}$ and the following three utility functions:

x	$u(x)$	$u^*(x)$	$u^{**}(x)$
9	1.00	5.00	1.00
0	0.20	2.60	0.04
−1	0.00	2.00	0.00

With function u prospect $\mathbb{P}_A = \{.05, .20, .75\}$ is preferred to prospect $\mathbb{P}_B = \{.06, .10, .84\}$, since $E_A u = .05(1.0) + .20(0.2) = .09$ while $E_B u = .06(1.0) + .10(0.2) = .08$. Utility $u^* = 2 + 3u$ gives the same ranking, since $E_A u^* = 2 + 3Eu = 2.27 > E_B u^* = 2.24$. Therefore, there is nothing to choose between these two functions. However, even though $u^{**} = u^2$ ranks the *prizes* in the same way as u and u^*, it reverses the ranking of $\mathbb{P}_A$ and $\mathbb{P}_B$, since $E_A u^{**} = .058 < E_B u^{**} = .064$.

Note that the equivalence of u and $u^* = a + bu$ ($b > 0$) as utilities implies that we can rescale a utility function by any positive constant and recenter it *arbitrarily* without changing predictions about preferences. This is comparable to the essential arbitrariness of measuring temperature by Fahrenheit, Celsius, or Kelvin scales—or for that matter, by F^2, C^2, or K^2—when the only purpose is to judge whether one thing is hotter than another. On the other hand, the necessity part of the uniqueness property means that nonlinear transformations of the utility function are not permitted. While nonlinear transformations do preserve the ranking of the *prizes*, they do not preserve the ranking of all *prospects* over those prizes. This is merely a consequence of the linearity property of mathematical expectation—the familiar facts that $Eg(X) = g(EX)$ when $g(X)$ has the form $a + bX$ but that $Eg(X)$ does not always equal $g(EX)$ when g is nonlinear. (For example, $EX^2 > (EX)^2$ whenever X has positive variance.) For this reason von Neumann–Morgenstern utility is often called *cardinal* utility. That distinguishes it from the *ordinal* utility that is used in economic theory to describe people's choices among baskets of consumption goods. Since (in the usual view) no uncertainty is involved there, only the ranking of the baskets is important.

7.3   APPLYING EU THEORY

The expected-utility theorem provides a compelling model for characterizing *rational* choices among uncertain prospects. If one adopts the conventional rationality paradigm of economics, the theorem becomes the basis for models of how individuals actually do make such choices. The potential for application to portfolio theory and finance generally should be apparent: a portfolio (i.e., a particular mix of assets) is itself a stake in an uncertain future level of wealth. Different allocations of current wealth W_0 among available assets correspond to different probability distributions over future wealth W_1. Choosing among portfolios thus amounts to choosing among probability distributions. For example, investing all of wealth in a riskless asset with sure one-period (simple) rate of return r amounts to choosing a *degenerate* distribution for W_1 such that $\Pr[W_1 = W_0(1 + r)] = 1$, while choosing a mix of risky stocks amounts to choosing a distribution with some positive dispersion.

We shall now see specifically how to apply the von Neumann–Morgenstern theorem to what is called *static* portfolio choice. These one-period models describe the behavior of a rational investor who must make a one-time decision about how to allocate wealth among assets; that is, investors choose assets at some time $t = 0$ according to their perceived values at some arbitrary future date $t = 1$, without regard to the decisions that might need to be made at any time in the future. This is the same setting that applied in the Markowitz framework. Models in which the need to make future decisions is recognized are called *dynamic* models; they will be treated in Chapters 12 and 13. At present, confining attention to the one-period case, we will see in principle how a person's utility function for wealth could be determined, and from this how it could be possible to determine a person's beliefs (subjective probability distributions) about assets' future returns. We will then (1) see what properties the function $u(W)$ must have in order for predictions about choices to fit commonly observed behavior, (2) look at some examples of specific functions that have these properties, and (3) examine some general implications for choices among prospects generally and portfolios specifically. Finally, we will see that choosing the prospect with highest expected utility is not always consistent with Markowitz' mean–variance criterion, but that there are some not-implausible conditions under which the two theories yield similar predictions.

7.3.1 Implementing EU Theory in Financial Modeling

We have proved a version of the EU theorem that applies when prize space $\mathbb{X}$ is finite and when the prizes are not necessarily quantifiable. For the static portfolio application we can take the prizes to correspond to different—and obviously, quantifiable—levels of wealth. (Later, in the dynamic models, prizes will correspond to different levels of consumption of some representative good or market basket.) Fortunately, although the proof is harder, the theorem remains valid when the prize space is a continuum, such as $\mathbb{X} = \Re$ (the real numbers) or a subset thereof. Thus, in principle, we can consider prospects (portfolios) with infinitely many—even uncountably many—prizes (wealth payoffs).

For now, in keeping with the one-period focus of the Markowitz model, we consider an individual's utility to depend just on the level of wealth at the end of the investment "holding period;" that is, the individual chooses at $t = 0$ one of the possible *distributions* of end-of-period wealth W_1 that are attainable by some feasible composition of the available assets. Any particular set of proportions $p_1, ..., p_n$ of wealth invested in the n risky assets and proportion $p_0 = 1 - \sum_{j=1}^{n} p_j$ in the riskless asset yields a level of future wealth that depends on the realized rates of return. Thus, in symbols we have

$$W_1 = W_0 \left(1 + p_0 r + p_1 R_1 + \cdots + p_n R_n\right)$$

$$= W_0 \left[1 + r + \sum_{j=1}^{n} p_j \left(R_j - r\right)\right].$$

Applying the EU theorem, we view the investor as solving the problem

$$\max_{p_1, ..., p_n} Eu \left\{ W_0 \left[1 + r + \sum_{j=1}^{n} p_j \left(R_j - r\right)\right]\right\}.$$

Letting f be the joint PDF of the risky rates of return, this is expressed as

$$\max_{p_1, ..., p_n} \int \cdots \int u \left\{ W_0 \left[1 + r + \sum_{j=1}^{n} p_j \left(R_j - r\right)\right]\right\} f\left(r_1, ..., r_n\right) \cdot dr_1 \cdots dr_n.$$

$$(7.2)$$

In this very general setting it is unclear what, if anything, this implies about the individual's choices. To get any concrete predictions requires putting some structure on the investor's tastes, as embodied in the function u, and on the investor's beliefs about future returns, as embodied in the function f. Since our interest is not really in the behavior of any particular person but in that of the "typical" investor, it is necessary only to have a qualitative sense of what a typical person's utility function is like. Once we have that, the assumption of *rational expectations*—that beliefs accord pretty much with the facts—will give us a means of modeling f. However, it is indeed possible, in principle, actually to discover a *particular* person's utility function—assuming that the person obeys the von Neumann–Morgenstern axioms. Once the person's u is known, it is possible also to discover the person's beliefs. It will be instructive to see how these things could actually be done.

7.3.2 Inferring Utilities and Beliefs

We want to find the function u that actually characterizes a particular (rational) person's choices among risky prospects that affect wealth. This merely requires putting the continuity axiom into practice and observing choices among specific gambles for money. For example, fix two levels of wealth $w^* < w^{**}$ and set $u(w^*) = 0$ and $u(w^{**}) = 1$. For the moment w^{**} and w^* correspond to the top and bottom "prizes" x_1 and x_n in the development of the EU theorem. Values 1 and 0 can be assigned arbitrarily to the utilities of w^{**} and w^*, because the location and scale of u are arbitrary. Now pick some intermediate wealth level $w \in (w^*, w^{**})$. By the continuity axiom there are probabilities p and $1 - p$ such that the person is indifferent between having w for sure and getting w^{**} with probability p and w^* with probability $1 - p$. By trial and error the person should be able to determine such a unique value of p, which then corresponds to $u(w)$, the value of the utility function at w. Continuing in this way, the value of u at any level of wealth between w^* and w^{**} can be discovered. To extrapolate to some $w^+ > w^{**}$, get the person to name the value of p that makes having w^{**} for sure equivalent to getting w^+ with probability p and w^* with probability $1 - p$. Then p necessarily satisfies the condition $pu(w^+) + (1 - p)u(w^*) = u(w^{**})$, and this together with $u(w^*) = 0$ and $u(w^{**}) = 1$ give $u(w^+) = 1/p$. It should be easy now to figure out how to extrapolate to values of wealth $w^- < w^*$.

In the thought experiments just described the probabilities p, once decided on, would be objectively known. Thus, the person would know that some appropriate randomization device, such as a random-number generator on a computer, would determine whether the top prize was won. In the investment setting, however, objective probabilities of events are *not* known. For example, there is nowhere to look up the precise probability that the price of MSFT (Microsoft) will rise between 5% and 10% in the coming year, and different people would be apt to have different views. Nevertheless, implicit in a "rational" person's decision about buying MSFT are *subjective* probabilities of such events. In principle, one's subjective probability of any event A can actually be discovered once the utility function has been derived. Here is how. Propose the choice between a sure amount of cash C and the risky prospect that pays \$100 if event A occurs and nothing otherwise. By trial and error, find the precise value of C such that the person is indifferent between the cash and the risk. Then, if the person's current wealth is W_0, it follows that

$$u(W_0 + C) = \mathbb{P}(A)\, u(W_0 + 100) + [1 - \mathbb{P}(A)]u(W_0),$$

where $\mathbb{P}(A)$ is the subjective probability of event A. Knowing the value of u at all three points, one simply solves the equation for $\mathbb{P}(A)$:

$$\mathbb{P}(A) = \frac{u(W_0 + C) - u(W_0)}{u(W_0 + 100) - u(W_0)}.$$

It was mentioned in the introduction to the chapter that some economists distinguish between risk and uncertainty. The former applies when objective probabilities are known; the latter, when they are not. Thus, the investment problem would be considered a decision under uncertainty, while choosing the bet to place on a coin flip would be a decision under risk. While the distinction is often ignored in financial modeling, there is some evidence that people actually do respond differently in the two situations. A famous example was presented by Daniel Ellsberg (1961) and is known as the *Ellsberg paradox*. There are two urns. A subject is told that urn A contains a 50–50 mix of red and blue balls and that urn B contains only red and/or blue balls—but in *unknown* proportion; that is, urn B could have all red, all blue, or any mix in between. There is a reward for drawing a red ball. The person gets just one draw and must choose from which urn to draw. In effect, the probability of drawing red from A is known objectively to be $\frac{1}{2}$, so that whether one gets red or blue from urn A is a "toss-up." But since there is no reason to think that urn B has more of one color than the other, the same should be true there as well. In other words, from considerations of symmetry the *subjective* probability of red from urn B should be $\frac{1}{2}$ also, so it would seem that a rational individual would be indifferent as to the choice of urn. In fact, those confronted with such a choice almost uniformly choose to draw from urn A. This suggests that people find choices with unknown probabilities to be inherently riskier.[21]

7.3.3 Qualitative Properties of Utility Functions

What might the "typical" (rational) person's utility function for wealth actually look like? Mere armchair reasoning gives us some basic clues. First, we can safely assume that for all but self-flagellating monks u is an increasing function of wealth; i.e., that more wealth is preferred to less, all else equal. Do we draw the function then as a straight line with positive slope, or does it have some curvature?

Consider the implication of having a utility function that is an increasing *affine* function of wealth: $u(W_1) = a + bW_1$ with $b > 0$. Since $Eu(W_1) = a + bEW_1 = u(EW_1)$, such a person would be indifferent between having a risky wealth level with a given expected value and having that expected level of wealth for sure. For example, the person would be indifferent between winning or losing $1000 on the flip of a coin and not taking the gamble, between having a 50–50 chance of winning $1,000,000 or nothing and getting a gift of $500,000, etc. A person who is indifferent among all prospects with the same expected value is said to be *risk neutral*. While people do sometimes take fair gambles and even choose to play at *unfavorable* odds, there are often other considerations involved, such as entertainment value, being a good sport, or assuming an image of boldness. In such cases the prize space may be

---

[21] If someone gave you such a choice, might you not be a little suspicious that the person had filled urn A with blue balls only, so as to reduce your chance of winning? Might the experiment be more informative if the winning color were announced *after* the urn had been chosen?

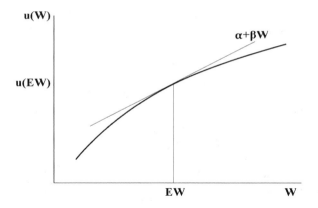

Figure 7.1   Illustration of Jensen's inequality.

thought to have some dimension other than the monetary one alone, so that—win or lose—there is some benefit to playing the game. There may be elements of this in the investment setting, also; for example, the attraction of appearing or considering oneself to be an insightful stock picker, of boosts to self-esteem from the occasional success at long odds, of being right when others are wrong, and so on. On the other hand, if such considerations played large enough in most people's minds to offset the worry about losses, there would be little diversification. As with the expectations models considered in Chapter 5, we would see people putting all wealth into the one venture with highest expected payoff, even if it had the highest risk.

If we really think that most people are risk averse in their important financial decisions, how should we model the "typical" investor's utility function? Risk aversion means that one always prefers a sure payoff (i.e., wealth level) to a risky prospect having the same expected payoff. Thus, u must be such that $Eu\left(W\right) < u\left(EW\right)$. To find functions with this property we can look to *Jensen's inequality* from probability theory. As we saw in Section 3.4.1, this tells us that a function u such that $Eu\left(W\right) < u\left(EW\right)$ for all risks is necessarily strictly *concave*. The proof that concave functions (and only those) have this property is easy. If (and only if) u is strictly concave on a support $\mathbb{W}$ of W, then at any point on the graph of the function a *dominating* line can be drawn (not necessarily uniquely) that touches the function there and lies on or above it elsewhere in $\mathbb{W}$, as in Fig. 7.1. Given a risk with expected value EW, choose intercept α and slope β such that the line $\alpha + \beta W$ touches u at the point $\left(EW, U\left(EW\right)\right)$ and never cuts the function. Since $\left(EW, U\left(EW\right)\right)$ lies on the line, we have $u\left(EW\right) = \alpha + \beta EW$. But $\alpha + \beta W > u\left(W\right)$ for other points in $\mathbb{W}$ implies that $E\left(\alpha + \beta W\right) > Eu\left(W\right)$ and therefore that $u\left(EW\right) > Eu\left(W\right)$.

7.3.4 Measures of Risk Aversion

It is apparent from daily experience that different people have different degrees of risk aversion. They even seem to have different levels of risk aversion at different times of their lives. Moreover, all else equal, a person's risk aversion might well change with changes in wealth, holding age and other circumstances constant. However, if we are to compare degrees of risk aversion objectively, we need some way to measure it. Since risk aversion is bound up with concavity of u, a natural idea is to look at the second derivative, u''. Clearly, this must be negative (assuming that u is differentiable), but can one properly gauge the degree of risk aversion from its absolute value (i.e., from $-u''$)? The answer is no, because of the fact that u has arbitrary scale. Recall that if function u describes preferences, then so does $u^* = a + bu$ for *arbitrary* $b > 0$, so it is clear that a scale-invariant measure is needed. There are two important ones in the literature, both from an influential paper by John Pratt (1964).

Both measures relate to the concept of *risk premium*. The *absolute* risk premium π_a is defined implicitly by $u\left(W_0 - \pi_a\right) = Eu\left(W_1\right)$. Here, W_1 is supposed to be the level of wealth attained on taking a "fair" bet having expected value zero, so that $EW_1 = W_0$. The quantity $-\pi_a$ is the *certainty equivalent* of the risk, with π_a itself being the sure amount of present wealth that one would willingly give up to avoid it. Clearly, $\pi_a = 0$ corresponds to risk neutrality. Pratt also considers a *relative* risk premium, π_r, equal to the *proportion* of wealth that would be given up to avoid the risk. This is defined implicitly by $u\left(W_0 - W_0\pi_r\right) = Eu\left(W_1\right)$, so that $\pi_r = \pi_a/W_0$. In general, the values of π_a and π_r depend on the initial wealth level and on the nature of the risk itself and not just on the function u, and that lessens their usefulness in making global comparisons of degrees of risk aversion. However, consider the following location–scale-invariant quantities that are defined in terms of u and w alone:

1. The coefficient of *absolute* risk aversion:

$$\rho_a\left(w\right) \equiv -\frac{u''\left(w\right)}{u'\left(w\right)} = -\frac{d\ln u'\left(w\right)}{dw}.$$

2. The coefficient of *relative* risk aversion:

$$\rho_r\left(w\right) \equiv w\rho_a\left(w\right).$$

Pratt proved that of two utility functions u_1 and u_2 such that $\rho_{a1}\left(w\right) > \rho_{a2}\left(w\right)$ for *all* positive w, function u_1 would have higher absolute risk premium than would u_2 for *all* risks and all initial wealth levels. Likewise, if $\rho_{r1}\left(w\right) > \rho_{r2}\left(w\right)$ for all levels of wealth, u_1 would imply higher π_r than u_2. These two measures have come to be the standard ones for analyzing models of utility functions.

7.3.5 Examples of Utility Functions

Suppose that we want to model a person whose *absolute* risk premium is invariant with wealth. For such a person, the amount willingly paid to get rid of a fixed fair risk (e.g., to win or lose $1000 on a coin flip) would be the same no matter how poor or wealthy he was. Pratt's theorem shows that such a person would have a constant coefficient of absolute risk aversion. In other words, u would be such that $\rho_a\left(w\right) = \rho_a$ for all

$w > 0$. Since $u''(w)/u'(w) = d\ln u'(w)/dw$, the only utility functions having this property are the solutions to the differential equation $d\ln u'(w)/dw = -\rho_a$. These solutions are all of the form $u(w) = a - be^{-\rho_a w}$, where a is arbitrary and $b > 0$. To put in the simple form in which one usually sees this *exponential* utility function, multiply by $1/b$ and add $1 - a/b$ to get the canonical form

$$u(w) = 1 - e^{-\rho_a w}.$$

Of course, the constant "1" in front has no bearing on the choices implied by this function but it has the virtue of setting $u(0) = 0$ as a point of reference. A nice feature of the exponential model is that expected utility is just an affine function of the moment-generating function of the random variable W. Thus, if W has density $f(\cdot)$ and if $\mathcal{M}_W(\zeta) \equiv Ee^{\zeta W} = \int e^{\zeta w} f(w) \cdot dw$ exists for values of ζ in some interval about the origin, then $Eu(W) = 1 - \mathcal{M}_W(-\rho_a)$. Many common families of distributions have MGFs, and many of these are of tractable form. It is thus easy to express and work with expected utilities when the various risky prospects belong to such a family. We will see an example of this in the next section. If W is supported on $[0, \infty)$—meaning that none of the choices under consideration permits insolvency— then $\Pr(0 \le e^{-\rho_a W} \le 1) = 1$ for all $\rho_a \ge 0$, and so (Laplace transform) $Ee^{-\rho_a W}$ is necessarily finite.

While the constant absolute risk aversion utility is a convenient model from a mathematical standpoint, it seems likely that absolute risk aversion is, for most people, not really strictly constant with respect to wealth. It is often thought that wealthy people would generally pay *less* in dollar amounts to avoid a given dollar risk, since the consequence of a fixed loss would be less severe. Common models of utilities with *decreasing* absolute risk aversion are

$$u(w) = \ln w, w > 0$$
$$u(w) = \frac{w^{1-\gamma} - 1}{1 - \gamma}, \gamma \ge 0, \gamma \ne 1, w \ge 0.$$

Obviously, neither of these is applicable when the choice set includes distributions for which $\Pr(W < 0) > 0$. Both forms have constant *relative* risk aversion— $\rho_r = 1$ for the logarithmic form and $\rho_r = \gamma$ for the power form; for this reason they are commonly referred to as "CRRA utilities." l'Hôpital's rule shows that $\lim_{\gamma \to 1}(w^{1-\gamma} - 1)/(1 - \gamma) = \ln w$, so that the logarithmic form can be considered a special case of the power form if the latter is defined at $\gamma = 1$ by continuity. Notice that utility is linear in w in the borderline case $\gamma = 0$, which is the case of risk neutrality, while utility is increasing and concave for all $\gamma > 0$.

7.3.6 Some Qualitative Implications of the EU Model

Having acquired the necessary background, we can now see what can be deduced about the portfolios a "rational" individual would choose by imposing weak but plausible conditions on the individual's tastes and subjective distributions of assets' returns. As to tastes, we suppose (1) that the domain of u includes the positive real numbers and (2) that u is strictly increasing, strictly concave, and at least twice differentiable there. Some of the results also require that coefficient of absolute risk aversion ρ_a decreases (strictly) as wealth increases. At this level of generality not much can be inferred about behavior in the general $(n + 1)$-asset setting of (7.2), so let us suppose that the individual's choices at $t = 0$ are limited to portfolios comprising default-free unit bonds maturing at $t = 1$ and a single risky asset. To make this seem a bit more realistic, think of the risky asset as an index fund of stocks. Now, initial wealth W_0 constrains the choices to numbers b and s of unit bonds and shares of the fund, respectively, that satisfy $W_0 = bB + sS_0$. Here $B \equiv B(0, 1) \in (0, 1)$ and $S_0 > 0$ are the assets' market prices, which are considered to be unaffected by the individual's own choices. Since the bonds have unit value at maturity, the portfolio (b, s) leads to end-of-period wealth

$$W_1 = b + sS_1 = \frac{W_0}{B} + s\left(S_1 - \frac{S_0}{B}\right),$$

whose distribution is governed by that of the index fund's uncertain future value, S_1. In accordance with the principle of limited liability, we suppose that $\Pr(S_1 \geq 0) = 1$, and for added realism we assume also that $\Pr(a \leq S_1 < b) > 0$ for all intervals $[a, b) \subset [0, \infty)$. Thus, S_1 is nonnegative, can take values arbitrarily close to zero, and is unbounded above. Clearly, we also want expected utility to be real and finite for all feasible choices of s. For certain utilities meeting conditions (1) and (2) the restriction $E|u(W_1)| < \infty$ will require the solvency constraint that $0 \leq s \leq W_0/S_0$. This rules out short sales and buying on margin, either of which could lead to negative values of wealth. Also, if $\Pr(S_1 = 0) > 0$, CRRA utilities with $\gamma \geq 1$ would require the strict inequality $s < W_0/S_0$. Finally, we assume that the fund's expected future value is finite: $ES_1 < \infty$.

We first consider under what condition the investor will choose to hold a positive quantity of the risky fund. Thus, putting $\mathcal{E}_u(s) \equiv Eu[W_0/B + s(S_1 - S_0/B)]$, we want to know under what condition $\mathcal{E}_u(s) > \mathcal{E}_u(0)$ for all sufficiently small, positive s. Now, assuming that

$$\mathcal{E}'_u(0+) \equiv \lim_{s \downarrow 0} s^{-1}[\mathcal{E}_u(s) - \mathcal{E}_u(0)]$$

exists, we have $[\mathcal{E}_u(s) - \mathcal{E}_u(0)]/s > 0$ for all $s \in (0, \varepsilon)$ and some $\varepsilon > 0$ if and only if $\mathcal{E}'_u(0+) > 0$. In fact, as shown below, the right-hand derivative $\mathcal{E}'_u(0+)$ does exist under the conditions stated above and is given by $\mathcal{E}'_u(0+) = u'(W_0/B)(ES_1 - S_0/B)$. Accordingly, under the stated conditions the investor will take *some* position in the fund so long as its expected return exceeds that of the bond; i.e., as long as $E(S_1/S_0) > 1/B$. The interesting aspect of this conclusion is that it holds without explicit restriction on the fund's risk characteristics (nor does it require diminishing absolute risk aversion).

To show that $\mathcal{E}'_u(0+)$ exists, we must verify that the right-hand derivative of expected utility with respect to s equals the expectation of the derivative. As explained in Section 2.3.3, this interchange of operations is valid if $|du(W_1)/ds|$

is bounded in a neighborhood of the origin by a function of finite expectation.[22] Boundedness is established as follows. Since $\Pr(S_1 \geq 0) = 1$, it follows that $W_1 \geq (W_0 - \varepsilon S_0) B > 0$ with probability one whenever $0 \leq s \leq \varepsilon < W_0/S_0$. Now by assumption u' exists and is positive, continuous, and strictly decreasing on $(0, \infty)$, so $u'(W_1) \leq u'[(W_0 - \varepsilon S_0)/B]$ for $s \in [0, \varepsilon]$. Therefore

$$E\left|u'(W_1)\left(S_1 - \frac{S_0}{B}\right)\right| = Eu'(W_1)\left|S_1 - \frac{S_0}{B}\right|$$
$$\leq u'\left(\frac{W_0 - \varepsilon S_0}{B}\right) E\left|S_1 - \frac{S_0}{B}\right|$$
$$\leq u'\left(\frac{W_0}{B}\right)\left(ES_1 + \frac{S_0}{B}\right) < \infty.$$

Thus, dominated convergence implies

$$\mathcal{E}'_u(0+) = E\lim_{s\downarrow 0} s^{-1}\left\{u\left[\frac{W_0}{B} + s\left(S_1 - \frac{S_0}{B}\right)\right] - u(W_0/B)\right\}$$
$$= Eu'\left(\frac{W_0}{B}\right)\left(S_1 - \frac{S_0}{B}\right)$$
$$= u'\left(\frac{W_0}{B}\right)\left(ES_1 - \frac{S_0}{B}\right).$$

We can now develop some qualitative implications about how the demand for shares varies with the individual's attitude toward risk and with respect to initial wealth and the assets' market prices. Note that these are exercises in *comparative statics*, meaning that we look at how optimal choices would vary respect to certain variables when all other features of the model remain the same. In other words, we compare alternative optima rather than consider how the individual might actually move from one to another. For this we suppose that the optimization problem has an *interior* solution; that is, that the optimal number of shares, s^u say, satisfies the first-order condition

$$\mathcal{E}'_u(s^u) = Eu'(W_1)\left(S_1 - \frac{S_0}{B}\right) = 0 \tag{7.3}$$

for some $s^u \in (0, W_0/S_0)$. That such an extremum is indeed an optimum is confirmed on noting that

$$\mathcal{E}''_u(s) = Eu''(W_1)\left(S_1 - \frac{S_0}{B}\right)^2 < 0 \tag{7.4}$$

for all allowed s when the individual is globally risk averse (i.e., when $u''(w) < 0$ for all $w \geq 0$). Of course, condition (7.4) implies that $\mathcal{E}'_u(s)$ is strictly decreasing for all $s \in (0, W_0/S_0)$ and, in particular, at the optimal number of shares s^u.

---

[22]For a proof see Theorem 16.8 of Billingsley (1995).

1. **Changing risk aversion**. Let g be a concave, increasing function on $\Re$ with at least two continuous derivatives, and put $v(w) \equiv g\left[u(w)\right]$. It is easy to see that $-v''(w)/v'(w) > -u''(w)/u'(w)$ for $w \geq 0$, so that v is more risk averse than u. Putting $\mathcal{E}_v(s) \equiv Ev(W_1) = Eg\{u\left[W_0/B + s\left(S_1 - S_0/B\right)\right]\}$, we have $\mathcal{E}_v'(s) = Eg'\left[u(W_1)\right]u'(W_1)\left(S_1 - S_0/B\right)$, with $\mathcal{E}_v(s^v) = 0$ at the optimal number of shares under v. The question is how s^v relates to s^u, the optimal value for function u. Expanding g' about $u(W_0/B)$, the derivative $\mathcal{E}_v'(s)$ can be written as

$$g'\left[u\left(\frac{W_0}{B}\right)\right]Eu'(W_1)\left(S_1 - \frac{S_0}{B}\right) + Eg''(\bar{u})u'(W_1)^2\left(S_1 - \frac{S_0}{B}\right)^2,$$

where $\bar{u}$ is between $u(W_1)$ and $u(W_0/B)$. If we evaluate this at s^u, the first term vanishes by virtue of (7.3), giving $\mathcal{E}_v'(s^u) = Eg''(\bar{u})u'(W_1)^2$ $\left(S_1 - S_0/B\right)^2 < 0$. Since $\mathcal{E}_v'(s)$—like $\mathcal{E}_u'(s)$—is strictly decreasing in s, we conclude that $\mathcal{E}_v'(s)$ intersects the s axis (and so equals zero) at a point $s^v < s^u$. Thus, given that $s^u \in (0, W_0/S_0)$ initially, the individual would choose to hold fewer shares of the risky asset upon becoming more averse to risk.

2. **Changing initial wealth**. This and the remaining implications do require that absolute risk aversion be strictly decreasing in wealth. (We are now done with the alternative function v and can focus again on the generic u.) Differentiating $\mathcal{E}_u'(s^u)$ implicitly with respect to W_0 gives the equality

$$0 = Eu''(W_1)\left[\frac{1}{B} + \frac{\partial s^u}{\partial W_0}\left(S_1 - \frac{S_0}{B}\right)\right]\left(S_1 - \frac{S_0}{B}\right)$$

and the following solution for $\partial s^u/\partial W_0$:

$$\frac{\partial s^u}{\partial W_0} = \frac{Eu''(W_1)\left(S_1 - S_0/B\right)}{-BEu''(W_1)\left(S_1 - S_0/B\right)^2}. \tag{7.5}$$

Since the denominator is positive, the sign of $\partial s^u/\partial W_0$ is that of the numerator. Putting $\rho_a(w) \equiv -u''(w)/u'(w)$, the numerator is

$$Eu''(W_1)\left(S_1 - \frac{S_0}{B}\right) = -E\rho_a(W_1)u'(W_1)\left(S_1 - \frac{S_0}{B}\right)$$

$$= E\left[\rho_a\left(\frac{W_0}{B}\right) - \rho_a(W_1)\right]u'(W_1)\left(S_1 - \frac{S_0}{B}\right),$$

where the second equality follows from (7.3) and the fact that $\rho_a(W_0/B)$ is $\mathcal{F}_0$-measurable (known at time 0). But the signs of $\rho_a(W_0/B) - \rho_a(W_1)$ and $S_1 - S_0/B$ are the same when $s^u > 0$, since ρ_a is decreasing and $W_1 - W_0/B$ and $S_1 - S_0/B$ themselves have the same sign. We conclude that $Eu''(W_1)(S_1 - S_0/B) > 0$, so that $\partial s^u/\partial W_0$ is positive when $s^u \in (0, W_0/B]$.

3. **Changing stock price** S_0. By varying S_0 while holding fixed the distribution of S_1, we effect a multiplicative change in the fund's return per dollar invested. Again differentiating (7.3) implicitly gives

$$0 = Eu''(W_1) \left[\frac{\partial s^u}{\partial S_0} \left(S_1 - \frac{S_0}{B} \right) - \frac{s^u}{B} \right] \left(S_1 - \frac{S_0}{B} \right) - \frac{Eu'(W_1)}{B}$$

and the solution

$$\frac{\partial s^u}{\partial S_0} = \frac{s^u Eu''(W_1)(S_1 - S_0/B) + Eu'(W_1)}{BEu''(W_1)(S_1 - S_0/B)^2}.$$

Because the numerator is positive and the denominator is negative, we conclude that $\partial s^u / \partial S_0 < 0$; thus, all else equal, the demand for the risky asset diminishes as its price rises and its total return stochastically declines.

4. **Changing bond price** B. We first consider what happens if realizations of $S_1/S_0 - 1/B$ do not vary with B, so that the distribution of S_1/S_0 shifts along with $1/B$. In this case it is easy to see that

$$\frac{\partial s^u}{\partial B} = -\frac{W_0}{B} \frac{\partial s^u}{\partial W_0} < 0. \tag{7.6}$$

Since we ordinarily think of bonds and shares as substitutes, it seems odd at first that the demand for shares would move counter to the price of the bond. However, this is more easily understood if one expresses B in terms of the average continuously compounded spot rate, as $B = e^{-r}$, and writes (7.6) as

$$\frac{\partial s^u}{\partial r} = \frac{\partial s^u}{\partial \ln W_0} > 0.$$

This shows that when the distribution of $S_1/S_0 - 1/B$ is invariant, the change in the spot rate acts like a change in wealth. By contrast, suppose now that realizations of S_1/S_0 alone are invariant to B. In this case, so long as $s^u S_0 < W_0$, we find that

$$\frac{\partial s^u}{\partial r} = \frac{\partial s^u}{\partial \ln W_0} + \frac{\partial s^u}{\partial \ln S_0}. \tag{7.7}$$

Thus, so long as some bonds are held, the total effect of the change in r consists of a positive wealth effect and a negative substitution effect, and the sign of the whole is indeterminate. Of course, the wealth component is not present if $s^u S_0 = W_0$, and in that case $\partial s^u / \partial r$ is unambiguously negative.

Results such as these indicate that the EU theory does yield general, plausible, and testable implications about an investor's behavior. Indeed, it has become for economists the almost universally accepted way to characterize decisions under uncertainty.[23]

---

[23]One such application that relates closely to what we have just seen is that of Miles Kimball (1990). Considering an individual who faces stochastic labor income and must decide how much to save, Kimball

7.3.7 Stochastic Dominance

Suppose we consider just two portfolios, whose implied distributions for end-of-period wealth are represented by CDFs F_A and F_B. According to the EU theorem, there is for each "rational" individual a utility function u such that the person's choice will be indicated by the sign of $E_A u(W) - E_B u(W) = \int u \cdot dF_A - \int u \cdot dF_B$. That is, F_A is *weakly* preferred to F_B (preferred or equivalent) if and only if $E_A u(W) - E_B u(W) \geq 0$. Of course, this might be positive for one individual, indicating a preference for F_A, but negative for another person with different tastes. One may ask, however, whether there are conditions on F_A and F_B such that *all* "rational" individuals in some particular class would weakly prefer F_A. In that case we would say for that class that F_A *stochastically dominates* F_B. Such questions were first addressed in the economics literature by James Quirk and Rubin Saposnik (1962), Josef Hadar and William Russell (1969, 1971), and Giora Hanoch and Haim Levy (1969), and they have inspired many subsequent studies. The concept of dominance is appealing from a normative standpoint, since one who knows himself to be in a class for which F_A is known to dominate $F_B, F_C, \ldots$ would not have to consider the latter prospects at all. We describe the two most basic concepts, known as *first-order* and *second-order* stochastic dominance. These pertain, respectively, to the class of individuals whose utilities are strictly increasing in wealth and to the subclass of these whose utilities are also strictly concave. For brevity we write $F_A D_1 F_B$ if F_A is weakly preferred by all who prefer more wealth to less, and we write $F_A D_2 F_B$ if F_A is weakly preferred by the subclass of *risk averters*. Of course, since this is a subclass, it follows that $F_A D_1 F_B$ implies $F_A D_2 F_B$.

Some general restrictions are needed before anything can be shown, since we cannot pair all utilities with all distributions. We assume that all utilities $\{u\}$ being considered are defined at least on $(0, \infty)$ and that the distributions $\{F\}$ are such that $F(0-) = 0$ and $\int_{[0,\infty)} |u| \cdot dF < \infty$ for each admissible u. Note that if u is unbounded below on $(0, \infty)$, then the finiteness of the integral requires $F(0) = \Pr(W = 0) = 0$. If u is bounded below, we set $u(0) = 0$ without loss of generality. In either case $\int_{[0,\infty)} u \cdot dF = \int_{(0,\infty)} u \cdot dF$.

1. We show first that $F_A D_1 F_B$ if and only if $F_B(w) \geq F_A(w)$ for each $w \geq 0$. For this the class of utilities can include all increasing, right-continuous functions. This allows there to be one or more special thresholds of wealth w' such that $u(w') - u(w'-) > 0$. Of course, CDFs F_A and F_B are also monotone and right continuous, but we assume that both are continuous at any point where u is discontinuous. Under these conditions, the usual integration-by-parts formula applies[24] to give $\int_{(a,b]} u(w) \cdot dF(w) = u(b) F(b) - u(a) F(a) - \int_{(a,b]} F(w) \cdot du(w)$ for $F \in \{F_A, F_B\}$.

shows that a measure of the risk sensitivity of optimal precautionary saving—a quantity that he calls "prudence"—plays a role like that of the coefficient of risk aversion in financial applications. Comparative-static implications about saving choices can then be deduced, much as we have done here for an investor's choice of portfolios.

[24]See Theorem 18.4 in Billingsley (1995).

(a) To prove sufficiency, assume that $\Delta\left(w\right) \equiv F_B\left(w\right) - F_A\left(w\right) \geq 0$ for all $w \geq 0$, fix some $n \in \aleph$, and write

$$\int_{(n-1,n]} u \cdot dF_A - \int_{(n-1,n]} u \cdot dF_B \qquad (7.8)$$

$$= -u\left(n\right)\Delta\left(n\right) + u\left(n^{-1}\right)\Delta\left(n^{-1}\right) + \int_{(1/n,n]}\Delta \cdot du.$$

By right continuity the second term on the right goes to $u\left(0\right)\Delta\left(0\right) = 0$ as $n \to \infty$, and by monotone convergence the last term goes to $\int_{(0,\infty)}\Delta \cdot du$. The first term is

$$u\left(n\right)\left[F_A\left(n\right) - F_B[n]\right] = u\left(n\right)\left[1 - F_B(n)\right] - u\left(n\right)\left[1 - F_A\left(n\right)\right],$$

and we need to see what happens to this as $n \to \infty$. Now for $F \in \{F_A, F_B\}$ we have

$$\int_{(0,\infty)} |u| \cdot dF - \int_{(0,n]} |u| \cdot dF = \int_{(n,\infty)} |u\left(w\right)| \cdot dF\left(w\right)$$

$$\geq \int_{(n,\infty)} |u(n)| \cdot dF\left(w\right)$$

$$= |u\left(n\right)|\left[1 - F\left(n\right)\right] \geq 0,$$

and since the left side goes to zero as $n \to \infty$ (as $|u|$ is integrable), so does the right side. Thus, the term $-u\left(n\right)\Delta\left(n\right)$ on the right of (7.8) vanishes in the limit. Dominated convergence then implies

$$E_A u\left(W\right) - E_B u\left(W\right) = \int_{[0,\infty)}\left(F_B - F_A\right) \cdot du, \qquad (7.9)$$

which is nonnegative under the hypothesis $\Delta \equiv F_B - F_A \geq 0$.

(b) For necessity, suppose that there is a w' such that $F_B(w') < F_A\left(w'\right)$. By right continuity, there are $\delta > 0$ and $\varepsilon > 0$ such that $\Delta(w) \equiv F_B\left(w\right) - F_A(w) < -\delta$ for $w \in \left[w', w' + \varepsilon\right]$. Now pick one of our increasing, integrable functions u that happens to have a density, $u' = du/dw$, and for $n \in \aleph$ put

$$u_n\left(w\right) = u(w)\mathbf{1}_{[0,w')}\left(w\right) + \left[u\left(w\right) + n\left(w - w'\right)\right]\mathbf{1}_{[w',w'+\varepsilon]}\left(w\right)$$
$$+ \left[u\left(w\right) + n\varepsilon\right]\mathbf{1}_{(w'+\varepsilon,\infty)}\left(w\right).$$

(This just makes u_n increase more rapidly than u where $F_B < F_A$ but leaves it in the class of increasing functions.) Then $u'_n\left(w\right) = u'\left(w\right) + n\mathbf{1}_{[w',w'+\varepsilon]}\left(w\right)$ and $E_A u_n\left(W\right) - E_B u_n\left(W\right)$ equals

$$\int_0^\infty \Delta(w)u'\left(w\right) \cdot dw + n\int_{w'}^{w'+\varepsilon}\Delta(w) \cdot dw$$

$$< \int_0^\infty \Delta(w)u'\left(w\right) \cdot dw - n\delta\varepsilon,$$

which is negative for sufficiently large n.

2. A necessary and sufficient condition for $F_A D_2 F_B$ is that

$$\int_0^w [F_B(t) - F_A(t)] \cdot dt \geq 0 \ \forall \ w > 0; \qquad (7.10)$$

that is, the accumulated area between F_B and the horizontal axis up to any w is at least as large as that for F_A. Of course, this holds if $F_B(w) \geq F_A(w)$ for all w, but it is clearly a weaker condition. While F_B must indeed start out above F_A, the curves could nevertheless cross one or more times thereafter. Since our utility class now includes only functions that are strictly concave, there can no longer be discontinuities in u. On the other hand, nothing need be assumed about differentiability. We make use of the fact that if $u : \Re^+ \to \Re$ is concave then for any $w_0 \geq 0$ there is a least one pair of constants α_0, β_0 such that $u(w_0) = \alpha_0 + \beta_0 w_0$ and such that $u(w) \leq \alpha_0 + \beta_0 w$ for all $w \geq 0$. In other words, when u is concave there is *at each point* at least one dominating line that touches u there and lies above u everywhere else. This fact was used in the proof of Jensen's inequality in Section 7.3.3 and illustrated in Fig. 7.1. (At a point where u is differentiable the tangent line would be the unique dominating line.) When u is strictly increasing, each such dominating line will necessarily have positive slope.

(a) Here is a simple proof that (7.10) implies $F_A D_2 F_B$. We assume that F_A does cross above F_B at some point, else there is first-order dominance and nothing left to prove. Let $w_0 = \inf \{w : F_A(w) \geq F_B(w)\}$ be the "crossing point," as in Fig. 7.2. For the moment we assume that there are no further crossings, so that $\Delta(w) \equiv F_B(w) - F_A(w) \leq 0$ for all $w \geq w_0$. (The CDFs need not be continuous, and w_0 itself could be a point of left discontinuity for either or both.) Let $\alpha_0 + \beta_0 w_0$ be a dominating line for u at w_0, and pick $w_1 > w_0$ arbitrarily. Then from (7.9) $E_A u(W) - E_B u(W)$ equals

$$\int_{[0,w_0)} \Delta \cdot du + \int_{[w_0,w_1)} \Delta \cdot du + \int_{[w_1,\infty)} \Delta \cdot du.$$

Now the concavity of u implies that, for any positive g

$$\int_{[0,w_0)} g(t) \cdot du(t) \geq \int_{[0,w_0)} g(t) \cdot d(\alpha_0 + \beta_0 t) = \beta_0 \int_{[0,w_0)} g(t) \cdot dt$$

and that $\int_{[w_0,w_1)} h(t) \cdot du(t) \geq \beta_0 \int_{[w_0,w_1)} h(t) \cdot dt$ for any negative h. Since $\Delta > 0$ on $[0, w_0)$ and $\Delta \leq 0$ on $[w_0, w_1)$ it follows that

$$E_A u(W) - E_B u(W) > \beta_0 \int_{[0,w_1)} \Delta \cdot dt + \int_{[w_1,\infty)} \Delta \cdot du.$$

The first term is nonnegative for each w_1 by hypothesis, and the second term converges to zero as $w_1 \to \infty$, so it follows that $E_A u(W) -$

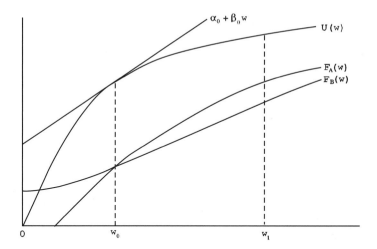

Figure 7.2   Second-order dominance of F_A over F_B.

$E_B u (W) \geq 0$. Now if the CDFs do subsequently cross so that $F_B > F_A$ somewhere beyond w_0, then F_A is even more attractive, and so the conclusion holds a fortiori.

(b) As in the first-order case necessity can be proved by constructing a specific concave increasing utility for which $E_A u < E_B u$ when (7.10) does not hold. This is left as an exercise.

7.4   IS THE MARKOWITZ INVESTOR RATIONAL?

Clearly, the Markowitz prescription/description of choosing portfolios on the basis of mean and variance is a far more limited theory than expected utility, which can apply to choices among prospects over spaces containing prizes that are not representable as numbers—trips to Disneyland, dates with a celebrity, and so on. But even within its narrow scope, might the Markowitz theory be inconsistent with rational behavior? If we accept that rational people obey the von Neumann–Morgenstern axioms and therefore behave *as if* they choose among risky prospects the one with highest expected utility, might it be irrational always to act in accordance with the Markowitz mean–variance dominance condition?

Indeed, following the Markowitz prescription is *not* always rational. The mean–variance dominance condition states that *all* individuals would prefer investment A

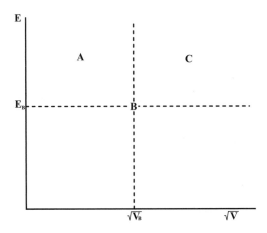

Figure 7.3   The mean–variance dominance condition.

to investment B *if and only if*

$$E_A W \geq E_B W, \sqrt{V_A W} \leq \sqrt{V_B W}, \qquad (7.11)$$

where at least one of these inequalities is strict. To get a visual interpretation of this, we can plot portfolios A, B, C in the $\left(\sqrt{V}, E\right)$ plane (i.e., with standard deviation $\sqrt{V}$ on the horizontal axis and mean E on the vertical). Then, if A lies to the northwest of B as in Fig. 7.3, (7.11) implies that *everyone* would prefer A to B, which means that A dominates B and hence that (7.11) is *sufficient* for dominance. On the other hand, if C is to the northeast of B *some* people might prefer B and some might prefer C. This means that neither dominates the other, and hence that (7.11) is *necessary* for dominance. Thus, in Markowitz' view condition (7.11) is both necessary and sufficient. In fact, though, if we accept the rationality postulates of expected-utility theory, (7.11) is neither necessary *nor* sufficient. Moreover, that is true even if we restrict to those rational people who prefer more wealth to less and want to avoid risk—and, of course, to prospects whose payoffs have finite moments. We have just seen that, under the rationality postulates, all wealth-loving risk-averse individuals would prefer distribution F_A to distribution F_B if and only if these satisfy condition (7.10) for second-order dominance. Now (7.10) does imply that $E_A W \geq E_B W$, since (recalling relation (3.6)), $\int_0^\infty (F_B - F_A) \cdot dw = \int_0^\infty [(1 - F_A) - (1 - F_B)] \cdot dw = E_A W - E_B W$. However, (7.10) has no explicit connection to variance; indeed, it can hold even for distributions one or both of which has infinite second moment.

Hanoch and Levy (1969) provide the following example to show that (7.11) is not necessary for dominance. For discrete prize space $\mathbb{X} = \{0, 1, 2\}$ and initial wealth $W_0 = 0$, prospect C offers $\mathbb{P}_C (0) = .65, \mathbb{P}_C (1) = .25, \mathbb{P}_C (2) = .10$, while prospect

B assigns to winnings $0, 1, 2$ the probabilities $.86, .03, .11$, respectively. Then if one plots the CDFs, simple geometry shows that

$$\int_0^w [F_B(t) - F_C(t)] \cdot dt = \begin{cases} .21w, & 0 \le w < 1 \\ .21 - .01(w-1), & 1 \le w < 2 \\ .20 & w \ge 2 \end{cases}.$$

This is nonnegative for all $w \ge 0$, so $F_C D_2 F_B$. However, condition (7.11) is not satisfied, since $E_C W_1 = .45 > E_B W_1 = .25$ while $V_C = .4475 > V_B = .4075$.

Another example shows that (7.11) is not sufficient for dominance. Prospect A delivers $W_1 = 1$ with probability 0.80 and $W_1 = 100$ with probability 0.20. Prospect B delivers $W_1 = 10$ with probability 0.99 and $W_1 = 1000$ with probability 0.01. The means are $E_A W_1 = 20.8$ and $E_B W_1 = 19.9$, while the standard deviations are $\sqrt{V_A W_1} = \sqrt{1568}$ and $\sqrt{V_B W_1} = \sqrt{9703}$. Thus, A does dominate B in the Markowitz sense. However, a wealth-loving, risk-averse person with $u(w) = 2\sqrt{w}$ (the CRRA utility with $\gamma = \frac{1}{2}$) would choose B, since

$$E_A\left(2\sqrt{W_1}\right) = .80 \cdot 2\sqrt{1} + .20 \cdot 2\sqrt{100} = 5.60$$

$$E_B\left(2\sqrt{W_1}\right) = .99 \cdot 2\sqrt{10} + .01 \cdot 2\sqrt{1000} \doteq 6.89.$$

Of course, these particular examples of payoff distributions are hardly representative of those that one encounters in the investment setting. Are there any special circumstances under which the Markowitz criterion *is* consistent with expected-utility theory? Indeed, there are such circumstances. One way of making the criterion consistent with EU is to impose severe restrictions on both utilities and returns. For example, suppose that risky rates of return $\{R_j\}_{j=1}^n$ over the fixed holding period are jointly normal and that there is a riskless asset with sure one-period rate r. In that case the terminal wealth $W_1 = W_0\left[1 + r + \sum_{j=1}^n p_j(R_j - r)\right]$ achieved by a portfolio $\mathbf{p} = (p_1, p_2, ..., p_n)'$ with $\mathbf{p}'\mathbf{p} > 0$ is also normally distributed, since it is an affine function of the $\{R_j\}_{j=1}^n$. For an investor with constant absolute risk aversion, expected utility is then $Eu(W_1) = 1 - \mathcal{M}_W(-\rho_a)$, where $\mathcal{M}_W$ is the moment-generating function. Now recall that a generic random variable X that is normally distributed with mean μ and variance σ^2 has MGF $\mathcal{M}_X(\zeta) \equiv E e^{\zeta X} = e^{\zeta\mu + \zeta^2\sigma^2/2}$ for any real ζ. It follows then that expected utility under the joint condition of exponential utility and jointly normal returns is

$$Eu(W_1) = 1 - \exp\left\{-\rho_a\left[EW_1 - \frac{\rho_a}{2}VW_1\right]\right\}.$$

This is maximized with respect to $\mathbf{p}$ simply by maximizing the linear function of mean and variance within the square brackets. If the vector of rates of return has mean μ and covariance matrix Σ, then it is not hard to see that the solution is $\mathbf{p} = (\rho_a W_0)^{-1}\Sigma^{-1}(\mu - r\mathbf{1}) = \rho_r^{-1}\Sigma^{-1}(\mu - r\mathbf{1})$, where $\mathbf{1}$ is a vector of units and ρ_r is the coefficient of *relative* risk aversion. For example, if there is just one risky asset with rate of return $R \sim N(\mu, \sigma^2)$, then the optimal proportional investment is

$p = (\mu - r) / (\rho_r \sigma^2)$. Note the pleasing implication that this increases with μ and decreases with r, σ^2, and relative risk aversion. Somewhat more generally, it suffices to restrict choices to prospects whose distributions for terminal wealth fall into a class called "elliptical," which includes (but is not limited to) the normal distribution.[25] However, such justifications of mean–variance theory are obviously very restrictive and essentially ad hoc.

Perhaps the most compelling argument for mean–variance analysis is that it can serve as a useful *approximation* to rational decision criteria in a realistic investment setting when holding periods are of moderate length and utilities are differentiable. The idea first appeared in a paper by Paul Samuelson (1970), but here is a simpler and more compelling version based on later work by Lawrence Pulley (1981). Let $R_{\mathbf{p}\tau} = 1 + r_\tau + \sum_{j=1}^{n} p_j (R_{j\tau} - r_\tau)$ be the rate of return over the period $[0, \tau]$ of a portfolio that puts proportions $\{p_j\}_{j=1}^{n}$ of initial wealth into n risky assets and proportion $1 - \sum_{j=1}^{n} p_j$ into an asset with sure rate of return r_τ. The τ subscripts remind us that the holding period now is of explicit length τ. As usual, the investor's problem is to choose vector $\mathbf{p} = (p_1, p_2, ..., p_n)'$ to solve $\max_{\mathbf{p}} Eu(W_\tau) = \max_{\mathbf{p}} Eu[W_0(1 + R_{\mathbf{p}\tau})]$. Using Taylor's theorem, expand function u out to second order about $W_\tau = W_0$ and add a remainder term of third order to represent $Eu(W_\tau)$ as

$$E\left[u(W_0) + u'(W_0) W_0 R_{\mathbf{p}\tau} + \frac{u''(W_0)}{2} W_0^2 R_{\mathbf{p}\tau}^2 + \frac{u'''(W_\tau^*)}{3!} W_0^3 R_{\mathbf{p}\tau}^3\right].$$

Here, the W_τ^* in the remainder is some unspecified point between W_τ and W_0, and Taylor's theorem shows that *some* such point exists that makes the left and right sides precisely equal so long as u''' is continuous. Pushing E through and recognizing that only $R_{\mathbf{p}\tau}$ and W_τ^* are random variables give for $Eu(W_\tau)$ the exact expression

$$u(W_0) + u'(W_0) W_0 \mu_{\mathbf{p}\tau} + \frac{u''(W_0)}{2} W_0^2 (\mu_{\mathbf{p}\tau}^2 + \sigma_{\mathbf{p}\tau}^2) + E\frac{u'''(W_\tau^*)}{3!} W_0^3 R_{\mathbf{p}\tau}^3,$$

where $\mu_{\mathbf{p}\tau} = ER_{\mathbf{p}\tau}$ and $\mu_{\mathbf{p}\tau}^2 + \sigma_{\mathbf{p}\tau}^2 = (ER_{\mathbf{p}\tau})^2 + VR_{\mathbf{p}\tau} = ER_{\mathbf{p}\tau}^2$. Of course, we have to assume that u and the joint distribution of $\mathbf{R}_\tau \equiv (R_{1\tau}, R_{2\tau}, ..., R_{n\tau})'$ are such that the last term is finite for all $\mathbf{p}$. Given that it is, the question now is how this expression would behave as τ, the length of the holding period, approaches zero. Typically, we find that mean rates of return and variances are roughly in proportion to the length of the holding period. For example, if we use annual rates of return ($\tau = 1$ year) to calculate estimates of means and variances, we will get values equal to about 12 times those from monthly rates ($\tau = \frac{1}{12}$). Thus, we say that means and variances are $O(\tau)$—of "order" τ as $\tau \to 0$. On the other hand, higher moments tend to be proportional to higher powers of τ and so are $o(\tau)$—of order *less than* τ. For example, $E|R_{\mathbf{p}\tau}^3|$ is typically $O(\tau^{3/2}) = o(\tau)$, which means that it declines *faster* than either $\mu_{\mathbf{p}\tau}$ or $\sigma_{\mathbf{p}\tau}^2$ as $\tau \to 0$. If the utility function is well enough behaved, as our standard models are, then the term involving $Eu'''(W_\tau^*) W_0^3 R_{\mathbf{p}\tau}^3$ becomes negligible

---

[25]For a discussion of this concept, see Ingersoll (1987), pp. 104–113.

relative to the remaining terms when τ is sufficiently small. Moreover, since $\mu_{\mathbf{p}\tau}$ is of order τ, its square is $o(\tau)$, so that the $\mu_{\mathbf{p}\tau}^2$ term can also be neglected. This gives as an *approximation*

$$Eu(W_\tau) \doteq u(W_0) + u'(W_0)W_0\mu_{\mathbf{p}\tau} + \frac{u''(W_0)}{2}W_0^2\sigma_{\mathbf{p}\tau}^2.$$

Now factor out $u'(W_0)W_0$ from the second and third terms to get

$$Eu(W_\tau) \doteq u(W_0) + u'(W_0)W_0\left[\mu_{\mathbf{p}\tau} + \frac{u''(W_0)W_0}{2u'(W_0)}\sigma_{\mathbf{p}\tau}^2\right]$$
$$= u(W_0) + u'(W_0)W_0\left(\mu_{\mathbf{p}\tau} - \frac{\rho_r}{2}\sigma_{\mathbf{p}\tau}^2\right),$$

where the last expression follows from the definition of the coefficient of relative risk aversion ρ_r. Now since W_0 is fixed and known, to maximize $Eu(W_\tau)$ with respect to the portfolio weights requires only that one maximize the factor $\mu_{\mathbf{p}\tau} - \rho_r\sigma_{\mathbf{p}\tau}^2/2$, a simple linear function that increases in the mean and decreases in the variance. Subject to the approximation error, the expected-utility maximizer would be indifferent among all portfolios having the same value of $\mu_{\mathbf{p}\tau} - \rho_r\sigma_{\mathbf{p}\tau}^2/2$. In other words, indifference curves drawn in the $(\sigma_{\mathbf{p}\tau}^2, \mu_{\mathbf{p}\tau})$ plane would be straight lines with slope $\rho_r/2 > 0$. In the $(\sigma_{\mathbf{p}\tau}, \mu_{\mathbf{p}\tau})$ plane they would be positively sloped *convex* curves, such as depicted in Fig. 5.1. In other words, the mean–variance dominance condition does hold *approximately* when holding periods are sufficiently short, when utilities are well behaved, and when moments of returns behave as they appear to do. The "approximately" optimal portfolio—which we might better call "asymptotically" optimal—is just the same as in the special case of exponential utility and normal returns; i.e., $\mathbf{p} = \rho_r^{-1}\Sigma^{-1}(\mu - r\mathbf{1})$ or, when there is just one risky asset, $p = (\mu - r)/(\rho_r\sigma^2)$.

How short must holding periods be in practice before the approximation becomes acceptable? If one actually solves the two problems $\max_{\mathbf{p}} Eu(W_\tau)$ and $\max_{\mathbf{p}}(\mu_{\mathbf{p}\tau} - \rho_r\sigma_{\mathbf{p}\tau}^2/2)$ for various universes of common stocks, various reasonable utility functions, and various holding periods—estimating moments and expected utilities from historical averages—one finds that the two alternative maximizing portfolios are usually almost indistinguishable, even when τ is as long as one year. Thus, while one can certainly construct examples of risky prospects in which the mean–variance condition fails, it does seem to be a reasonable guide in selecting portfolios of common stocks.

EXERCISES

7.1 An individual with current cash wealth $W_0 = 10$ (in $1000s, say) has the opportunity to choose among risky prospects that pay uncertain amounts X taking values on the set $\mathbb{X} = \{4, 3, 0\}$ (also in $1000s). After the play the individual's wealth will be $W_0 + X$. The individual *claims* to obey the rational-choice axioms of von Neumann and Morgenstern. Setting $u(10) = 0$ and

$u(14) = 1$ arbitrarily, determine $u(13)$ assuming that the individual claims to be indifferent between prospects $\mathbb{P}_A = \{0, 1, 0\}$ and $\mathbb{P}_B = \{.80, 0, .20\}$. Suppose now that the individual—still maintaining that $\mathbb{P}_A \sim \mathbb{P}_B$—states that $\mathbb{P}_C = \{.20, 0, .80\} \succ \mathbb{P}_D = \{0, .25, .75\}$ (i.e., that $\mathbb{P}_C$ is strictly preferred). Determine whether these preferences are in fact consistent with the EU axioms.

7.2 The prize space is $\mathbb{X} = \{5000, 1000, 0, -100\}$ (in \$US). Here are five prospects:

$$\mathbb{P}_A = \{.15, .15, .10, .60\}$$
$$\mathbb{P}_B = \{.20, .00, .00, .80\}$$
$$\mathbb{P}_C = \{.00, .00, 1.0, .00\}$$
$$\mathbb{P}_D = \{.40, .00, .00, .60\}$$
$$\mathbb{P}_E = \{.00, 1.0, .00, .00\}.$$

An individual who obeys the EU axioms states that $\mathbb{P}_E \sim \mathbb{P}_D \succ \mathbb{P}_C \sim \mathbb{P}_B$. Determine how $\mathbb{P}_A$ fits in the ranking.

7.3 The prize space is $\mathbb{X} = \{6000, 3000, 0\}$ (in \$US). Of the prospects

$$\mathbb{P}_A = \{.45, .00, .55\}$$
$$\mathbb{P}_B = \{.00, .90, .10\}$$
$$\mathbb{P}_C = \{.001, .00, .999\}$$
$$\mathbb{P}_D = \{.00, .002, .998\}$$

an individual states that $\mathbb{P}_B \succ \mathbb{P}_A$ and $\mathbb{P}_C \succ \mathbb{P}_D$. Show that these preferences violate the substitution axiom. (*Hint*: Show that $\mathbb{P}_A$ is equivalent to a compound prospect that offers a .90 chance of playing the game $\{.50, .00, .50\}$ and a .10 chance of the "game" (actually a sure thing) $\{.00, .00, 1.0\}$. Show that $\mathbb{P}_C$ is equivalent to another compound prospect involving the same two games but with different probabilities of getting to play them. Likewise, resolve $\mathbb{P}_B$ into .90 and .10 chances of playing two games and $\mathbb{P}_D$ into chances of playing the same two games.)

7.4 An individual who obeys the EU axioms has utility function for wealth equal to $u(W_1) = 1 - \exp(-W_1/1000)$ and current wealth $W_0 = \$1000$. Find the *certainty equivalents* of each of the prospects in Exercise 7.2; that is, the certain sum of money such that the individual would be indifferent between receiving that sum (an amount that could be negative) and playing the game.

7.5 The prize space is $\mathbb{X} = \{999, 99, 9, 0\}$ (in \$US), and there are prospects

$$\mathbb{P}_A = \{.00, .20, .00, .80\}$$
$$\mathbb{P}_B = \{.01, .00, .99, .00\}.$$

Which would a person choose who always preferred prospects with higher mean given the variance and prospects with lower variance given the mean?

Which would a person choose who was an expected-utility maximizer with initial wealth $W_0 = 1$ and utility $u(W) = \ln W$?

7.6 An EU maximizer will allocate wealth $W_0 = 1$ between a safe asset yielding sure rate of return r and a risky asset yielding rate R (a random variable) over the next period. Given that fraction p of wealth is put into the risky asset and that $u(W)$ is the individual's increasing and concave utility function, the problem is to solve

$$\max_p Eu\left[1 + r + p(R - r)\right].$$

Assuming that u is differentiable and (to avoid technicalities) *bounded,* show that when $ER > r$ the individual will always invest some positive fraction of wealth in the risky asset. (*Hint*: Differentiate Eu with respect to p, and—as the boundedness feature suffices to authorize—reverse the order of differentiation and expectation. Then show that the resulting expression is positive at $p = 0$, and draw the appropriate conclusion.)

7.7 Verify expressions (7.6) and (7.7).

EMPIRICAL PROJECT 3

This project involves applying expected-utility theory to portfolio choice. The file Monthly1977–2007 in folder Project 3 on the FTP site contains annualized percentage rates of return on one-month CDs (certificates of deposit) and monthly (dividend- and split-adjusted) prices of Ford (F) and General Motors (GM) stock, all as of the beginning of each month. For this project we will measure time in months, taking $t = 0$ to correspond to the beginning of January 1977, $t = 1$ to the beginning of February1977, and so on up to $t = 372$ for the beginning of January 2008. You will use the empirical distribution of the monthly returns from these assets to determine the portfolio that maximizes the expected utility of wealth at the end of a holding period of one month, taking $u(W) = \ln W$ as the utility function. Let r_t be the one-month CD rate over the month beginning at t and $\dot{R}_t^F = P_{t+1}^F/P_t^F - 1 - r_t$ and $\dot{R}_t^{GM} = P_{t+1}^{GM}/P_t^{GM} - 1 - r_t$ be the simple *excess* rates of return over the month for F and GM. (Note that r_t would have been known to an investor at t but that $\dot{R}_t^F$ and $\dot{R}_t^{GM}$ would not.) Standing at $t = 372$ and using the available 31 years of historical data, you will find (1) the value p^F that maximizes $\hat{E}\ln\left(1 + r_t + p^F \dot{R}_t^F\right)$, (2) the value p^{GM} that maximizes $\hat{E}\ln\left(1 + r_t + p^{GM}\dot{R}_t^{GM}\right)$, and (3) the value $\mathbf{p} = \left(p^F, p^{GM}\right)'$ that maximizes $\hat{E}\ln(1 + R_{\mathbf{p}t})$, where $R_{\mathbf{p}t} \equiv r_t + p^F \dot{R}_t^F + p^{GM}\dot{R}_t^{GM}$. Here in each case, $\hat{E}$ is the estimate of expected value based on the 31-year sample. In each of the three cases you will compare the exact solution with the solution to mean–variance problem $\max_{\mathbf{p}}(\hat{\mu}_{\mathbf{p}t} - \rho_r \hat{\sigma}_{\mathbf{p}t}^2/2)$, using the coefficient of relative risk aversion ρ_r that corresponds to log utility and empirical estimates $\hat{\mu}_{\mathbf{p}t}$ and $\hat{\sigma}_{\mathbf{p}t}^2$ of the mean and variance of the portfolio's rate of return. Here are the specific steps to follow:

1. Convert the annualized percentage rates $(\%)_t$ on CDs to monthly simple rates, as $r_t = (\%)_t / 1200$.

2. Calculate the simple monthly rates of return for each of the two stocks and the excess rates $\dot{R}_t^F$ and $\dot{R}_t^{GM}$.

3. Calculate the estimates of expected log returns per dollar for given values of p^F, p^{GM}, and $\mathbf{p} = \left(p^F, p^{GM} \right)'$ as

$$\hat{E} \ln \left(1 + r_t + p^F \dot{R}_t^F \right) = \frac{1}{372} \sum_{t=0}^{371} \ln \left(1 + r_t + p^F \dot{R}_t^F \right) \tag{7.12}$$

$$\hat{E} \ln \left(1 + r_t + p^{GM} \dot{R}_t^{GM} \right) = \frac{1}{372} \sum_{t=0}^{371} \ln \left(1 + r_t + p^{GM} \dot{R}_t^{GM} \right) \tag{7.13}$$

$$\hat{E} \ln \left(1 + R_{\mathbf{p}t} \right) = \frac{1}{372} \sum_{t=0}^{371} \ln \left(1 + R_{\mathbf{p}t} \right). \tag{7.14}$$

4. Optimize the pertinent $\hat{E}$ value with respect to the portfolio weight or weights. It is feasible to do this by trial and error by evaluating at sequences of values of p^F, p^{GM}, and $\mathbf{p}$, but using an optimizing routine in Matlab® or other software is much faster.

5. In each case, once the optimal portfolio is found, multiply the corresponding maximum value of expected utility by 12 and record the result. This is the annualized, continuously compounded expected rate of return of the optimal portfolio.

6. Estimate the means, variances, and covariances of the excess rates of return on F and GM, using the usual unbiased estimators,

$$\frac{1}{372} \sum_{t=0}^{371} \left(\dot{R}_t^F, \dot{R}_t^{GM} \right)'$$

for mean vector $\hat{\mu} = (\hat{\mu}^F, \hat{\mu}^{GM})'$ and

$$\frac{1}{371} \sum_{t=0}^{371} \left[\begin{array}{cc} \left(\dot{R}_t^F - \hat{\mu}^F \right)^2 & \left(\dot{R}_t^F - \hat{\mu}^{GM} \right) \left(\dot{R}_t^{GM} - \hat{\mu}^{GM} \right) \\ \left(\dot{R}_t^F - \hat{\mu}^{GM} \right) \left(\dot{R}_t^{GM} - \hat{\mu}^{GM} \right) & \left(\dot{R}_t^{GM} - \hat{\mu}^{GM} \right)^2 \end{array} \right]$$

for covariance matrix $\hat{\Sigma}$.

7. Express $\hat{\mu}_{\mathbf{p}t} = \hat{E} R_{\mathbf{p}t}$ and $\sigma_{\mathbf{p}t}^2 = \hat{V} R_{\mathbf{p}t}$ for arbitrary portfolio vector $\mathbf{p}$ and find the solutions of $\max_{\mathbf{p}} (\hat{\mu}_{\mathbf{p}t} - \rho_r \hat{\sigma}_{\mathbf{p}t}^2 / 2)$ in each of the three cases. (Obviously, in case (1) $\mathbf{p} \equiv \left(p^F, 0 \right)'$ and $R_{\mathbf{p}t} = r_t + p^F \dot{R}_t^F$, while in case (2) $\mathbf{p} \equiv \left(0, p^G \right)'$

Table 7.1

Case	Method	p^F	p^{GM}	$12\hat{E}\ln(1+R_{\mathbf{p}t})$
(1)	EU	.6843	0	.0875
(1)	MV	.6615	0	.0875
(2)	EU	0	.2711	.0685
(2)	MV	0	.2684	.0685
(3)	EU	.8973	-.3333	.0903
(4)	MV	.8473	-.3083	.0902

and $R_{\mathbf{p}t} = r_t + p^{GM}\dot{R}_t^{GM}$.) This requires no optimization, but in case (3) it does require inverting a 2×2 matrix or by some other means solving a pair of simultaneous equations.

8. Plug each of the mean–variance solutions into the corresponding summations (7.12)–(7.14) and multiply by 12 to get the annualized, continuously compounded expected rate of return of the mean–variance approximation to the optimal portfolio.

The results you find should be close to those in Table 7.1. (In each case the proportion of wealth held in the CD is unity minus the proportion(s) in the stock(s).) Notice that the optimal and approximate portfolio compositions are very close and that the corresponding annualized rates of return are almost identical. The negative values of p^{GM} in the solutions for case (3) indicate that GM would be sold short.

Note that although these are the correct solutions to $\max \hat{E}\ln(1 + R_{\mathbf{p}t})$ based on the *empirical* distribution of F and GM returns over the past 31 years, they cannot be consistent with the objective $\max E\ln(1 + R_{\mathbf{p}t})$ on the basis of a more accurate model of the *actual* distribution. This is because $\dot{R}_t^{GM}$ has no definite upper bound in reality, and a short position would lead to insolvency and $\ln(1 + R_{\mathbf{p}t}) = -\infty$ with positive probability. Thus, in implementing a viable portfolio strategy one would need to constrain both p^F and p^{GM} to be nonnegative. For the same reason one would need the constraint $p^F + p^{GM} \le 1$, since sufficiently low realizations of the returns would lead to insolvency if shares were purchased on margin. We will return to this issue in Empirical Project 5 (Chapter 13), where we simulate the performance of various portfolio strategies over time.

CHAPTER 8

OBSERVED DECISIONS UNDER RISK

If we accept that the axioms of expected-utility theory describe rational behavior, and if we further maintain that intelligent, self-interested individuals simply *must* be rational, then expected-utility theory becomes a guide to *positive* or *descriptive* theories. Much of the theoretical work that has been done in economics and finance since the 1950s is in fact based on these joint assumptions. This has been guided, in large part, by Milton Friedman's influential essay "The methodology of positive economics" in Friedman (1953). There he argues persuasively that theories should be judged not by their *assumptions,* but by their implications or *predictions.* In his view, all models are merely convenient tools that help us organize, distill, and make sense of the complexity of the world around us. If we find that a model is useful in generating predictions, then that is all we can ask of it. For example, the usual supply-demand dichotomy in economics is such an organizing tool that gives us an immediate sense of what will happen when costs go up or incomes go down or prices of competing products rise, and so forth. The model is useful even if we cannot put complete confidence in the underlying framework of competitive firms whose managers conscientiously and resolutely strive to maximize profits and fully informed, self-interested consumers who relentlessly optimize. Friedman offers the compelling example of the expert pool player, whose approach to a particular shot can be predicted by calculating the vector forces imparted by the cue ball on the object of impact. According to Friedman, if we want to know what the pool player will do, our model should be based on the physics of the process. In other words, we should suppose that the pool player behaves *as if* he makes the requisite calculations and applies the vector force that the laws of physics deem appropriate to the objective. The "as if" stipulation is crucial. It is not necessary to claim that the player actually makes the calculations; presumably, long experience coupled with the quick feedback that promotes learning and error correction would produce the same result.

There are, however, some opposing views. For one, it has been pointed out that more realistic assumptions lead to broader, more encompassing, and more useful theories. For example, a model of how feedback and error correction actually operate to improve one's ability at pool would extend the theory even to novices and enable one to make predictions about the rate of improvement of play. Moreover—and this is especially relevant in economics—the "implications" of theories are not always easy to test, since (at least at the level of the market) it is well nigh impossible to carry out controlled experiments that make it possible to distinguish among separate, and often conflicting, influences. Great reliance inevitably has to be placed on statistical methods, which themselves rely on untested—often untestable—assumptions about the data. For example, in the CAPM tests, the effect on the cross-sectional Fama–MacBeth regressions produced by sampling errors in the betas depends on the statistical model for errors in the time-series regression of returns of individual securities on the market return.

For the issue now at hand, how can we judge whether financial models should be based on expected-utility theory when a long chain of other assumptions are necessary for the predictions—frictionless markets, "rational" expectations, etc.? As it happens, the modeling of decision making under uncertainty has not been left to economists alone. This has been for many years an active field in psychology, and psychologists have carried out carefully controlled experiments to judge the adequacy of the expected-utility (EU) theory and some competing ones. More recently, experimental economists have made valuable contributions as well. Let us now look at some of the main findings.

8.1   EVIDENCE ABOUT CHOICES UNDER RISK

8.1.1   Allais' Paradox

One of the first to cast doubts on the applicability of the EU theorem was French economist and 1988 Nobel laureate Maurice Allais. Allais (1953) posed to various individuals hypothetical choices of the following sort (here substituting dollars for French francs):

1. Prospect A gives a gain of $1 million for sure. Prospect B gives $5 million with probability .10, $1 million with probability .89, and nothing with probability .01. One must choose either A or B.

2. Prospect C gives \$1 million with probability .11 and nothing with probability .89. Prospect D gives \$5 million with probability .10 and nothing with probability .90. One must choose either C or D.

Allais noted that in situation 1 most people say that they prefer A to B, evidently considering that the certainty of getting \$1 million outweighs the chance of getting an extra \$4 million at the risk of coming away empty handed. In situation 2 most people choose D, the \$5 million top prize dominating the \$1 million offered by C and the probabilities looking about the same. As Allais pointed out, these modal choices—A over B and D over C—violate the EU theorem. To see this, take as the prize space $\mathbb{X} = \{5, 1, 0\}$ (in \$ millions) and define the prospects as $\mathbb{P}_A = \{0, 1, 0\}$, $\mathbb{P}_B = \{.10, .89, .01\}$, $\mathbb{P}_C = \{0, .11, .89\}$, $\mathbb{P}_D = \{.10, 0, .90\}$. Set $u(5) = 1$ and $u(0) = 0$ as an arbitrary choice of location and scale. Then, under the EU hypothesis $\mathbb{P}_A \succ \mathbb{P}_B$ implies $u(1) > .10 + .89u(1)$ or $u(1) > \frac{10}{11}$. On the other hand, $\mathbb{P}_D \succ \mathbb{P}_C$ implies $.10 > .11u(1)$ or $u(1) < \frac{10}{11}$. Obviously, there is no *function* $u : \mathbb{X} \to \Re$ that can satisfy these two conditions. Intuitively, the preference for $\mathbb{P}_A$ over $\mathbb{P}_B$ means that, starting with $\mathbb{P}_B$, one would trade off the 10% chance of \$5 million and 1% chance of nothing in order to get an additional 11% chance of \$1 million; while the preference for $\mathbb{P}_D$ over $\mathbb{P}_C$ means just the opposite, since going from D to C means sacrificing an 11% chance at \$1 million to get the \$5 million-or-nothing gamble.

8.1.2 Prospect Theory

Many other examples of this sort were put forth during the 1950s, but economists downplayed their importance. One reason for this was that EU theory provides such a cogent, systematic, logical way of modeling behavior, with nothing else to supplant it apart from Markowitz' seemingly ad hoc mean–variance approach. Another reason is that economists considered that test subjects would not give much thought to *hypothetical* choices of the sort that Allais and others posed, since they have no real incentive for careful reflection. Also, after Allais most of the experimental work in this area appeared in the psychology literature, into which few economists were willing to venture. However, the 1979 *Econometrica* article by psychologists Daniel Kahneman and Amos Tversky finally forced many economists to recognize the serious challenges to EU theory. As did Allais, Kahneman and Tversky reported the results of choices made by various subjects, mostly university students and faculty. Although the choices were still hypothetical, they were simple enough and realistic enough that readers could compare their own hypothetical decisions with those of the experimental subjects. Thus, the evidence for anomalous behavior was extremely compelling. Moreover, Kahneman and Tversky offered an alternative theory—*prospect theory*—that showed how humans' limited ability to process information could account for it.

8.1.2.1 *The Certainty Effect* Among the anomalies that Kahneman and Tversky documented are more paradoxes of the Allais type, but involving sums of money that people deal with in everyday life. For example, with $\mathbb{X} = \{4000, 3000, 0\}$ (in dollars, say) they find that $\mathbb{P}_B = \{0, 1, 0\}$ is usually preferred to $\mathbb{P}_A = \{.80, 0, .20\}$ and that $\mathbb{P}_C = \{.20, 0, .80\}$ is preferred to $\mathbb{P}_D = \{0, .25, .75\}$. The preference

for $\mathbb{P}_B$ over $\mathbb{P}_A$ is consistent with risk aversion, in that one prefers the certainty of 3000 to a gamble with higher expected payoff, $E_A X = 3200$. Again, however, there is no utility function of the prizes that is consistent with the modal preferences between the two pairs of prospects. This is because $\mathbb{P}_B \succ \mathbb{P}_A$ implies $u(3000) > .89$ (taking $u(4000) = 1$), while $\mathbb{P}_C \succ \mathbb{P}_D$ implies $.20 > .25u(3000)$ or $u(3000) < .20/.25 = .80$. This simpler example makes it easy to see just which axiom of EU theory is violated. $\mathbb{P}_B \succ \mathbb{P}_A$ means that prize $x_2 = 3000$ is preferred to the gamble $\{.80, 0, .20\}$. By the substitution axiom, $\mathbb{P}_D$ should be preferred to a prospect that substitutes for the .25 chance of winning x_2 a .25 chance of playing the $\{.80, 0, .20\}$ game. The result of this substitution is, however, the prospect $\{0 + .25(.80), 0, .75 + .25(.20)\} = \{.20, 0., .80\} = \mathbb{P}_C$. The modal choice, made by people of arguably above-average intellectual skills, is therefore in violation of the critical substitution axiom of von Neumann and Morgenstern. Kahneman and Tversky (1970) consider this violation an example of what they call the "certainty" effect, which is that the *certainty* of winning, as in the sure gain of 3000 in $\mathbb{P}_B$, has some special significance to people. Specifically, increments in probability that lead to certainty are considered more important or more understandable than increments that yield probabilities less than unity. Likewise, a decrement that reduces a probability to zero has more significance than if the final result is still positive. Prospect theory accounts for this by processing probabilities through a weight function that gives unit weight to values 0 and 1 but overweights small risks and underweights probabilities close to but below unity.

8.1.2.2 *The Reflection Effect*

In the single-period portfolio models we have considered the application of EU theory involves *integrating* portfolio gains and losses into initial wealth before they are evaluated by the decision maker. An investor is supposed to regard the prospective wealth level $W_1 = W_0(1 + R_p)$ that results from choosing portfolio p in the same way, regardless of whether it involves a gain from a relatively low initial wealth ($R_p > 0$) or a loss ($R_p < 0$) from a larger W_0. In other words, only the final result matters, not the starting point.

The following experiment brings this assumption of wealth integration into sharp question. Recall the fact that $\mathbb{P}_A = \{.80, 0, .20\}$ was preferred to $\mathbb{P}_B = \{0, 1, 0\}$ and that $\mathbb{P}_C = \{.20, 0, .80\}$ was preferred to $\mathbb{P}_D = \{0, .25, .75\}$ on the prize space $\mathbb{X} = \{4000, 3000, 0\}$. Now modify the prize space and choices symmetrically as follows. Take $\mathbb{X}' = \{0, -3000, -4000\}$, with $\mathbb{P}'_A = \{.20, 0, .80\}$, $\mathbb{P}'_B = \{0, 1, 0\}$, $\mathbb{P}'_C = \{.80, 0, .20\}$, $\mathbb{P}'_D = \{.75, .25, 0.\}$. All we have done is to switch the signs of the "prizes" to make them losses instead of gains. Now one finds that the modal choice is $\mathbb{P}'_A$ over $\mathbb{P}'_B$ and $\mathbb{P}'_D$ over $\mathbb{P}'_C$. Again, there is no utility function that is consistent with the choice, since preference for $\mathbb{P}'_A$ implies (taking $u(0) = 1$ and $u(-4000) = 0$) that $.20 > u(-3000)$ while preference for $\mathbb{P}'_D$ implies $.75 + .25u(-3000) > .80$ or $u(-3000) > .20$. Moreover, if $\mathbb{P}_B \succ \mathbb{P}_A$ was consistent with risk *aversion*, then $\mathbb{P}'_A \succ \mathbb{P}'_B$ is consistent with risk *preference*. Kahneman and Tversky infer that people respond differently to prospects of gains and prospects of loss, which indicates that people consider outcomes *in relation to their current wealth*. Moreover, they conclude that people are risk averse when considering gains and risk taking when considering

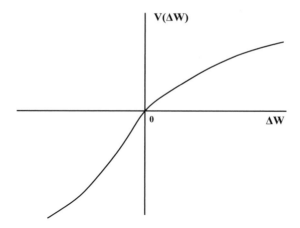

Figure 8.1 The Kahneman–Tversky value function.

losses. Their theory proposes a "value" function that has *changes* in wealth as its argument and is concave for positive changes, convex for negative changes, and steeper in the loss direction than in the gain direction, as in Fig. 8.1. The last feature accommodates the usual aversion that people show for bets involving symmetric gains and losses.

8.1.3 Preference Reversals

The apparent violation of the substitution axiom clearly challenges the EU theory, and the failure to integrate gains and losses is at odds with the way the theory has usually been applied in finance. The latter failure is not so important in one-period models in which people take gambles or make investments without regard to later events that transpire after the results are known. In other words, not much would be lost by dispensing with wealth integration and applying an expected-utility model using the Kahneman and Tversky value function. However, this would be much harder to manage in the multiperiod models that we discuss later, in which people are supposed to optimize a utility function of lifetime consumption. Even more serious than these difficulties, however, would be violations of something so fundamental as the *transitivity* of preferences. That would challenge even the basic ordinal utility theory that underlies models of consumers' behavior. In fact, there is now a substantial body of experimental evidence that does indicate that people—even smart people— systematically make intransitive choices.

This is indicated through a phenomenon known to psychologists as *preference reversals*. Here is a brief illustration. As usual, there is a prize space of monetary gains and losses, such as $\mathbb{X} = \{14, 3, -1\}$, consisting of a small loss, a moderate gain, and a much larger gain. Two alternative prospects are considered. The first, called a "$ bet," offers relatively low probability of the large gain and a high probability of the small loss. The second prospect, called a "P bet," offers a higher probability of the moderate gain and low probability of the loss, with no chance at the best prize. For example, for some low probability p (on the order of .05, say) we have $\mathbb{P}_\$ = \{p, 0, 1 - p\}$ and $\mathbb{P}_P = \{0, 3p, 1 - 3p\}$. The $\mathbb{P}_\$$ and $\mathbb{P}_P$ prospects are configured so that the *expected* payoff of the $ bet is larger (as $E_\$ X = 14p - 1(1 - p) = 15p - 1$ versus $E_P X = 12p - 1$ in the example). Several such pairs of bets are presented to the subject, with somewhat different prize spaces and p values, and the choices are indicated in each case. Most subjects almost invariably take the P bets, as is consistent with risk aversion.

Next, the subject is presented with the same set of prospects, one at a time and in random order, but is not asked to choose between bets of the two different types. For each prospect, whether it is P or $\$$, the subject is told that he "owns" it—i.e., that he will get to play the game and take whatever is won or lost. However, there is an alternative: the subject is allowed to *sell* the bet for some amount of money instead of playing it. An ingenious scheme is used to determine the absolute *minimum* selling price that the subject is willing to accept, eliminating the incentive to hold out for more. The subject then gets the dollar amount that is named instead of playing the game. These selling prices are supposed to be the *certainty equivalents* of the bets—the certain amounts of money that are equivalent to the prospects. Let these be denoted c_P and $c_\$$ for the two types of bets. Then, it is clear that $c_P \sim \mathbb{P}_P$ and $c_\$ \sim \mathbb{P}_\$$, so that, by transitivity (and the assumption that more money is preferred to less), $\mathbb{P}_P \succ \mathbb{P}_\$$ implies $c_P > c_\$$. However, the results are generally the opposite—minimum selling prices for the $ bets are usually greater than those for the P bets, a clear violation of transitivity.

Considering the importance of such findings for economics, and being a bit skeptical of evidence from researchers outside their own field, Cal Tech economists David Grether and Charles Plott (1979) undertook their own experiments, using elaborate controls to avoid every possible rational explanation for making intransitive choices. For example, to give subjects the incentive to choose carefully, they required them to actually play the games or to sell the bets for real money. Efforts were made to control for changes in tastes due to income effects arising from the gains and losses during the course of the experiment. The probability numbers were made more tangible by using a physical apparatus to determine outcomes, such as by spinning a pointer on a graduated circular scale. Subjects were questioned carefully to be sure they understood the instructions and the consequences of their choices. The researchers' findings are summed up in the following conclusion to their paper (p. 634):

> Needless to say the results we obtained were not those expected when we initiated this study. Our design controlled for all the economic-theoretic explanations of the phenomenon which we could find. The preference reversal phenomenon which is inconsistent with the traditional statement of preference theory remains.

Grether and Plott's was not the only experiment with careful controls and real motivation for subjects to make good choices. A fascinating paper by psychologists Sarah Lichtenstein and Paul Slovic (1973) described an experiment conducted in a Las Vegas casino. A space was made available for them to offer real P and $

bets to people who had come to the casino to gamble. Presumably, these were individuals who were used to thinking probabilistically and to putting their own cash at risk. In fact, interviews with those who chose to participate showed that most were better educated than the run-of-the-mill gambling crowd, having careers as engineers, lawyers, airline pilots, and so forth. Nevertheless, many of these experienced people made the "mistake" of choosing P bets over $ bets while selling P bets for less.

8.1.4   Risk Aversion and Diminishing Marginal Utility

In such experimental settings as those we have just described subjects almost universally display an aversion to risk, even when the stakes at risk amount to just a few dollars. Of course, within the EU paradigm, explaining any such strong aversion to risk requires that the utility function for wealth be strongly concave—that is, that marginal utility of wealth decrease rapidly. For example, if a "rational" individual with initial wealth W_0 refuses a 50–50 chance of winning $11 versus losing $10, we must conclude that $u(W_0 + 11) - u(W_0) < u(W_0) - u(W_0 - 10)$. But if u' decreases perceptibly over such a small range of wealth, how could even *huge* potential gains tempt the individual to risk modest losses? In other words, if people generally are so timid as to refuse these insignificant favorable bets, how could they ever be induced to play the stock market, to buy "junk" bonds, or to start up a business? Indeed, expected-utility maximizers could *not* be so induced according to an insightful analysis by Matthew Rabin (2000). Assuming only that the utility function u is defined, increasing, and concave over the relevant range of wealth, Rabin proves that demonstrated aversion to small but favorable risks implies aversion even to highly favorable risks that produce significant but still modest losses. For example, a person who at any initial level of wealth refuses an even chance to win $110 and lose $100 would also refuse an even chance to win $2,090 and lose just $800; and *no* finite gain could persuade the same person to take a 50–50 chance of losing $1000!

How could these perverse-seeming prospective choices be explained within the EU paradigm? Well, there are some ways to do it. For one thing, we do not know that the person who declines the 50–50 chance of +$110 versus -$100 *could* ever be induced to risk losing $1000. (Thus far, taxpayers have been unwilling to pony up enough money to fund research at this scale.) Thus, perhaps the experimental subjects who refuse the small bets do *not* play the market or otherwise take significant financial risks. After all, we already know that the throngs of salivating gamblers who flock to the casinos are in a class by themselves, so perhaps the explanation for Rabin's paradox just lies in the inhomogeneity of the crowd. Or perhaps we could salvage the rationality of choices under risk by allowing for some *irrationality* in assessments of risk. For example, the experiments all involve probabilities that are objectively known, whereas one might persuade oneself that one's skill as a stock picker skews the odds of market success strongly in one's favor. Besides, if the goal is just to explain the stock market, it is not irrelevant that a significant share of investment funds are moved around by professional managers who gamble with someone *else's* money. As well, some of the risks that individuals take in financial markets are at least partially hedged—either directly through options and short sales or indirectly by being played off against human capital (present value of labor income), housing, private businesses, and other nonfinancial assets. Finally, as we shall see when we discuss dynamic models in Chapter 12, it is inappropriate to try to explain commitments to long-term investments in a simple one-period framework. One's decisions about taking gambles with immediate payoffs or losses depend on how current resources will be used to finance consumption over the lifetime. Extreme aversion to the "timeless" risks proposed in experiments might be explained if the subjects cared far more about the present than about the future. Such a short-term focus is implied by models with *hyperbolic* discount rates for future consumption, such as those by David Laibson (1997) and Christopher Harris and Laibson (2001).

Although these counter arguments are not entirely without force, Rabin's own proposed explanation is far simpler and at least as compelling; namely, that people just plain hate to *lose*. In other words, the mere fact of losing is distasteful and is to some extent independent of the consequence for future consumption. Such an extrinsic influence on decisions can be accommodated within the EU paradigm in a very natural way (see Exercise 8.3). Nevertheless, the rather ad hoc kinked value function in prospect theory accomplishes the same thing and does explain at least qualitatively the paradoxes that Rabin raises.

8.2 TOWARD "BEHAVIORAL" FINANCE

Sadly, in the mainstream theoretical economics literature there still seems to be little acknowledgment of these challenges to the EU paradigm. Expected-utility maximization is still the core assumption underlying microeconomic models that take account of uncertainty. Indeed, even macroeconomists nowadays try to explain economic aggregates through models that start with rational, optimizing agents who behave "as if" they solve complicated dynamic programming problems, work out mathematical expectations under complicated models of price dynamics, and respond instantly to changes in technology, institutions, and the legal codes. There are several reasons why acknowledgment of contrary evidence has been so slow:

1. Some economists argue that the findings by their experimentalist colleagues and by psychologists are unpersuasive. They note that subjects have little incentive to make good decisions, since the rewards—if there are any at all—typically amount to no more than a few dollars. Moreover, test subjects are often students and other novices rather than people whose decision making skills have been sharpened through long experience.

2. Theories of behavior based on alternatives to expected utility have thus far been rather ad hoc. They may explain particular deviations from the rationalist paradigm but are incapable of extension to more general decision problems. For all its flaws, the rational-optimizing approach does provide a systematic way of approaching a broad variety of empirical issues. In a sense, it provides a kind of glue that holds together the different branches of economics.

3. Philosopher of science Thomas Kuhn (1970) points out that scientific progress typically occurs in fits and starts. To mark the beginning of a cycle, a discipline accepts some new "paradigm," a world view that offers a way to understand or "explain" the phenomena of principal interest. For a time, in their enthusiasm for the new idea, scientists rush to provide confirmatory or supporting data, but eventually the tide turns and inconsistencies come to light. Still, even after the weight of evidence has become ponderous, there is no general rejection of the paradigm until some compelling successor theory is found and the cycle can begin anew. In Kuhn's view this is not just an instance of the human failing known as *cognitive dissonance*—not liking to have to rethink one's views. Rather, it reflects the vital need for some clear way of organizing what we see around us so as to make some sense of all the chaos.

4. Finally, there is a less charitable explanation, which again is not limited to economists. Researchers in any field build up vested interests in particular lines of research that they have followed over the years. Professional reputation and status are often tied to the success of some theory long promoted by one's arguments and supported by one's empirical findings. Having become recognized as the experts in their particular fields, these self-interested individuals are the ones to whom new papers that challenge the orthodox views are sent to be reviewed for publication.

Fortunately, there has been some change of view in finance. Theorists in the new field of "behavioral" finance do in fact allow for certain deviations from "rational" behavior and thereby try to explain some of the empirical findings that seem to contradict the standard theory. Many have finally put aside EU theory as a descriptive model and turned to ideas from prospect theory or alternatives to characterize humans' decisions.[26] Robert Shiller's (2003) survey article gives an introduction to the growing literature and documents how the mounting evidence against rationality has initiated a paradigm shift in finance. One very influential contribution of this sort, by Nicholas Barberis, Ming Huang, and Tano Santos (2001), is discussed in Chapter 14. We shall see there that loss-averse behavior by investors can explain many of the qualitative features of broad indexes of stock prices, interest rates, and macroeconomic variables that are missing from dynamic models based on standard EU theory.

---

[26]For a good laugh—and some good analysis, too—read the article by Rabin and Richard Thaler (2001). You may agree with them that economists' stubborn defense of EU is similar to a pet store owner's efforts to persuade a customer that a recently purchased parrot is not really dead.

EXERCISES

8.1 An individual with current wealth W_0 chooses among risky prospects the one that maximizes the expected value of the utility or "value" function

$$u\left(W_1\right) = \begin{cases} e^{W_1 - W_0} - 1, W_1 \leq W_0 \\ 1 - e^{W_0 - W_1}, W_1 \geq W_0 \end{cases} . \tag{8.1}$$

Find the certainty equivalents c_A, c_B, c_C of the following prospects over prize space $\mathbb{X} = \{-1, 0, 1\}$ (in \$1000 US, say), where each prospect corresponds to a different distribution of a random variable X and $W_1 = W_0 + X$:

$$\mathbb{P}_A = \{.50, .50, .00\}$$
$$\mathbb{P}_B = \{.00, .50, .50\}$$
$$\mathbb{P}_C = \{.50, .00, .50\} .$$

On the basis of the characteristics of the utility function and the certainty equivalents, describe generally how the individual reacts toward financial risk.

8.2 An individual with utility function (8.1) owns a house whose value equals half his current wealth (i.e., $W_0/2$) and has no other income-producing investments. Housing prices being currently stable, the individual foresees wealth as being just the same next period unless fire or storm should destroy the house, an event to which he assigns the probability $p << 1$. An insurance company offers a policy that pays $W_0/2$ if the house should be destroyed. The policy costs $pW_0/2$. Would the individual buy it?

8.3 Suppose that one who must choose among risky prospects cares not only about the level of future wealth that would result but also about whether the *decision* produces a favorable or an unfavorable result. For example, suppose one's self-esteem depends to some degree on the outcome of one's decisions—whether one makes a winning choice or a losing choice. Such a person might still be "rational" in the sense of obeying the axioms of EU theory, but the usual models for utility functions would not satisfactorily represent the person's choices since the "prizes" have a nonpecuniary component. As a specific example of a model that does incorporate such sentiments, suppose that we combine the conventional utility function $1 - e^{-\alpha W_1}$ with an "esteem" function that adds $e^{\beta(W_1 - W_0)} - 1$ for $W_1 \leq W_0$ and $1 - e^{\gamma(W_0 - W_1)}$ for $W_1 > W_0$. Thus, the esteem function registers the effect of gains and losses on the pride or shame that results from one's decisions. Adding the two functions together gives

$$u\left(W_1\right) = \begin{cases} -e^{-\alpha W_1} + e^{\beta(W_1 - W_0)}, W_1 \leq W_0 \\ 2 - e^{-\alpha W_1} - e^{\gamma(W_0 - W_1)}, W_1 \geq W_0 \end{cases} .$$

As does (8.1), this specifically links choices to the initial level of wealth, but it has very different implications for decision making under risk. Analyze the following properties of this function under the condition that $\alpha \geq \beta > \gamma > 0$: **(a)** the sign of the first derivative for $W_1 > W_0$ and for $W_1 < W_0$; **(b)** the values of

$$\lim_{h \to 0} \frac{u\left(W_0 + h\right) - u\left(W_0\right)}{h}$$

$$\lim_{h \to 0} \frac{u\left(W_0\right) - u\left(W_0 - h\right)}{h},$$

which are the right- and left-hand derivatives at $W_1 = W_0$; **(c)** the sign of the second derivative at the various values of W_1 where it exists. Explain the implications of these properties for choices involving **(i)** symmetric risks that offer equal chances of comparable gains and losses; **(ii)** chances of gain with no risk of loss; **(iii)** risks of small loss with no chance of gain; and **(iv)** risks of large loss with no chance of gain.

CHAPTER 9

DISTRIBUTIONS OF RETURNS

The portfolio chosen by the "rational" maximizer of expected utility depends not just on the individual's tastes, as expressed in the utility function, but also on the person's beliefs about the probabilistic behavior of assets' returns. Thus, in a one-period model the decision maker needs to know the joint distribution of assets' returns over the holding period. Even to carry out the Markowitz prescription of choosing portfolios with high mean and low variance requires knowledge of the first two moments. To model a person's behavior, the researcher must know both the person's tastes *and* the person's beliefs. Of course, just as tastes differ, different people obviously have different ideas about the likely future values of different assets. Still, one might at least hope that the *typical* conception—what the financial press calls the "consensus" belief in regard to forecasts of analysts—is correct *most* of the time. On this basis theorists have considered it reasonable to assume that the investor knows the "true" model for the joint distribution of returns. In economics this has come to be known as the assumption of *rational expectations*. In finance it is generally called the *efficient markets* hypothesis, the view being that agents make efficient use of current information in anticipating future events and that this information is, therefore, fully reflected in market prices. We shall see in Chapter 15 that there are now serious doubts about whether markets really respond rationally and objectively to information, just as there are doubts about whether people make rational decisions, but for the time being we will swallow the doubts and see where the models take us. Keep in mind that people are not assumed to know what future rates of return will be, but merely that they know the right model for the joint probability distribution. To proceed in drawing inferences about what the typical investor does, the researcher then needs also to know the "right" model.

9.1 SOME BACKGROUND

This chapter covers some of the standard models for marginal distributions of continuously compounded rates of return of individual common stocks. Several questions need to be answered by way of motivation:

1. What do we mean by a stock's "continuously compounded" rate of return, and why focus on continuous rates instead of the simple one-period rates dealt with heretofore?

2. Why model rates of return anyway instead of *prices*?

3. Why consider *marginal* distributions of rates of return instead of *conditional* distributions given past returns (or other current information) or *joint* distributions of returns over many periods?

We can deal with the first question quickly, since the concepts should be familiar from the discussion of interest rates in Chapter 4. Answers to the last two questions are interconnected.

Thus far our treatment of static portfolio models has dealt just with assets' one-period simple rates of return, $R = (P_1 + D_1)/P_0 - 1$, where P_0 and P_1 are initial and terminal prices and D_1 is the cash value of whatever is paid or accrues during the period. For brevity, we refer to any such amount as a *dividend*. In one-period models it was not necessary to specify the length of the holding period, so we have just measured time in holding period units. However, if we are to model the behavior of returns over time or to compare the returns of different securities, there must be a standard time unit, and in finance this is customarily taken to be one year. This means that for the present purpose our notation must indicate the time intervals involved, corresponding to the notation for spot and forward rates of interest. Also, just as for interest rates, it will be convenient to work with continuously compounded rates of return, for reasons explained below. Thus, with time measured in years, we let $R_{t-\tau,t} \equiv \tau^{-1} \ln \left[(P_t + D_t)/P_{t-\tau} \right] = \tau^{-1} \left[\ln \left(P_t + D_t \right) - \ln P_{t-\tau} \right]$ represent the average continuously compounded rate over the interval from $t - \tau$ to t. Of course, this can be converted back to total return per dollar just by exponentiating, as $(P_t + D_t)/P_{t-\tau} = \exp \left(\tau R_{t-\tau,t} \right)$.

Note that if the asset in question is a discount unit bond maturing at t, then $R_{t-\tau,t}$ corresponds to the average spot rate,

$$r(t - \tau, t) = \tau^{-1} \ln \left[\frac{B(t,t)}{B(t - \tau, t)} \right]$$

$$= -\tau^{-1} \ln B(t - \tau, t).$$

Now, if an initial investment of \$1 were made at $t = 0$ and held for n periods, each of length τ, with dividends reinvested at the end of each period, then the total n-period return per dollar would be

$$\left(\frac{P_\tau + D_\tau}{P_0} \right) \left(\frac{P_{2\tau} + D_{2\tau}}{P_\tau} \right) \cdot \dots \cdot \left(\frac{P_{n\tau} + D_{n\tau}}{P_{(n-1)\tau}} \right) = \prod_{i=1}^{n} e^{\tau R_{(i-1)\tau, i\tau}}$$

$$= \exp \left(\tau \sum_{i=1}^{n} R_{(i-1)\tau, i\tau} \right).$$

Putting $T \equiv n\tau$, the average continuously compounded rate over the period $[0, T]$ would be found by taking the logarithm and dividing by T, as

$$R_{0,T} = n^{-1} \sum_{i=1}^{n} R_{(i-1)\tau, i\tau}.$$

This illustrates the ease of working with these continuous rates—while returns over long periods are found *multiplicatively* from the simple rates over subperiods, they are found *additively* from continuously compounded rates. Note, however, that when simple rates of return are close to zero, as they are with high probability when measured over short intervals of time, they are roughly proportional to the continuous versions. For example, let $R \equiv (P_\tau + D_\tau)/P_0 - 1$ represent the simple rate from $t = 0$ to $t = \tau$ and $R_{0,\tau}$ the continuous version. Then $\tau R_{0,\tau} = \ln(1 + R)$, and a Taylor expansion of the natural log gives

$$\tau R_{0,\tau} = R - \frac{1}{2(1 + R^*)^2} R^2 = R + o(R),$$

where $|R^*| < |R|$ and $o(R)/R \to 0$ as $|R| \to 0$. For example, with $\tau R_{0,\tau} = .10$ we have $|\ln(1.10) - .10| \doteq 4.7 \times 10^{-3}$ as the difference between continuous and simple rates, while with $\tau R_{0,\tau} = .01$ we have $|\ln(1.01) - .01| \doteq 5 \times 10^{-5}$. Thus, when we analyze the behavior of rates over fairly short periods of time, what we learn about continuous rates applies roughly to simple rates also.

Note: We will want to work exclusively with continuously compounded rates for the time being, so "continuous" should be understood whenever "rate of return" appears without qualification. Since the discussion pertains to the distributions of these rates, we will use uppercase to indicate their status as random variables and lowercase to denote particular realizations. Thus, we write $\Pr(R_{t-\tau,t} \le r)$ to represent the value of the CDF of the average (annual) rate $R_{t-\tau,t}$ at the point r, adding a subscript to distinguish realizations at different times, as $\Pr(R_{t-\tau,t} \leq r_t, R_{t,t+\tau} \leq r_{t+\tau})$. Finally, we will express one-period total returns per dollar, $(P_{t+\tau} + D_{t+\tau})/P_t$, just as $P_{t+\tau}/P_t$. Implicitly, this means that end-of-period prices (but not the *initial* prices) have been "dividend adjusted."[27]

We now take up the other issues listed at the beginning of this section, starting with question 3: Why consider marginal distributions $f(r)$ instead of conditional or joint distributions $f(r_t \mid r_{t-\tau}, r_{t-2\tau}, ...)$ or $f(r_t, r_{t-\tau}, ...)$? Modeling the marginal distribution is a logical place to begin for two reasons: (1) it does not get us into the more complicated task of modeling interdependence of returns over time, and (2) for many years it was thought that rates over nonoverlapping periods are essentially independent. In that case *only* the marginal distribution would matter, since $f(r_t \mid r_{t-\tau}, r_{t-2\tau}, ...) = f(r_t)$ and $f(r_t, r_{t-\tau}, ...) = f(r_t) \cdot f(r_{t-\tau}) \cdot \cdots$. In the next chapter we shall look at some of the empirical evidence and theoretical arguments that supported independence, but we will also see later on that new ways of looking at the data now lead us to reject that hypothesis. For the same reason—not having to model the dependence structure—the present focus is on marginal distributions of returns of individual stocks, rather than on the joint distribution for a *collection* of stocks. We have already briefly encountered factor models that do account for interdependence, and such models will be explored later in more detail. For now, though, the focus is on marginal distributions of one-period rates for individual stocks. However, we shall have something to say about how these distributions change with the length of the period over which returns are measured.

Finally, we can address question 2: Why model rates of return instead of prices? This has to do with the degree to which the two quantities exhibit a property known as *stationarity*. A time series $\{X_t\}_{t=1,2,...}$ is said to be *strictly stationary* if the joint distribution of $X_t, X_{t+1}, ..., X_{t+n}$ is, for each $n \geq 1$, independent of t. In other words, the same model holds for $X_t, X_{t+1}, ..., X_{t+n}$ as for $X_{t+m}, X_{t+m+1}, ..., X_{t+m+n}$. In particular, taking $n = 1$, this implies that the marginal distributions of $X_t, X_{t+1}, ...$ are all the same. Clearly, the definition of stationarity implies that moments of a stationary process (e.g., mean and variance) do not change with time. However, it is clear from looking at any stock chart that a *price* series $\{P_t\}$ is *not* a stationary process. Instead of bouncing back and forth randomly around some fixed typical value, prices wander up and down without evident indication of clustering. A histogram of prices looks nothing like one of the distributions one sees in statistics, where frequency is high about some central value and drops off rapidly in the tails. On the other hand, rates of return do appear to be more nearly stationary, as one can see by comparing the histograms of split-adjusted prices and returns for Microsoft (MSFT) in Figs. 9.1 and 9.2. Of course, if we have a model for the marginal distribution of $R_{t,t+\tau}$, and if rates in different periods are independent, then we automatically have a model for the

---

[27] Stocks begin to trade "*ex* dividend"— without dividend—shortly before the cash payment is made to shareholders. One who purchases on or after the *ex* date does not receive the dividend, and so the share price typically drops by about that amount when the stock "goes *ex.*" Much the same thing happens to the price of a registered coupon bond, for which the buyer pays the accrued interest. "Adjusting price for dividends" just means adding the cash payment back to the price to offset the artifactual decline.

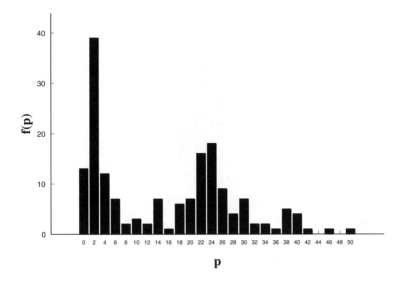

Figure 9.1   Histogram of monthly prices, MSFT, 1990–2005.

distribution of next period's dividend-adjusted price, $P_{t+\tau} = P_t e^{\tau R_{t,t+\tau}}$, *conditional on* P_t. Also, while simple and continuous rates over short periods behave much the same, the latter do appear to be more nearly stationary. This is another part of the answer to question 1 above.

9.2   THE NORMAL/LOGNORMAL MODEL

A model for rates of return that has long been the benchmark in finance theory is the Gaussian or normal distribution. If $R_{t,t+\tau}$, the average rate over the interval from t to $t + \tau$, is normal with mean μ_τ and variance σ_τ^2 for each t, and if rates in nonoverlapping intervals are independent, then the logarithm of total return over n periods, $\tau \sum_{t=0}^{(n-1)\tau} R_{t,t+\tau} \equiv \tau \sum_{j=1}^{n} R_{(j-1)\tau,j\tau}$, is the sum of i.i.d. normals. It is therefore distributed as $N\left(n\tau\mu_\tau, n\tau^2\sigma_\tau^2\right)$. Thus, the model has the property that *logs* of total returns over periods of any length are in the same family of distributions. What does this imply about the model for total return itself and for the conditional distribution of a future price?

We need to recall some basic facts from Section 3.8. If $Z \sim N(0,1)$ (standard normal), and if μ and $\sigma > 0$ are constants, then $X = \mu + \sigma Z$ is distributed as $N\left(\mu, \sigma^2\right)$, with PDF

$$f_X(x) = \frac{1}{\sqrt{2\pi\sigma^2}} \exp\left[-\frac{(x-\mu)^2}{2\sigma^2}\right], x \in \Re.$$

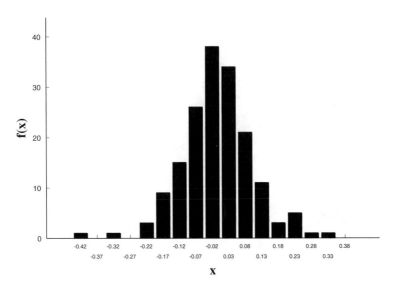

Figure 9.2   Histogram of monthly rates of return, MSFT, 1990–2005.

The random variable $Y \equiv e^X$ is then distributed as *lognormal* with parameters μ and σ^2 (in symbols, $Y \sim LN\left(\mu, \sigma^2\right)$), and its PDF is

$$f_Y\left(y\right) = \frac{1}{y\sqrt{2\pi\sigma^2}} \exp\left[-\frac{\left(\ln y - \mu\right)^2}{2\sigma^2}\right] \mathbf{1}_{(0,\infty)}\left(y\right).$$

Note that while f_X is symmetric about μ with infinitely long tails in both directions, f_Y is right-skewed and supported on the *positive* reals. Figures 9.3 and 9.4 illustrate the shapes of normal and lognormal distributions with $\mu = .10, \sigma^2 = .04$.

Applying now to rates of return, we see that the normality of $R_{t,t+\tau}$ implies the *log*normality of the one-period total return per dollar, $e^{\tau R_{t,t+\tau}}$. Moreover, the dividend-adjusted price at $t + \tau$ is also lognormal conditional on current price P_t, since $\ln P_{t+\tau} = \ln P_t + \tau R_{t,t+\tau} \sim N\left(\ln P_t + \tau\mu_\tau, \tau^2\sigma_\tau^2\right)$ implies $P_{t+\tau} \sim LN$ $\left(\ln P_t + \tau\mu_\tau, \tau^2\sigma_\tau^2\right)$. Standing at time t and knowing the current price, this would be the model for assigning probabilities to next period's dividend-adjust price. Note that the positivity of lognormal random variables is consistent with the limited-liability feature of most traded securities, in that $\Pr\left(P_{t+\tau} < 0 \mid P_t\right) = 0$. On the other hand, the lognormal model does not allow price to go to zero, which is something that does happen when US firms dissolve under provisions of Chapter 7 bankruptcy laws.

Let us see now how the model can be used in practice. Suppose, starting at $t = 0$, that we have $P_0 = 10$ and we take $R_{0,1}$ to be normal with mean $\mu = .1$, variance $\sigma^2 = .04$, and standard deviation $\sigma = .2$. Again letting P_1 represent the dividend-adjusted price at $t = 1$, how would we find $\Pr\left(P_1 > 12.0\right)$? This is done

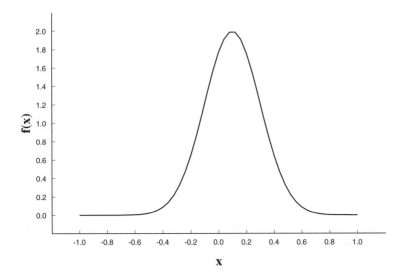

Figure 9.3   PDF of N(.10,.04).

by converting lognormal to normal, then normal to standard normal, by the following steps:

$$
\begin{aligned}
\Pr\left(P_1 > 12.0\right) &= \Pr\left(\ln P_1 > \ln 12\right) \\
&= \Pr\left(\ln P_0 + R_{0,1} > \ln 12\right) \\
&= \Pr\left(R_{0,1} > \ln 12 - \ln 10\right) \\
&= \Pr\left(.10 + .2Z > \ln 1.2\right) \\
&= \Pr\left(Z > \frac{\ln 1.2 - .10}{.20} \doteq .41\right) \\
&\doteq 0.34.
\end{aligned}
$$

Now, how would we find the moments of P_1? Recall that if $X \sim N\left(\mu, \sigma^2\right)$ its moment-generating function is $\mathcal{M}_X\left(\zeta\right) = Ee^{\zeta X} = e^{\zeta\mu + \zeta^2\sigma^2/2}$ for any real ζ. Since $Y = e^X$ is distributed as $LN\left(\mu, \sigma^2\right)$, it follows that the kth moment of Y about the origin is

$$
EY^k = Ee^{kX} = \mathcal{M}_X\left(k\right) = e^{k\mu + k^2\sigma^2/2}.
$$

In particular, the mean of Y (the case $k = 1$) is $EY = e^{\mu + \sigma^2/2}$ and the variance is $VY = EY^2 - \left(EY\right)^2 = e^{2\mu + \sigma^2}\left(e^{\sigma^2} - 1\right)$. Since $P_1 \sim LN\left(\ln 10 + .10, .04\right)$

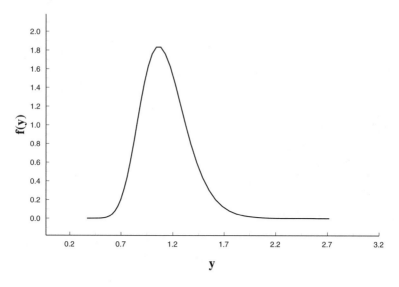

Figure 9.4   PDF of LN(.10,.04).

conditional on $P_0 = 10$, we have $EP_1 = \exp\left(\ln 10 + .1 + .04/2\right) = 10e^{.12} \doteq 11.27$.

Adopting the normal model for annual rates of return and assuming these rates to be independent over time automatically provides a model for total returns and prices after longer periods. This is because $\sum_{t=1}^{T} R_{t-1,t} \sim N\left(T\mu, T\sigma^2\right)$ when $\{R_{t-1,t}\}$ are independent and identically distributed (i.i.d.) as $N\left(\mu, \sigma^2\right)$. Thus, continuing the preceding example with one-period rates distributed as $N\left(.10, .04\right)$, consider the implications for the value of a position at $t = 5$ that begins at $t = 0$ with a single share worth $P_0 = 10$. Assuming reinvestment of dividends, the value is $V_5 = 10 \cdot \exp\left(\sum_{t=1}^{5} R_{t-1,t}\right)$, which is distributed as $LN\left[\ln 10 + 5\left(.10\right), 5\left(.04\right)\right]$. Thus

$$\Pr\left(V_5 > 12.0\right) = \Pr\left(Z > \frac{\ln 1.2 - .50}{\sqrt{.20}} \doteq -.71\right) \doteq .76$$

and

$$EV_5 = \exp\left(\ln 10 + .5 + .2/2\right) = 10e^{.6} \doteq 18.22.$$

Clearly, the normal/lognormal model is convenient. What reason is there to think that it actually might describe the data? The most obvious suggestion comes from the central-limit theorem (CLT). Recall that if random variables $\{X_j\}_{j=1}^{\infty}$ are i.i.d. with mean μ and finite variance σ^2, then the partial sum $\sum_{j=1}^{n} X_j$ has mean $n\mu$ and variance $n\sigma^2$, and the average value, $n^{-1}\sum_{j=1}^{n} X_j$, has mean μ and variance σ^2/n. The CLT tells us that the distribution of the *standardized* version of either of these, namely

$$Z_n \equiv \frac{\sum_{j=1}^n X_j - n\mu}{\sqrt{n\sigma^2}} \equiv \frac{n^{-1}\sum_{j=1}^n X_j - \mu}{\sigma/\sqrt{n}},$$

converges to standard normal, in the sense that $\Pr(Z_n \le z) \to \Pr(Z \le z)$ as $n \to \infty$, for any real number z. This means that when n is large, $n^{-1}\sum_{j=1}^n X_j$ itself is approximately normal with mean μ and variance σ^2/n. Applying this to rates of return, suppose that we divide the time interval $[0, t]$ into n equal subintervals of length t/n and (for brevity) let $R_{t_j} \equiv R_{(j-1)t/n, jt/n}$ be the average continuously compounded return over the jth such subinterval. Now the average rate over $[0, t]$ is just the average of the rates over the subintervals: $R_{0,t} = n^{-1}\sum_{j=1}^n R_{t_j}$. Thus, if the random variables $\{R_{t_j}\}_{j=1}^n$ are i.i.d. with finite variance, although not normally distributed, the CLT indicates that rates over moderately long periods should be at least *approximately* normal. For example, although the normal might not be a good model for *daily* rates of return—and we shall see that it is not—it might reasonably approximate the distribution of monthly or annual rates.

How does it work in practice? Figure 9.5 is a repeat of Fig. 9.2 with a fitted normal PDF superimposed. As before, the histogram pertains to monthly rates of return of MSFT, and the smooth curve is the PDF of a normal distribution with mean and variance equal to the sample mean (.0199) and variance (.0110) of MSFT rates of return over the same time period. From the figure it appears that the normal model for monthly rates of return is at least plausible. However, a closer look reveals some hint of discrepancies; in particular, the frequency of outliers in the tails—particularly in the *left* tail. For example, in 2 of the 169 months of the sample MSFT's price fell by 28% or more, while the probability of this under the normal model is less than half the proportion $\frac{2}{169}$. Empirical Project 4 at the end of Chapter 10 gives an opportunity to examine the issue more carefully, applying statistical tests to judge the fit of the normal distribution to rates of return measured at various frequencies. We generally find that the fit deteriorates as we move from monthly to weekly to daily to intraday data. Recall that a common measure of tail thickness is the coefficient of *kurtosis*, $\kappa_4 = \mu_4/\sigma^4 = E(X - \mu)^4 / (VX)^2$. Any normal random variable has $\kappa_4 = 3.0$ regardless of mean and variance, but sample estimates of kurtosis for high-frequency returns of financial assets are almost always greater than 3.0. Moreover, the data often produce negative estimates of coefficients of skewness, $\kappa_3 = \mu_3/\sigma^3$.

9.3 THE STABLE MODEL

If random variables $\{Z_j\}_{j=1}^\infty$ are i.i.d. as $N(0, 1)$ and if $S_n \equiv \sum_{j=1}^n Z_j$ is the partial sum, then $S_n \sim N(0, n)$, so that S_n has the same distribution as $\sqrt{n}Z_1$. In other words, the sum of n i.i.d. standard normals has the same distribution as that of each component scaled by $\sqrt{n}$. Extending this to arbitrary normals, if $\{X_j\}_{j=1}^\infty$ are i.i.d. as $N(\mu, \sigma^2)$, then $S_n \sim N(n\mu, n\sigma^2)$, so S_n has the same distribution as $\sqrt{n}X_1 + (n - \sqrt{n})\mu$. Thus, when $\mu \ne 0$ a centering constant as well as a scale factor is needed to make the distribution of each component match that of the sum.

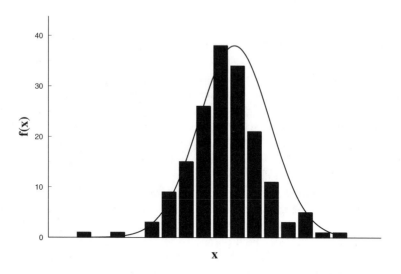

Figure 9.5   Histogram of MSFT rates of return and fitted normal.

Independent and identically distributed random variables with the property that $\sum_{j=1}^{n} X_j$ is distributed as $a_n X_1 + b_n$ for certain constants a_n, b_n are said to have a *stable* distribution. We have just seen that this holds for $a_n = \sqrt{n}$ and $b_n = (n - \sqrt{n})\mu$ when $\{X_j\}$ are i.i.d. as $N\left(\mu, \sigma^2\right)$. Thus, the normal distribution is stable, but it is not the only member of the stable family. For all such stable random variables the appropriate scaling factor a_n happens to be of the form $n^{1/\alpha}$ for some constant $\alpha \in (0, 2]$. For the normal distribution we have $\alpha = 2$, while the Cauchy distribution, another member of the stable class, has $\alpha = 1$. The Cauchy PDF is

$$f(x) = \frac{1}{\pi \left\{ 1 + \left[(x - m)/s \right]^2 \right\}}, x \in \Re,$$

where m and $s > 0$ are centering and scaling parameters. Although m and s correspond to μ and σ for the normal model, they are not mean and standard deviation, because this distribution possesses no finite moments of positive integer order. The problem is that the Cauchy PDF falls off toward zero only *quadratically*, so that even though $\int_{-\infty}^{\infty} f(x) \cdot dx$ does converge (to unity, obviously), the integral of the *product* of x and $f(x)$ does not. One can see how thick the tails are by comparing the plots of standard normal and standard Cauchy PDFs ($m = 0$ and $s = 1$) in Fig. 9.6. Other stable random variables with $1 < \alpha < 2$ do have first moments, but they, too, lack variances. Thus, the normal is the only stable distribution with finite variance, and of course it possesses moments of *all* positive integer orders.

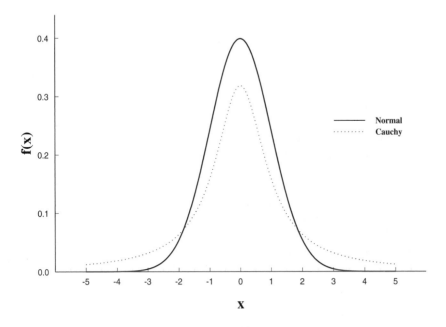

Figure 9.6   Standard normal and standard Cauchy PDFs.

While the normal distribution is the limiting form of appropriately standardized (scaled and centered) sums of i.i.d. random variables with finite variance, it turns out that appropriately standardized sums of i.i.d. variables with *infinite* variances have limiting forms in the nonnormal stable class. Thus, if distributions of rates of return over very short intervals really have such thick tails, then stable distributions might be good models for rates of return over longer periods. This fact and the observation that distributions of high-frequency returns of common stocks, commodity futures, etc., do have very thick tails prompted Benoit Mandelbrot (1963) to propose non-normal stable distributions as generic models. While all the rage for several years, and despite Mandelbrot's continuing efforts to promote them (e.g., Mandelbrot and Richard Hudson (2004)), these models have fallen out of favor, for two reasons: (1) the nonexistence of moments is obviously inconvenient for doing analysis—in particular, there is no hope of rescuing the Markowitz theory, even as an approximation, when rates of return follow nonnormal stable laws, and even expected utility would fail to exist without severe restrictions on the functional form of utility; and (2) since sums of i.i.d. stable variates retain the same characteristic exponent α no matter how many variables are summed, we should see tails that are just as thick in low-frequency returns (monthly, annually) as in high-frequency data (daily or intraday), and this is simply not the case.

9.4 MIXTURE MODELS

As it happens, there is an elegant way to account for the thick-tailedness of marginal distributions of high-frequency rates of return without having to forego variances and to ignore the apparent approach to normality over increasingly long intervals of time. Let $f(x; \theta)$ be any PDF with parameter $\theta \in \Theta$, where Θ is the space of allowable values. For example, if f is the normal PDF with mean zero and variance θ, then Θ is the positive real numbers. Now pick two different values, θ_1 and θ_2 in Θ, and use some random scheme for sampling from the two distributions having these different values of θ. For example, we could draw from $f(\cdot; \theta_1)$ if a coin turns up heads and from $f(\cdot; \theta_2)$ if it turns up tails. The effect is that θ has itself become a random variable with the distribution $\Pr(\theta = \theta_1) = \frac{1}{2} = \Pr(\theta = \theta_2)$. If we repeat indefinitely the experiment of flipping the coin and making the draw, the distribution of values that are drawn will look like neither $f(\cdot; \theta_1)$ nor $f(\cdot; \theta_2)$. In fact, it is a sort of *mixture* of the two f's. When parameter θ controls the scale of the distribution, the mixture typically has thicker tails than does either component and a higher coefficient of kurtosis.

Two specific mixture models have been prominent in modeling assets' returns; both are mixtures of normals having different variances:

1. James Press (1967) proposed the *compound-events model*, in which the logarithm of an asset's total return over a span τ is the sum of a normal random variable plus a Poisson-distributed number of other (independent) normals; i.e.,

$$\tau R_{t-\tau,t} = Y_\tau + \sum_{j=0}^{N_\tau} X_j. \tag{9.1}$$

Here (a) $Y_\tau \sim N\left(\mu\tau, \sigma^2\tau\right)$; (b) N_τ is independent of Y_τ and is distributed as Poisson with parameter $\lambda\tau > 0$ (so that $\Pr(N_\tau = n) = (\lambda\tau)^n e^{-\lambda\tau}/n!$ for $n = 0, 1, 2, ...$); (c) $\{X_j\}_{j=1}^\infty$ are i.i.d. as $N(\eta, \gamma^2)$ and independent of N_τ and Y_τ; and (d) we put $X_0 \equiv 0$ to handle the event $N_\tau = 0$. Conditional on N_τ it is clear that $\tau R_{t-\tau,t}$ is just a sum of independent normals; specifically, the conditional distribution of $\tau R_{t-\tau,t}$ is normal with mean $E(\tau R_{t-\tau,t} \mid N_\tau) = \mu\tau + N_\tau\eta$ and variance $V(\tau R_{t-\tau,t} \mid N_\tau) = \sigma^2\tau + N_\tau\gamma^2$. Thus, as Poisson variate N_τ takes on different realizations we simply mix normals of different means and variances. Recalling relation (3.8), the tower property of conditional expectation, we find the unconditional mean of $R_{t-\tau,t}$ to be

$$\begin{aligned}
ER_{t-\tau,t} &= E\left[E\left(R_{t-\tau,t} \mid N_\tau\right)\right] \\
&= E\left(\mu + \tau^{-1}N_\tau\eta\right) \\
&= \mu + \lambda\eta.
\end{aligned}$$

For the variance, the relation (3.9) between unconditional variance and conditional moments gives

$$VR_{t-\tau,t} = E\left[V\left(R_{t-\tau,t} \mid N_\tau\right)\right] + V\left[E\left(R_{t-\tau,t} \mid N_\tau\right)\right]$$

$$= E\left(\tau^{-1}\sigma^2 + \tau^{-2}N_\tau\gamma^2\right) + V\left(\mu + \tau^{-1}N_\tau\eta\right) \tag{9.2}$$

$$= \left[\sigma^2 + \lambda\left(\eta^2 + \gamma^2\right)\right]\tau^{-1}. \tag{9.3}$$

The coefficients of skewness and kurtosis work out to be

$$\kappa_3 = \frac{\lambda\eta\left(\eta^2 + 3\gamma^2\right)}{\sqrt{\tau}\left[\sigma^2 + \lambda\left(\eta^2 + \gamma^2\right)\right]^{3/2}} \tag{9.4}$$

$$\kappa_4 = 3 + \frac{\lambda\left(\eta^4 + 6\eta^2\gamma^2 + 3\gamma^4\right)}{\tau\left[\sigma^2 + \lambda\left(\eta^2 + \gamma^2\right)\right]^2}. \tag{9.5}$$

The sign of κ_3 is that of η, and $\kappa_4 - 3$ is always positive, but both tend to zero as time interval τ increases. Thus, the moments look increasingly like those of the normal distribution as $\tau \to \infty$. Indeed, putting $\tau = m\delta$ for some positive integer m and positive δ, we have

$$R_{t,t+\tau} = m^{-1}\sum_{j=1}^{m} R_{t+(j-1)\delta,t+j\delta}.$$

Assuming the components to be independent, $R_{t,t+\tau}$ is thus the average of m i.i.d. random variables with finite variance. Thus, by the central-limit theorem the *standardized* continuously compounded average rate of return over the span τ converges in distribution to $N\left(0,1\right)$ as $\tau \to \infty$.

2. Peter Praetz (1972) and Robert Blattberg and Nicholas Gonedes (1974) proposed the Student t family as models for rates of return.[28] We are familiar with the Student model from statistics as the distribution of the ratio of centered sample mean to its sample standard deviation in samples from normal populations: the ratio $(\bar{X} - \mu)/(S/\sqrt{n})$. This function of sample statistics and the population mean has, in fact, the distribution of the ratio of a standard normal to the square root of an independent chi-squared random variable divided by its degrees of freedom. In particular, if $Z \sim N\left(0,1\right)$ and $U \sim \chi^2\left(k\right)$ (chi-squared with parameter k), and if Z and U are independent, then $T_k \equiv Z/\sqrt{U/k}$ has Student's t distribution with k degrees of freedom. Parameter k governs the shape, including dispersion and tail thickness. The distribution is symmetric about the origin. Random variable T_1 actually has the standard Cauchy distribution, but T_k approaches the standard normal as $k \to \infty$. Positive integer moments $E\left|T_k\right|^m$ are finite only for $m \le k - 1$. As this implies, the distribution does possess a variance when $k \ge 3$. The coefficient of kurtosis

---

[28]"Student" was the pseudonym of William Sealy Gosset, whose 1908 *Biometrika* paper presented the first derivation of the t distribution.

equals $3 + 6/(k-4)$ for $k > 4$, which clearly approaches the normal value 3.0 as $k \to \infty$. In fact, k does not have to be an integer, but any positive real number will do. To see in what sense the t distribution is a *mixture* of normals, let $Y \equiv 1/\sqrt{U/k}$, where $U \sim \chi^2(k)$ and $k > 0$. Then the random variable $T_k = ZY$ has a normal distribution with mean zero and variance Y^2, conditional on Y. In effect, a random draw from a chi-squared distribution determines the variance of a normal distribution. To model rates of return, we add location and scale parameters and allow the degrees of freedom to depend on the time interval, as $R_{t-\tau,t} = \mu_\tau + s_\tau T_{k_\tau}$. Here the mean μ_τ, the scale parameter s_τ, and the degrees of freedom k_τ would be specified so as to fit the data at hand. Any such specification that sends k_τ to infinity as $\tau \to \infty$ ensures that the standardized variable $(R_{t-\tau,t} - \mu_\tau)/s_\tau$ converges in distribution to the normal.

9.5   COMPARISON AND EVALUATION

There are really just three considerations in evaluating a model: (1) the accuracy of its predictions in specific applications, (2) the breadth of its range of successful applications, and (3) the ease of using it to obtain predictions. Here are some specific things that we would like from a model of assets' returns in discrete time.

- The model should apply to and give good predictions of rates of return over periods of arbitrary length.

- The model should generalize to the multivariate case to allow modeling the joint behavior of returns of many assets.

- The model for the joint distribution of returns of many assets should lead to a tractable characterization of rates of return for portfolios.

None of the discrete-time models we have discussed scores well on all these points, although, as we shall see later, some of the problems disappear when we work in continuous time. Here are the assessments based on how well the discrete-time models meet these goals.

1. When average rates of return $\{R_{t-\tau,t}\}$ over periods of length τ are jointly normally distributed (which they are when they are independent with normal marginals), then average rates over longer periods are also normally distributed. This is a nice feature of the normal model. It implies that to model rates over periods of different length we can simply adjust the parameters without having to adopt another *family* of models. The stable laws and the compound-events model also have this feature, but the Student model generally does not; sums even of i.i.d. Student variables are no longer in the Student family. On the other hand, the normal model does not have thick enough tails to give a good description of high-frequency data, and the stable models do not explain the

marked decline in tail thickness as the time interval grows. The compound-events model does account—at least qualitatively—for tails that are thick at high frequencies and thinner at lower frequencies.

2. The normal model generalizes readily to the multivariate case. Specifically, if $\{R_{j,t-\tau,t}\}_{j=1}^{n}$ represent the rates for assets $j \in \{1, 2, ..., n\}$, and if these are modeled as multivariate normal, then the marginal distribution of the rate for each individual asset remains in the normal family. Such generalized versions of the stable and mixture models exist also, but they are much harder to use.

3. As applied to continuously compounded rates of return, none of these models leads to a simple characterization of rates of return of *portfolios*. The dividend-adjusted total return of portfolio $\mathbf{p} = (p_1, p_2, ..., p_n)'$ from t to $t + \tau$ is $\sum_{j=1}^{n} p_j \exp(\tau R_{j,t,t+\tau})$. If the continuous rates are multivariate normal, then total one-period returns $\{\exp(\tau R_{j,t,t+\tau})\}_{j=1}^{n}$ are multivariate *lognormal*. No problem there. However, although arbitrary linear functions of multivariate normals are still normal, linear functions of multivariate lognormals are *not* lognormal. In other words, we must adopt an entirely different family of models when we go from individual assets to portfolios. Likewise, linear functions of log-stable and mixture models for total returns do not remain in the respective families. In this situation the best one can do is to use the approximation that simple rates are probably close to continuously compounded rates when the period is short. In that case, under the normal model the simple rates are approximately multivariate normal and the corresponding portfolio rate is also approximately univariate normal. The approximations do become exact as the time interval goes to zero and we collapse to a model in continuous time. Whether the same "closure" property holds for multivariate stable and mixture models depends on just how these are generalized to the multivariate case.

EXERCISES

9.1 Let X be a random variable with moment-generating function $\mathcal{M}(\zeta)$, and let $\mathcal{L}(\zeta) \equiv \ln \mathcal{M}(\zeta)$ represent the cumulant-generating function. Show that

(a) $\mathcal{L}'(0) \equiv d\mathcal{L}(\zeta)/d\zeta|_{\zeta=0} = EX$.

(b) $\mathcal{L}''(0) = VX$.

(c) $\mathcal{L}^{(3)}(0)/\mathcal{L}''(0)^{3/2} = \kappa_3$, the coefficient of skewness.

(d) $\mathcal{L}^{(4)}(0)/\mathcal{L}''(0)^{2} = \kappa_4 - 3$, the "excess" kurtosis.

9.2 Use the results of the previous exercise to verify expressions (9.3) and (9.5). Why can the moments of the Student model not be found in this way?

9.3 Two assets have *simple* monthly rates of return R_1 and R_2. These are distributed as bivariate normal with means $\mu_1 = .010$ and $\mu_2 = .005$, variances $\sigma_1^2 = .04$

and $\sigma_2^2 = .01$, and correlation $\rho_{12} = -.50$. Let $R_p \equiv pR_1 + (1 - p) R_2$ be the rate of return of a portfolio in which proportion p of wealth is invested in asset 1. Find

(a) The minimum-variance portfolio; i.e., the value p_* such that $VR_p \geq VR_{p_*}$ for all p.

(b) The probability that $R_{p_*} > .10$.

(c) The value p_0 that maximizes the investor's expected utility of next month's wealth W_1 if present wealth is $W_0 = \$100$ and the utility function is $u(W_1) = 1 - \exp(-W_1/100)$.

(d) The probability that $R_{p_0} > .10$.

9.4 The total return of a security over the period $[0, \tau]$ is $P_\tau/P_0 = e^{\tau R_{0,\tau}}$, where $R_{0,\tau}$ is modeled as in (9.1). Show that P_τ/P_0 has CDF

$$\Pr\left(\frac{P_\tau}{P_0} \leq r\right) = \sum_{n=0}^{\infty} \Phi\left(\frac{\ln r - \mu\tau - n\eta}{\sqrt{\sigma^2\tau + n\gamma^2}}\right) \frac{(\lambda\tau)^n e^{-\lambda\tau}}{n!} \qquad (9.6)$$

for $r > 0$, where $\Phi(z) \equiv \Pr(Z \leq z)$ is the standard normal CDF, and $\Pr(P_\tau/P_0 \leq r) = 0$ for $r \leq 0$. (Hint: Apply the law of total probability, expression (3.1).)

9.5 The logarithm of a stock's total return per dollar over $[0, \tau]$ is to be modeled as $\tau R_{0,\tau} = \mu\tau + \sigma\sqrt{\tau}T_{k\tau}$ for certain constants μ, σ, k, where $T_{k\tau}$ is distributed as Student with $k\tau$ degrees of freedom. Measuring time in years and taking the number of trading days per year as 252, we have $\tau = 1/252$ for intervals of one day. In a long sample of observed *daily* data we estimate the mean, variance, and kurtosis of $\tau R_{0,\tau}$ as $\hat{E}\tau R_{0,\tau} = .0004$, $\hat{V}\tau R_{0,\tau} = .0001$, and $\hat{\kappa}_4 = 6.0$, respectively. Determine the values of μ, σ, and k that fit the moments. From these values estimate the mean, variance, and kurtosis of $R_{0,1}$ (the stock's average continuously compounded rate of return over one full year).

CHAPTER 10

DYNAMICS OF PRICES AND RETURNS

In the previous chapter we considered models for marginal distributions of continuously compounded rates of return of individual assets. When rates in different periods of time are i.i.d., we know that *joint* distributions are just products of marginals, so under the i.i.d. assumption knowledge of the marginals suffices to explain the joint behavior of returns over many periods. In addition, average rates over many periods are just averages of i.i.d. random variables, and their distributions are easily determined under some of the models, such as the normal/lognormal. We will now look at things from a dynamic perspective to see how prices actually evolve over time. We begin by inspecting the empirical evidence about whether rates of return in nonoverlapping periods are really independent. We then survey some of the discrete- and continuous-time models for prices and judge their consistency with these empirical findings and those of Chapter 9.

10.1 EVIDENCE FOR FIRST-MOMENT INDEPENDENCE

If random variables X, Y are statistically independent, then knowing the realization of one tells us nothing about what can happen to the other; i.e., for any x, y the conditional probability $\Pr(Y \leq y \mid X = x)$ is the same as the marginal probability

$\Pr\left(Y \leq y\right)$. On the other hand, when X, Y are dependent, these probabilities may well differ. We shall use the term *first-moment dependence* to describe situations in which the conditional mean of one variable (the first moment) depends on the realization of the other. We are familiar with this idea through the linear regression model, in which $E\left(Y \mid X\right)$ has the linear form $\alpha + \beta X$ with $\beta \neq 0$. More generally, we could have $E\left(Y \mid X\right) = g\left(X\right)$ for some function g. Of course, in the case of independence we would have simply $E\left(Y \mid X\right) = EY$, not involving X at all. Always in the linear case, and often more generally, first-moment dependence can be detected (in large enough samples) by estimating the covariance between X and Y, $\sigma_{XY} = E\left(X - \mu_X\right)\left(Y - \mu_Y\right)$; that is, since

$$E\left(X - \mu_X\right)\left(Y - \mu_Y\right) = E\left(X - \mu_X\right)Y - E\left(X - \mu_X\right)\mu_Y = E\left(X - \mu_X\right)Y$$

we have in the linear case

$$\begin{aligned}
\sigma_{XY} &= E\left[\left(X - \mu_X\right)E\left(Y \mid X\right)\right] \\
&= E\left[\left(X - \mu_X\right)\left(\alpha + \beta X\right)\right] \\
&= \alpha E\left(X - \mu_X\right) + \beta E\left[X\left(X - \mu_X\right)\right] \\
&= \beta \sigma_X^2.
\end{aligned}$$

Thus, the first (and easiest) way to spot dependence in rates of return in different periods is to look for nonzero covariances. Equivalently, since the coefficient of correlation is proportional to the covariance, we can simply see whether rates in nonoverlapping periods, $R_{t-\tau,t}$, $R_{t,t+\tau}$, ..., are correlated. Since we refer to correlations of a variable with its own past or future values, they are called *autocorrelations*. Here is a sample of what we see. Figures 10.1 and 10.2 show autocorrelation functions of monthly ($\tau \doteq \frac{1}{12}$ years) and daily ($\tau \doteq \frac{1}{252}$) rates of return of the S&P 500 index for long time samples, and Fig. 10.3 shows autocorrelations for daily rates of MSFT. The dashed horizontal lines represent approximate 2σ confidence bands that indicate statistical significance of the individual sample correlations. As one can see, there is not much evidence of linear dependence. Knowing the past record of rates of return seems to be of little help in predicting future rates over the next few days or months—at least if one restricts to linear models.

One often sees a different picture, however, in higher-frequency data, such as returns over spans of an hour, a few minutes, or from one trade to the next. For example, Fig. 10.4 shows that there is statistically and economically significant negative correlation at one lag in transaction-to-transaction price changes for Dow Chemical Corp. common stock (ticker symbol DOW), as calculated for the 1794 trades on the NYSE on January 2, 2002. This *seems* to indicate that one could profit, on average, by buying when price had ticked down on the last trade or selling when it had ticked up. In fact, this pattern is an artifact of what is called *bid-asked bounce*. It arises because "market" buy orders that come to the exchange—orders to trade right away at the best available price—typically execute at the price that is asked by a dealer or specialist, while market orders to sell typically trade at the price that the market maker bids. The spread between bid and asked is a compensation received by market

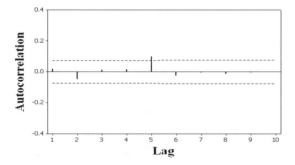

Figure 10.1   Autocorrelations, S&P monthly rates of return, 1950–2008.

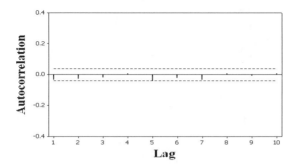

Figure 10.2   Autocorrelations, S&P daily rates of return, 1998–2008.

makers for standing ready to buy and sell immediately. To see the effect, suppose that a specialist (the individual or firm at the exchange that presides over trading in the particular stock) quotes DOW at 34.00 bid and 34.05 asked, standing ready to buy up to 1000 shares at the bid price and to sell 1000 at the asked. A market buy order for fewer than 1000 shares would then hit the "tape" at 34.05. If followed by another small buy order, and if the quotes had not changed, there would be another trade at the same price, for a recorded zero rate of return. However, the very first sell order would come in at 34.00, leading to a reported downtick or negative rate of return. Following this, price would tick back up with the next buy order. Thus, over periods in which bid-asked quotes remain the same, the sequence of prices might look like $AABBBABBAA\cdots$, where A and B are asked and bid. The sequence of price changes would then look like $0, -S, 0, 0, +S, -S, 0, +S, 0$, where $S = A - B$ is the spread. The price reversals interspersed by no-change events would show up statistically as negative autocorrelations at one lag, which is precisely what the data show.

A moment's thought shows that there is no way to extract profit from this bid-asked bounce, on average. For example, to attempt to exploit the downtick caused by

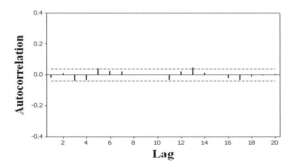

Figure 10.3 Autocorrelations, MSFT daily rates of return, 1998–2008.

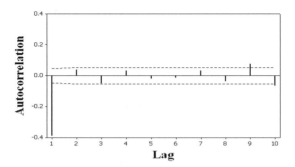

Figure 10.4 Autocorrelations, DOW trade-to-trade price changes, January 2, 2002.

a market sell order, one would have to place a market buy order. That order would then ordinarily be executed at the same higher asked price that prevailed before the downtick. If the shares were bought at this price and then unloaded right away, they would typically fetch the bid price, leading to a "profit" of $-S$. Figure 10.5 shows that the effect of bid-asked bounce for DOW on this date was no longer statistically significant when price was recorded at intervals of only 5 minutes, a period during which the quotes would typically have changed and price would have drifted outside the initial interval $[B, A]$. In general, just how long it takes for the effects of bid-asked bounce to vanish depends on how actively traded the stock is.

In high-frequency data for stock indexes, such as the S&P 500 or NASDAQ index, one often sees small *positive* autocorrelations. Indeed, this often shows up in daily data, although it does not appear in the time sample used to generate Fig. 10.2 for the S&P. While such positive autocorrelation would be consistent with a slow diffusion of information, at least *some* of it is due to what is called the *asynchronous trading effect*. Here is what causes it. The index price that is recorded for any time t is calculated from the most recent *trade* prices of the component stocks—that is, from prices on their last trades *prior to* t. If news of common influence arrives just before

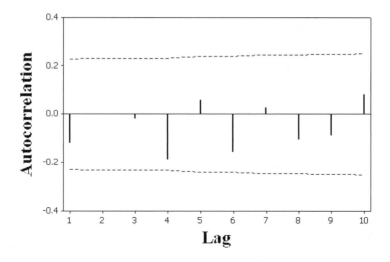

Figure 10.5   Autocorrelations, DOW 5-minute price changes, January 2, 2002.

t, the stocks that happen to trade after the news but before t will ordinarily reflect it at t, but the effect cannot show up in prices of other stocks until they next trade. Stocks of smaller companies are typically more thinly traded, and so trade prices of stocks of large-cap companies often do lead those of smaller firms. Of course, it can sometimes go the other way, depending on when orders to trade happen to come in. In any case, the fact that trades in different stocks are not synchronized can cause the index price at $t + \tau$ to be slightly correlated with the price at t. The effect would be present even if markets were so informationally efficient that bid-asked quotes of all stocks moved simultaneously to reflect relevant news.

It is wise to keep in mind also that high-frequency returns of common stocks are extremely noisy, as is indicated by the thick tails in the marginal distributions. This often causes statistical results for a particular time sample to be dominated by a few outliers. In such applications the hypothesis tests and confidence limits produced by standard statistical software, such as those for the autocorrelations in the preceding figures, are often a poor guide to the actual statistical significance of the results. Nevertheless, the persistent failure of researchers to find useful linear predictive models for assets' returns over moderate periods gives us some confidence that there is little first-moment dependence in data of moderate to high frequency. The story may be different over very long intervals corresponding to market cycles. Indeed, we will see in Section 15.2.3 that there is credible evidence for predictability at very long horizons—a phenomenon that seems to mirror the excessive volatility of prices relative to the fundamental factors that are thought to drive them.

10.2   RANDOM WALKS AND MARTINGALES

The limited indication of first-moment dependence was thought for many years to have confirmed the *random-walk hypothesis*. Expressed in log form, the hypothesis is that logarithms of dividend-corrected prices behave as random walks with trends; e.g., as

$$\ln P_t = \mu\tau + \ln P_{t-\tau} + \sqrt{\tau}u_t, t = \tau, 2\tau, ...,$$

where the shocks $\{u_t\}$ are i.i.d. with means equal to zero. The idea is that the shocks are purely unpredictable, and so the changes in log price from $t - \tau$ to t, $\ln P_t - \ln P_{t-\tau} = \ln (P_t/P_{t-\tau})$, bounce randomly back and forth about their mean value $\mu\tau$, which just depends on the length of the period. Of course, this is precisely the model in which continuously compounded rates of return are i.i.d., since $\tau R_{t-\tau,t} = \ln (P_t/P_{t-\tau})$ when P_t is dividend-corrected. In this case $E\tau R_{t-\tau,t} = \mu\tau$ and $V\tau R_{t-\tau,t} = \tau V u_t = \sigma^2\tau$, say.

It is now recognized that the strict random-walk hypothesis is untenable, in that the changes in log price are *not* strictly independent. While there continues to be little evidence against first-moment independence except at very long horizons, we now know that there is substantial dependence of higher order. Specifically, we find that the variance of $\tau R_{t-\tau,t} = \ln (P_t/P_{t-\tau})$ is highly predictable from the past record. One can see that in Figs. 10.6 and 10.7, which show autocorrelations of *squared* daily rates of return of the S&P index and of MSFT stock. These give strong indication of positive autocorrelation out to 20 lags. Thus, a big move in the index on one day—whether positive or negative—tends to be followed by unusually large moves for many days thereafter. Likewise, unusually calm markets tend to persist as well. Evidence of this *volatility clustering* has been found in virtually all markets and time periods. Of course, knowing today's volatility is of no help in predicting the direction of tomorrow's price change, so it does not provide obvious profitable trading opportunities in the asset itself. On the other hand, predicting a stock's future volatility can be of great value in dealing with options on the stock—an issue we shall discuss in detail later on.

That stock prices follow random walks was thought for many years to be equivalent to the hypothesis that markets are "efficient." We discuss the efficient-markets hypothesis at length in Chapter 15, but for now we can think of it as asserting that assets' prices always embody all the relevant public information about their underlying values. It is not immediately obvious what this implies about the intertemporal behavior of prices, but it was usually taken to mean that no use of publicly available information can produce profits in excess of the "appropriate" compensation for risk bearing; that is, there should be no special model or trading system that consistently produces "good deals." In very simple terms, the idea was that prices change in response to *news*, and by definition news is itself unpredictable. Of course, if logs of prices follow random walks, then their changes certainly are unpredictable—that is, one has no way of knowing in advance whether a future rate will be below or above its long-run average. However, the discovery of volatility persistence forced economists to rethink the matter and recognize that efficiency does not actually require full in-

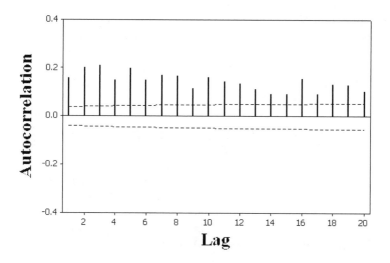

Figure 10.6   Autocorrelations, squared S&P daily rates of return, 1998–2008.

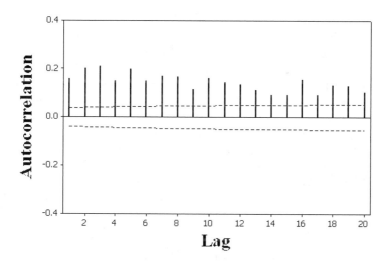

Figure 10.7   Autocorrelations, squared MSFT daily rates of return, 1998–2008.

dependence. The successor idea, which caught on during the 1970s, was that in efficient markets (appropriately scaled or discounted) prices behave as *martingales*. This turns out to be a nice way of formalizing the idea of first-moment independence, but it requires some explanation.

A discrete-time stochastic process $\{X_t\}_{t=0}^{\infty}$ is a martingale *adapted* to evolving "information sets" $\{\mathcal{F}_t\}_{t=0}^{\infty}$ if (1) EX_t exists and is finite (i.e., $E|X_t| < \infty$) for each t and (2) $E(X_t \mid \mathcal{F}_{t-1}) = X_{t-1}$ for $t \geq 1$. Here, the information set $\mathcal{F}_{t-1}$ on which the expectation is conditioned includes the entire history of the process, $X_0, X_1, ..., X_{t-1}$, and anything else that is relevant to know about X_t at $t-1$. More formally, $\mathcal{F}_{t-1}$ is the σ field of events that we know either to have occurred or not to have occurred at t. We require these information sets to grow over time, so that $\mathcal{F}_t$ contains all of $\mathcal{F}_{t-1}$ plus (at least) the knowledge of X_t itself. Condition (1) of the definition just guarantees that the conditional expectations can be found. The critical condition (2) is called the *fair-game property*, which gets its name from the following scenario.

Suppose that one starts at time $t = 0$ with some amount of capital X_0 and plays a sequence of \$1 bets, one per unit time. The payoff on the tth bet is Y_t, and the game is "fair" in the sense that the expected value of Y_t is always zero, regardless of what has happened in the past; that is, $E(Y_t \mid \mathcal{F}_{t-1}) = 0$. In other words, the loss event $Y_t = -1$ is balanced by just the right chance of gain. Then $X_t = X_{t-1} + Y_t$, and $E(X_t \mid \mathcal{F}_{t-1}) = X_{t-1}$ for each $t \in \{1, 2, ...\}$. In particular, $E(X_1 \mid \mathcal{F}_0) = X_0$ and $E(X_2 \mid \mathcal{F}_1) = X_1$, so

$$E(X_2 \mid \mathcal{F}_0) = E[E(X_2 \mid \mathcal{F}_1) \mid \mathcal{F}_0] = E(X_1 \mid \mathcal{F}_0) = X_0, \qquad (10.1)$$

and so on, with

$$E(X_t \mid \mathcal{F}_0) = X_0 \qquad (10.2)$$

for all $t \geq 0$. The first equality in (10.1) follows from the tower property of conditional expectation. As explained in Section 3.4.2, the tower property implies that the expected value of a quantity conditional on information $\mathcal{F}_{t-1}$—whatever it is—equals the expected value of the expectation conditional on *more* information $\mathcal{F}_t$. To express this more intuitively, it means that our best guess—our *forecast*—given what we know today equals our best guess of what our forecast will be tomorrow. Thus, (10.2) tells us that, *on average* in infinitely many repetitions of t plays of a fair game, one would end up with the same capital one had at the outset. But even more is true, for suppose that the amount bet on the play at each time $t-1$ can vary and depend in any way on $\mathcal{F}_{t-1}$. Letting b_{t-1} be the number of dollars bet, we have $X_t = X_{t-1} + b_{t-1}Y_t$. But since $b_{t-1} \in \mathcal{F}_{t-1}$—that is, since b_{t-1} is $\mathcal{F}_{t-1}$-*measurable,* so that we know the value of b_{t-1} at $t-1$—we still have $E(X_t \mid \mathcal{F}_{t-1}) = X_{t-1} + b_{t-1}E(Y_t \mid \mathcal{F}_{t-1}) = X_{t-1}$ and $E(X_t \mid \mathcal{F}_0) = X_0$. This tells us that there is no feasible betting strategy that can bias the game in one's favor.[29]

Clearly, martingales are relevant for (fair) games of chance, but what is their connection to prices of financial assets? Returning to the random-walk model, namely

$$\ln P_t = \mu\tau + \ln P_{t-\tau} + \sqrt{\tau}u_t, t = \tau, 2\tau, ...,$$

---

[29]Strictly, we can say that the martingale result holds for $\{X_t\}$ if the bet sizes are *bounded*. This requirement rules out *doubling strategies* that could ultimately bias the game in one's favor. For example, if one could keep doubling the bet following each loss, all past losses would be recouped and a net gain obtained on the first winning play. For such strategies always ultimately to succeed, however, one must have unlimited resources; for example, if \$1 is bet on each play, the total loss after n losing plays would be $2^n - 1$.

we can exponentiate both sides to get

$$P_t = e^{\mu\tau} P_{t-\tau} e^{\sqrt{\tau} u_t}$$

$$= \left(e^{\mu\tau} E e^{\sqrt{\tau} u_t} \right) P_{t-\tau} \left(\frac{e^{\sqrt{\tau} u_t}}{E e^{\sqrt{\tau} u_t}} \right)$$

$$= e^{\nu\tau} P_{t-\tau} U_t,$$

where $U_t \equiv e^{\sqrt{\tau} u_t} / E e^{\sqrt{\tau} u_t}$ and $e^{\nu\tau} \equiv e^{\mu\tau} E e^{\sqrt{\tau} u_t}$. Now replace the assumption that the $\{u_t\}$ are independent with the *weaker* one that $E\left(U_t \mid \mathcal{F}_{t-\tau}\right) = 1$ for all possible information sets $\mathcal{F}_{t-\tau}$. (This means that $E(e^{\sqrt{\tau} u_t} \mid \mathcal{F}_{t-\tau})$ equals unconditional expectation $E e^{\sqrt{\tau} u_t}$ for all $\mathcal{F}_{t-\tau}$.) It then follows algebraically that $P_{t-\tau} = e^{-\nu\tau} E\left(P_t \mid \mathcal{F}_{t-\tau}\right)$ for each $t \in \{\tau, 2\tau, ...\}$ or, equivalently, that $P_t = e^{-\nu\tau} E\left(P_{t+\tau} \mid \mathcal{F}_t\right)$ for each $t \in \{0, \tau, 2\tau, ...\}$. This says that current price equals the discount factor $e^{-\nu\tau}$ (a constant depending just on time interval τ) times the expectation of next period's price given available information—and that this is true in each period and regardless of the current situation and past events. In other words, the market sets current price equal to the best guess of next period's price, discounted to allow for growth at average continuously compounded rate ν. This means that one who concludes on the basis of any public information that the current price is either too low or too high will be wrong on average. The current price simply reflects everything that is relevant for determining expected future price.

Still, $\{P_t\}_{t=0}^{\infty}$ clearly does not have the fair-game property, so it is not yet a martingale; however, we can turn $\{P_t\}_{t=0}^{\infty}$ into a martingale very easily. Let $X_t \equiv e^{-\nu t} P_t$ be the time t price discounted back to some arbitrary time origin. Then

$$E\left(X_t \mid \mathcal{F}_{t-\tau}\right) = e^{-\nu t} E\left(P_t \mid \mathcal{F}_{t-\tau}\right)$$

$$= e^{-\nu t} e^{\nu\tau} P_{t-\tau}$$

$$= e^{-\nu(t-\tau)} P_{t-\tau}$$

$$= X_{t-\tau}.$$

This in turn implies that $E\left(X_t \mid \mathcal{F}_0\right) = X_0$ for each $t \in \{\tau, 2\tau, ...\}$. Thus, appropriately *discounted* prices are indeed martingales when those prices are based on "rational" expectations of *future* prices. This fact was first pointed out long ago by Paul Samuelson (1965), at which time it was thought to be both remarkable and reassuring.

Not surprisingly, we have learned a few things since then. Samuelson's logical proof is unassailable given the hypothesis that actual prices *are* based on "rational" expectations of future prices, but this is not necessarily the case. Whether prices discounted by known discount factors are close to martingales in reality is an empirical issue. We do know now that they *need* not be so in efficient markets, if by "efficiency" we mean that prices reflect all current public information and that people behave "rationally." This fact will emerge later in the course of discussing dynamic pricing models, where we will see that the appropriate discount factor may well be *stochastic*. When it is, there need be no simple martingale property for rescaled prices themselves.

10.3   MODELING PRICES IN CONTINUOUS TIME

Thus far, we have considered how prices behave only in discrete time, focusing on simple and continuously compounded average rates of return, as $R_t = (P_t + D_t) / P_{t-\tau} - 1$ and $R_{t-\tau,t} = \tau^{-1} \ln(1 + R_t) = \tau^{-1} \ln(P_t / P_{t-\tau})$ (with P_t dividend corrected). To study models of portfolio choice in continuous time and models for pricing options and other derivatives, it is necessary to get away from discrete-time units and think of prices as evolving continuously. The primary building blocks for such models in finance have been Brownian motions and Poisson processes, which are particular forms of *stochastic processes*. In general, a stochastic process $\{X_t\}$ is simply a *family* of random variables indexed by t. We would write $\{X_t\}_{t=0}^{\infty}$ or $\{X_t\}_{t \in \{0,1,2,\ldots\}}$ or $\{X_t\}_{t \in \{0,\tau,2\tau,\ldots\}}$ for a process that evolves at discrete instants in time and $\{X_t\}_{t \geq 0}$ for one that evolves continuously.

A little conceptual background is needed. We are used to thinking in terms of a single random variable or a set of finitely many random variables whose values are determined by the outcome of some random experiment. For example, suppose that we plan to draw at random from an urn containing balls marked with various real numbers. We think of the set of balls as our sample space Ω, of which ω is some generic outcome—some particular but unspecified ball. Then, if $X(\omega)$ represents the number on ball ω, we have a function or mapping $X : \Omega \to \Re$ from the outcomes to the real numbers. Of course, such a function is what we call a *random variable*. When we perform the experiment—make the draw—we observe outcome ω and the numerical value or *realization* $x = X(\omega)$. Now consider a modification of the experiment. The plan now is to draw at random from the urn n times, with replacement. The sample space is now all the ordered *sequences* of n balls. A single $\omega = (\omega_1, \omega_2, \ldots, \omega_n)$ that represents some particular such sequence now generates the realization of a sequence $X_1, X_2, \ldots, X_n$ of random variables, where X_1 takes the value $X(\omega_1)$ of the number on the first ball, X_2 takes the value $X(\omega_2)$ of the number on the second ball, and so on. Thus, a single *outcome* of the experiment of n draws generates an entire sequence of random variables. Now extend these ideas to a continuous-time stochastic process. We may think again of some random experiment (performed, perhaps, by nature) that generates an entire *track* or *sample path* $\{X_t\}_{t \geq 0}$ starting from some initial point X_0. This track is the realization of a set of random variables on the continuum $[0, +\infty)$, of which by a particular time $T > 0$ we will have seen only the first segment $\{X_t\}_{0 \leq t \leq T}$. Different outcomes of the experiment produce different tracks from some initial value X_0. The expression $\Pr(X_t \leq x)$ can be interpreted as the proportion of all the *possible* tracks—of which there might be an infinite number—whose values at t did not exceed x. The expression EX_t can be interpreted as the average of the values of the possible tracks at t. In this thought experiment all the possible tracks whose proportions and average values are calculated must start under the same initial conditions, represented by an information set $\mathcal{F}_0$, which necessarily includes initial value X_0. Thus, $\Pr(X_t \leq x)$ and EX_t mean the same things as $\Pr(X_t \leq x \mid \mathcal{F}_0)$ and $E(X_t \mid \mathcal{F}_0)$. Likewise, for any $s \in [0, t)$ the expressions $\Pr(X_t \leq x \mid \mathcal{F}_s)$ and $E(X_t \mid \mathcal{F}_s)$ represent the proportions of events

$X_t \leq x$ and the average values of X_t conditional on the history of the process up through s (and whatever else is known at s). Of course, $\Pr(X_t \leq x \mid \mathcal{F}_t)$ is just unity or zero accordingly as $X_t \leq x$ or $X_t > x$, and $E(X_t \mid \mathcal{F}_t)$ is just X_t itself.

We are now ready to look at the specific models for stochastic processes that are most often used in finance. The Poisson process and its compound generalization are the easiest to understand, so we begin there.

10.3.1 Poisson and Compound-Poisson Processes

A Poisson process, denoted $\{N_t\}_{t\geq0}$, is one of a family of so-called *point processes*. Initialized to zero at time 0, the value of a point process at $t > 0$ is merely a count of a certain class of events that have occurred up through t. The *track* of a realization of a point process on $[0, \infty)$ is thus a step function that starts at 0 and moves up by one unit at each point in time at which an event in the relevant class occurs. Examples of phenomena that could be modeled by such processes include the number of arrivals of customers in the local Walmart store, the number of impacts by meteorites in the continental United States, detections of neutrinos in a physics experiment, landings of aircraft at an airport, and occurrences of trades in IBM on the NYSE. For Poisson processes in particular we specify that the random number of events up through t has a Poisson distribution with mean $\lambda t > 0$; that is, the probability mass function for N_t ($t > 0$) as of time 0 is assumed to be

$$\Pr(N_t = n) = \Pr(N_t = n \mid \mathcal{F}_0) = \frac{(\lambda t)^n e^{-\lambda t}}{n!}$$

for $n \in \{0, 1, 2, ...\}$. To abbreviate, we write $N_t \sim \text{Poisson}(\lambda t)$. The parameter $\lambda = E N_t / t$ is called the *intensity* of the process, since it is the mean rate at which events occur through time. More generally, the increment $N_t - N_s$ in the process from time s to any later time t also has a Poisson distribution with mean (and variance) $\lambda(t - s)$. Moreover, increments of the process over nonoverlapping periods—as from q to r and from s to t for $q < r \leq s < t$—are statistically independent. We thus have the following defining properties of a Poisson process $\{N_t\}_{t\geq0}$:

1. $N_0 = 0$

2. $N_t - N_s \sim \text{Poisson}[\lambda(t - s)]$ for $0 \leq s < t$

3. $N_t - N_s$ independent of $N_r - N_q$ for $0 \leq q < r \leq s < t$.

The familiar fact that sums of independent Poisson random variables are also Poisson distributed implies that

$$(N_r - N_q) + (N_t - N_s) \sim \text{Poisson}[\lambda(t - s) + \lambda(r - q)].$$

In particular, dividing interval $[0, t]$ into subintervals, as $0 = t_0 < t_1 < t_2 < \cdots < t_{n-1} < t_n = t$, we have

$$N_t = \sum_{j=1}^{n}(N_{t_j} - N_{t_{j-1}}) \sim \text{Poisson}\left[\lambda \sum_{j=1}^{n}(t_j - t_{j-1}) = \lambda t\right].$$

Thus, the distribution of N_t is *infinitely divisible*, in the sense that it can be viewed as the sum of arbitrarily many independent Poisson variates.

Finally, consider the increment of a Poisson process over the interval $[t, t + \Delta t]$. Property 2 implies that $\Delta N_t \equiv N_{t+\Delta t} - N_t \sim$ Poisson $(\lambda \Delta t)$, so

$$\Pr(\Delta N_t = n) = \frac{(\lambda \Delta t)^n e^{-\lambda \Delta t}}{n!} 1_{\{0,1,2,\dots\}}(n).$$

Exercise 10.1 will show that $\Pr(\Delta N_t = 0) = 1 - \lambda \Delta t + o(\Delta t)$, that $\Pr(\Delta N_t = 1) = \lambda \Delta t + o(\Delta t)$, and that $\Pr(\Delta N_t = n) = o(\Delta t)$ for $n > 1$. Thus, over very short (i.e., infinitesimal) intervals we could say that the process can take *at most* one jump. Expressed in differential form, this is

$$\Pr(dN_t = 0) = 1 - \lambda \cdot dt$$
$$\Pr(dN_t = 1) = \lambda \cdot dt$$
$$\Pr(dN_t = n) = 0, n > 1.$$

Thus, Poisson processes move along through time and at any particular point in time either do not change at all or else undergo a *unit* incremental change.

By themselves, Poisson processes are clearly pretty limited in what they can describe, but they can easily be generalized. Consider the following thought experiment. Starting from $t = 0$, let us plan to track Poisson process $\{N_t\}$ and use it to generate something new. The first time there is a jump, we will take a draw from some probability distribution, represented by density $f(x)$, and obtain a realization of a random variable X_1. At the next jump we will take another (independent) draw from the same distribution and observe the realization of X_2; and so on. At time t there will have been N_t jumps and, provided $N_t > 0$, a sequence $X_1, X_2, \dots, X_{N_t}$ of i.i.d. random variables. Now define a new process whose value at t is equal to the sum of the X's obtained up to that time; that is, we set $S_t = \sum_{j=1}^{N_t} X_j$ if $N_t > 0$ and $S_t = 0$ otherwise, or else we handle both cases by putting $S_t = \sum_{j=0}^{N_t} X_j$ and $X_0 \equiv 0$. The result is a *compound-Poisson* process. In Chapter 9 we encountered a slightly extended discrete-time version of such a process in the compound-events model of Press (1967). There $\ln(P_t/P_{t-\tau}) = Y_\tau + \sum_{j=0}^{N_\tau} X_j$ for $t = \tau, 2\tau, \dots$, with Y_τ and the $\{X_j\}_{j=1}^\infty$ as independent normals and $N_\tau \sim$ Poisson $(\lambda \tau)$. Here N_τ was regarded as counting the significant news shocks affecting the value of a stock during $(t - \tau, t]$, each of which makes a random, normally distributed contribution to the log of total return. The extension was to introduce Y_τ as an additional component that represents the "normal" variation in price. Extending this idea further to continuous time, we could set $\ln(P_t/P_0) = \mu t + \sigma W_t + \sum_{j=1}^{N_t} X_j$, where $\{W_t\}_{t \geq 0}$ is a *Brownian motion*. We will apply this same *jump–diffusion* process in Chapter 21 as we seek to model arbitrage-free prices of stock options, but now it is time to describe Brownian motions themselves.

10.3.2   Brownian Motions

A particle of dust, starting from an origin $(0, 0, 0)$ in three-dimensional (x, y, z) space and moving about randomly in the air, is found to have x coordinate W_t in t units of time, where t is some positive integer. Suppose that we model $W_t - W_{t-1}$, the net change over the unit interval from $t - 1$ to t, as standard normal. Moreover, let us assume that the net changes over all such unit intervals are independent. Then, starting at $W_0 = 0$, the particle will be at a distance $W_t = W_1 + (W_2 - W_1) + \cdots + (W_t - W_{t-1})$ from the origin after t units of time. Since this is the sum of t independent standard normals, we have $W_t \sim N(0, t)$. The *Brownian motion* process $\{W_t\}_{t \geq 0}$ is merely a continuous-time version of such a discrete-time process with independent normal increments. Specifically, it has these defining properties:

1. $W_0 = 0$ (process starts at the origin)

2. $W_t - W_s \sim N(0, t - s)$ for $s < t$ (increments are normal with variance equal to the time span)

3. $W_t - W_s$ independent of $W_r - W_q$ for $q < r \leq s < t$ (increments over nonoverlapping periods are independent).

In this *standard* form, the process is obviously pretty special, but, like the Poisson process, it is easily generalized. For example, process $\{X_t = \sigma W_t\}$ has increments over $[t, t + \Delta t]$ distributed as $N(0, \sigma^2 \Delta t)$, while $\{Y_t = \mu t + \sigma W_t\}$ has increments distributed as $N(\mu \Delta t, \sigma^2 \Delta t)$. The version $\{Y_t\}_{t \geq 0}$ is called Brownian motion with *drift* or *trend*, as if a steady wind were adding a deterministic impetus to the particle. One such form of Brownian motion (BM) was in fact proposed by Albert Einstein (in one of his three famous 1905 papers) as a model for the position of a particle subject to random molecular vibration in a surrounding medium, such as air or water. In fact, Brownian motion is named after the Scottish botanist Robert Brown, who first reported seeing this seemingly random motion of suspended particles under a microscope. It was Einstein who attributed the phenomenon to molecular vibration and thus pointed to the first solid evidence of what was then the "particle theory" of matter. Even earlier than Einstein's application was that of French graduate student Louis Bachelier (1900), who in his doctoral dissertation used BM to model speculative prices. As we shall see, it is still commonly used for that purpose today. Brownian motions are also called *Wiener processes* in honor of American probabilist Norbert Wiener, who worked out many of their mathematical properties.

Since we shall be doing lots of work with models based on BMs, it is important to get a feel for these properties and develop some shortcuts for calculations. First, note that property 2 implies that $W_t - W_s$ has the same distribution as $Z\sqrt{t - s}$, where $Z \sim N(0, 1)$; that is, both are normal with mean zero and variance $t - s$. Thus, one calculates $\Pr(W_t - W_s \leq w)$ for some real number w as

$$\Pr\left(Z\sqrt{t - s} \leq w\right) = \Pr\left(Z \leq \frac{w}{\sqrt{t - s}}\right) = \Phi\left(\frac{w}{\sqrt{t - s}}\right),$$

where Φ is the standard normal CDF. In other words, for probability calculations with BMs one simply converts to the appropriate function of a standard normal. The conversion also helps in finding moments. For example, (1) $E(W_t - W_s)^2 = (t-s)EZ^2 = t - s$, (2) $E(W_t - W_s)^4 = (t-s)^2 EZ^4 = 3(t-s)^2$, and (3) $E|W_t - W_s| = \sqrt{t-s}E|Z| = \sqrt{t-s}\sqrt{2/\pi}$. The last follows because

$$
\begin{aligned}
E|Z| &= \int_{-\infty}^{\infty} \frac{|z|}{\sqrt{2\pi}} e^{-z^2/2} \cdot dz = 2\int_0^{\infty} \frac{z}{\sqrt{2\pi}} e^{-z^2/2} \cdot dz \\
&= \sqrt{\frac{2}{\pi}} \int_0^{\infty} d(-e^{-z^2/2}) = \sqrt{\frac{2}{\pi}} \left[\lim_{z\to\infty} \left(-e^{-z^2/2}\right) - \left(-e^{-0}\right) \right] \\
&= \sqrt{\frac{2}{\pi}} (0+1).
\end{aligned}
$$

Also, *odd* moments of increments to BMs vanish, since odd moments of standard normals are zero; e.g., $E(W_t - W_s)^{2j-1} = (t-s)^{j-1/2} EZ^{2j-1} = 0$ for $j = 1, 2, \ldots$. Next, property 3 implies that for $q < r \leq s < t$

$$
(W_t - W_s) + (W_r - W_q) \sim Z_1\sqrt{t-s} + Z_2\sqrt{r-q} \sim N(0, t-s+r-q),
$$

where Z_1, Z_2 are i.i.d. as $N(0,1)$. Extending, if $0 = t_0 < t_1 < \cdots < t_{n-1} < t_n = t$, then

$$
\sum_{j=1}^n (W_{t_j} - W_{t_{j-1}}) \sim \sum_{j=1}^n Z_j \sqrt{t_j - t_{j-1}} \sim N\left[0, \sum_{j=1}^n (t_j - t_{j-1}) = t\right].
$$

Thus, as for Poisson variate N_t, the distribution of W_t is infinitely divisible.

While the defining properties of standard BMs make calculations straightforward, they do have some strange and counterintuitive implications.

1. Process $\{W_t\}_{t\geq 0}$ is continuous, but it is *nowhere* differentiable. That is, the track of an evolving BM has no jumps or breaks, but it is so jagged or fuzzy that a tangent line cannot be drawn at any point. It is not difficult to see that these properties follow from the definition. First, consider the change in the value of a BM from t to $t + \Delta t$, the random variable $\Delta W_t \equiv W_{t+\Delta t} - W_t$. Since $\Delta W_t \sim Z\sqrt{\Delta t}$ we have for any $\varepsilon > 0$

$$
\begin{aligned}
\Pr(|\Delta W_t| > \varepsilon) &= \Pr\left(\left|\frac{\Delta W_t}{\sqrt{\Delta t}}\right| > \frac{\varepsilon}{\sqrt{\Delta t}}\right) \\
&= \Pr\left(|Z| > \frac{\varepsilon}{\sqrt{\Delta t}}\right).
\end{aligned}
$$

As $\Delta t \to 0$ this approaches $\Pr(|Z| > +\infty) = 0$,[30] confirming that there is zero probability of a discontinuity at any given t. Next, consider $\Delta W_t / \Delta t$, the

[30]Recall from Section 2.2.2 that the *monotone* property of probability (and of measures generally) justfies this interchange of limits: $\lim_{\Delta t \to 0} \Pr\left(|Z| > \varepsilon/\sqrt{\Delta t}\right) = \Pr\left(|Z| > \lim_{\Delta t \to 0} \varepsilon/\sqrt{\Delta t}\right)$.

average *rate* of change of a BM over t to $t + \Delta t$. Letting B be an arbitrarily large positive number, we obtain

$$\Pr\left(\left|\frac{\Delta W_t}{\Delta t}\right| > B\right) = \Pr\left(\left|\frac{\Delta W_t}{\sqrt{\Delta t}}\right| > B\sqrt{\Delta t}\right) = \Pr\left(|Z| > B\sqrt{\Delta t}\right),$$

which approaches unity as $\Delta t \to 0$. Thus, the "instantaneous" rate of change of the process at any t is almost surely unbounded, confirming that the derivative simply does not exist.

2. The length of a *track* of a BM over any positive interval of time is almost surely...*infinite*. Let L_t be the length of a track that starts at the origin and has the value W_t at t. Referring to Fig. 10.8, one can see that $L_t \geq +\sqrt{W_t^2 + t^2}$, which is the length of the *shortest possible* path from $(0, 0)$ to (t, W_t). Thus, for $t > 0$ we have $L_t > +\sqrt{W_t^2} = |W_t|$. Now break $[0, t]$ into n pieces each of length t/n, and let $t_j = jt/n$ be the jth intermediate point, with $t_0 = 0$ and $t_n = t$. The length of the track on the interval from t_{j-1} to t_j is at least $\left|W_{t_j} - W_{t_{j-1}}\right|$, and the total length on $[0, t]$ is the sum of lengths on the subintervals, so we also have $L_t > \sum_{j=1}^{n}\left|W_{t_j} - W_{t_{j-1}}\right|$. The jth term is distributed as $|Z_j|\sqrt{t/n}$, so for each n we have $L_t > \sqrt{t/n}\sum_{j=1}^{n}|Z_j|$. Now $E|Z_j| = \sqrt{2/\pi}$, so for any $B > 0$

$$\Pr(L_t > B) > \Pr\left(\sqrt{\frac{t}{n}}\sum_{j=1}^{n}|Z_j| > B\right)$$

$$= \Pr\left[\sqrt{\frac{t}{n}}\left(\sum_{j=1}^{n}|Z_j| - n\sqrt{\frac{2}{\pi}}\right) > B - \sqrt{\frac{2tn}{\pi}}\right].$$

The expression on the left of the last inequality is proportional to $1/\sqrt{n}$ times a sum of n i.i.d. random variables with mean zero and finite variance—the terms $\left\{|Z_j| - \sqrt{2/\pi}\right\}$. By the central-limit theorem this weighted sum converges in distribution as $n \to \infty$ to a normal random variable, and since the expression on the right of the inequality approaches $-\infty$, the probability necessarily approaches unity. But no probability can exceed unity, so it follows that $\Pr(L_t > B) = 1$ for any B. Thus, with probability one, the total length of a track followed by a BM over any span of time—no matter how short—is indeed *infinite*!

The implication of property 2 is that one who attempts to lay down a piece of string to match the track of a BM over any interval of time will almost certainly require an infinitely long piece. Of course, any ordinary (i.e., *rectifiable*) curve $\{g(s)\}_{0 \leq s \leq t}$ has *finite* length over a finite span. We consider such ordinary curves in the plane to be one-dimensional objects. On the other hand, we consider the interiors of closed figures in the plane, such as circles and polygons, to be of dimension 2. The paths

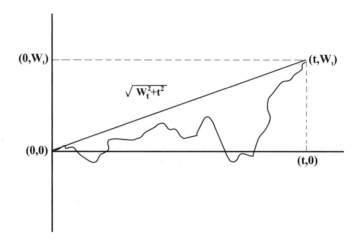

Figure 10.8   Track of a Brownian motion over $[0, t]$.

of Brownian motions are of neither sort, and so are considered to have *fractional* or "fractal" dimension.

We shall need to be aware of one more curious property of BMs. Notice that what gives rise to property 2 is the fact that

$$\lim_{n \to \infty} \sum_{j=1}^{n} \left| W_{t_j} - W_{t_{j-1}} \right| = +\infty.$$

For any function g the limit of sums like $\sum_{j=1}^{n} \left| g\left(t_j\right) - g\left(t_{j-1}\right) \right|$ as $n \to \infty$ and $\left(t_j - t_{j-1}\right) = t/n \to 0$ is called the *variation* of g over $[0, t]$. Thus, property 2 holds because a BM has *unbounded* variation. Now instead of summing the absolute values of the increments, suppose that we sum the *squares* of the increments to a BM, as $\sum_{j=1}^{n} \left(W_{t_j} - W_{t_{j-1}}\right)^2$. The limit of such sums as $n \to \infty$ and $\left(t_j - t_{j-1}\right) \to 0$ is called the *quadratic variation* over $[0, t]$ and is denoted $\langle W \rangle_t$. As t increases, we can think of $\{\langle W \rangle_t\}_{t \geq 0}$ as an evolving *process*, just as the BM itself is. We are now ready for property 3:

3. The quadratic variation of a BM over $[0, t]$ equals t; therefore, the quadratic-variation process $\{\langle W \rangle_t\}_{t \geq 0}$ is *nonvanishing* and *predictable*. This result is a consequence of the *strong law of large numbers*. As we recall from Section 3.7, the SLLN states that the average of n i.i.d. random variables with mean μ converges to μ as $n \to \infty$. To see how this applies, replace $\left(W_{t_j} - W_{t_{j-1}}\right)^2$ with $Z_j^2 \left(t_j - t_{j-1}\right) = Z_j^2 t/n$ in the sum. Here the $\{Z_j\}_{j=1}^{n}$ are i.i.d. as

$N(0, 1)$ and so $EZ_j^2 = VZ_j = 1$ for each j. Then

$$\langle W \rangle_t = \lim_{n \to \infty} \sum_{j=1}^{n} \left(W_{t_j} - W_{t_{j-1}} \right)^2 = \lim_{n \to \infty} \frac{t}{n} \sum_{j=1}^{n} Z_j^2 = t.$$

Notice that this last property of BMs also sets them apart from ordinary differentiable functions, which have *vanishing* quadratic variations. Thus, for a differentiable function h with $|h'| \leq B < \infty$ on $[0, t]$ we would have

$$\langle h \rangle_t = \lim_{n \to \infty} \sum_{j=1}^{n} [h(t_j) - h(t_{j-1})]^2 = \lim_{n \to \infty} \sum_{j=1}^{n} \left[h'(t_j^*)(t_j - t_{j-1}) \right]^2,$$

where (from the mean value theorem) t_j^* is some point between t_{j-1} and t_j. Given that $|h'|$ is bounded, we then have

$$\langle h \rangle_t = \lim_{n \to \infty} \frac{t^2}{n^2} \sum_{j=1}^{n} h'(t_j^*)^2 \leq t^2 \lim_{n \to \infty} \frac{B^2}{n} = 0.$$

These properties of BMs are more than just curiosities, for the lack of smoothness that gives rise to them has profound implications for the temporal behavior of *functions* of a BM. Just as calculus provides the tools needed to understand the evolution of smooth functions—their slopes, their lengths, and the areas they bound—we must look to *stochastic* calculus to fathom such properties of functions of BMs. We begin that study in the next chapter.

10.3.3 Martingales in Continuous Time

In discrete-time settings we have already encountered sequences of expanding information sets (σ fields) $\{\mathcal{F}_t\}_{t=0}^{\infty}$ with the property that $\mathcal{F}_t \subseteq \mathcal{F}_{t+1}$, and we have defined discrete-time processes that are martingales with respect to such sequences. Both concepts extend to continuous time. Thus, consider information that evolves continuously as $\{\mathcal{F}_t\}_{t \geq 0}$, with $\mathcal{F}_s \subseteq \mathcal{F}_t$ for $s < t$. We call such a sequence a *filtration*, since it is indeed like a sequence of increasingly fine filters that pick up more and more information as we proceed through time. A stochastic process $\{X_t\}_{t \geq 0}$ is said to be *adapted* to $\{\mathcal{F}_t\}_{t \geq 0}$ if at each t the information $\mathcal{F}_t$ includes the history of the process up to and including time t. Now, paralleling the definition in discrete time, an $\{\mathcal{F}_t\}_{t \geq 0}$-adapted process $\{X_t\}_{t \geq 0}$ is a continuous-time $\{\mathcal{F}_t\}_{t \geq 0}$ martingale if (1) $E|X_t| < \infty$ for all t and (2) $E(X_t \mid \mathcal{F}_s) = X_s$ for $s \leq t$.

EXERCISES

10.1 Let X_j be the total winning after j plays in a coin-tossing game in which the player wins \$1 when the (fair) coin comes up heads and loses \$1 when the coin comes up tails. For $j \in \{1, 2, ...\}$ let $\mathcal{F}_j$ represent the information about what faces turned up on all plays up through j, and let $\mathcal{F}_0$ (the initial information before play begins) consist of the knowledge that $X_0 = 0$. Find (a) $E\left(X_{10}|\mathcal{F}_9\right)$, (b) $E\left(X_1|\mathcal{F}_0\right)$, (c) $E(X_{10}|\mathcal{F}_0)$, and (d) $E\left[E\left(X_{10}|\mathcal{F}_9\right)|\mathcal{F}_0\right]$.

10.2 Consider a game similar to that in Exercise 10.1 but involving repeated rolls of a fair 6-sided die instead of repeated flips of a fair coin. A player wins \$1 if any of the numbers 1,2,...,5 turns up and loses \$$L$ if a 6 is obtained. As before, let X_j be the total winning after j plays starting with $X_0 = 0$, with $\mathcal{F}_j$ as the knowledge of what happened on all plays up through j. Find L such that $\{X_j\}_{j=0}^{\infty}$ is a martingale adapted to $\{\mathcal{F}_j\}_{j=0}^{\infty}$.

10.3 An investor starts off with capital W_0 and invests it in the stock of ABC company, leaving it invested period after period. In each period $j \in \{1, 2, ...\}$ each dollar so invested returns $1 + R_j$ dollars, where $\{R_j\}_{j=1}^{\infty}$ are i.i.d. random variables satisfying $\Pr\left(R_j > -1\right) = 1$ and $ER_j = \mu > 0$. Find (a) $E\left(W_{10}|\mathcal{F}_9\right)$, (b) $E\left(W_{10}|\mathcal{F}_0\right)$, (c) $E\left[E\left(W_{10}|\mathcal{F}_9\right)|\mathcal{F}_0\right]$, and (d) a quantity b such that the process $\left\{b^j W_j\right\}_{j=0}^{\infty}$ is a martingale adapted to $\{\mathcal{F}_j\}_{j=0}^{\infty}$.

10.4 Put $Y_0 = 0$ and let $\{Y_j\}_{j=1}^{\infty}$ be independent random variables, each distributed as $N\left(\mu, \sigma^2\right)$. For $n \in \{0, 1, 2, ...\}$ let $S_n \equiv \sum_{j=0}^n Y_j$ and insist that information $\mathcal{F}_n$ include knowledge of $\{Y_j\}_{j=0}^n$.

(a) Find a sequence of constants $\{a_n\}_{n=0}^{\infty}$ such that $\{D_n = S_n - a_n\}_{n=0}^{\infty}$ is a martingale adapted to $\{\mathcal{F}_n\}_{n=0}^{\infty}$ and such that $D_0 = 0$.

(b) Find a sequence of constants $\{b_n\}_{n=0}^{\infty}$ such that $\left\{G_n = b_n e^{S_n}\right\}_{n=0}^{\infty}$ is an $\{\mathcal{F}_n\}_{n=0}^{\infty}$ martingale with $G_0 = 1$.

10.5 A price process $\{P_t\}_{t\in\{\tau, 2\tau,...\}}$ evolves over periods of length τ as $P_t = P_{t-\tau} e^{(\mu - \sigma^2/2)\tau + \sigma\sqrt{\tau}Z_t}$, where $\{Z_t\}_{t\in\{\tau, 2\tau,...\}}$ are i.i.d. as standard normal and P_0 is a finite, $\mathcal{F}_0$-measurable constant. Determine the appropriate discount rate ν that makes discounted process $\{e^{-\nu t} P_t\}$ a martingale adapted to evolving history $\{\mathcal{F}_t\}_{t\in\{0, \tau, 2\tau,...\}}$.

10.6 $\{N_t\}_{t\geq 0}$ is an $\{\mathcal{F}_t\}_{t\geq 0}$-adapted Poisson process with intensity λ. Find (deterministic) processes $\{\mu_t\}$ and $\{\nu_t\}_{t\geq 0}$ such that (a) $\{X_t \equiv N_t - \mu_t\}_{t\geq 0}$ and (b) $\left\{Y_t \equiv e^{N_t}/\nu_t\right\}_{t\geq 0}$ are $\{\mathcal{F}_t\}_{t\geq 0}$ martingales.

10.7 $\{W_t\}_{t\geq 0}$ is an $\{\mathcal{F}_t\}_{t\geq 0}$-adapted Brownian motion. Find (deterministic) processes $\{\mu_t\}_{t\geq 0}$, $\{\nu_t\}_{t\geq 0}$, and $\{\beta_t\}_{t\geq 0}$ such that (a) $\{X_t = W_t - \mu_t\}_{t\geq 0}$, (b) $\left\{Y_t = e^{W_t}/\nu_t\right\}_{t\geq 0}$, and (c) $\left\{Q_t \equiv W_t^2 - \beta_t\right\}_{t\geq 0}$ are $\{\mathcal{F}_t\}_{t\geq 0}$ martingales.

10.8 Given an initial price $P_0 \in \mathcal{F}_0$, we model the logarithm of a security's total return over $[0, t]$ as $tR_{0,t} = \ln\left(P_t/P_0\right) = \mu t + \sigma W_t + \sum_{j=0}^{N_t} X_j$, where (1) $\{W_t\}_{t\geq 0}$ and $\{N_t\}_{t\geq 0}$ are independent BM and Poisson processes adapted to evolving information $\{\mathcal{F}_t\}_{t\geq 0}$, (2) $X_0 \equiv 0$, and (3) $\{X_j\}_{j=1}^{\infty}$ are i.i.d. as $N\left(0, \sigma^2\right)$, independent of $\{W_t\}_{t\geq 0}$ and $\{N_t\}_{t\geq 0}$. Find a deterministic process $\{\alpha_t\}_{t\geq 0}$ such that $\{P_t/\alpha_t\}_{t\geq 0}$ is an $\{\mathcal{F}_t\}_{t\geq 0}$ martingale.

EMPIRICAL PROJECT 4

In this project you will examine the marginal distributions of rates of return of two common stocks and the S&P index and judge the extent to which the first two moments of these are predictable from past values. The Project 4 folder at the FTP site contains three files: Monthly1977–2007.xls, Daily1977–2007.xls, and Intraday.xls. The first two files have monthly and daily closing prices, respectively, for Ford (F) and General Motors (GM) common stock and for the S&P 500 index over the period 1977–2007. The intraday file has prices for F and GM at intervals of 5 minutes for a period of one month (June 1997). In each series the price corresponding to time t is the price recorded for the last trade at or before t. With these data we can study the behavior of returns at moderate, high, and very high frequencies. The exercise will illustrate the difficulty of drawing reliable statistical conclusions from data with such extreme outliers as we find in stock returns. All of the statistical work here can be done with common statistical software packages, such as Minitab$^{\circledR}$.

The first task is to convert the price data to continuously compounded average rates of return as $R_{t-\tau,t} = \tau^{-1}\ln\left(P_t/P_{t-\tau}\right)$, where τ is the relevant time interval for the data at hand. (Actually, scale factor τ^{-1} is irrelevant in the computations to follow, so one can just work with $\ln\left(P_t/P_{t-\tau}\right) = \tau R_{t-\tau,t}$ instead of $R_{t-\tau,t}$.) The properties of the returns data can now be studied, as follows:

1. Time-series properties.

    (a) To look for evidence of first-moment dependence, for each security cal-
    culate the autocorrelation function of rate of return out to three lags for
    monthly data and out to five lags for daily and intraday data. In other
    words, calculate the correlation between $R_{t-\tau,t}$ and each of $R_{t-2\tau,t-\tau}$,
    $R_{t-3\tau,t-2\tau}, ..., R_{t-(k-1)\tau,t-k\tau}$, where $k = 3$ for monthly data and $k = 5$
    for the high-frequency returns. For each series calculate a "portman-
    teau" test of serial dependence, such as the Box–Pierce statistic, $BP =$
    $n\sum_{j=1}^{k}\hat{\rho}_j^2$, or the Ljung–Box statistic, $LB = n\left(n+2\right)\sum_{j=1}^{k}\hat{\rho}_j^2/\left(n-j\right)$,
    where $\{\hat{\rho}_j\}_{j=1}^{k}$ are the sample autocorrelation coefficients from a series
    of n observations and k is the maximum number of lags. Under the null
    hypothesis that the data are serially independent (and the maintained hy-
    pothesis that they are identically distributed with finite second moment)
    each of these statistics converges in distribution to chi-squared with k
    degrees of freedom as $n \to \infty$, but the approximation for any given large
    n is more accurate for LB. Large values of BP and LB are consistent
    with the alternative hypothesis that there is some form of serial depen-
    dence; thus H_0 would be rejected at level α if the statistic exceeds the
    upper-α quantile of $\chi^2\left(k\right)$. With monthly data you will find that none of
    the autocorrelations exceeds 0.10 in absolute value and that none of the
    LB statistics rejects H_0 at level $\alpha = .05$. With daily data there are small
    but—in this large sample—statistically significant correlations for S&P
    and F, the LB statistics for both leading to rejection at the .05 level. The

results are thus generally consistent with those depicted for the S&P and MSFT in Figs. 10.1, 10.2, and 10.3 in the text (obtained for a different time period). With the intraday data the story for F and GM is very different from that at lower frequencies, and different also from the findings for DOW in Fig. 10.5. Specifically, autocorrelations at one lag for both F and GM are large negative and highly statistically significant. Evidently, the effect of bid-asked bounce had not washed out at 5-minute intervals for these stocks during this period.

(b) To determine whether the statistically significant autocorrelation in daily returns could feasibly be used to predict the future behavior of prices, regress the current daily rate of return $R_{t-\tau,t}$ for each security on past values $\left\{ R_{t-(j-1)\tau,t-j\tau} \right\}_{j=1}^{5}$ and calculate the R^2 statistic, which indicates the proportion of sample variance in response variable $R_{t-\tau,t}$ that is "explained" by variation in the predictors. You will find that in each case less than .5% of the variation is so explained, indicating that bets based on the predictions would pay off with about the same frequency as bets on flipping a coin.

(c) To look for evidence of cross-security predictive power, with both daily and intraday data regress the rate of return for each of F and GM on its own first lag and on the first lag of the other; that is, regress $R_{t-\tau,t}^{F}$ on $R_{t-2\tau,t-\tau}^{F}$ and $R_{t-2\tau,t-\tau}^{GM}$, and then regress $R_{t-\tau,t}^{GM}$ on the same variables. With daily data do this for τ corresponding to one day; with intraday data, for τ corresponding to 5 minutes. With daily data you will find that the lagged value of the other company's return has no significant predictive power. However, in the intraday data the lagged GM return does have significant predictive power for current F. There are two possible explanations. One is due to asynchronous trading. GM's stock is more widely held than F's and it trades more actively. Thus, the last trade for GM within a given 5-minute interval usually occurs *after* the final trade for F. When this happens GM's price can reflect news relevant to auto stocks that is not picked up in the price of F until the next 5-minute interval. This can occur even if the bid and asked quotes for F and GM always moved simultaneously, in which case there would be no way to profit by trading F in response to moves in the price of GM. The other possibility is that bid and asked quotes for F really do respond more slowly to news than do those of GM, on average. This is might well be true, because more investors own and "follow" GM's stock, and if it really were true, it would be a genuine instance of market inefficiency. To know, we would need to look for systematic lags in bid and asked quotes rather than in the trade prices. Of course, there is no reason that causality could not run in both directions, since F's quotes might sometimes respond more quickly than GM's, even if they did not do so on average.[31]

---

[31] For more evidence on leads and lags in high-frequency price moves, see Epps (1979).

(d) To look for evidence of second-moment dependence, we will see whether the sample data exhibit the usual volatility clustering. For the individual stocks we will also look for a relation between the second moment of return and the initial price. This so-called *leverage effect* would be consistent with cross-sectional evidence that returns of high-priced stocks—typically those of larger companies—are less volatile than returns of low-priced stocks. With monthly data for the S&P regress the *square* of the current month's rate of return on the squares of the rates in the three previous months. Do the same for monthly data on F and GM, but add as predictor variable the inverse of the initial price $P_{t-\tau}^{-1}$. At this monthly frequency you should find little evidence of volatility clustering; that is, whatever linkage there is among second moments usually seems to last for less than a month. Neither of the regressions for the individual stocks gives evidence of a leverage effect, which would produce a positive coefficient on $P_{t-\tau}^{-1}$. With daily data, regress the current squared rate for the S&P on its five lags, and for F and GM include the inverse price as a sixth regressor. Now there is strong evidence in all three series that variations from mean levels of volatility tend to persist for several days, but for this sample period only GM shows any link between volatility and initial price.

2. Marginal distributions. To assess whether the normal distribution can reasonably represent marginal distributions of continuously compounded returns at the various frequencies, calculate the sample coefficients of skewness and kurtosis and apply the Anderson–Darling test for normality to all eight series (monthly and daily for S&P, F, and GM and intraday for F and GM). The sample skewness and kurtosis are calculated as

$$\hat{\kappa}_3 = S_\tau^{-3} \sum_{t=1}^{n} \left(R_{t-\tau,t} - \bar{R}_\tau\right)^3$$

$$\hat{\kappa}_4 = S_\tau^{-4} \sum_{t=1}^{n} \left(R_{t-\tau,t} - \bar{R}_\tau\right)^4,$$

where $\bar{R}_\tau$ and S_τ are the sample mean and standard deviation. The Anderson–Darling statistic measures the discrepancy between the CDF of the sample and that of a normal distribution with the same mean and variance. To calculate it, arrange the standardized sample of rates of return $\left(R_{t-\tau,t} - \bar{R}_\tau\right)/S_\tau$ in rank order, from smallest to largest. Let x_j represent the jth of these ranked values and put $z_j \equiv \Phi(x_j)$, where Φ is the standard normal CDF. Then the Anderson–Darling statistic is

$$AD = n^{-1} \sum_{j=1}^{n} (2j-1) \ln\left[z_j\left(1 - z_{n-j+1}\right)\right] - n.$$

Rejection of H_0: "data are i.i.d. as normal" occurs for large values of AD; specifically, with samples of the sizes used here H_0 can be rejected at about level .05 if $AD > .75$. (Minitab carries out the Anderson–Darling test and supplies a probability value automatically.) With monthly data you should find that the sample kurtosis somewhat exceeds 3.0 (the value for the normal distribution) for all three series and that the Anderson–Darling test shows the normality hypothesis to be tenable only for F. With high-frequency data the evidence against normality is far stronger, with values of AD above 50 for all three daily series and above 100 for the two intraday series.

CHAPTER 11

STOCHASTIC CALCULUS

As stochastic processes whose increments are normally distributed and hence unbounded below, Brownian motions themselves are obviously not good models for prices of limited-liability assets, but with a little work we can find functions of BMs that do have reasonable properties. However, building these models and analyzing how functions of a BM behave over time require new concepts of integration and differentiation. These new concepts and the associated tools were developed largely during the 1940s–1960s. Collectively, they are called *stochastic calculus* or *Itô calculus*, after Japanese probabilist Kiyosi Itô, one of the principal contributors. While much of the theory is very advanced, the fundamental ideas are easily accessible, and one can quickly learn how to apply these concepts properly. To get started, one should review the discussions of Stieltjes integrals and stochastic convergence in Sections 2.3 and 3.6. With these concepts fresh in mind, we can see how to define integrals with respect to Brownian motions and other stochastic processes of infinite variation.

11.1 STOCHASTIC INTEGRALS

11.1.1 Itô Integrals with Respect to a Brownian Motion (BM)

Let $\{W_t\}_{t\geq 0}$ be a standard BM and $\{h_t\}_{t\geq 0}$ be a process that is adapted to the same evolving information or filtration $\{\mathcal{F}_t\}_{t\geq 0}$ as $\{W_t\}_{t\geq 0}$ itself. Thus, the value of h at t can depend on time itself and/or on the value of the BM at any time *at or before* t. We want to construct an integral of h_t *with respect to* BM; namely, $I(0,t) \equiv \int_0^t h_s \cdot dW_s$. One can think of this as something like a Stieltjes integral, but in this case the value of h is weighted not by an increasing, deterministic function, but by the stochastic (i.e., random, unpredictable) increment to the BM. Also, unlike for Stieltjes integrals generally, we do not have to specify whether the interval of integration is open or closed since weight function $\{W_t\}$ is continuous. As in the development of Riemann–Stieltjes integrals we will consider approximating sums of the form $I_n(0,t) = \sum_{j=1}^n h_{t_j^*}\left(W_{t_j} - W_{t_{j-1}}\right)$, where $0 = t_0 < t_1 < \cdots < t_{n-1} < t_n \equiv t$. However, another—and *crucial*—difference now is that we steadfastly fix each point t_j^* to be at the lower boundary of its interval: $t_j^* = t_{j-1}$. As we shall see, this will confer on the Itô integral a vital property that accounts largely for its usefulness in finance. Since increments over time to the BM are stochastic (and since function h may be as well), the sum in $I_n(0,t)$ is clearly a random variable as of the initial time 0. We define the Itô integral as the *probability limit* of the sum as $n \to \infty$:

$$\int_0^t h_s \cdot dW_s = P \lim_{n\to\infty} \sum_{j=1}^n h_{t_{j-1}}\left(W_{t_j} - W_{t_{j-1}}\right). \tag{11.1}$$

■ **EXAMPLE 11.1**

To see what this looks like for a specific integrand, take $h_t = W_t$. Then (recalling that $t_n \equiv t$ and $W_{t_0} = W_0 = 0$), we obtain

$$I_n(0,t) = \sum_{j=1}^n W_{t_{j-1}}\left(W_{t_j} - W_{t_{j-1}}\right) = \sum_{j=1}^n W_{t_{j-1}}W_{t_j} - \sum_{j=1}^n W_{t_{j-1}}^2$$

$$= \sum_{j=1}^n W_{t_{j-1}}W_{t_j} - \frac{1}{2}\left(\sum_{j=1}^n W_{t_j}^2 + \sum_{j=1}^n W_{t_{j-1}}^2 - W_t^2\right)$$

$$= \frac{W_t^2}{2} - \frac{1}{2}\sum_{j=1}^n \left(W_{t_j} - W_{t_{j-1}}\right)^2.$$

Taking limits and recalling the definition of quadratic variation, we have

$$\mathbb{P}\lim_{n\to\infty} I_n(0,t) = \frac{W_t^2}{2} - \frac{\langle W\rangle_t}{2} = \frac{W_t^2}{2} - \frac{t}{2}.$$

Since $W_t^2 \sim tZ^2$ and $Z^2 \sim \chi^2(1)$, we see that $\int_0^t W_s \cdot dW_s$ is distributed as t times a linear function of a chi-squared random variable with one degree of freedom.

Four points deserve emphasis:

1. The result $\int_0^t W_s \cdot dW_s = W_t^2/2 - t/2$ from Itô calculus is quite different from what conventional calculus would imply for a differentiable function $h(t)$. For example, putting $h(0) = 0$ to correspond to $W_0 = 0$, the usual Riemann integration formula would give $\int_0^t h(s) \cdot dh(s) = \int_0^t h(s) h'(s) \cdot ds = h(t)^2/2$. As we shall see in more detail in discussing stochastic differentials below, the extra term $-t/2$ arises from the erratic nature of the nondifferentiable process $\{W_t\}_{t \geq 0}$, whose variation is infinite and whose *quadratic* variation is nonvanishing. Typically, as in the example, we shall find that integration formulas from Itô calculus differ from those in the calculus inherited from Newton and Leibnitz.

2. BMs—and integrable functions of BMs—are themselves *Riemann* integrable with respect to time, because realizations (tracks) of BMs are continuous functions. This means that nothing new is involved in integrating them. For example, given the realization $\{W_s(\omega)\}_{0 \leq s \leq t}$ of a sample path of Brownian motion and an integrable function h, we could calculate $\int_0^t W_s(\omega) \cdot ds$ and $\int_0^t h[W_s(\omega)] \cdot ds$ as ordinary Riemann integrals, since both quantities are $\mathcal{F}_t$-measurable and hence known at t. However, we do have to recognize that from the *time* 0 perspective, equipped merely with information $\mathcal{F}_0$, integrals $\int_0^t W_s \cdot ds$ or $\int_0^t h(W_s) \cdot ds$ of the *random* process $\{W_s\}_{0 \leq s \leq t}$ are themselves random variables.

3. Suppose that our integrand process $\{h_t\}_{t \geq 0}$ is a *deterministic*, integrable function of time alone. Thus, the entire process is $\mathcal{F}_0$-measurable—known as of time 0. Then for each n our approximating sum in (11.1) is $\sum_{j=1}^n h_{t_{j-1}} \left(W_{t_j} - W_{t_{j-1}} \right)$. Clearly, as a weighted sum of independent normals, this is itself normally distributed. The mean and variance of the approximating sum are

$$E \sum_{j=1}^n h_{t_{j-1}} \left(W_{t_j} - W_{t_{j-1}} \right) = \sum_{j=1}^n h_{t_{j-1}} E \left(W_{t_j} - W_{t_{j-1}} \right)$$

$$= 0$$

$$V \sum_{j=1}^n h_{t_{j-1}} \left(W_{t_j} - W_{t_{j-1}} \right) = \sum_{j=1}^n h_{t_{j-1}}^2 V \left(W_{t_j} - W_{t_{j-1}} \right)$$

$$= \sum_{j=1}^n h_{t_{j-1}}^2 \left(t_j - t_{j-1} \right).$$

Letting $n \to \infty$ to turn the approximating sum into the integral, the expected value remains equal to zero, while the variance converges to $\int_0^t h_s^2 \cdot ds$. As this suggests, we have that $\int_0^t h_s \cdot dW_s \sim N\left(0, \int_0^t h_s^2 \cdot ds\right)$ when $\{h_t\}$ is an integrable, deterministic process.

4. When integrand process $\{h_t\}$ is not deterministic (but still adapted to $\{\mathcal{F}_t\}$), it is still true that $E \int_0^t h_s \cdot dW_s = 0$ (see below), but now $V \int_0^t h_s \cdot dW_s = E \int_0^t h_s^2 \cdot ds$. However, in this stochastic case the normality result no longer holds.

11.1.2 From Itô Integrals to Itô Processes

Given some arbitrary $\mathcal{F}_0$-measurable initial value X_0, setting $X_t \equiv X_0 + \int_0^t h_s \cdot dW_s$ and thinking of this as a function of the upper limit of integration give rise to a new stochastic *process*, $\{X_t\}_{t \geq 0}$. Its properties make it a good building block for models of assets' prices in continuous time.

Let us see what the salient properties are. First, like its progenitor $\{W_t\}_{t \geq 0}$, the derived process $\{X_t\}_{t \geq 0}$ is continuous, as can be seen from the relation

$$
\begin{aligned}
\lim_{\Delta t \to 0} (X_{t+\Delta t} - X_t) &= \lim_{\Delta t \to 0} \int_t^{t+\Delta t} h_s \cdot dW_s \\
&\doteq h_t \cdot \lim_{\Delta t \to 0} (W_{t+\Delta t} - W_t) \\
&= h_t \cdot 0.
\end{aligned}
$$

Second, and very importantly, it has the fair-game property of martingales. Specifically, since information $\mathcal{F}_t$ at time $t \leq u$ includes the history of the process itself up to t, we have

$$
\begin{aligned}
E (X_u \mid \mathcal{F}_t) &= E \left[X_0 + \int_0^u h_s \cdot dW_s \mid \mathcal{F}_t \right] \\
&= X_0 + \int_0^t h_s \cdot dW_s + E \left[\int_t^u h_s \cdot dW_s \mid \mathcal{F}_t \right] \\
&\doteq X_t + \sum_{j=1}^n E \left[h_{t_{j-1}} \left(W_{t_j} - W_{t_{j-1}} \right) \mid \mathcal{F}_t \right],
\end{aligned}
$$

where $t_j = t + j (u - t) / n \geq t$ for $j \geq 0$. But by the tower property of conditional expectation each term of the sum equals

$$
E \left[h_{t_{j-1}} E \left(W_{t_j} - W_{t_{j-1}} \mid \mathcal{F}_{t_{j-1}} \right) \mid \mathcal{F}_t \right] = 0,
$$

so $E (X_u \mid \mathcal{F}_t) = X_t$ for $u \geq t$. Note that evaluating integrand process $\{h_t\}_{t \geq 0}$ at t_{j-1} rather than at t_j or at some point in between is crucial for this fair-game property unless $\{h_t\}_{t \geq 0}$ happens to be deterministic. Finally, the processes $\{h_t\}$ that we shall encounter will be such that $E \int_0^t h_s^2 \cdot ds < \infty$, and under this condition we also have $E |X_t| < \infty$ for each t, which makes $\{X_t\}_{t \geq 0}$ a genuine $\{\mathcal{F}_t\}_{t \geq 0}$-adapted martingale in continuous time.

The expression $X_{t+\Delta t} - X_t = \int_t^{t+\Delta t} h_s \cdot dW_s \doteq h_t \cdot (W_{t+\Delta t} - W_t) \equiv h_t \cdot \Delta W_t$ suggests a useful *stochastic differential* form as a shorthand expression for the increment to X_t over a "short" interval:

$$dX_t = h_t \cdot dW_t.$$

Thus, the change dX_t over the next instant *from t* equals the increment to the BM times an $\mathcal{F}_t$-measurable factor—one whose value is known at t. As such, it has zero expected value given information $\mathcal{F}_t$, a fact that accounts for the fair-game property of the integral. The weighting factor h_t is called the *diffusion coefficient*.

That processes built up directly from Itô integrals are martingales is very useful mathematically. More importantly for finance, the unpredictability of their increments is consistent with the "first-moment independence" that assets' prices seem to display over short-to-medium-term horizons. However, the fair-game property per se is inconsistent with the fact that prices typically drift upward over time. To capture this feature, we can simply add a trend term and define a new process as

$$X_t = X_0 + \int_0^t g_s \cdot ds + \int_0^t h_s \cdot dW_s, \qquad (11.2)$$

where $\{g_t\}_{t \geq 0}$ can be any Riemann-integrable process that is adapted to $\{\mathcal{F}_t\}_{t \geq 0}$. In differential form this is

$$dX_t = g_t \cdot dt + h_t \cdot dW_t.$$

In this context we refer to g_t as the *instantaneous drift rate* or just as the *drift term*. Now the increment to X_t consists of a part that is predictable given information $\mathcal{F}_t$ and a part that is not. Processes of this general form can provide attractive models for prices of assets and for interest rates. Here are some examples:

1. Setting $g_t = \mu X_t$, $h_t = \sigma X_t$, where μ and $\sigma > 0$ are constants, we have

$$dX_t = \mu X_t \cdot dt + \sigma X_t \cdot dW_t.$$

   This process, in which the drift term and the diffusion coefficient are both proportional to the level of the process, is called a *geometric* Brownian motion. It is the model on which the Black–Scholes–Merton theory of option pricing is based, and, as such, it is still considered a sort of benchmark in continuous-time finance. The constant $\sigma = h_t/X_t$ in the diffusion coefficient is called the *volatility parameter*. We will make extensive use of the geometric BM model in succeeding chapters.

2. Setting $g_t = \mu - \nu X_t$ and $h_t = \sigma$ gives

$$dX_t = (\mu - \nu X_t) \cdot dt + \sigma \cdot dW_t.$$

When $\nu > 0$ the process is *mean reverting*, in that values of $X_t > \mu/\nu$ make the drift term negative, which helps to push the process back toward μ/v, while if $X_t < \mu/\nu$, the process has positive average trend. This model and variants with other diffusion coefficients have been widely used to characterize interest rates.

3. Setting $g_t = \mu X_t$ and $h_t = \sigma X_t^\gamma$ for $0 < \gamma < 1$ gives the *constant-elasticity-of-variance* (CEV) *model*. The volatility is now $h_t/X_t = \sigma X_t^{\gamma-1}$, which is decreasing in X_t. This feature corresponds to the leverage effect that sometimes appears in continuously compounded rates of return of common stocks—the fact that volatility tends to decrease as the price rises and the issuing firm's ratio of debt value to equity value declines.

11.1.3 Quadratic Variations of Itô Processes

We have seen that the quadratic variation of a BM is the nonvanishing and predictable process

$$\langle W \rangle_t = \lim_{n \to \infty} \sum_{j=1}^{n} \left(W_{t_j} - W_{t_{j-1}} \right)^2 = t. \tag{11.3}$$

It will be important also to work with the quadratic-variation process $\{\langle X \rangle_t\}_{t \geq 0}$ of the more general Itô process $dX_t = g_t \cdot dt + h_t \cdot dW_t$. Again breaking interval $[0, t]$ into n pieces each of length t/n, the definition is[32]

$$\langle X \rangle_t = P \lim_{n \to \infty} \sum_{j=1}^{n} \left(X_{t_j} - X_{t_{j-1}} \right)^2.$$

To evaluate, focus on the generic term $X_{t_j} - X_{t_{j-1}}$, whose square can be approximated as

$$\left[g_{t_{j-1}} \frac{t}{n} + h_{t_{j-1}} \left(W_{t_j} - W_{t_{j-1}} \right) \right]^2$$

$$= \frac{g_{t_{j-1}}^2 t^2}{n^2} + \frac{2 g_{t_{j-1}} h_{t_{j-1}} t}{n} \left(W_{t_j} - W_{t_{j-1}} \right) + h_{t_{j-1}}^2 \left(W_{t_j} - W_{t_{j-1}} \right)^2.$$

Since $W_{t_j} - W_{t_{j-1}}$ is distributed as $\sqrt{t/n}$ times a standard normal random variable, the first two terms approach zero *faster* than t/n as $n \to \infty$, and so their sum over $j \in \{1, 2, ..., n\}$ simply vanishes in the limit. With a bit of work one can see that the sum of the third terms converges in probability to $\int_0^t h_s^2 \cdot ds$. Thus, the quadratic-variation process for $dX_t = g_t \cdot dt + h_t \cdot dW_t$ is $\langle X \rangle_t = \int_0^t h_s^2 \cdot ds$. Of course, unless h is a constant or some other deterministic function, this is no longer *predictable*. The differential form, which is what we shall need, is simply

$$d \langle X \rangle_t = h_t^2 \cdot dt.$$

---

[32]Here we must specify *probability* limit, because, unlike the sum in (11.3), the convergence is not necessarily *strong* or *almost sure*. Recall the definitions in Section 3.6.

Thus, for the geometric BM process $dX_t = \mu X_t \cdot dt + \sigma X_t \cdot dW_t$ we have $d\langle X \rangle_t = \sigma^2 X_t^2 \cdot dt$, a result that will be of great importance for future developments.

11.1.4   Integrals with Respect to Itô Processes

Since $\{W_t\}_{t \geq 0}$ is of the form (11.2) with $X_0 = 0$, $g_t = 0$, and $h_t = 1$, BMs are themselves Itô processes. Just as we integrated $\{h_t\}_{t \geq 0}$ with respect to a BM and obtained the more general $\{X_t\}_{t \geq 0}$, we can integrate another suitable process $\{f_t\}_{t \geq 0}$ with respect to $\{X_t\}_{t \geq 0}$. As before, this gives us yet another Itô process. Taking $dX_t = g_t \cdot dt + h_t \cdot dW_t$ or, in integral form, $X_t = X_0 + \int_0^t g_s \cdot ds + \int_0^t h_s \cdot dW_s$, and letting Y_0 be $\mathcal{F}_0$-measurable but otherwise arbitrary, set

$$Y_t = Y_0 + \int_0^t f_s \cdot dX_s = Y_0 + \int_0^t f_s g_s \cdot ds + \int_0^t f_s h_s \cdot dW_s.$$

In differential form, this is

$$dY_t = f_t g_t \cdot dt + f_t h_t \cdot dW_t.$$

The ability to generate new Itô processes from simpler ones in this way affords great flexibility in modeling many of the erratic, unpredictable processes observed in nature and in markets. By regarding f, g, h, and W as vectors, we will see later that a process like $\{Y_t\}$ can represent the value of a *portfolio* when X_t itself represents a vector of assets' prices.

11.2   STOCHASTIC DIFFERENTIALS

We have seen that Itô processes can be built up as stochastic integrals of suitable functions with respect to a BM, as

$$X_t = X_0 + \int_0^t g_s \cdot ds + \int_0^t h_s \cdot dW_s,$$

and that there corresponds to this the shorthand *differential* form $dX_t = g_t \cdot dt + h_t \cdot dW_t$. Intuitively, we can think of this differential form as representing the "instantaneous" change in the process, which consists of a part that is known at t, $g_t \cdot dt$, and an unpredictable part, $h_t \cdot dW_t$. Suppose now that we want to represent the instantaneous change of a suitable *function* of an Itô process, $f(X_t)$. The appropriate expression for the stochastic differential $df(X_t)$ comes from what is called *Itô's formula* or *Itô's lemma*. This is a fundamental tool for analyzing and developing stochastic models in continuous time.

To ease into this, let us consider some things that should be familiar from a first course in calculus. Let g be a function with continuous first derivative g', and represent the change in g over the interval $[0, t]$ as the sum of the changes over the n subintervals $\{[t_{j-1}, t_j]\}_{j=1}^n$, where $t_j - t_{j-1} = t/n$ for each j. Thus,

$$g(t) - g(0) = \sum_{j=1}^n [g(t_j) - g(t_{j-1})] = \sum_{j=1}^n g'(t_j^*)(t_j - t_{j-1}),$$

where the last expression comes from the mean-value theorem, with $t_j^* \in (t_{j-1}, t_j)$. Now letting $n \to \infty$ and recalling the definition of the Riemann integral, we see that

$$g(t) - g(0) = \int_0^t g'(s) \cdot ds, \qquad (11.4)$$

which is essentially the content of the fundamental theorem of calculus. The corresponding differential form is just $dg(t) = g'(t) \cdot dt$. While this differential form and its interpretation as "instantaneous change" are useful aids to intuition, it is well to keep in mind that they have rigorous meaning only through the integral expression (11.4). Notice that if g also had a continuous *second* derivative, we could have used Taylor's theorem to expand $g(t_j)$ out to the quadratic term, as $g(t_j) = g(t_{j-1}) + g'(t_{j-1})(t_j - t_{j-1}) + \frac{1}{2}g''(t_j^*)(t_j - t_{j-1})^2$. However, since the extra term would have been proportional to t^2/n^2 and since the sum of n such terms would have vanished in the limit, we would have gotten precisely the same result as (11.4). Thus, to determine the limit of the sum as $n \to \infty$, we need to carry the expansion only as far as the term of order $1/n$, since the remaining terms of order *less than* $1/n$—represented collectively as $o(n^{-1})$—would be irrelevant.

Now let us go one step farther within the domain of ordinary calculus. Introduce another function f that is also continuously differentiable, and consider the composite function $f[g(\cdot)]$. Again, the total change from 0 to t equals the sum of changes on subintervals: $\sum_{j=1}^n \{f[g(t_j)] - f[g(t_{j-1})]\}$. Now expand $f[g(t_j)]$ about $g(t_{j-1})$ to get[33]

$$f[g(t_j)] - f[g(t_{j-1})] = f'[g(t_j^*)][g(t_j) - g(t_{j-1})],$$

and further expand g about t_{j-1} to get

$$f[g(t_j)] - f[g(t_{j-1})] = f'[g(t_j^*)] g'(t_j^{**})(t_j - t_{j-1}).$$

As before, we can stop the expansion with the $(t_j - t_{j-1})$ term, because terms of order $(t_j - t_{j-1})^2$ and higher are $o(n^{-1})$ and will vanish when we add them up and take limits. Now, summing the terms for $j \in \{1, 2, ..., n\}$ and letting $n \to \infty$ give

$$f[g(t)] - f[g(0)] = \int_0^t f'[g(s)] g'(s) \cdot ds.$$

In differential form this is $df[g(t)] = f'[g(t)] g'(t) \cdot dt$, which is the familiar *chain rule* for differentiating composite functions. Intuitively, what makes this formula valid is that the instantaneous change in $g(t)$ is of order dt. We will now see that a different formula is required for functions of BMs—and for Itô processes generally.

---

[33]Note that since g is continuous, there is a t_j^* between t_{j-1} and t_j that corresponds to any g^* between $g(t_{j-1})$ and $g(t_j)$.

11.3   ITÔ'S FORMULA FOR DIFFERENTIALS

11.3.1   Functions of a BM Alone

With $\{W_t\}_{t\geq 0}$ as a standard BM, consider the change $f(W_t) - f(W_0)$ in the value of a function f over the interval $[0, t]$. One can think of the $\{W_t\}$ process as corresponding to the continuous function g in the chain rule; however, things will work differently because $\{W_t\}$ is not differentiable. Indeed, because its realizations are so rough, we will now require f to have a greater degree of smoothness than before; specifically, f must now have at least *two* continuous derivatives. To make it easy to extend the result to functions f with more than one argument, we will indicate derivatives from here on with subscripts instead of primes, as $f_W = f'$ and $f_{WW} = f''$.

Proceeding as before, the total change in f is

$$f(W_t) - f(W_0) = \sum_{j=1}^{n} \left[f\left(W_{t_j}\right) - f\left(W_{t_{j-1}}\right) \right].$$

Now use Taylor's formula to expand f about $W_{t_{j-1}}$ to terms of the *second* order, writing $f\left(W_{t_j}\right)$ as

$$f\left(W_{t_{j-1}}\right) + f_W\left(W_{t_{j-1}}\right)\left(W_{t_j} - W_{t_{j-1}}\right) + \frac{f_{WW}\left(W_{t_j^*}\right)}{2}\left(W_{t_j} - W_{t_{j-1}}\right)^2,$$

so that total change $f(W_t) - f(W_0)$ equals

$$\sum_{j=1}^{n} f_W\left(W_{t_{j-1}}\right)\left(W_{t_j} - W_{t_{j-1}}\right) + \sum_{j=1}^{n} \frac{f_{WW}\left(W_{t_j^*}\right)}{2}\left(W_{t_j} - W_{t_{j-1}}\right)^2. \quad (11.5)$$

As we send n off to infinity, the first term on the right converges to the stochastic integral $\int_0^t f_W(W_s) \cdot dW_s$, as per the construction we have already seen. What is new is the behavior of the second term. This term *does not vanish* as $n \to \infty$, unlike the second-order term in the expansion of $f[g(t)]$ for a differentiable g. One can see why at once on replacing $W_{t_j} - W_{t_{j-1}}$ with $Z_j\sqrt{t_j - t_{j-1}} = Z_j\sqrt{t/n}$ and $\left(W_{t_j} - W_{t_{j-1}}\right)^2$ with $Z_j^2 t/n$, where $Z_j \sim N(0,1)$. Here the *second-order* term in ΔW_t is converging to zero at the same $1/n$ rate as did the *first-order* term in $\Delta g(t)$, and the sum of n such terms does not vanish, in general. But if the term does not go to zero, what limiting value does it have? From what we have seen about the predictability of the quadratic variation of a BM, one might suppose that we could just replace each factor $\left(W_{t_j} - W_{t_{j-1}}\right)^2$ in (11.5) by $t_j - t_{j-1}$ as we send n off to infinity. That is not quite correct, for after all the change in the BM is *stochastic*. Still, the effect turns out to be the same. To see it, write the second term as

$$\sum_{j=1}^{n} \frac{f_{WW}\left(W_{t_j^*}\right)}{2}(t_j - t_{j-1}) + \sum_{j=1}^{n} \frac{f_{WW}\left(W_{t_j^*}\right)}{2}\left[\left(W_{t_j} - W_{t_{j-1}}\right)^2 - (t_j - t_{j-1})\right],$$

which comes just from adding and subtracting $t_j - t_{j-1}$ from $\left(W_{t_j} - W_{t_{j-1}} \right)^2$ in each summand. The first term converges to the Riemann integral $\int_0^t f_{WW} (W_s) \cdot ds$, and with a bit of work one can see that the second term converges in probability to zero. Putting everything together leaves us with Itô's integral formula:

$$f (W_t) - f (W_0) = \int_0^t f_W (W_s) \cdot dW_s + \int_0^t \frac{f_{WW} (W_s)}{2} \cdot ds. \qquad (11.6)$$

Since the predictable quadratic variation $\langle W \rangle_t$ is just equal to t, we can write this in compact differential form as

$$df (W_t) = f_W \cdot dW_t + \frac{f_{WW}}{2} \cdot d \langle W \rangle_t. \qquad (11.7)$$

For what we do later it is helpful to commit to memory the differential form (11.7). It reminds us of the key point that to represent "instantaneous" changes in a (smooth) function of a BM, we need a second-order term involving the function's second derivative.

■ **EXAMPLE 11.2**

With $f (W_t) = W_t^2$ we have $f_W (W_t) = 2W_t$, $f_{WW} (W_t) = 2$, and $dW_t^2 = 2W_t \cdot dW_t + dt$. In integral form (recall that $W_0 = 0$) this is $W_t^2 = 2 \int_0^t W_s \cdot dW_s + \int_0^t ds = 2 \int_0^t W_s \cdot dW_s + t$. Thus, the stochastic integral is

$$\int_0^t W_s \cdot dW_s = \frac{W_t^2}{2} - \frac{t}{2}.$$

We worked this out earlier from the definition of the integral, but here we have done it more easily by finding a function that differentiates to the integrand using differentiation formula (11.7). In the process we have found that $X_t = \left(W_t^2 - t \right) / 2$ solves the stochastic differential equation (SDE) $dX_t = W_t \cdot dW_t$ subject to the initial condition $X_0 = 0$.

■ **EXAMPLE 11.3**

With $f (W_t) = e^{W_t}$ we have $f_W (W_t) = f_{WW} (W_t) = e^{W_t} = f (W_t)$ and $de^{W_t} = e^{W_t} (dW_t + dt/2)$. The integral form is $e^{W_t} - e^{W_0} = e^{W_t} - 1 = \int_0^t e^{W_s} \cdot dW_s + \frac{1}{2} \int_0^t e^{W_s} \cdot ds$. This shows that $X_t = e^{W_t} - \frac{1}{2} \int_0^t e^{W_s} \cdot ds$ solves s.d.e. $dX_t = e^{W_t} \cdot dW_t$ subject to $X_0 = 1$.

11.3.2 Functions of Time and a BM

Now consider a function $f (t, W_t)$ that depends directly on time as well as on a BM. Again, we require that f be twice continuously differentiable in W_t, but only one continuous derivative with respect to t is needed. The derivatives—now *partial* derivatives—are denoted f_t, f_W, and f_{WW}. The extended version of Itô's formula is

$$f(t, W_t) - f(0, W_0) = \int_0^t f_t(s, W_s) \cdot ds + \int_0^t f_W(s, W_s) \cdot dW_s$$
$$+ \int_0^t \frac{f_{WW}(s, W_s)}{2} \cdot d\langle W \rangle_s,$$

or in differential form

$$df(t, W_t) = f_t \cdot dt + f_W \cdot dW_t + \frac{f_{WW}}{2} \cdot d\langle W \rangle_t$$
$$= \left(f_t + \frac{f_{WW}}{2} \right) \cdot dt + f_W \cdot dW_t.$$

■ **EXAMPLE 11.4**

For $f(t, W_t) = e^{tW_t}$ we have $f_t = W_t e^{tW_t}$, $f_W = t e^{tW_t}$, $f_{WW} = t^2 e^{tW_t}$ and

$$de^{tW_t} = e^{tW_t} \left(W_t + \frac{t^2}{2} \right) \cdot dt + e^{tW_t} t \cdot dW_t.$$

11.3.3 Functions of Time and General Itô Processes

In applications to continuous-time finance we will want to be able to express instantaneous changes in functions of assets' prices—e.g., functions representing values of portfolios and values of derivative securities such as puts and calls. Since the prices themselves will be modeled as more general Itô processes than BMs—e.g., as *geometric* BMs, we shall need further extensions of Itô's formula. Fortunately, things follow the same patterns as before.

Consider first a function $f(t, X_t)$ of time and a single Itô process, $X_t = X_0 + \int_0^t g_s \cdot ds + \int_0^t h_s \cdot dW_s$. The change in f over $[0, t]$ is then

$$f(t, X_t) - f(0, X_0) = \int_0^t f_t(s, X_s) \cdot ds + \int_0^t f_X(s, X_s) \cdot dX_s$$
$$+ \int_0^t \frac{f_{XX}(s, X_s)}{2} \cdot d\langle X \rangle_s,$$

or in differential form

$$df(t, X_t) = f_t \cdot dt + f_X \cdot dX_t + \frac{f_{XX}}{2} \cdot d\langle X \rangle_t.$$

Using $dX_t = g_t \cdot dt + h_t \cdot dW_t$ and $d\langle X \rangle_t = h_t^2 \cdot dt$, this becomes

$$df(t, X_t) = \left(f_t + f_X g_t + \frac{f_{XX}}{2} h_t^2 \right) \cdot dt + f_X h_t \cdot dW_t.$$

Here are two important applications.

■ **EXAMPLE 11.5**

Let $\{X_t\}_{t\geq 0}$ be a geometric BM, with $dX_t = \mu X_t \cdot dt + \sigma X_t \cdot dW_t$ and $X_0 > 0$ given, and assume for the moment that $\Pr(X_t > 0) = 1$ for each t. Putting $f(t, X_t) = \ln X_t$ (not depending directly on t), we have $f_t = 0$, $f_X = X_t^{-1}$, $f_{XX} = -X_t^{-2}$, and

$$d\ln X_t = \frac{1}{X_t} \cdot dX_t - \frac{1}{2X_t^2} \cdot d\langle X\rangle_t$$

$$= \mu \cdot dt + \sigma \cdot dW_t - \frac{\sigma^2}{2} \cdot dt$$

$$= \left(\mu - \frac{\sigma^2}{2}\right) \cdot dt + \sigma \cdot dW_t.$$

In integral form this is

$$\ln X_t - \ln X_0 = \int_0^t \left(\mu - \frac{\sigma^2}{2}\right) \cdot ds + \sigma \int_0^t dW_s$$

$$= \left(\mu - \frac{\sigma^2}{2}\right) t + \sigma W_t.$$

This shows that the solution to SDE $dX_t = \mu X_t \cdot dt + \sigma X_t \cdot dW_t$ (subject to the initial condition) is $X_t = X_0 e^{(\mu - \sigma^2/2)t + \sigma W_t}$ and that $\ln(X_t/X_0) \sim N\left[(\mu - \sigma^2/2)\, t, \sigma^2 t\right]$. Thus, future values of a geometric BM process are distributed as lognormal—which with $X_0 > 0$ justifies the assumption that $X_t > 0$.

■ **EXAMPLE 11.6**

Let $f(t, X_t) = \ln X_t$ with $X_0 > 0$ as before, but suppose that $\{X_t\}_{t\geq 0}$ evolves as $dX_t = \mu_t X_t \cdot dt + \sigma_t X_t \cdot dW_t$, where *processes* $\{\mu_t\}_{t\geq 0}$ and $\{\sigma_t\}_{t\geq 0}$ are time varying but $\mathcal{F}_0$-measurable—known at time 0—and Riemann integrable. Then

$$d\ln X_t = \left(\mu_t - \frac{\sigma_t^2}{2}\right) \cdot dt + \sigma_t \cdot dW_t$$

and

$$\ln X_t - \ln X_0 = \int_0^t \left(\mu_s - \frac{\sigma_s^2}{2}\right) \cdot ds + \int_0^t \sigma_s \cdot dW_s$$

$$\sim N\left[\int_0^t \left(\mu_s - \frac{\sigma_s^2}{2}\right) \cdot ds, \int_0^t \sigma_s^2 \cdot ds\right].$$

Thus, the lognormality result continues to hold when time-varying processes $\{\mu_t\}_{t\geq 0}$ and $\{\sigma_t\}_{t\geq 0}$ are deterministic.

Next, let $f(t, X_t, Y_t)$ depend on *two* Itô processes, $X_t = X_0 + \int_0^t g_s \cdot ds + \int_0^t h_s \cdot dW_s$ and $Y_t = Y_0 + \int_0^t a_s \cdot ds + \int_0^t b_s \cdot dW'_s$. Here $\{W_t, W'_t\}_{t\geq 0}$ are standard BMs, but we allow them to be dependent by setting covariance $C(W_t, W'_t) = EW_t W'_t$ equal to ρt, where $\rho = C(W_t, W'_t)/\sqrt{VW_t \cdot VW'_t} = \rho t/\sqrt{t \cdot t}$ is the coefficient of correlation. The covariance between W_t and W'_t also equals the (predictable) quadratic *co-variation* over $[0, t]$. This is defined as $\langle W, W' \rangle_t \equiv \lim_{n\to\infty} \sum_{j=1}^n (W_{t_j} - W_{t_{j-1}})(W'_{t_j} - W'_{t_{j-1}})$, with $t_j = jt/n$ as usual. Regarding $\langle W, W' \rangle_t$ as a function of t, we have the quadratic-covariation *process* $\{\langle W, W' \rangle_t\}_{t\geq 0}$ with differential form $d\langle W, W' \rangle_t = \rho \cdot dt$. The corresponding covariation process for X_t and Y_t is $\{\langle X, Y \rangle_t\}_{t\geq 0}$, where $\langle X, Y \rangle_t = \rho \int_0^t h_s b_s \cdot ds$ and $d\langle X, Y \rangle_t = \rho h_t b_t \cdot dt$.

When the quadratic covariation between Itô processes does not vanish (i.e., when $\rho \neq 0$), then it will also make a contribution to $df(t, X_t, Y_t)$. Specifically, when f has continuous partial derivatives f_t, f_{XX}, f_{YY}, and f_{XY}, we have

$$df(t, X_t, Y_t) = f_t \cdot dt + f_X \cdot dX_t + f_Y \cdot dY_t + \frac{f_{XX}}{2} \cdot d\langle X \rangle_t$$

$$+ \frac{f_{YY}}{2} \cdot d\langle Y \rangle_t + f_{XY} \cdot d\langle X, Y \rangle_t$$

$$= \left(f_t + f_X g_t + f_Y a_t + \frac{f_{XX}}{2} h_t^2 + \frac{f_{YY}}{2} b_t^2 + f_{XY} h_t b_t \rho \right) \cdot dt$$

$$+ f_X h_t \cdot dW_t + f_Y b_t \cdot dW'_t.$$

This includes as a special case the situation that $\{Y_t\}_{t\geq 0}$ is *deterministic*, for then $b_t \equiv 0$, $dY_t = a_t \cdot dt$, and

$$df(t, X_t, Y_t) = \left(f_t + f_X g_t + f_Y a_t \frac{f_{XX}}{2} h_t^2 \right) \cdot dt + f_X h_t \cdot dW_t.$$

Finally, generalizing to functions of the m Itô processes

$$\{dY_{jt} = a_{jt} \cdot dt + b_{jt} \cdot dW_{jt}\}_{j=1}^m,$$

where $\langle W_j, W_k \rangle_t = \rho_{jk} t$ (with $\rho_{jk} = 1$ for $j = k$), we have

$$df(t, Y_{1t}, ..., Y_{mt}) = \left(f_t + \sum_{j=1}^m f_{Y_j} a_{jt} + \sum_{j=1}^m \sum_{k=1}^m \frac{f_{Y_j Y_k}}{2} b_{jt} b_{kt} \rho_{jk} \right) \cdot dt$$

$$+ \sum_{j=1}^m f_{Y_j} b_{jt} \cdot dW_{jt}.$$

EXERCISES

11.1 Evaluate $\int_{[0,1]} (1+x) \cdot dF(x)$ and $\int_{(0,1]} (1+x) \cdot dF(x)$ when **(a)** $F(x) = x$, **(b)** $F(x) = 1 - e^{-x}$, and **(c)** $F(x) = 0$ for $x < 0$, $F(x) = .25$ for $0 \leq x < .5$, $F(x) = .75$ for $.5 \leq x < 1$, $F(x) = 1$ for $x \geq 1$.

11.2 In defining Itô integral $\int_0^t h_s \cdot dW_s$ in (11.1), we insisted that the integrand h in the approximating sums be evaluated at the *left* endpoint of each time interval $[t_{j-1}, t_j]$. To see the significance, modify the construction of $\int_0^t W_s \cdot dW_s$ in Example 11.1 as $P \lim_{n\to\infty} \sum_{j=1}^n W_{t_j} \left(W_{t_j} - W_{t_{j-1}}\right)$ and show that this equals $W_t^2/2 + t/2$ instead of $W_t^2/2 - t/2$. Which of the *processes* $\left\{W_t^2/2 + t/2\right\}_{t\geq 0}$ and $\left\{W_t^2/2 - t/2\right\}_{t\geq 0}$ is a martingale? What happens if we define $\int_0^t W_s \cdot dW_s$ as

$$P \lim_{n\to\infty} \sum_{j=1}^n \left(\frac{W_{t_j} + W_{t_{j-1}}}{2}\right) \left(W_{t_j} - W_{t_{j-1}}\right)?$$

(This is known as the *Stratonovich* construction.)

11.3 $\{X_t\}_{t\geq 0}$ is an Itô process evolving as $dX_t = g_t \cdot dt + h_t \cdot dW_t$, where $\{W_t\}_{t\geq 0}$ is a Brownian motion, and $\{M_t\}_{t\geq 0}$ is a deterministic process evolving as $dM_t = M_t r \cdot dt$ for some $r \neq 0$. Apply Itô's formula to find df for each of the following functions f: **(a)** $f(X_t) = X_t^2$, **(b)** $f(X_t) = \exp(X_t^2)$, **(c)** $f(X_t) = \ln(X_t)$, **(d)** $f(t, X_t) = \exp(\theta t + X_t^2)$, **(e)** $f(t, X_t) = tX_t^{-1}$, and **(f)** $f(M_t, X_t) = M_t X_t$.

11.4 Taking $g_t = g$ and $h_t = h$ (time-invariant constants), **(a)** find a process $\{X_t\}$ with initial value X_0 that solves the SDE $dX_t = g \cdot dt + h \cdot dW_t$; **(b)** find $E(X_t \mid \mathcal{F}_s)$ for any $s \in [0, t]$; **(c)** determine whether $\{X_t\}$ is a martingale.

11.5 Taking $g_t = gX_t$ and $h_t = hX_t$, **(a)** find a process $\{X_t\}$ with initial value X_0 that solves the SDE $dX_t = gX_t \cdot dt + hX_t \cdot dW_t$; **(b)** find $E(X_t \mid \mathcal{F}_s)$ for any $s \in [0, t]$; **(c)** determine whether $\{X_t\}$ is a martingale.

11.6 Let $x(t)$ and $y(t)$ be differentiable, deterministic functions of time; $\{X_t\}_{t\geq 0}$ be an Itô process evolving as $dX_t = g_t \cdot dt + h_t \cdot dW_t$; and $\{Y_t\}_{t\geq 0}$ be an Itô process evolving as $dY_t = a_t \cdot dt + b_t \cdot dW_t'$, where $d\langle W, W'\rangle_t = \rho \cdot dt$. Find **(a)** $d[x(t)y(t)]$, **(b)** $d[x(t)X_t]$, and **(c)** $d[X_t Y_t]$.

11.7 $\{X_t\}_{t\geq 0}$ evolves as $dX_t = \mu X_t \cdot dt + \sigma X_t \cdot dW_t$, $\{M_t = M_0 e^{rt}\}_{t\geq 0}$ is deterministic, and $\{Y_t\}_{t\geq 0}$ evolves as $dY_t = \nu Y_t \cdot dt + \gamma Y_t \cdot dW_t'$, where $d\langle W, W'\rangle_t = \rho \cdot dt$. Find dX_t^* and dX_t^{**} if $X_t^* \equiv X_t/M_t$ and $X_t^{**} \equiv X_t/Y_t$.

CHAPTER 12

PORTFOLIO DECISIONS OVER TIME

We now have the essential tools for modeling prices of assets and functions of those prices in discrete and continuous time. In this chapter we return to the discrete-time setting to further prepare for applications in dynamic modeling.

In the general framework of expected-utility theory we characterized portfolio choice at $t = 0$ as the solution to optimization problem

$$\max_{\mathbf{p}} Eu\left(W_1\right) = \max_{\mathbf{p}} Eu\left[W_0\left(1 + R_{\mathbf{p}}\right)\right],$$

where $R_{\mathbf{p}} \equiv r_0 + \sum_{j=1}^{n} p_j \left(R_j - r_0\right)$ is the (simple) one-period rate of return from a portfolio with proportions $p_1, ..., p_n$ of initial wealth W_0 in n risky assets and proportion $p_0 = 1 - \sum_{j=1}^{n} p_j$ in a riskless asset paying sure rate r_0. To solve the problem, one must (1) propose a joint distribution $f\left(\mathbf{r}\right) \equiv f\left(r_1, ..., r_n\right)$ for risky rates $\mathbf{R} \equiv \left(R_1, ..., R_n\right)'$, (2) work out the integral

$$Eu\left[W_0\left(1 + R_{\mathbf{p}}\right)\right] = \int u\left\{W_0\left[1 + r_0 + \sum_{j=1}^{n} p_j\left(r_j - r_0\right)\right]\right\} f\left(\mathbf{r}\right) \cdot dr_1 \cdots dr_n$$

to express expected utility as a function of the $\{p_j\}$, then (3) maximize the resulting expression subject to the applicable constraints. Explicit solutions are not easy to come by, but we have seen one example. When utility has the exponential form $u(W_1) = 1 - \exp(-\rho_a W_1)$ and rates of return are jointly normally distributed with mean vector μ and covariance matrix Σ, the (unconstrained) solution for $\mathbf{p}$ is $\rho_a^{-1} \Sigma^{-1} (\mu - r_0 \mathbf{1})$. Thus, in this case the portfolio depends on means, variances, and covariances only, as in the Markowitz framework.

By the late 1960s economists were beginning to think more broadly about the portfolio problem as just one facet of an individual's decision making. The view was that what really motivates people is not wealth per se, but the opportunities for consumption that wealth affords. Given some current level W_0, one must decide not only how to allocate investment shares among assets but also what portion to consume now and what portion to invest. Moreover, it was realized by the early 1970s that a person's current optimal consumption–investment decision would have to take into account that more decisions of the same sort would be made all through life and would affected by what is done today. In other words, if the current decision is to be truly optimal, it must be seen as merely the first step in a problem of *lifetime* portfolio choice. We will need to make one easy extension of the one-period framework before we can see how to frame portfolio choice as a truly *dynamic* problem.

12.1 THE CONSUMPTION–INVESTMENT CHOICE

Consider the problem of choosing consumption at just two points in time—"now" (the beginning of the investment holding period) and "later" (the end of the period). We might think of posing the problem for a "rational" agent as $\max_{c_0, c_1} u(c_0, c_1)$, where the form of utility function u registers the trade-offs that the person is willing to make between current and future consumption. Since (for now) the individual does not look beyond $t = 1$, we can assume that all the wealth W_1 that is then available will be consumed at that time; thus, c_1 is implicitly determined once W_1 is known. However, the value of W_1 depends not just on the initial consumption choice c_0, but also on the payoffs of the risky investments made at $t = 0$. Since W_1 is therefore unknown initially, the problem needs to be reformulated in terms of *expected* utility, as $\max Eu(c_0, W_1)$. Maximization is with respect to the two things that can be chosen at $t = 0$; namely, c_0 and portfolio vector $\mathbf{p}_0$. Thus, with c_0 dollars devoted to current consumption and $W_0 - c_0$ invested, the problem is

$$\max_{c_0, \mathbf{p}_0} Eu(c_0, W_1) = \max_{c_0, \mathbf{p}_0} Eu\left[c_0, (W_0 - c_0)\left(1 + R_{\mathbf{p}_0, 1}\right)\right]$$

subject to $0 \leq c_0 \leq W_0$. (Time dating is required for the dynamic problem that follows, so we use $R_{\mathbf{p}_0, 1}$ to represent the rate of return realized between $t = 0$ and $t = 1$ for portfolio $\mathbf{p}_0$.)

■ **EXAMPLE 12.1**

Suppose that utility has the Cobb-Douglas form $u(c_0, c_1) = c_0^\alpha c_1^\beta$ with $\alpha > \beta > 0$ and $\alpha + \beta < 1$. To simplify, let there be just two assets. One asset is riskless, offering rate of return r_0 known at $t = 0$, and the other is risky with rate of return R_1. The problem is then

$$\max_{c_0, p_0} E c_0^\alpha W_1^\beta = \max_{c_0, p_0} E \left\{ c_0^\alpha (W_0 - c_0)^\beta [1 + r_0 + p_0 (R_1 - r_0)]^\beta \right\}$$

$$= \max_{c_0} c_0^\alpha (W_0 - c_0)^\beta \cdot \max_{p_0} E [1 + r_0 + p_0 (R_1 - r_0)]^\beta$$

subject to $0 \le c_0 \le W_0$. (Whether p_0 is constrained does not matter for the present purpose.) Notice that the consumption and investment decisions are separable with this utility function, in that the solutions for each of p_0 and c_0 can be made independently of the other. Of course, the optimal p_0 depends on the distribution of R_1, on r_0, and on the constant β. The coefficient of relative risk aversion for W_1^β is $1 - \beta$, so β is a measure of risk "tolerance," with $\beta = 1$ corresponding to risk neutrality. Letting p_0^* be the optimal portfolio share and setting $\theta_0 \equiv E [1 + r_0 + p_0^* (R_1 - r_0)]^\beta$, we are left with the consumption problem $\max_{c_0} c_0^\alpha (W_0 - c_0)^\beta \theta_0$. The unconstrained optimal value $c_0^* = [\alpha / (\alpha + \beta)] W_0$ satisfies the constraint, and so the maximum value of expected utility as of $t = 0$ is

$$v_0 (W_0) \equiv \max_{c_0, p_0} E u (c_0, c_1) = \frac{\alpha^\alpha \beta^\beta}{(\alpha + \beta)^{\alpha+\beta}} W_0^{\alpha+\beta} \theta_0.$$

Notice in the example that the maximum value of expected utility depends just on positive constants and the $\mathcal{F}_0$-measurable initial level of wealth, which is why we represent it as $v_0 (W_0)$. This *value function* represents the real current *value* of wealth as an instrument for producing utility; that is, it shows the utility that can be derived from wealth by allocating wealth *optimally* between consumption and investment. This value function is what one would use to decide among *timeless* prospects that would pay off instantly, before the consumption and investment decisions were to be made. It is precisely this function that would be used by gamblers in a casino or by subjects in experiments like those described in Chapters 7 and 8. Since $v_0 (W_0)$ in the example is simply a positive constant times $W_0^{\alpha+\beta}$, the agent would choose among such timeless prospects the one that maximizes $E W_0^{\alpha+\beta}$. The restriction $\alpha + \beta < 1$ makes the agent averse to these timeless risks.

12.2   DYNAMIC PORTFOLIO DECISIONS

In the framework just considered the individual is supposed to think about consumption now and consumption later. Nevertheless, it is only at $t = 0$ that pertinent decisions must be made, since the decision to set $c_1 = W_1$ at $t = 1$ is essentially automatic. Thus, in making the current decision there is no need to recognize that future actions will depend on what is decided now, so the framework is still not *dynamic*. In reality, though, people do need to make many consumption–investment decisions over their lifetimes, and the decision made at any one stage influences the *constraints* imposed on future decisions.[34] It does so because today's decision determines (or at least influences) the value of wealth in later periods, which thereby limits the amount that can later be consumed and invested. In general, as in the game of chess, it is not optimal to think just one move ahead, and we must expand our decision framework to take this into account.

Here is an example that illustrates both the need to think ahead and a general approach to solving dynamic problems. Suppose that we want to get from point S (start) to point F (finish) in Fig. 12.1 in the least possible travel time. We allow for two overnight stops in between and must travel between stops by straight-line compass marches. Objective F sits at the top of a hill with steep access from the west and a gentle slope on the south. Of course, uphill marches are much slower than those on a level. Now the straight path $SABF$ is the shortest in terms of distance, and rest point A could be reached comfortably in one day. However, continuing from A by the direct route would require two long, arduous days of climbing. Choosing a route like $SABF$ is an example of the mistake one makes when one fails to look beyond the first move. A brief inspection of the map shows many other feasible routes, such as $SCDF$, that might have more promise. How could one go about finding the route that is optimal, in the sense of requiring the least actual marching time?

At first, the problem seems utterly mind-boggling. How do we pick a first stop C, knowing that the subsequent choice of D would depend on where C turns out to be? The key to solving the problem is to tackle it in reverse. Note that from the second stop D there is no further decision to be made, given that each day's march must follow a straight path. The travel time from D on to F would clearly be fixed—a function of the position of D. But D would have been chosen from the first stop C, and at that point there would have been infinitely many stops D from which to choose. How would one make that choice? Clearly, the best choice would be the one that minimizes the *total* time to get from C to D and then from D on to F. That is, we could figure the time to get from C to any point D and the time to get from there to F, and then pick the D that minimizes the sum. That would give the best route from C to F. But that best route from C would depend on the *position* of C, so how does one choose C? It should now be obvious. For any C we can calculate the time to get there from S and then the shortest time from there to F, so once again we just choose the C that minimizes the total time. As we will see, the optimal dynamic consumption–investment problem is solved in exactly this backward-recursive manner.

12.2.1 Optimizing via Dynamic Programming

We can characterize the multiperiod consumption–investment decision as follows. Standing at time $t = 0$, one wants to maximize expected utility from lifetime consumption, $Eu\,(c_0, c_1, ..., c_T)$, where T is the last year of life. (In more advanced and realistic applications we could let T be a random variable and could incorporate the

---

[34]Indeed, if tastes adapt to experience, current decisions may even influence future objectives. Modeling the evolution of preferences is beyond our present scope, but for an example of such a model, see the paper by John Campbell and John Cochrane (1999).

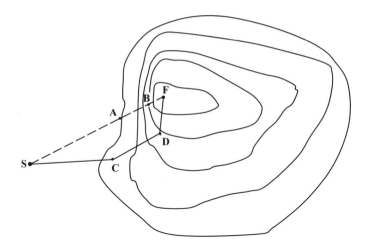

Figure 12.1 Possible routes from start to finish (S to F).

goal of leaving a bequest to heirs, but we focus on the basics here.) The immediate decisions are to choose how much of current wealth W_0 to consume now, c_0, and how to allocate among available assets the remaining $W_0 - c_0$ that is to be invested. In making these choices one recognizes that the decision made today will affect the total amount of resources available at $t = 1$, which will constrain the consumption–investment decisions to be made then. This is just as in the map example, where the choice of stopping point C constrains the choice of subsequent routes. In turn, the result of the $t = 1$ decision will constrain the decision at $t = 2$, and so on. However, nothing can actually be decided at $t = 0$ besides initial consumption c_0 and portfolio $\mathbf{p}_0 = (p_{10}, ..., p_{n0})'$, since the result of these decisions is not fully predictable. It is not fully predictable because wealth W_1 is determined not *just* by the choice of c_0 and $\mathbf{p}_0$ but also by the uncertain payoffs to be obtained from the risky investments. Thus, unlike the deterministic situation in the map example, where one could determine optimal points C and D at the outset, in the investment setting it is usually not possible at $t = 0$ to determine explicitly the optimal values of $c_1, \mathbf{p}_1, c_2, \mathbf{p}_2, ...$ What *can* be determined in advance is the optimal *policy*. This is the optimal allocation and investment choice at each t *contingent on* each possible attainable level of wealth W_t; that is, what can be done at $t = 0$ is to determine the optimal functions $c_t(W_t), \mathbf{p}_t(W_t)$ for each t. Finding the optimal policy is possible through an application of what is called the *principle of optimality*:

> The optimal decision at any stage t subject to the constraints at t will be based on the assumption that all future decisions are optimal with respect to the constraints then faced.

Application of the principle proceeds by a process known as *backward induction* or *dynamic programming*. Suppose that we are at t with past consumption and investment decisions having already been made and having (somehow) produced the current wealth W_t. An optimal decision would make best use of this resource and produce some value

$$\max_{c_t, \mathbf{p}_t} E_t u \left(c_0, ..., c_{t-1}, c_t, c_{t+1}, ..., c_T \right)$$

subject to the constraints

$$0 \leq c_t \leq W_t, \ p_{0t} + \mathbf{p}_t' \mathbf{1} \equiv \sum_{j=0}^{n} p_{jt} = 1. \tag{12.1}$$

Here for brevity $E_t \left(\cdot \right) \equiv E(\cdot \mid \mathcal{F}_t)$ denotes expectation conditional on what is known at t, and $\mathbf{1}$ is just a vector of units. Also, to simplify things on the first pass, let us assume that the n vectors of rates of return $\{\mathbf{R}_t\}_{t=1}^{T}$ in different periods are independent and identically distributed. Under this condition the optimal $c_t = c_t \left(W_t \right)$ and $\mathbf{p}_t = \mathbf{p}_t \left(W_t \right)$ depend just on current wealth, so the maximum expected value of lifetime consumption would be some function $v_t \left(c_0, ..., c_{t-1}, W_t \right)$ of current wealth and past consumption alone. But if we behave optimally at each future period, then next period's decision would be made in the same general way to produce the highest possible value derivable from W_{t+1}:

$$v_{t+1} \left(c_0, ..., c_t, W_{t+1} \right) = \max_{c_{t+1}, \mathbf{p}_{t+1}} E_{t+1} u \left(c_0, ..., c_t, c_{t+1}, ..., c_T \right).$$

Following the principle of optimality and proceeding at t as if optimal decisions will be made henceforth, the problem at t reduces to

$$\max_{c_t, \mathbf{p}_t} E_t v_{t+1} \left(c_0, ..., c_{t-1}, c_t, W_{t+1} \right).$$

Here $c_0, ..., c_{t-1}$ are predetermined, and the choices are subject to the constraints (12.1) and $W_{t+1} = \left(W_t - c_t \right) \left(1 + R_{\mathbf{p}_t, t+1} \right)$, where

$$R_{\mathbf{p}_t, t+1} = r_t + \sum_{j=1}^{n} p_{jt} \left(R_{j,t+1} - r_t \right) = r_t + \mathbf{p}_t' \left(\mathbf{R}_{t+1} - r_t \mathbf{1} \right)$$

is the one-period simple rate of return on portfolio $\mathbf{p}_t$. This now looks much like the one-period consumption–investment problem that we have already seen how to solve, but there is actually a critical difference. In the one-period setting we knew the form of the function $u \left(c_0, W_1 \right)$ whose expected value we wanted to maximize, but how can we determine the form of the function v_{t+1} that confronts us now? What makes a solution possible is that we can in fact determine the value function at each stage by starting at the end and working *backward*.

Here is how. At $t = T$ the decision is the trivial one of putting $c_T = W_T$. This makes terminal value $v_T \left(c_0, ..., c_{T-1}, W_T \right) = u \left(c_0, ..., c_{T-1}, W_T \right)$ a *known* function. Of course, W_T is determined from the $T - 1$ decision as $\left(W_{T-1} - c_{T-1} \right) \left(1 + R_{\mathbf{p}_{T-1}, T} \right)$, so at $t = T - 1$ one solves, subject to the usual constraints,

$$\max_{c_{T-1}, \mathbf{p}_{T-1}} E_{T-1} v_T \left[c_0, ..., c_{T-2}, c_{T-1}, \left(W_{T-1} - c_{T-1} \right) \left(1 + R_{\mathbf{p}_{T-1}, T} \right) \right].$$

The optimal values of c_{T-1} and $\mathbf{p}_{T-1}$, once found, will be functions of W_{T-1} (and, as explained below, other features that characterize the "state" at $T-1$). Plugging the optimal values into objective function $E_{T-1}v_T(c_0, ..., c_{T-1}, W_T)$ produces value function $v_{T-1}(c_0, \ldots, c_{T-2}, W_{T-1})$, which is the basis for the $T-2$ decision: $\max_{c_{T-2},\mathbf{p}_{T-2}} E_{T-2}v_{T-1}(c_0, ..., c_{T-2}, W_{T-1})$. Continuing the process, one eventually arrives at the appropriate value function for the current decision and solves $\max_{c_0,\mathbf{p}_0} E_0 v_1(c_0, W_1)$. Plugging the optimal $c_0(W_0)$, $\mathbf{p}_0(W_0)$ into $E_0 v_1(c_0, W_1)$ would in turn yield the function $v_0(W_0)$ that would be used to evaluate timeless gambles (prospects) whose payoffs are to be known before the first consumption–investment decision must be made. Again, this would be the function that characterizes choices by (rational) casino gamblers and subjects in research experiments.

If rates of return in different time periods are dependent, then the "state" variables on which the value function depends at each stage will include the realized returns up to that stage in addition to the current wealth. This makes any numerical solution more complicated but offers no conceptual difficulty. However, the backward-recursive strategy would not be applicable in a situation in which current decisions affected assets' future returns, so that possibility does have to be excluded.

■ EXAMPLE 12.2

Let us extend Example 12.1 to allow for consumption at dates $t = 0, 1, 2$, again assuming that there is just one risky asset. At $t = 0$ we solve

$$\max_{c_0,p_0} E_0 u(c_0, c_1, c_2) = \max_{c_0,p_0} E_0 \left(c_0^\alpha c_1^\beta c_2^\gamma \right),$$

where $\alpha > \beta > \gamma > 0$ and $\alpha + \beta + \gamma < 1$. Again, the $t = 2$ decision is automatic—spend everything we have—so the value function relevant for $t = 1$ is $v_2(c_0, c_1, W_2) \equiv u(c_0, c_1, W_2) = c_0^\alpha c_1^\beta W_2^\gamma$. The problem at $t = 1$ is then $\max_{c_1,p_1} E_1 c_0^\alpha c_1^\beta W_2^\gamma$ or, since the multiplicative form separates investment and consumption decisions,

$$c_0^\alpha \max_{c_1} c_1^\beta (W_1 - c_1)^\gamma \cdot \max_{p_1} E_1 [1 + r_1 + p_1 (R_2 - r_1)]^\gamma.$$

Let θ_1 be the value of $E_1 [1 + r_1 + p_1 (R_2 - r_1)]^\gamma$ attained at the optimal p_1^*. (This might well depend on the realized return R_1, but it does not depend on W_1.) Then the optimal $t = 1$ consumption just maximizes $c_0^\alpha c_1^\beta (W_1 - c_1)^\gamma \theta_1$. Plugging in the solution $c_1 = [\beta/(\beta + \gamma)] W_1$ gives the optimal value function as of $t = 1$:

$$v_1(c_0, W_1) = \frac{\beta^\beta \gamma^\gamma}{(\beta + \gamma)^{\beta+\gamma}} c_0^\alpha W_1^{\beta+\gamma} \theta_1.$$

We are now ready to choose c_0, p_0. Again, separability of the consumption and investment decisions reduces the problem to

$$\max_{c_0} c_0^\alpha (W_0 - c_0)^{\beta+\gamma} \cdot \max_{p_0} E_0 \left\{ [1 + r_0 + p_0 (R_1 - r_0)]^{\beta+\gamma} \theta_1 \right\}. \qquad (12.2)$$

If random variables R_1 and R_2 are independent, then θ_1 is itself a constant independent of R_1 and can be factored out of the expectation. Otherwise it will be a random variable as of $t = 0$ and cannot be separated out. In either case there would be an optimal p_0^* and an optimal consumption $c_0^* = [\alpha/(\alpha + \beta + \gamma)] W_0$. Finally, plugging c_0^* into (12.2) shows that the appropriate utility function for evaluating timeless prospects is proportional to $W_0^{\alpha+\beta+\gamma}$. Notice that we start out consuming a small fraction of wealth and gradually increase it over time, as

$$c_0 = \frac{\alpha}{\alpha + \beta + \gamma} W_0$$

$$c_1 = \frac{\beta}{\beta + \gamma} W_1$$

$$c_2 = \frac{\gamma}{\gamma} W_2.$$

Notice also that the coefficient of relative risk aversion—one minus the sum of the exponents on W—increases over time, as $1 - \alpha - \beta - \gamma$ for risks that are resolved immediately to $1 - \beta - \gamma$ at stage 0 to $1 - \gamma$ at stage 1. This does seem consistent with how people actually behave as they age, although one cannot expect that this simple model really explains that observation.

A critical omission from the situation in the example is the labor income that the typical consumer could use to help finance consumption and investment. As individuals age and the present values of their remaining streams of labor income diminish, they have fewer opportunities to recoup and offset losses from investments. This is thought to be one reason that older people adopt more conservative investment styles. Dynamic models of this sort could be—and have been—extended to accommodate labor income, private and government-sponsored retirement plans, and other opportunities and constraints faced in real life.

12.2.2   A Formulation with Additively Separable Utility

In developing models of human behavior over time, economists often model lifetime utility as having the additive form

$$u(c_0, c_1, ..., c_T) = \sum_{t=0}^{T} U_t(c_t),$$

where each U_t is a concave, increasing function. The time subscript allows for consumption at different future dates to be valued differently. While obviously very special, the additive form often makes possible explicit solutions for optimal consumption and portfolios. A common specification for period t utility is the CRRA form $U_t(c_t) = \delta^t \left(c_t^{1-\gamma} - 1 \right) / (1 - \gamma)$, where $\delta \in (0, 1)$ is a discount factor and γ is the coefficient of relative risk aversion. For any such specification of U_t we can apply the principle of optimality to write

$$v_t(c_0, ..., c_{t-1}, W_t) = \max_{c_t, \mathbf{p}_t} E v_{t+1}(c_0, ..., c_t, W_{t+1}),$$

where $v_t(c_0, ..., c_{t-1}, W_t)$ is the value function at stage t of an optimal policy and $0 \le c_t \le W_t$. By starting the backward solution at T, we get as terminal value function $v_T(c_0, ..., c_{T-1}, W_T) = \sum_{t=0}^{T-1} U_t(c_t) + U_T(W_T)$. Moving to $t = T - 1$, we find $v_{T-1}(c_0, ..., c_{T-2}, W_{T-1})$ as the solution to

$$\sum_{t=0}^{T-2} U_t(c_t) + \max_{c_{T-1}, \mathbf{p}_{T-1}} [U_{T-1}(c_{T-1}) + E_{T-1} U_T(W_T)]. \qquad (12.3)$$

With such additive utilities it is helpful to separate the value function at any stage t into two parts, as $v_t(c_0, ..., c_{t-1}, W_t) = \sum_{\tau=0}^{t-1} U_\tau(c_\tau) + J_t(W_t)$. At the last stage we have $J_T(W_T) = U_T(W_T)$, so expression (12.3) becomes

$$v_{T-1}(c_0, ..., c_{T-2}, W_{T-1}) = \sum_{t=0}^{T-2} U_t(c_t) + \max_{c_{T-1}, \mathbf{p}_{T-1}} U_{T-1}(c_{T-1}) + E_{T-1} J_T(W_T).$$

In general, for arbitrary $t \in \{0, 1, ..., T-1\}$, we obtain

$$v_t(c_0, ..., c_{t-1}, W_t) = \sum_{\tau=0}^{t-1} U_\tau(c_\tau) + \max_{c_t, \mathbf{p}_t} \{U_t(c_t) + E_t J_{t+1}(W_{t+1})\},$$

where we take the summation to be zero when $t = 0$.

■ **EXAMPLE 12.3**

Again take $T = 2$ and set $U_t(c_t) = \delta^t c_t^{1-\gamma}$ for some $\gamma \in (0, 1)$. At the terminal stage we have $J_2(W_2) = \delta^2 W_2^{1-\gamma}$ and $v_2(c_0, c_1, W_2) = c_0^{1-\gamma} + \delta c_1^{1-\gamma} + J_2(W_2)$. Backing up to $t = 1$, we find $v_1(c_0, W_1)$ by solving

$$c_0^{1-\gamma} + \delta \max_{c_1} \left\{ c_1^{1-\gamma} + \delta(W_1 - c_1)^{1-\gamma} \cdot \max_{p_1} E_1 [1 + p_1(R_2 - r_1)]^{1-\gamma} \right\}$$

for the optimal c_1 and p_1 and plugging them in. As in the Cobb–Douglas example, p_1^* can be found without regard to c_1, but c_1^* *will* now depend on p_1^*. Setting $\theta_1 \equiv E_1 [1 + p_1^*(R_2 - r_1)]^{1-\gamma}$, we find that $c_1^* = \lambda_1 w_1$, where

$$\lambda_1 \equiv \frac{(\theta_1 \delta)^{-1/\gamma}}{1 + (\theta_1 \delta)^{-1/\gamma}}.$$

Substituting gives for $v_1(c_0, W_1)$

$$c_0^{1-\gamma} + \delta \left[\lambda_1^{1-\gamma} + \theta_1 \delta (1 - \lambda_1)^{1-\gamma} \right] W_1^{1-\gamma} \equiv c_0^{1-\gamma} + J_1(W_1).$$

Proceeding, we find c_0 and p_0 by maximizing

$$c_0^{1-\gamma} + E_0 J_1 \{(W_0 - c_0)[1 + p_0(R_1 - r_0)]\}.$$

Putting $\phi \equiv \delta \left[\lambda_1^{1-\gamma} + \theta_1 \delta (1 - \lambda_1)^{1-\gamma} \right]$, assuming the returns R_1 and R_2 to be independent, and letting $\theta_0 \equiv E_0 \left[1 + p_0^* (R_1 - r_0) \right]^{1-\gamma}$, we get $c_0^* = \lambda_0 W_0$, with

$$\lambda_0 \equiv \frac{(\theta_0 \phi)^{-1/\gamma}}{1 + (\theta_0 \phi)^{-1/\gamma}}.$$

EXERCISES

12.1 Suppose that lifetime utility is given by $u(c_0, c_1, c_2) = \ln c_0 + \delta \ln c_1 + \delta^2 \ln c_2$, where $0 < \delta < 1$ is a discount factor that accounts for "impatience." Assuming a single risky asset, the problem at stage $t = 1$ is to choose c_1 and p_1 to maximize

$$E_1 v_2 (c_0, c_1, W_2) = \ln c_0 + \delta \ln c_1 + \delta^2 E_1 \ln \left\{ (W_1 - c_1) \left[1 + p_1 (R_2 - r_1) \right] \right\}.$$

Show, first, that the optimal portfolio decisions at $t = 0$ and $t = 1$ are just the same if R_1, R_2 are identically distributed and if interest rates r_0, r_1 at $t = 0$ and $t = 1$ are equal. In other words, although there is not enough information to find their common value, show that $p_0^* = p_1^*$. Next, show that the optimal initial consumption is $c_0^* = W_0 / (1 + \delta + \delta^2)$, where W_0 is the initial wealth.

12.2 Initial wealth is $W_0 = 1000$ and lifetime utility is $u(c_0, c_1, c_2) = c_0 + \delta c_1 + \delta^2 c_2$ with $\delta = 0.7$. One can choose each period between a riskless asset and a risky asset with one-period rates of return $r_0 = r_1 = .10$ and R_1, R_2 i.i.d. as $N(\mu = .25, \sigma^2 = .50)$. Express the optimal initial consumption and portfolio, c_0^* and p_0^*, assuming that consumption and portfolio in each period are constrained as $0 \le c_t \le W_t$ and $0 \le p_t \le 1$. How much will actually be consumed at $t = 1$ and $t = 2$?

12.3 For the problem in Example 12.2 set $\alpha = \beta = \gamma = 1$; take $r_0 = r_1 = r$; and let R_1, R_2 be independent with $ER_1 = ER_2 = \mu > r$. Find the optimal portfolio and consumption at $t = 0$ assuming that at each t the portfolio and consumption decisions are constrained as $0 \le p_t \le 1$ and $0 \le c_t \le W_t$.

12.4 For the problem in Example 12.2 set $\alpha = \beta = \gamma = 1$; take $r_0 = r_1 = 0$; and let R_1, R_2 be jointly normally distributed with zero means, unit variances, and correlation $\rho = .5$. Find the optimal portfolio and consumption at $t = 0$ given that at each t the portfolio and consumption decisions are constrained as $0 \le p_t \le 1$ and $0 \le c_t \le W_t$. (Hint: Under the stated assumptions, $E_1 R_2 = E(R_2 \mid R_1) = \rho R_1$.)

CHAPTER 13

OPTIMAL GROWTH

Let us suppose that one has a choice between two investments. One is a riskless bond; the other is shares in a hedge fund. At the end of each period beginning at $t = 0, 1, 2, \ldots$ the bond pays sure rate of return $r_t = .10$, while the hedge fund offers simple rates of return $R_{t+1} = 1.00$ and $R_{t+1} = -.60$ with equal probability, period after period. Thus, at time t the rate paid by the bond during $(t, t + 1]$ is a known value, while the rate offered by the fund is a random variable with $\Pr(R_{t+1} = 1.00) = \Pr(R_{t+1} = -.60) = .5$. Which of these investments looks more promising? One notes that the expected rate of return of the hedge fund is an impressive 20% per annum—precisely double the rate on the bond. Thinking in terms of a long-term commitment, it would seem logical to apply the law of large numbers. Thus, making a \$1 investment in the hedge fund in each of T years would produce average rate of return $\bar{R}_T \equiv T^{-1} \sum_{t=0}^{T-1} R_{t+1}$, which converges as $T \to \infty$ to $ER_{t+1} = .20$. Of course, the average rate of return from the bond would just be .10. Certainly, the hedge fund looks like the better deal.

However, a little reflection might give one pause. In running an investment portfolio the real test is how value grows over time when funds are left there to grow. One starts at $t = 0$ by investing some amount V_0, holds the investment for a period at some simple rate of return R_1, and thus attains net value $V_1 = V_0 (1 + R_1)$

Quantitative Finance. By T.W. Epps
Copyright © 2009 John Wiley & Sons, Inc.

at $t = 1$. Assuming no additions or withdrawals, this amount is then reinvested, giving net value $V_2 = V_1 (1 + R_2) = V_0 (1 + R_1)(1 + R_2)$ at $t = 2$, and so on, with $V_T = V_0 \prod_{t=0}^{T-1} (1 + R_{t+1})$ after T periods. Thus, portfolio values grow by the *multiplicative* process of compounding rather than by an additive process. The appropriate summary statistic for average performance is therefore not the *arithmetic* mean of rates of return but the *geometric* mean of total returns, or $\sqrt[T]{V_T/V_0} = \sqrt[T]{\prod_{t=0}^{T-1} (1 + R_{t+1})}$. Just as earning a constant rate $\bar{R}_T = T^{-1} \sum_{t=0}^{T-1} R_{t+1}$ in each of T periods would produce the given *sum* of rates $\sum_{t=0}^{T-1} R_{t+1} = T\bar{R}_T$, earning a constant average rate $\sqrt[T]{V_T/V_0} - 1$ in each of T periods would produce the same *product* of one-period total returns $\prod_{t=0}^{T-1} (1 + R_{t+1})$—and therefore the same terminal value. Thus, the geometric mean is the more appropriate summary statistic for investment performance. An even better such statistic, since it does not depend on the arbitrary length of the holding periods, is the mean *continuously compounded* rate, $R_{0,T} = T^{-1} \ln (V_T/V_0)$. Of course, this could be expressed in terms of the simple returns as $T^{-1} \sum_{t=0}^{T-1} \ln (1 + R_{t+1})$. (In either case, if time is measured in years the result is an *annualized* rate.) $R_{0,T}$ does summarize the T-period experience, since investing V_0 for T periods at the constant rate $R_{0,T}$ with continuous compounding would yield the realized terminal wealth; that is,

$$V_0 e^{T R_{0,T}} = V_0 \exp \left[T \cdot T^{-1} \ln \left(\frac{V_T}{V_0} \right) \right] = V_0 \cdot \frac{V_T}{V_0} = V_T.$$

Applying the law of large numbers again, assuming that simple rates $R_1, R_2, \ldots$ are independent and identically distributed (and so have the same distribution as R_1), gives as the "long-run" average rate

$$R_{0,\infty} = \lim_{T \to \infty} T^{-1} \sum_{t=0}^{T-1} \ln (1 + R_{t+1}) = E \ln (1 + R_1).$$

Calculating this expected value would afford a reasonable way to assess an investment that would be maintained until some distant horizon, such as the date of retirement.

Let us see how the hedge fund compares with the riskless bond according to this standard. The value for the bond is simply $\ln (1.10) \doteq .0953$, while that for the hedge fund is

$$E \ln (1 + R_1) = .5 \ln (1 + 1) + .5 \ln (1 - .6) \doteq -.1116.$$

Thus, investing period after period in the hedge fund would, on average, cause wealth to *decline* by about 11% per unit time, whereas investing in the bond would lead to steady growth at a continuously compounded rate exceeding 9.5%. How can this be? Very simply, with the hedge fund one "wins"—gets the good return—about half the time and one "loses" about half the time. A win doubles one's capital, but a loss reduces it by *more* than half. For example, the table below shows how an initial investment of $1 would fare from a sequence of 10 annual investments with an equal number of wins and losses. (Note that the final result would be the same regardless of the order in which these occurred.)

t	R_t	V_t
1	1.00	2.00
2	1.00	4.00
3	−.60	1.60
4	1.00	3.20
5	1.00	6.40
6	−.60	2.56
7	−.60	1.024
8	1.00	2.048
9	−.60	0.8192
10	−.60	0.32768

13.1 OPTIMAL GROWTH IN DISCRETE TIME

To put what we have just seen in more formal terms, suppose that one starts with capital V_0 and plans to invest for a "large" number of periods T. We assume that the account is tax free and that transactions are costless. In the United States, dividends and capital gains in retirement plans run through traditional individual retirement accounts (IRAs), 401(k) accounts, and 403(b) accounts are in fact tax free until the date T at which funds are to be withdrawn. Funds in "Roth" IRAs and accounts held by nonprofit organizations, such as university endowments, are not taxed at all. Moreover, in these days of online trading transaction costs are indeed small in proportion to account value, so the assumptions are not unrealistic in many applications. Although it would be easy to allow for planned fractional withdrawals or contributions, we assume for simplicity that there are to be none such before time T. The goal, let us say, is to choose investments at each $t \in \{0, 1, ..., T - 1\}$ so as to make V_T as large as possible. Of course, investment returns are always unpredictable unless one sticks to holding government bonds to maturity, so finding a strategy that attains the goal of *maximizing* the random variable V_T seems impossible. Indeed, it is. Yet we will see that it *is* possible to find a strategy that is *asymptotically* optimal, in the sense that its *long-run* average continuously compounded rate of return, $\lim_{T \to \infty} T^{-1} \ln (V_T/V_0)$, is larger than that of any other strategy.

The example with the bond and hedge fund points the way. In building a portfolio, suppose that we can choose among all combinations of a riskless bond and n risky assets. During each period from t to $t+1$ the bond pays a rate r_t that is known at t, while the risky assets have unpredictable simple rates of return $\{R_{j,t+1}\}_{j=1}^n$. Although "unpredictable," these do have a known joint probability distribution, $f_t(r_1, r_2, ..., r_n)$. Although not strictly necessary for the result, we simplify by assuming rates in different periods to be independent. (What *is* necessary is that the dependence be weak enough for a law of large numbers to apply. Of course, rates of different assets in a *given* period may well covary, so long as they are not *perfectly* correlated.) A portfolio chosen at time t is a vector $\mathbf{p}_t = (p_{0t}, p_{1t}, ..., p_{nt})'$ with $\sum_{j=0}^n p_{jt} = 1$. Its rate of return during t to $t+1$ is

$$R_{\mathbf{p}_t,t+1} = p_{0t}r_t + \sum_{j=1}^{n} p_{jt}R_{j,t+1} = r_t + \sum_{j=1}^{n} p_{jt}\left(R_{j,t+1} - r_t\right).$$

We want to choose $\mathbf{p}_t$ in each period so as to maximize the long-run growth rate of wealth:

$$\lim_{T\to\infty} \frac{1}{T} \ln\left(\frac{V_T}{V_0}\right) = \lim_{T\to\infty} \frac{1}{T} \ln \prod_{t=0}^{T-1} \left(1 + R_{\mathbf{p}_t,t+1}\right)$$

$$= \lim_{T\to\infty} \frac{1}{T} \sum_{t=0}^{T-1} \ln\left(1 + R_{\mathbf{p}_t,t+1}\right).$$

To see how to do this, we need a trivial extension of Kolmogorov's strong law of large numbers (SLLN), as it was stated in Section 3.7. If random variables $\{X_t\}$ are i.i.d. with mean μ, then the standard version of the SLLN tells us that $T^{-1}\sum_{t=1}^{T} X_t \to \mu$. Equivalently, since $\{X_t - \mu\}$ are also i.i.d. and $E\left(X_t - \mu\right) = 0$, it follows that $\lim_{T\to\infty} T^{-1}\sum_{t=1}^{T}(X_t - \mu) = 0$. Now suppose the $\{X_t\}$ are still independent but that they (may) have different expected values, $EX_t = \mu_t$. Then since $E\left(X_t - \mu_t\right) = 0$ for each t, we still have $0 = \lim_{T\to\infty} T^{-1}\sum_{t=1}^{T}\left(X_t - \mu_t\right)$. Now writing

$$\frac{1}{T}\sum_{t=1}^{T} X_t = \frac{1}{T}\sum_{t=1}^{T}\left(X_t - \mu_t\right) + \frac{1}{T}\sum_{t=1}^{T}\mu_t,$$

and taking limits shows that

$$\lim_{T\to\infty} \frac{1}{T}\sum_{t=1}^{T} X_t = \lim_{T\to\infty} \frac{1}{T}\sum_{t=1}^{T} \mu_t.$$

Thus, if the average of the $\{\mu_t\}$ does have a limiting value, then so will the average of the $\{X_t\}$, and the two will coincide. Of course, nothing guarantees that the limit will be finite, but that is of no concern to us here.

Now apply this result to the continuously compounded returns from our portfolio, taking $X_t \equiv \ln\left(1 + R_{\mathbf{p}_t,t+1}\right)$ and $\mu_t \equiv E_t\left[\ln\left(1 + R_{\mathbf{p}_t,t+1}\right)\right]$, where $E_t\left(\cdot\right) \equiv E\left(\cdot \mid \mathcal{F}_t\right)$ represents expectation conditional on what is known at t:

$$\lim_{T\to\infty} \frac{1}{T} \ln\left(\frac{V_T}{V_0}\right) \equiv \lim_{T\to\infty} \frac{1}{T}\sum_{t=0}^{T-1} \ln\left(1 + R_{\mathbf{p}_t,t+1}\right)$$

$$= \lim_{T\to\infty} \frac{1}{T}\sum_{t=0}^{T-1} E_t \ln\left(1 + R_{\mathbf{p}_t,t+1}\right).$$

Clearly, then, to maximize the asymptotic growth rate of wealth we should choose portfolio $\mathbf{p}_t$ in each period so as to maximize

$$E_t \ln (1 + R_{\mathbf{p}_t,t+1}) = E_t \ln \left[1 + r_t + \sum_{j=1}^{n} p_{jt} (R_{j,t+1} - r_t)\right]. \qquad (13.1)$$

Since the value of capital at $t+1$ is $V_{t+1} = V_t (1 + R_{\mathbf{p}_t,t+1})$ and V_t is $\mathcal{F}_t$-measurable, maximizing (13.1) is equivalent to maximizing $\ln V_t + E_t \ln (1 + R_{\mathbf{p}_t,t+1}) = E_t \ln V_{t+1}$. Thus, the prescription for optimal growth is just to proceed as though the utility function for next period's wealth is logarithmic.

As an example of how to do this, return to the case of the bond and the hedge fund, but suppose now that we are able to divide wealth between them in any proportions. Although full investment in the bond dominates full investment in the fund, some blend of the two might produce a better result in the long run. Let us see. With wealth share p_t in the fund and $1 - p_t$ in the bond, the one-period portfolio rate of return is now $R_{p_t,t+1} = .10 + p_t (R_{t+1} - .10)$. We want to choose p_t to solve

$$\max_{p_t} E \ln (1 + R_{p_t,t+1}) = \max_{p_t} E \ln [1.10 + p_t (R_{t+1} - .10)]$$
$$= \max_{p_t} [.5 \ln (1.10 + .90 p_t) + .5 \ln (1.10 - .70 p_t)].$$

(Because the $\{R_t\}$ are i.i.d., there is no need for a time subscript on E.) Differentiating with respect to p_t and equating to zero give the first-order condition

$$0 = \frac{.9}{1.1 + .9 p_t} - \frac{.7}{1.1 - .7 p_t}$$

and the solution $p_t \doteq .1746$. Since the second derivative is negative for all p_t, we have indeed found the function's maximum. And since the rates of return are i.i.d. in our example, this optimal investment proportion is the same for each t. Plugging the solution into $E \ln (1 + R_{p_t,t+1})$ gives as the maximum continuously compounded rate of growth the value

$$\frac{1}{2} \ln [1.10 + .1746 (.90)] + \frac{1}{2} \ln [1.10 - .1746 (.70)] \doteq .1032,$$

versus .0953 for investment in the bond alone. Figure 13.1 shows the result of a simulation that compares investments in the bond, the hedge fund, and the optimal combination of the two over 50 periods, starting at $t = 0$ with $1 in each.

One can get some intuitive feel for how it is possible to gain by adding a "bad" asset to a good one. Each period, given the returns from the previous period, we *rebalance* the portfolio to keep proportion $p_t = .1746$ in the hedge fund. Doing this means that we must sell off some of the hedge fund when it has done well in the previous period and boosted our wealth, using the proceeds to buy more of the bond. Having done so, the next loss that the fund produces will not hurt so badly. By contrast, when the fund has done poorly, we shift capital out of bonds and *into* the fund and so benefit

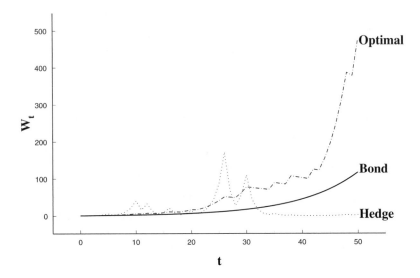

Figure 13.1 Value at t for three alternative investment strategies.

from the next good outcome. The result is a little like trying to ride waves at the beach while blindfolded and unable to see what is coming. By constantly paddling gently toward shore, we catch many of the inbound surges and don't wear ourselves out battling the back flow.

13.2 OPTIMAL GROWTH IN CONTINUOUS TIME

It was easy to solve the growth-optimal problem in the example, because there was just one risky asset and because its rate of return could have just two possible values. But what about real-world investments with many assets and rates of return having (essentially) continuous distributions? The computational problem becomes much harder when distributions of returns are more realistic, even when there just a single risky asset. To see this, let us modify the example by adopting a fairly plausible model for the per-unit value of the risky asset; namely, the lognormal model that corresponds to geometric Brownian motion in continuous time. Letting S_t be the value of the risky "stock" at t, assume that the "instantaneous" change from t to $t + dt$ is $dS_t = \mu S_t \cdot dt + \sigma S_t \cdot dW_t$, where $\{W_t\}_{t \geq 0}$ is a Brownian motion and μ and $\sigma > 0$ are constants. Then (recalling Example 11.5), we have the solution $S_t = S_0 e^{(\mu - \sigma^2/2)t + \sigma W_t}$ when the initial value is $S_0 > 0$. Since

$$\ln S_t = \ln S_0 + \left(\mu - \frac{\sigma^2}{2}\right)t + \sigma W_t \sim N\left[\ln S_0 + \left(\mu - \frac{\sigma^2}{2}\right)t, \sigma^2 t\right],$$

the stock's future price itself is distributed as lognormal.

We can also invest in a default-free, discount bond that matures and pays \$1 at t. Its initial value is $B(0,t) = e^{-tr(0,t)}$, where $r(0,t)$ is the average spot rate to t. Now suppose that we invest capital V_0 at time 0 by buying b_0 units of the bond and s_0 units of stock. The capital value at future date t is then

$$V_t = s_0 S_t + b_0 = s_0 S_0 e^{(\mu - \sigma^2/2)t + \sigma W_t} + b_0.$$

The cost at time 0 is $V_0 = s_0 S_0 + b_0 B(0,t)$, so we have

$$b_0 = \frac{V_0 - s_0 S_0}{B(0,t)} = (V_0 - s_0 S_0)\, e^{tr(0,t)}$$

and

$$
\begin{aligned}
V_t &= s_0 S_t + (V_0 - s_0 S_0)\, e^{tr(0,t)} \\
&= (s_0 S_0) \frac{S_t}{S_0} + (V_0 - s_0 S_0)\, e^{tr(0,t)} \\
&= V_0 \left[\left(\frac{s_0 S_0}{V_0} \right) \frac{S_t}{S_0} + \left(1 - \frac{s_0 S_0}{V_0} \right) e^{tr(0,t)} \right] \\
&= V_0 \left[p_0 e^{(\mu - \sigma^2/2)t + \sigma W_t} + (1 - p_0)\, e^{tr(0,t)} \right],
\end{aligned}
$$

where $p_0 \equiv s_0 S_0 / V_0$ is the proportion of capital invested in the stock. Finding the growth-optimal value of p_0 requires solving

$$\max_{p_0 \in [0,1]} E \ln \left(\frac{V_t}{V_0} \right) = \max_{p_0 \in [0,1]} E \ln \left[p_0 e^{(\mu - \sigma^2/2)t + \sigma W_t} + (1 - p_0)\, e^{tr(0,t)} \right].$$

or

$$e^{tr(0,t)} \max_{p_0 \in [0,1]} E \ln \left\{ 1 + p_0 \left[e^{(\mu - r(0,t) - \sigma^2/2)t + \sigma W_t} - 1 \right] \right\} \qquad (13.2)$$

for given values of μ, σ^2, $r(0,t)$, and t. (Note that the restriction $p_0 \in [0,1]$ is necessary to keep $\Pr(V_t > 0) = 1$.) While it is very easy to get a numerical solution using the computer, there is, in general, no simple analytical formula for the optimal p_0, because there is no simple (closed-form) expression for the expected value of an affine function of a lognormal random variable.[35] Of course, adding more risky assets just makes the problem all the more difficult.

---

[35]An exception is when $\mu - r(0,t) - \sigma^2/2 = 0$, in which case the solution is $p_0 = .5$, regardless of the values of σ and t. To see this (1) put $\mu = r(0,t) - \sigma^2/2$; (2) write the expectation in (13.2) in terms of normal CDF Φ, as $\int_{-\infty}^{\infty} \ln \left[1 + p_0 \left(e^{\sigma \sqrt{t} \cdot z} - 1 \right) \right] \cdot d\Phi(z)$; (3) break the range of integration into $(-\infty, 0)$ and $(0, \infty)$; (4) change variables as $z \to -z$ in the integral over $(-\infty, 0)$; and (5) use the facts that $\Phi(-z) = 1 - \Phi(z)$ and $\ln x + \ln y = \ln(xy)$ to reduce the expectation to

$$\int_0^{\infty} \ln \left\{ 1 + 2p_0 (1 - p_0) \left[\cosh \left(\sigma \sqrt{t} \cdot z \right) - 1 \right] \right\} \cdot d\Phi(z).$$

Since $\cosh(x) > 1$ for $x > 0$, the integrand is maximized for *each* $z > 0$ at $p_0 = .5$, and likewise the integral. Note that the result does not require the distribution of $X_t \equiv \ln(S_t/S_0) - tr(0,t)$ to be normal. It holds for any distribution F such that $F(x) = 1 - F(-x)$ for all $x > 0$ and such that $\int |\ln[1 + p_0(e^x - 1)]| \cdot dF(x) < \infty$ for $p_0 \in [0,1]$.

This is the sort of problem that becomes much more tractable in continuous time, and it provides our first application of Itô's famous formula. Suppose now that, standing at time t, we want to choose proportion p_t of capital to place in the stock and proportion $1 - p_t$ to invest in a money-market fund that earns interest continuously at the short rate. For the moment we take the short rate to have constant value r. The stock's current price is S_t, and the current value of a unit of the money fund is $M_t = M_0 e^{rt}$. We make the investment choice knowing that at $t + dt$ we will rebalance the portfolio, selling some of one asset and using the proceeds to buy more of the other, but neither adding nor withdrawing funds. That means that the portfolio will be *self-financing*. Given that we hold s_t units of stock and m_t units of the money fund at t, the change in capital from t to $t + dt$ is[36]

$$
\begin{aligned}
dV_t &= d\left(s_t S_t + m_t M_t\right) \\
&= s_t \cdot dS_t + S_t \cdot ds_t + m_t \cdot dM_t + M_t \cdot dm_t \\
&= \left[s_t \cdot dS_t + m_t \cdot dM_t\right] + \left[S_t \cdot ds_t + M_t \cdot dm_t\right].
\end{aligned}
$$

Here, the quantity in the first pair of brackets represents the effect of changes in the assets' prices, while what is inside the second pair of brackets represents the effect of portfolio adjustments. However, this latter effect will be precisely zero, since purchases of one asset must be financed by sales of the other, so only the price effect matters. Expressing the instantaneous changes in the individual prices, we then have

$$
\begin{aligned}
dV_t &= s_t\left(\mu S_t \cdot dt + \sigma S_t \cdot dW_t\right) + m_t M_t r \cdot dt \\
&= V_t\left[\frac{s_t S_t}{V_t}\left(\mu \cdot dt + \sigma \cdot dW_t\right) + \frac{m_t M_t}{V_t} r \cdot dt\right] \\
&= V_t p_t\left(\mu \cdot dt + \sigma \cdot dW_t\right) + V_t\left(1 - p_t\right) r \cdot dt \\
&= V_t\left[r + p_t\left(\mu - r\right)\right] \cdot dt + V_t p_t \sigma \cdot dW_t.
\end{aligned}
$$

This is simply an Itô process of the same general sort as that followed by the stock itself. The quadratic variation of such a process is $\langle V \rangle_t = \int_0^t V_s^2 p_s^2 \sigma^2 \cdot ds$, and so $d\langle V \rangle_t = V_t^2 p_t^2 \sigma^2 \cdot dt$. Applying Itô's formula to $\ln V_t$, we have for the "instantaneous" change in the logarithm of capital

$$
\begin{aligned}
d\ln V_t &= \frac{dV_t}{V_t} - \frac{d\langle V \rangle_t}{2V_t^2} \\
&= \left[r + p_t\left(\mu - r\right)\right] \cdot dt + p_t \sigma \cdot dW_t - p_t^2 \frac{\sigma^2}{2} \cdot dt \\
&= \left[r + p_t\left(\mu - r\right) - p_t^2 \frac{\sigma^2}{2}\right] \cdot dt + p_t \sigma \cdot dW_t. \tag{13.3}
\end{aligned}
$$

Now, the expected value of $d\ln V_t$ is just the first term, since the change in the BM has mean zero, and maximizing this first term with respect to p_t gives the simple

---

[36]As quantities s_t and m_t are under our control, they are of finite variation. This accounts for the absence of interaction term $d\langle s_t, S_t \rangle$ in the expression for $d\left(s_t S_t\right)$.

solution $p_t = (\mu - r)/\sigma^2$. Recall from Section 7.4 that this is precisely the "short holding period" solution we found for a utility with relative risk aversion $\rho_r = 1$ when expected utility is approximated as a function of mean and variance. We now see that this approximation becomes exact in continuous time when interest rates and prices are modeled in this fairly realistic way.

Now plugging $p_t = (\mu - r)/\sigma^2$ into $Ed \ln V_t$ and putting $\theta \equiv (\mu - r)/\sigma$ give

$$r + \left(\frac{\mu - r}{\sigma^2}\right)(\mu - r) - \left(\frac{\mu - r}{\sigma^2}\right)^2 \frac{\sigma^2}{2} = r + \frac{\theta^2}{2}$$

as the maximum expected growth rate of capital and $d \ln V_t = (r + \theta^2/2) \cdot dt + \theta \cdot dW_t$ as the actual instantaneous change in $\ln V_t$. Given the initial condition $V_0 > 0$, the solution to this SDE is $V_t = V_0 \exp\left[(r + \theta^2/2)t + \theta W_t\right]$, showing that the future value of capital is also lognormally distributed. Moreover, the average continuously compounded rate of return out to horizon T is

$$R_{0,T} = T^{-1} \ln\left(\frac{V_T}{V_0}\right) = r + \frac{\theta^2}{2} + \theta \frac{W_T}{T}, \tag{13.4}$$

and since W_T has the distribution of $\sqrt{T}Z$ where $Z \sim N(0, 1)$, we confirm that the long-run average continuously compounded rate of return is $R_{0,\infty} = r + \theta^2/2$.

It is trivial to generalize these results to allow the short rate and the mean drift rate and volatility of $\{S_t\}_{t \geq 0}$ to be time varying, so long as their paths $\{r_t, \mu_t, \sigma_t\}_{0 \leq t}$ are considered to be integrable and $\mathcal{F}_0$-measurable (known at $t = 0$). In this case expression (13.4) becomes

$$R_{0,T} = T^{-1}\left[\int_0^T \left(r_t + \frac{\theta^2}{2}\right) \cdot dt + \int_0^T \theta_t \cdot dW_t\right],$$

still distributed as lognormal, and $R_{0,\infty} = \lim_{T \to \infty} T^{-1} \int_0^T (r_t + \theta_t^2/2) \cdot dt$ (assuming that the limit exists). Moreover, as Exercise 13.3 indicates, the results generalize to accommodate portfolios of arbitrarily many risky assets. The quantity θ is usually referred to as the *Sharpe ratio* of the risky asset. Exercise 13.5 shows that when there are n risky assets the optimal portfolio itself has Sharpe ratio $\theta = \sqrt{(\mu - r\mathbf{1})' \Sigma^{-1} (\mu - r\mathbf{1})}$, where μ and $\mathbf{1}$ are n vectors of mean rates of return and units, respectively, and Σ is the $n \times n$ covariance matrix (assumed to be nonsingular). In that multiasset, time-invariant setting the maximum long-run rate of return is still $R_{0,\infty} = r + \theta^2/2$. We can thus summarize the prescription for optimal growth in continuous time as follows: *Always hold the portfolio with the largest Sharpe ratio.*

13.3 SOME QUALIFICATIONS

The idea of maximizing the growth of capital certainly does seem appealing, and its implementation—at least if one could really trade frequently enough and cheaply enough and knew the right model for prices—is relatively straightforward. Still, there are several cautions to bear in mind. The first is a technical one that relates to the continuous-time solution. In the case of a single risky asset the optimal investment proportion $p_t = (\mu_t - r_t)/\sigma_t^2$ at any t (allowing r, μ, and σ to be time varying) could (in one's opinion at the time) take any value whatsoever, including values less than zero (corresponding to short sales of the stock) and values greater than unity (corresponding to borrowing, or shorting the money fund). Likewise, in the case of n assets the optimal $\mathbf{p}_t = \Sigma_t^{-1}(\mu_t - r_t\mathbf{1})$ could have negative components or be such that $\mathbf{p}_t'\mathbf{1} > 1$. Why is this a problem? In continuous time, when the stock follows a continuous Itô process, such positions involve no chance of financial ruin in the "infinitesimal" time until the next rebalancing. But in reality, since no one can rebalance continuously, it is possible that the stocks' prices could increase or decline sufficiently between rebalances as to reduce capital to zero. For example, if one borrows cash to finance purchases of stocks, a sufficient decline in their prices would make one unable to repay the loan. Likewise, a sufficient increase in the price of any one stock would make one unable to buy it back and cover a short position. In practice, therefore, one would need to constrain $\mathbf{p}_t$ to be nonnegative with $\mathbf{p}_t'\mathbf{1} \leq 1$ in order to ensure that the portfolio's value remains positive. Clearly, no portfolio for which $\Pr(V_t \leq 0) > 0$ could be growth optimal, since $\ln V_t \to -\infty$ as $V_t \to 0$. On the other hand, such a constrained solution would not be optimal in continuous time, nor is there reason to think that it correctly represents the discrete-time solution. Exercise 13.6 will make clear that discrete- and continuous-time solutions sometimes do diverge in fundamental ways.

Two other issues are more fundamental, applying to both continuous- and discrete-time versions of the optimal-growth theory. Recall that the guarantee of optimal growth holds only *asymptotically*; that is, the prescription is certain to deliver the fastest average growth of capital only in the longest of long runs. The simulated performance shown in Fig. 13.1 illustrates that it can take a very long time indeed for the "optimal" portfolio to break away from the pack of contenders. One who manages to follow the discipline religiously for even 30 years *may* find that his retirement account would have done better if kept in a mundane stock index fund.

Finally, there is the related issue that the goal of growth optimality may appear more appealing than it should—as it apparently did to some academic researchers in the early years after its discovery. The possible misconception can be put in the form of a question. "If (almost) everyone considers himself to be better off with more money than with less, wouldn't a portfolio strategy that delivers the highest possible growth of capital be the best for (almost) everyone?" The fallacy here is made clear on rephrasing the question in economists' terms: "Since (almost) everyone's utility function is increasing in future wealth, shouldn't (almost) everyone adopt the function $u(W_t) = \ln W_t$?" Remember that the purpose of investing is ultimately to finance consumption, and that different—but fully rational—people simply have different tolerances for unplanned fluctuations in what they consume. The risk/return trade-offs willingly made by an individual with log utility might be totally unacceptable to a "rational" individual with different tastes.

One way to judge the risk level of a particular portfolio strategy is through simulations like that used to produce Fig. 13.1. One sees in such simulations, and even in the single track of the figure, that the growth-optimal strategy can lead to pretty choppy wealth levels in the short and even the intermediate runs. In other words, the long-run optimal solution can produce a *lot* of volatility in the short run. To persist in a program that may lead to substantial fluctuations in wealth calls for stronger nerves—read lower risk aversion—than many of us have. One can see this directly, without simulations, through simple thought experiments. For example, suppose at the flip of a coin that you could either lose half of everything you own or else double it. Would you want to take that gamble, or even be indifferent? If you are indifferent at age 21, would you be apt to be at 65? If the answer is "no," then you should not like the idea of growth optimality, because the calculation

$$\ln\left(W_0 + c\right) = \frac{1}{2}\ln\left(\frac{W_0}{2}\right) + \frac{1}{2}\ln\left(2W_0\right)$$
$$= \ln W_0 + \frac{1}{2}\left(-\ln 2\right) + \frac{1}{2}\ln 2$$
$$= \ln W_0$$

shows that the certainty equivalent c of such a deal is precisely zero when utility is logarithmic.

EXERCISES

13.1 You have a gambling stake $V_0 > 0$ and will undertake a series of plays of a game, thereby generating a sequence of values $V_1, V_2, \ldots$. At each play of the game you can bet any fraction $f \in [0, 1]$ of the current stake, and the result of the bet will determine what you have at the next play. The game is such that each dollar that is bet returns \$2 with probability .6 and \$0 with probability .4. If you intend to play the game indefinitely, and if you want to maximize the long-run average growth of your gambling stake, what is the optimal fraction f_j of wealth to bet on the jth play, for $j \in \{0, 1, 2, \ldots\}$?

13.2 You can make repeated investments in two assets, the joint distribution of whose simple rates of return in any one period, R_A and R_B, is tabulated below. (To interpret the table, both rates will take the value $-.5$ with probability .3, R_A will take the value $-.5$ and R_B the value 1.0 with probability .2, and so on.)

$R_B\backslash R_A$	$-.5$	1.0
$-.5$	.3	.2
1.0	.2	.3

Letting $R_p = pR_A + (1 - p)R_B$ be the rate of return on a portfolio, find the growth-optimal value of p and the corresponding continuously compounded growth rate.

13.3 An investor with current capital V_t can choose among among a money-market fund, whose value per unit evolves as $dM_t = rM_t \cdot dt$, and two stocks, whose share prices move as $dS_{1t} = \mu_1 S_{1t} \cdot dt + \sigma_1 S_{1t} \cdot dW_{1t}$ and $dS_{2t} = \mu_2 S_{2t} \cdot dt + \sigma_2 S_{2t} \cdot dW_{2t}$. Brownian motions $\{W_{1t}, W_{2t}\}_{t\geq 0}$ have $EW_{1t} W_{2t} = \rho t$, with $|\rho| < 1$. The value of a self-financing portfolio comprising s_{1t} and s_{2t} shares of the two stocks and m_t units of the money fund evolves as

$$dV_t = s_{1t} \cdot dS_{1t} + s_{2t} \cdot dS_{2t} + m_t \cdot dM_t.$$

Find the growth-optimal numbers of units of these assets. What would happen if $\{W_{1t}, W_{2t}\}_{t\geq 0}$ were perfectly correlated ($|\rho| = 1$)?

13.4 An investor can choose among a riskless asset that earns a continuously compounded return at constant rate r and n risky assets with simple rates of return $\mathbf{R}_T \equiv (R_{1T}, R_{2T}, ..., R_{nT})'$ over $[0, T]$. Assume that $E\mathbf{R}_T = \mu T \equiv (\mu_1 T, \mu_2 T, ..., \mu_n T)'$; that

$$V\mathbf{R}_T = \Sigma T \equiv \begin{pmatrix} \sigma_1^2 T & \sigma_{12} T & \cdots & \sigma_{1n} T \\ \sigma_{21} T & \sigma_2^2 T & \cdots & \sigma_{2n} T \\ \vdots & \vdots & \ddots & \vdots \\ \sigma_{n1} T & \sigma_{n2} T & \cdots & \sigma_n^2 T \end{pmatrix},$$

where matrix Σ is nonsingular; and that $E|R_{jT}|^3 = o(T)$ as $T \to 0$ for each $j \in \{1, 2, ..., n\}$. (Thus, moments higher than the second approach zero faster than do the means and variances as the holding period becomes shorter.) Investing proportions $\mathbf{p} = (p_1, p_2, ..., p_n)'$ of initial wealth W_0 in the risky assets and proportion $p_0 = 1 - \sum_{j=1}^n p_j \equiv 1 - \mathbf{p}'\mathbf{1}$ in the safe one, the investor's wealth at T will be

$$\begin{aligned} W_T &= W_0 (1 + R_{\mathbf{p}T}) \\ &= W_0 \left[1 + p_0 (e^{rT} - 1) + \mathbf{p}'\mathbf{R}_T \right] \\ &= W_0 \left\{ e^{rT} + \mathbf{p}' \left[\mathbf{R}_T - (e^{rT} - 1)\mathbf{1} \right] \right\}. \end{aligned}$$

(a) Show that $E \ln W_T = \ln W_0 + rT + \mathbf{p}'(\mu - r\mathbf{1})T - \mathbf{p}'\Sigma\mathbf{p}T/2 + o(T)$.

(b) Show that the growth-optimal portfolio can be expressed as $\mathbf{p} = \Sigma^{-1}(\mu - r\mathbf{1}) + o(1)$, where $o(1) \to 0$ as $T \to 0$.

(c) Compare this short-holding-period approximation with the solutions to Exercise 13.3 as of $t = 0$.

13.5 Show that the Sharpe ratio of the growth-optimal portfolio in the n-asset, continuous-time setting is $\theta = \sqrt{(\mu - r\mathbf{1})' \Sigma^{-1} (\mu - r\mathbf{1})}$.

13.6 A European-style call option on a share of stock gives the holder the right to buy the stock at a predetermined "strike" price at a future date T, at which time the option expires and confers no further rights. When the price of the stock evolves as $dS_t = \mu S_t \cdot dt + \sigma S_t \cdot dW_t$ and the short rate of interest is known out to T, the unique current price of such an option in a frictionless, arbitrage-free market is given by the famous Black–Scholes–Merton formula (17.9), which is derived in Chapter 17. The formula expresses the price at time t as a function $C(S_t, T - t)$ of the current price of the stock and the option's remaining life, with the option's strike, the short rate, and the stock's volatility as parameters. Since $C(S_t, T - t)$ satisfies the smoothness requirements for Itô's formula, the option's "instantaneous" change in value can be represented by the stochastic differential $dC(S_t, T - t) = -C_{T-t} \cdot dt + C_S \cdot dS_t + C_{SS} \cdot d\langle S\rangle_t /2$, where the subscripts indicate partial derivatives (also depending on S_t and $T - t$) and where

$$-C_{T-t} = \frac{\partial C(S_t, T - t)}{\partial (T - t)} \cdot \frac{\partial (T - t)}{\partial t} = C_t.$$

Suppose that an investor who can rebalance continuously desires to create a self-financing, growth-optimal portfolio of call options and the money-market fund, as $V_t = c_t C(S_t, T - t) + m_t M_t$, where $dM_t = M_t r \cdot dt$. Express the appropriate values of c_t and m_t. What would happen if the investor desired to include the underlying stock as a third component of the portfolio?

EMPIRICAL PROJECT 5

This project involves simulating the performances of various portfolio selection strategies. You will determine in each of a sequence of $n = 336$ months the portfolio that maximizes the empirically estimated expected value of each of several constant-relative-risk-aversion (CRRA) utilities, calculate the total return from each such portfolio over the following month, and then determine the average continuously compounded rate of return of each strategy after n months. Expected utility will be estimated empirically using the appropriate mean–variance approximation, with sample means and variances calculated from the preceding 36 months of data.

Here are the specifics. The file GEPG1977-2007.xls in the Project 5 folder at the FTP site contains a monthly record of annualized percentage interest rates on one-month certificates of deposit (CDs), plus split- and dividend-adjusted prices of General Electric (GE) and Proctor and Gamble (PG) common stock. The interest rates and prices are quoted as of the beginning of each month, starting in January 1977 and ending in January 2008. Thus, there are 373 entries for each asset. The CDs will be regarded as the riskless asset in the portfolio problem. For this project we will measure time in months and regard $t = 1$ as corresponding to the beginning of January 1977, $t = 2$ as the beginning of February 1977, and so on up to $t = 372$ for the beginning of December 2007. Now proceed to do the following.

1. Convert the annualized percentage interest rate $(\%)_t$ of the CD at each $t \in \{1, 2, ..., 372\}$ to the simple rate of return over the month beginning at t, as $r_t = (\%)_t /1200$. Note that the CD rate for the month *beginning* at t would have been known at time t.

2. For each $t \in \{1, 2, \ldots, 372\}$ calculate the one-month *excess* rate of return on GE and PG for the month beginning at t, as $\dot{R}_t^{\text{GE}} = P_{t+1}^{\text{GE}}/P_t^{\text{GE}} - 1 - r_t$ and $\dot{R}_t^{\text{PG}} = P_{t+1}^{\text{PG}}/P_t^{\text{PG}} - 1 - r_t$. Note that these rates would *not* have been known at t since they depend on the prices at $t + 1$.

3. Starting at $t = 37$, calculate the sample means and the unbiased sample variances and covariance of the excess monthly rates of return of the two stocks using the data for the preceding 36 months. This will produce the moments needed for the portfolio decision to be made at $t = 37$, as

$$\frac{1}{36} \sum_{t=1}^{36} \left(\dot{R}_t^{\text{GE}}, \dot{R}_t^{\text{PG}} \right)'$$

for the mean vector $\hat{\mu}_{37} = (\hat{\mu}_{37}^{\text{GE}}, \hat{\mu}_{37}^{\text{PG}})'$ of *excess* rates of return and

$$\frac{1}{35} \sum_{t=1}^{36} \left[\begin{array}{cc} \left(\dot{R}_t^{\text{GE}} - \hat{\mu}_{37}^{\text{GE}} \right)^2 & \left(\dot{R}_t^{\text{GE}} - \hat{\mu}_{37}^{\text{GE}} \right) \left(\dot{R}_t^{\text{PG}} - \hat{\mu}_{37}^{\text{PG}} \right) \\ \left(\dot{R}_t^{\text{GE}} - \hat{\mu}_{37}^{\text{GE}} \right) \left(\dot{R}_t^{\text{PG}} - \hat{\mu}_{37}^{\text{PG}} \right) & \left(\dot{R}_t^{\text{PG}} - \hat{\mu}_{37}^{\text{PG}} \right)^2 \end{array} \right]$$

for their covariance matrix, $\hat{\Sigma}_{37}$. (Using 36 months for estimation is arbitrary, but choosing some such fixed number of periods as we advance the simulation from month to month puts each month's results on an equal footing.) Let $\mathbf{p}_t = \left(p_t^{\text{GE}}, p_t^{\text{PG}} \right)'$ represent the portfolio shares to be held at the beginning of month t and $R_{\mathbf{p}_t, t} = r_t + p_t^{\text{GE}} \dot{R}_t^{\text{GE}} + p_t^{\text{PG}} \dot{R}_t^{\text{PG}}$ represent the rate of return on portfolio $\mathbf{p}_t$ over the month commencing at t.

4. We will consider four CRRA utility functions:

$$u(W_t) = \ln W_{t-1} + \ln (1 + R_{\mathbf{p}_t, t}),$$

which has relative risk aversion $\rho_r = 1$, and

$$u(W_t) = \frac{W_{t-1}^{1-\rho_r} (1 + R_{\mathbf{p}_t, t})^{1-\rho_r}}{1 - \rho_r}, \rho_r \in \{.5, 2, 4\}.$$

Choosing portfolios $\mathbf{p}_t (\rho_r)$ that maximize the expected values of these utilities will constitute four different portfolio strategies. Now, for each such strategy— each $\rho_r \in \{.5, 1, 2, 4\}$—solve for $\mathbf{p}_{37} (\rho_r) = \left[p_{37}^{\text{GE}} (\rho_r), p_{37}^{\text{PG}} (\rho_r) \right]'$, which are the portfolio shares at $t = 37$. For this, recall from Section 7.4 that the mean–variance estimate of the expected utility of total return over the month is proportional to

$$E (1 + R_{\mathbf{p}_{37}, 37}) - \frac{\rho_r}{2} V R_{\mathbf{p}_{37}, 37} = 1 + r_{37} + \mathbf{p}_{37}' \mu_{37} - \frac{\rho_r}{2} \mathbf{p}_{37}' \Sigma_{37} \mathbf{p}_{37}.$$

Expressed in terms of estimated values $\hat{\mu}_{37}$ and $\hat{\Sigma}_{37}$, the *provisional* optimal portfolio for risk aversion ρ_r at $t = 37$ is

$$\mathbf{p}_{37} (\rho_r) = \rho_r^{-1} \hat{\Sigma}_{37}^{-1} \hat{\mu}_{37}. \tag{13.5}$$

However, since we have imposed no constraints on $\mathbf{p}_{37}(\rho_r)$, nothing guarantees that $1 + R_{\mathbf{p}_{37},37}$ will be nonnegative for all realizations of R_{37}^{GE} and R_{37}^{PG}, as it must be if the values of utility are always to exist as finite real numbers. Indeed, if either portfolio share were negative or if their sum were greater than unity, a sufficiently large price change in the succeeding period could lead to insolvency (negative wealth). To eliminate this possibility, use the constrained proportions

$$\tilde{p}_{37}^{GE}(\rho_r) = \max\left[p_{37}^{GE}(\rho_r),0\right]$$
$$\tilde{p}_{37}^{PG}(\rho_r) = \max\left[p_{37}^{PG}(\rho_r),0\right],$$

where $p_{37}^{GE}(\rho_r)$ and $p_{37}^{PG}(\rho_r)$ are the provisional values from (13.5). Further, if $\tilde{p}_{37}^{GE}(\rho_r) + \tilde{p}_{37}^{PG}(\rho_r) > 1$, replace each term by its value relative to the sum.

5. Now calculate the realized total return per dollar during the month beginning at $t = 37$ from each strategy $\rho_r \in \{.5, 1, 2, 4\}$, as

$$1 + R_{\tilde{\mathbf{p}}_{37}(\rho_r),37} = 1 + r_{37} + \tilde{p}_{37}^{GE}(\rho_r)\dot{R}_{37}^{GE} + \tilde{p}_{37}^{PG}(\rho_r)\dot{R}_{37}^{PG}.$$

6. Proceed in the same way for months beginning at $t = 38, 39, ..., 372$; that is, estimate sample mean vector μ_t and sample covariance matrix Σ_t from the previous 36 months of data, find the constrained optimal portfolio $\tilde{\mathbf{p}}_t(\rho_r)$ for each of the four levels of risk aversion, and calculate the realized total return of each portfolio over the succeeding month, $1 + R_{\tilde{\mathbf{p}}_t(\rho_r),t}$. Once this has been done for all 336 months from $t = 37$ to $t = 372$, find the continuously compounded *annualized* average rates of return, as

$$\bar{R}_{\tilde{\mathbf{p}}(\rho_r)} \equiv \frac{12}{336} \sum_{t=37}^{372} \ln\left[1 + R_{\tilde{\mathbf{p}}_t(\rho_r),t}\right].$$

The results should approximate those in the $q = 1.00$ row of Table 13.1 (the relevance of q will be explained), which also shows the average rates for each of the three individual assets. The results for $\rho_r = .5$ and $\rho_r = 1$ are essentially the same because the portfolio weights were in both cases usually bound by the constraints. Clearly, for this period, these assets, and this way of estimating the moments, the "optimal" strategies had nothing to offer relative to buying and holding either GE or PG. Part of the problem here is that the moment estimators, particularly of the means, are sensitive to recent fluctuations in returns and highly volatile. As we saw in Chapter 10, recent returns give poor indications of future *levels*, although they do help to forecast volatility. In any case, the effect of the noisy estimates of μ^{GE} and μ^{PG} is that the portfolio weights $\tilde{p}_t^{GE}$ and $\tilde{p}_t^{PG}$ tend to bounce back and forth between zero and unity. An alternative approach is to dampen the estimates of means at each t by bending them back toward the contemporary CD rate, as

$$\hat{\mu}_t(q) = q\hat{\mu}_t + (1 - q)r_t.$$

Table 13.1   Average rates of return of portfolios for four relative risk aversions and for individual assets

q	$\rho_r = .5$	$\rho_r = 1$	$\rho_r = 2$	$\rho_r = 4$	CD	GE	PG
1.00	.1470	.1470	.1427	.1328	.0638	**.1532**	.1513
0.75	.1448	.1464	.1440	.1328	.0638	**.1532**	.1513
0.50	.1515	.1520	.1516	.1340	.0638	**.1532**	.1513
0.25	.1585	**.1604**	.1584	.1369	.0638	.1532	.1513
0.10	.1597	**.1614**	.1580	.1357	.0638	.1532	.1513

The results for several values of q are shown in the remaining rows of the table; the top performance in each row appears in boldface. Although the results begin to look promising, one has to be cautious about the perils of "data mining;" that is, one could most likely find *some* way of estimating moments from historical data that made the optimized portfolios turn out better than buy and hold *during this period* and *for these particular securities*. As for any such experimental results, confidence can come only through *replication, replication, replication*.

CHAPTER 14

DYNAMIC MODELS FOR PRICES

We have modeled individuals as choosing optimal portfolios and levels of consumption given their individual levels of wealth, given market prices, and given their beliefs about assets' future values. Thus, the optimization problem for an individual implicitly determines the individual's demand *function* for assets, that is, the quantity to be held at each level of price. In equilibrium, the market price of any asset must be such that the aggregate of individuals' demands equals total market supply—the number of shares outstanding. By aggregating the demand functions for individuals, we thus obtain a model for the market-clearing price, given individuals' beliefs, tastes, and current endowments. By imposing *rational* expectations (i.e., assuming that individuals have worked out over time the true joint distribution of future prices), we can get a dynamic model that shows how prices behave over time as information evolves. From this exercise we can begin to learn about the connection between the informational efficiency of markets and the dynamics of prices; for example, whether the rationality of expectations is enough to make (discounted) prices martingales with respect to evolving information.

The first three sections of this chapter develop one such basic dynamic pricing model as a representative of (and introduction to) the class, then examine its implications for the cross section of returns and their behavior over time. Such models have been widely applied also to explain macroeconomic phenomena, such as general movements in the stock and money markets and their connection with aggregate consumption. The final section assesses the success of these models in such "macrofinance" applications.

14.1 DYNAMIC OPTIMIZATION (AGAIN)

Consider first the optimization problem in discrete time faced by some representative individual whose wealth at t is W_t and who must at that time choose current consumption and an investment portfolio. Available assets consist of n risky stocks priced at $\{S_{jt}\}_{j=1}^n$ and a default-free, one-period discount bond that returns \$1 at $t + 1$ and is now worth $B_t \equiv B(t, t+1) = e^{-r}$. (To simplify notation we just measure time in "holding periods" of unspecified length.) Current wealth must be allocated among consumption, stock, and bonds as $W_t = c_t + \sum_{j=1}^n s_{jt} S_{jt} + b_t B_t$, where $\{s_{jt}\}$ and b_t are the numbers of shares and units of bonds. At $t + 1$ the bonds mature and wealth is $W_{t+1} = \sum_{j=1}^n s_{jt}(S_{j,t+1} + D_{j,t+1}) + b_t$, where $\{D_{j,t+1}\}$ are the dividends paid during $(t, t+1]$. To get rid of one decision variable, apply the budget constraint to express b_t in terms of current wealth, consumption, and stock investment as $b_t = B_t^{-1} \left(W_t - c_t - \sum_{j=1}^n s_{jt} S_{jt} \right)$. This gives

$$W_{t+1} = \sum_{j=1}^n s_{jt} \left(S_{j,t+1} + D_{t+1} - B_t^{-1} S_{jt} \right) + B_t^{-1} \left(W_t - c_t \right).$$

Now adopt an additively separable form for lifetime utility

$$u\left(c_1, c_2, ..., c_T\right) = \sum_{t=1}^T U_t\left(c_t\right),$$

where U_t is increasing and concave for each t. Recall that the solution of the optimization problem at $t \in \{0, 1, ..., T-1\}$ can be expressed in terms of "value" functions that indicate the optimal levels of expected lifetime utility that can be achieved at a given time, given past decisions and the resources then available:

$$v_t\left(c_0, ..., c_{t-1}, W_t\right) = \max_{c_t, s_t} E_t v_{t+1}\left(c_0, ..., c_{t-1}, c_t, W_{t+1}\right). \tag{14.1}$$

(As usual, $E_t\left(\cdot\right) \equiv E\left(\cdot \mid \mathcal{F}_t\right)$ represents expectation conditional on time t information.) Now make use of the additive separability of lifetime utility to write

$$v_{t+1}\left(c_0, ..., c_{t-1}, c_t, W_{t+1}\right) = \sum_{\tau=0}^{t-1} U_\tau\left(c_\tau\right) + U_t\left(c_t\right) + J_{t+1}\left(W_{t+1}\right),$$

where J_{t+1} represents the maximum of expected utility of consumption from $t + 1$ *forward*, given the resources then available. Then, using a notation that dispenses with past consumptions, (14.1) becomes

$$J_t\left(W_t\right) = \max_{c_t, s_t} U_t\left(c_t\right) + E_t J_{t+1}\left(W_{t+1}\right). \tag{14.2}$$

Note for later reference that if we differentiate both sides of (14.2) with respect to W_t we get

$$J_t'(W_t) = E_t \left[J_{t+1}'(W_{t+1}) \frac{dW_{t+1}}{dW_t} \right] = B_t^{-1} E_t J_{t+1}'(W_{t+1}). \qquad (14.3)$$

Now differentiating the right side of (14.2) with respect to c_t and $\{s_{jt}\}_{j=1}^n$ and equating to zero give the $n + 1$ first-order conditions for a maximum of lifetime expected utility from t forward:

$$0 = U_t'(c_t) - B_t^{-1} E_t J_{t+1}'(W_{t+1}) \qquad (14.4)$$

$$0 = E_t J_{t+1}'(W_{t+1}) \left(S_{j,t+1} + D_{j,t+1} - B_t^{-1} S_{jt} \right), j \in \{1, 2, ..., n\}. \qquad (14.5)$$

Using (14.3), the first equation gives $U_t'(c_t) = J_t'(W_t)$ for each t, and so $U_{t+1}'(c_{t+1}) = J_{t+1}'(W_{t+1})$.

We now take these results in two directions. Although the main objective is to learn about the dynamics of prices, we shall first take a short detour to a target of opportunity and show how the framework described above can generate the venerable capital asset pricing model (CAPM) in a one-period setting.

14.2 STATIC IMPLICATIONS: THE CAPITAL ASSET PRICING MODEL

To simplify notation, let us assume that the market consists of a single representative individual, that the current date is time 0, and that time t is the end of the individual's holding period. The arguments will involve approximations that become increasingly accurate as $t \to 0$ and we approach the continuous-time solution. Putting $U_0'(c_0) = J_0'(W_0)$ in (14.4) and doing a little algebra give

$$\begin{aligned} B_0 &= E_0 \frac{J_t'(W_t)}{J_0'(W_0)} \\ &= E_0 \frac{J_0'(W_0) + J_0''(W_t^*)(W_t - W_0)}{J_0'(W_0)} \\ &= 1 + E_0 \left[\frac{J_0''(W_t^*)}{J_0'(W_0)} (W_t - W_0) \right] \end{aligned} \qquad (14.6)$$

for W_t^* between W_0 and W_t. Now, since our individual represents the entire market, W_0 must equal the value of the "market" portfolio that is featured in the CAPM. Thus, $W_t = W_0 (1 + R_{mt})$, where R_{mt} is the rate of return on the market portfolio between 0 and t. Moreover, when the period is short, we have $W_t^* \doteq W_0$, so Eq. (14.6) implies

$$B_0 \doteq 1 - \rho_r E_0 R_{mt},$$

where $\rho_r = -W_0 J_0''(W_0) / J_0'(W_0)$ is the coefficient of relative risk aversion derived from the optimal value function. Now work with (14.5):

$$\begin{aligned} 0 &= E_0 J_t'(W_t) \left(S_{jt} + D_{jt} - B_0^{-1} S_{j0} \right) \\ &= E_0 J_t'(W_t) \left(S_{jt} + D_{jt} \right) - B_0^{-1} S_{j0} E_0 J_t'(W_t) \\ &= E_0 J_t'(W_t) \left(S_{jt} + D_{jt} \right) - S_{j0} J_0'(W_0) \end{aligned}$$

using (14.3). Thus, with $1 + R_{jt} = (S_{jt} + D_{jt})/S_{j0}$ as the stock's total return during $[0, t]$, we have

$$
\begin{aligned}
1 &= E_0 \frac{J_t'(W_t)}{J_0'(W_0)} (1 + R_{jt}) \\
&= E_0 (1 - \rho_r R_{mt})(1 + R_{jt}) \\
&= E_0 (1 - \rho_r R_{mt}) \cdot E_0 (1 + R_{jt}) - \rho_r C(R_{jt}, R_{mt}) \\
&= B_0 (1 + E_0 R_{jt}) - \rho_r C(R_{jt}, R_{mt}),
\end{aligned}
$$

and thus

$$
1 + E_0 R_{jt} = B_0^{-1} [1 + \rho_r C(R_{jt}, R_{mt})]. \tag{14.7}
$$

Now form a value-weighted average across stocks to get a corresponding expression for the market portfolio. With $p_{mj} = s_{j0} S_{j0}/W_0$ as the jth asset's share of the market portfolio and $\sum_{j=1}^{n} p_{mj} = 1$, we have

$$
\begin{aligned}
1 + E_0 R_{mt} &= \sum_{j=1}^{n} p_{mj} (1 + E_0 R_{jt}) \\
&= B_0^{-1} \left[1 + \rho_r \sum_{j=1}^{n} p_{mj} C(R_{jt}, R_{mt}) \right] \\
&= B_0^{-1} [1 + \rho_r C(R_{mt}, R_{mt})] \\
&= B_0^{-1} [1 + \rho_r V R_{mt}].
\end{aligned}
$$

This gives an equilibrium relation between risk-aversion coefficient ρ_r and the first two moments of R_{mt}; namely, $\rho_r = [B_0 (1 + E_0 R_{mt}) - 1]/V R_{mt}$. Plugging into (14.7), setting $B_0^{-1} = e^{rt} \doteq 1 + rt$, and simplifying give the CAPM's famous *security market line* that relates assets' expected rates of return over $[0, t]$ to their betas, that is,

$$
\mu_{jt} = rt + (\mu_{mt} - rt)\beta_j,
$$

where $\mu_{jt} = E_0 R_{jt}$, $\mu_{mt} = E_0 R_{mt}$, and $\beta_j = C(R_{jt}, R_{mt})/V R_{mt}$. This is precisely the expression (6.5) that was derived from the purely static mean–variance model with $t = 1$. However, one must bear in mind that the result was obtained here as an approximation that applies when the holding period is short. In effect, what we have seen implies that the CAPM holds in continuous time in this general-equilibrium setting.

14.3   DYNAMIC IMPLICATIONS: THE LUCAS MODEL

Now we can return to (14.4) and (14.5) and obtain a truly dynamic model with implications for how price evolves over time. What follows is a simple version of an insightful general-equilibrium model by Robert Lucas (1978). To get the main idea with less notation we can specialize to the case of a single risky asset and start with the following version of (14.4) and (14.5):

$$0 = U_t'(c_t) - B_t^{-1} E_t J_{t+1}'(W_{t+1}) \tag{14.8}$$

$$0 = E_t J_{t+1}'(W_{t+1})(S_{t+1} + D_{t+1} - B_t^{-1} S_t). \tag{14.9}$$

The "stock" now is to be regarded as the general economic "factory" that produces a universally consumable good, the only one that our representative individual consumes. The "dividend" is the exogenously produced output of this factory, which follows some (Markovian) stochastic process, in which the *current* value is the only component of current information that is relevant for predicting the future, that is

$$\Pr(D_{t+1} \le d \mid \mathcal{F}_t) = \Pr(D_{t+1} \le d \mid D_t)$$

for all d. In equilibrium the dividend (which is perishable, supposedly) must be either consumed or discarded, and since our consumer likes to eat, we must have $c_t = D_t$ in each period. Also in equilibrium there is no net borrowing (just as there is no *net* borrowing in the real global economy), and the single "share" of the stock must be willingly held at the equilibrium price. Finally, in a true equilibrium our representative "rational" agent will have the dynamics of $\{D_t\}$ all figured out; that is, at each t his subjective distribution of D_{t+1} given information $\mathcal{F}_t$ will coincide with the true distribution. With this setup we can get a feel for both how price would be determined and how it would evolve over time.

To proceed, recall that (14.3) together with (14.8) give $U_t'(c_t) = J_t'(W_t)$ for each t, and so (14.9) becomes

$$\begin{aligned} 0 &= E_t U_{t+1}'(c_{t+1})(S_{t+1} + D_{t+1}) - B_t^{-1} S_t E_t U_{t+1}'(c_{t+1}) \\ &= E_t U_{t+1}'(c_{t+1})(S_{t+1} + D_{t+1}) - S_t U_t'(c_t), \end{aligned}$$

so that

$$S_t = E_t \left[\frac{U_{t+1}'(c_{t+1})}{U_t'(c_t)} \right] (S_{t+1} + D_{t+1}). \tag{14.10}$$

The bracketed quantity, known as the *stochastic discount factor* or *pricing kernel*, connects current price with future payoffs. It is stochastic in general because c_{t+1} is not known at t, but we can think of it as governing the relative value of consumption now and consumption later. It is, in fact, the marginal rate of substitution of future consumption for current consumption.

As we now explain, the intuition behind expression (14.10) is pretty compelling, although it makes more sense when there is more than one asset. Note first that concavity of the utility function implies that marginal utility declines with consumption. Now compare two different potential "factories." One of them produces especially high output in states of the world when consumption is also exceptionally high and low output when consumption is unusually low. Such a factory is not very useful to own—it just doesn't deliver when one really needs it. More valuable will be a factory that produces high output when there are poor prospects for consumption.

Thus, when consumption opportunities are positively correlated with factory output, the effect of the discount factor is to lower the current value of the factory. Of course, when there is just one asset, as was assumed here to simplify the argument, consumption is *necessarily* positively correlated with the factory's output, since they are one and the same. With n separate factories, however, consumption in equilibrium would equal the total of their outputs, and the output of any particular factory might well be negatively correlated with the total. The implication in that case is really the same idea as in the CAPM, where stocks whose returns are highly positively correlated with the market portfolio need to have higher expected returns—and therefore lower prices, all else equal—than do stocks whose returns are less highly correlated.

Two further points should be made about (14.10) in the n-factory setting. To emphasize that there is a multiplicity of assets, we express the pricing equation in two equivalent ways as

$$S_{jt} = E_t \delta_{t,t+1} S^+_{j,t+1} \tag{14.11}$$

$$1 = E_t \delta_{t,t+1} (1 + R_{j,t+1}) \tag{14.12}$$

for $j \in \{1, 2, ..., n\}$, where $\delta_{t,t+1}$ represents the discount factor, $S^+_{j,t+1} \equiv S_{j,t+1} + D_{j,t+1}$ is the dividend-adjusted price, and $R_{j,t+1}$ is asset j's one-period rate of return. The first point is that one and the same stochastic discount factor applies to each asset. Thus, if there are n assets, the joint conditional distribution of the n risky returns and the single discount factor provides all the information that is relevant for pricing everything in the market. A second point is related and leads to a new interpretation of (14.11) that will be of great importance in the theory of pricing derivative assets. Since (14.11) applies to all assets, it also applies to default-free discount bonds. Thus, if $B_t \equiv B(t, t+1)$ is the time t price of such a bond, then $B_{t+1} = 1$ and so $E_t \delta_{t,t+1} = B_t$. Thus, the expected value of the stochastic discount factor is the appropriate risk-free discount factor for *sure* $t+1$ claims. To see where this leads, put $\rho_{t,t+1} \equiv \delta_{t,t+1}/B_t$, write (14.11) as $S_{jt} = B_t E_t \rho_{t,t+1} S^+_{j,t+1}$, and note that $\rho_{t,t+1} > 0$ and $E_t \rho_{t,t+1} = 1$. We will see in Section 19.5 that taking the expectation of $\rho_{t,t+1} S^+_{j,t+1}$ amounts to taking the expectation of $S^+_{j,t+1}$ in a new probability measure. Thus, the asset's current equilibrium price is its expected future value in this new measure, discounted at the riskless rate of interest.

Returning to the one-asset setting, let us now recognize that the dividend of the sole factory is the only determinant of consumption. Thus, (14.10) really becomes

$$S_t = E_t \left[\frac{U'_{t+1}(D_{t+1})}{U'_t(D_t)} \right] (S_{t+1} + D_{t+1}). \tag{14.13}$$

This idealized setup gives us an insight into how the stock's price behaves over time. Since dividends are the only driving force in this model, and since the current dividend contains all the information there is about future dividends, the stock's price at any t *must* be a function of the current dividend; that is, there must be some *pricing function* s such that $S_t = s(D_t)$. By specifying utility and the dividend process, it is possible to deduce the form of this function. The simplest case is that utility is actually linear, so that the representative individual is risk neutral. In particular, suppose $U_t(D_t) = \delta^t D_t$

for some $\delta \in (0, 1)$. Then $U'_t = \delta^t$—in which case the "stochastic" discount factor is no longer stochastic—and so the model implies that $S_t = \delta E_t (S_{t+1} + D_{t+1})$. Now model dividends as $D_{t+1} = \theta D_t \varepsilon_{t+1}$ where $D_0 > 0$, $\theta > 0$, and the $\{\varepsilon_t\}$ obey $\varepsilon_t > 0$ and $E_t \varepsilon_{t+1} = 1$. Taking as a trial function $s(D_t) = qD_t$, we have

$$qD_t = \delta E_t (qD_{t+1} + D_{t+1}) = \delta (q + 1) \theta D_t.$$

Thus, the trial function with $q = \delta\theta / (1 - \delta\theta)$ does solve the equation, assuming $\delta\theta \neq 1$. In this case it is easy to see that process $\{S_t / \theta^t\}_{t=0,1,2,...}$ is a martingale adapted to evolving information (the filtration) $\{\mathcal{F}_t\}_{t=0,1,2,...}$. However, with nonlinear utility and a discount factor that really is stochastic, no simple rescaled version of price *need* have this martingale property, even though the representative individual is rational and fully informed about how the world works. For an example of a stochastic discount factor that does yield martingales, see Exercise 14.4.

14.4 ASSESSMENT

The Lucas model has been presented here as representative of a class of dynamic general-equilibrium models, to which there have been many contributions. What these models imply about the cross section of returns is both intuitive and appealing. The models have also served to show us that prices *can* have some first-moment predictability even in markets with rational, fully informed agents. Still, we must remember to judge models not by how they make us feel or for showing what is *possible*, but by how well they explain what we actually see. In most of the economics literature the focus has been on what the models imply at the macro level; specifically, about the behavior of the stock market as a whole and its relation to aggregate consumption and interest rates. Unfortunately, in such applications the standard rational-expectations, consumption-based, representative-agent models earn a grade somewhere between D and F. The incongruities between predictions and data—usually referred to in the economics/macrofinance literature as the asset-pricing "puzzles"—can be characterized through a sequence of interconnected questions:

- Why, over long periods, have stocks yielded such high average returns relative to default-free bonds?

- Why have riskless interest rates been so low?

- Why have prices of stocks been so volatile?

- Why have movements in stocks' prices been so loosely connected to movements in aggregate consumption?

We will first look at why the empirical observations seem at odds with the standard pricing models, next consider briefly some of the efforts to bring predictions into line, and then conclude with some general reflections on the entire enterprise.[37]

---

[37] For a detailed analysis of these issues and pointers to the extensive literature, see Cochrane (2001) and Campbell (2003).

14.4.1 The Puzzles

Focusing specifically on common stocks (equities), consider some broad portfolio of stocks, such as those traded on the NYSE or the 500 stocks in the S&P index. Multiplying both sides of (14.12) by p_{mj} (asset j's share of the value of this broad-market portfolio) and summing over j give a corresponding relation for the portfolio itself: $1 = E_t \delta_{t,t+1} (1 + R_{m,t+1})$. Robert Shiller (1982) and, in a more general setting, Lars Hansen and Ravi Jagannathan (1991) showed that this simple relation implies a rough upper bound on the portfolio's Sharpe ratio, $(ER_m - r)/\sqrt{VR_m}$. To see it, just average the conditional expectation $E_t \delta_{t,t+1} (1 + R_{m,t+1})$ over information sets to get a corresponding relation involving unconditional moments, then (dispensing with time notation) express the expectation of the product as the product of the means plus the covariance:

$$1 = E\delta (1 + R_m) = E\delta E (1 + R_m) + C (\delta, R_m).$$

Putting $E\delta = (1 + r)^{-1}$ (with r as the representative one-period simple interest rate implied by average one-period bond prices) and simplifying give $ER_m - r = -C (\delta, R_m) / E\delta$. The Schwarz inequality states that $C (\delta, R_m)^2 \leq V\delta \cdot VR_m$, and so it follows that

$$\frac{ER_m - r}{\sqrt{VR_m}} \leq \frac{|ER_m - r|}{\sqrt{VR_m}} \leq \frac{\sqrt{V\delta}}{E\delta}.$$

Thinking of the Sharpe ratio on the left as measuring an *equity premium*, we see that this is more or less bounded by the temporal coefficient of variation of the discount factor, $\delta_{t,t+1}$. Thus, if the equity premium has been historically high, this can be explained in the context of the model only if the volatility of the discount factor is high or if the mean is low. High volatility in the ratio of marginal utilities of aggregate consumption $U'_{t+1}(c_{t+1})/U'_t (c_t)$ would require either a lot of period-to-period variation in aggregate consumption itself, or else that the "representative" individual's marginal utility be quite sensitive to such variation in consumption— and since high sensitivity of U'_t means high values of $|dU'_t (c_t) /dc_t| = |U''_t (c_t)|$, this equates to high aversion to risk. Of course, a low value of $E\delta$ corresponds to low prices of discount, default-free bonds and, consequently, high risk-free rates of interest.

What do the numbers actually tell us? In fact, empirical estimates of the equity premium vary a lot depending on the sample period and the particular portfolio of assets being considered, but with post-1950 annual data for broad-based indexes of US common stocks we often see numbers on the order of .2–.5. John Campbell (2003) finds similar results in data for other developed countries. On the other hand, aggregate consumption in the developed economies has been relatively smooth over this period. This means that very high degrees of risk aversion would be needed to account for the volatility of stock prices. Keeping the bound in the 0.2 to 0.5 range with CRRA utility $U (c) = (c^{1-\gamma} - 1) / (1 - \gamma)$ would require values of relative risk aversion (γ) on the order of 20–50.[38] With more reasonable values of γ moving the bound high enough to match the data would require low values of $E\delta$, corresponding to a high rate of time preference and, accordingly, real interest rates that are much higher than we have observed.

As unaccountably high as the equity premium has been, it would have been more so (all else equal) had prices of stocks not been so volatile. Thinking of the pricing function $S_t = s\,(D_t)$ in the Lucas model, and regarding aggregate consumption c_t as the economy's dividend shows that volatility of prices should be proportional to that of consumption, which we know to have been relatively stable. Of course, the aggregate value of stocks is not the same as the market value of total wealth (as the Lucas model represents it), and so a direct comparison of the two volatilities is not really apt. But casting the model in terms of actual dividends and historical price–dividend ratios still does not account for the wild swings we see in the stock market.

Finally, there is the issue of the degree of comovement between stock returns and consumption. The compelling intuition that the consumption-based model brings to us is that assets should be priced according to how well or how poorly they hedge against variation in consumption. For this to be plausible as an explanation for the high cross-sectional variation in average returns, we should expect to see some significant actual correlation between returns and consumption in the time series. However, even allowing for the difficulty of measuring consumption, this correlation seems very weak. In quarterly US data for most of the period after World War II, Campbell (2003) finds a correlation between excess stock returns and consumption of only about .20. Values for the other developed economies are generally lower, and a few are even negative.

14.4.2 The Patches

There have been many attempts to modify the basic consumption-based pricing model to overcome the empirical contradictions. Robert Shiller's approach, which we discuss in the next chapter, has been to focus on expectations. In his view investors are influenced by psychological factors that produce cycles of optimism and pessimism that are not warranted by realistic prospects for assets' future cash flows. The other general approach has been to modify preferences alone while retaining the view that people's beliefs are more or less in line with reality. Two examples will serve. Campbell and Cochrane (1999) propose that the utility derived from a given level of consumption depends on how that level compares with an evolving norm. As consumption rises above the norm, risk aversion declines and price–dividend ratios rise and realized returns increase. When consumption falls toward the standard, less

---

[38] As a quick reality check, think of a one-period model with $U\,(W_1) = \left(1 - W_1^{-19}\right)/19$ and an investor who chooses between a riskless bond yielding $r = .05$ per annum and a risky stock with $\sqrt{VR} = \sigma = .2$ (roughly the recent annual volatility of the S&P 500). Using the short-holding-period approximation $p = (\mu - r)\,/\,(\gamma\sigma^2)$ for the proportion of wealth invested in the stock, the individual would put as much as half of wealth at risk only if $\mu \geq 0.45$. Yet, if we approximate R as $N\,(.45, .2^2)$, the probability of losing so much as a penny would then be only about 0.006.

is put at risk in the market and prices and realized returns fall. Thus, prices swing more widely than does consumption itself, and particular features of the model ensure that the outcome yields an equity premium and low interest rates. However, since consumption still drives behavior, there is still a counterfactually high association between consumption growth and returns. Barbaris et al. (2001) use some of the ideas from Kahneman-Tversky's prospect theory to get around this problem. In their model investors have an aversion to financial loss beyond what it signifies for their command over goods. Loss is measured relative to a norm that depends on recent experience, so that losses after a series of gains have less effect on risk aversion than do losses after losses. The result, like changing risk aversion in the Campbell–Cochrane model, is to amplify the swings in prices; but now movements in prices and consumption need not be in phase since risk aversion is no longer driven by consumption. There is a high equity premium, since the losses when they occur are especially painful. On the other hand, since default-free bonds held to maturity cannot produce losses, interest rates need not be high to induce people to hold them. Finally, the swings in prices generate a degree of predictability in returns over long horizons, such as does show up in broad-based stock indexes.

14.4.3   Some Reflections

One cannot but marvel at the ingenuity, technical skill, and boldness of the contributors to the macrofinance literature. Given the vast complexity and heterogeneity of financial markets, whose participants range from Wall Street "quants," to hedge-fund managers, to the local stockbroker, and on down to septuagenarian investment clubs, is it not remarkable that a model that extrapolates from the rational—or at least consistent—behavior of one or a few classes of optimizing, fully informed "agents" could have anything worthwhile to tell us about economic aggregates? We have already seen plenty of evidence that people do not consistently make optimal decisions under uncertainty, at least not as "optimal" is interpreted under the expected-utility theory—and we will see in the next chapter that there are times when markets seem to be driven by people who simply do not face reality. That we wonder at the inadequacy of predictions based on assumptions totally at odds with these observations is a little like a group of cavemen shaking their heads at the failure of a putative moon rocket made of bamboo and vines. Just as their wives might be yelling to them "Learn to make fire first," some might think it best for financial economists to concentrate for now on getting things right within a narrower scope. Of course, by building in enough complexity, one can always fit—the current term is "calibrate"—a model to any given set of facts. But, even if we can understand its mechanics, do we get from such a model any sense that it is more than just a fanciful, clever story—a kind of creation myth that is supposed to comfort us in our ignorance? Rather than turn out more of these, what if we try instead to make fire before shooting for the moon? What could we say about prices of at least *some* assets if we start from more basic, less controversial assumptions? What could we get merely by assuming, say, that people behave as we see them do in everyday life?

One thing that we can say with some confidence about those heterogeneous folks in the market is that they like money and will do just about anything that is legal (and many things that are not) to acquire it. Another thing that we can say is that most of them are neither totally stupid nor totally unaware of what goes on around them. Thus, even though most of them flunk exams on dynamic, stochastic optimization, when they come across $20 bills on the sidewalk, they—or at least those among them who are not economists—will stop and pick them up.[39] Can we get some useful predictions without assuming more than such as this? For the answer, see Part III.

EXERCISES

14.1 With $U_t(D_t) = \delta^t D_t$ and $D_{t+1} = \theta D_t \varepsilon_{t+1}$ as models for utility and dividends in the Lucas model, as in the text, show that $\{S_t/\theta^t\}_{t=0,1,2,\ldots}$ is an $\{\mathcal{F}_t\}_{t=0,1,2,\ldots}$ martingale.

14.2 Take $U_t(D_t) = \delta^t D_t$ as in the text, but model dividends so that $E_t D_{t+1} = \phi + \theta D_t$. Find a price function $s(D_t)$ that solves the Lucas model. Under what condition on the parameters would $\{S_t\}_{t=0,1,2,\ldots}$ be an $\{\mathcal{F}_t\}_{t=0,1,2,\ldots}$ martingale?

14.3 In the Lucas model take $U_t(D_t) = (1 - D_t^{-\gamma})/\gamma$ with $\gamma > 0$, and let the dividend evolve in discrete time as $\{D_t = D_0 e^{\mu t + \sigma W_t}\}_{t=0,1,2,\ldots}$, where $\{W_t\}_{t\geq 0}$ is a Brownian motion adapted to evolving information $\{\mathcal{F}_t\}_{t\geq 0}$. Find $r(t, t+1)$, the average continuously compounded spot rate over $[t, t+1]$. What restriction on the parameters is needed to make $r(t, t+1) > 0$?

14.4 With the same setup as in Exercise 14.3, find the price function $s(D_t)$ that solves (14.13) and verify that $s'(D_t) > 0$ under the parametric restrictions obtained in that exercise. Finally, find a discount factor θ such that $\{S_t/\theta^t\}_{t=0,1,2,\ldots}$ is a martingale.

14.5 Let $\{W_t\}_{t\geq 0}$ be a standard Brownian motion under the "natural" probability measure $\mathbb{P}$ that governs real-world events. Thus, $EW_t = W_0 = 0$, and $E(W_t \mid \mathcal{F}_s) = W_s$ for $s \leq t$, so that sample paths of $\{W_t\}_{t\geq 0}$ have no persistent trend. Now, if ν is some nonzero constant, it is clear that the trending process $\{\hat{W}_t \equiv W_t + \nu t\}_{t\geq 0}$ is *not* a standard BM under $\mathbb{P}$. Nevertheless, suppose that it were possible construct a new probability measure $\hat{\mathbb{P}}$ under which $\{\hat{W}_t\}$ *does* behave as a standard BM. How could this be done? It seems that it might be done just by judiciously altering the probabilities of sample paths. For example, if $\nu > 0$, we would want $\hat{\mathbb{P}}$ to assign lower probabilities than does $\mathbb{P}$ to

---

[39]I refer, of course, to the often-told story about the economist and friend who find on the sidewalk what appears to be a $20 bill. "Nah," says the economist, "If it was really 20 bucks, somebody would have already picked it up." Be assured that this story is almost certainly apocryphal; how many economists *have* friends?

bundles of sample paths that drift upward, and higher probabilities to those that drift downward. If done in just the right way, the positive average drift would thus be removed. In fact, we shall see in Chapter 19 that such a $\hat{\mathbb{P}}$ can indeed be constructed and that such *changes of measure* have great relevance in financial modeling. Using the same setup for the Lucas model as in Exercise 14.3, and letting $\hat{\mathbb{P}}$ be the measure under which $\{W_t + \nu t\}_{t \geq 0}$ is a standard BM, find the constant ν that would give $\{S_t\}_{t=0,1,2,\dots}$ the same one-period expected total return under $\hat{\mathbb{P}}$ as that of a default-free, one-period, discount bond.

CHAPTER 15

EFFICIENT MARKETS

Thus far, all our models for prices of assets have required some degree of rationality on the part of investors. Not only must they make rational decisions given their beliefs, but their beliefs themselves must correspond to a considerable degree with reality. Thus, in both the CAPM and the dynamic model of Robert Lucas current prices are based on peoples' expectations about the future, and we must assume that these expectations are in some sense "rational" if we are to relate the prices cranked out by the models to what we actually see in markets. In other words, individuals must correctly assess the joint distribution of future returns and/or whatever "state variables" there are that drive them. Another term for this rationality of beliefs is *informational efficiency*; that is, people are supposed to be able to acquire and digest relevant information rapidly and to use it to determine its implications for the evolution of assets' prices. The standard expression of this idea—as popularized by Burton Malkiel (2007) in his famous *Random Walk Down Wall Street*—is the statement that "prices fully reflect publicly available information."

A large part of the empirical literature in finance since the mid-1960s has been devoted to testing this hypothesis. Despite the reams of paper thus expended, the question as to whether markets are truly informationally efficient remains hotly debated. Why is the question so hard to answer? One reason is that any test of market efficiency

Quantitative Finance. By T.W. Epps

is necessarily a *joint* test of that hypothesis and a particular pricing model—that is, to determine whether prices today fully reflect current information, one must know what prices *should* be today, given what is known. Moreover, besides the uncertainty about the model, we must confront difficult statistical issues. Typically, whether a given statistical procedure is appropriate depends on other unverifiable assumptions, such as the joint distribution of key variables. Even when the assumptions are met, we still have to contend with sampling error and statistical identification problems that may confound the conclusions.

As if the scientific challenge were not enough, there is yet one other difficulty: The debate about market efficiency has been driven in part by polar political ideologies, with conservatives and liberals inclined to reach opposite conclusions based on their different world views. The former group, with their commitment to free trade, free domestic markets, and private enterprise, look with suspicion on any claim that markets might sometimes reach the wrong solution in determining such a fundamental quantity as the cost of capital. Political liberals, on the other hand, from their fixation on "corporate greed" and "exploitation" of the masses, are more apt to be receptive to notions that markets can sometimes fail. Here, as with so many other issues, there needs to be a clear distinction between the issues of what is and of what ought to be. The "what is" issue is a purely scientific one that can and should be approached dispassionately. If such study does lead us to conclude that markets are sometimes wrong—and not just in hindsight but wrong given the contemporary information—then there is the separate issue of what ought to be done about it. Solutions could include disseminating more and better information to shareholders, improving managerial incentives, stiffening corporate control, and more closely regulating specialists, dealers, and the technical aspects of trading. There is also the possibility of granting government the direct authority to influence prices, in hopes that "wise leaders" might avoid the mistakes made through the collective actions of the millions of people who stake their own money on their beliefs. There will likely be no consensus about the efficiency of markets so long as there are differing views about whether this ultimate policy "solution" lies at the bottom of a regulatory slippery slope.

In this chapter we look at some of the prominent methods and findings of empirical tests of market efficiency. These can be divided between static and dynamic procedures, much as pricing models themselves have been so categorized. Tests in the former class examine the speed and nature of the market's responses to one-time events that are presumed to affect valuations, while dynamic tests draw inferences from prices' persistent behavior over time. Because the range of methods used in event studies is narrower and the findings are less controversial, we begin the story with these.

15.1   EVENT STUDIES

Event studies illustrate some of the problems of testing market efficiency, but because of their limited aims they are considered among the most credible efforts. As efficiency tests, the objective is to assess how and how quickly markets react to certain specific events affecting firms, without taking a stand on what the appropriate *level* of reaction is. Of course, if one merely *assumes* market efficiency, then event studies can also be used to infer the value of the event from the change in the firm's market value. Indeed, in legal proceedings before the US Supreme Court event studies are now accepted as a means of assessing value. Examples of events to which the method has been applied include stock splits, earnings announcements, mergers and acquisitions, and changes in management, capital structure, and regulatory policy. Typically, the reactions are gauged by the effects of the *announcements* of these events on prices of companies' common stocks and marketable debt; that is, it is the timing of the announcement itself rather than the timing of the event that is considered relevant. Generally, the findings are that markets react very quickly to such announcements; indeed, price effects often show up well in advance, suggesting that those privy to inside information do often beat the market. John Campbell, Andrew Lo, and Craig MacKinlay (1997) give a thorough discussion of the technical aspects of event studies and the relevant statistical issues. Here we describe the most common methods and summarize one influential study.

15.1.1   Methods

The general approach taken in event studies is to gauge how the realized rates of return on a firm's stock behave before and after the announcement of some event. If the event is interpreted as a favorable signal on balance, then we should observe at least on the announcement day an "abnormal" positive return—one bigger than would be expected in the absence of the announcement. If markets react quickly, the effect should be short-lived; that is, price should adjust quickly to the information and thereafter revert to its "normal" behavior. Obviously, a major challenge in doing these studies is to determine what is normal. Another challenge is to distinguish the effect of the announcement from effects of other relevant events that occur on the same day. Various efforts have been made to meet these challenges.

Of the many ways that have been proposed for estimating a normal rate of return, three are most common. The simplest estimate of the normal rate on day t is the firm's own *average* rate over some period well before the announcement—$\bar{R} = T^{-1}\sum_{\tau=1}^{T} R_{t-k-\tau}$, say. For example, one might calculate the mean daily rate of return over an estimation period of several years ending some months *before* the announcement. Allowing a space of time between estimation period and announcement is supposed to eliminate the bias caused by trading in anticipation of the event, either by those with inside information or by astute outsiders. With this approach the estimate of normal return would have the same constant value, $\hat{R}_t = \bar{R}$, for each day during the entire period of study. Another estimate used in some studies is the average rate of return on the same day of a collection of similar firms, $\hat{R}_t = n^{-1}\sum R_{jt}$. Of course, these would be firms for which the event in question did not occur at any time near the study period. Finally, perhaps the most common procedure is to judge R_t in relation to a market index, such as the S&P 500. When there is not a history of prior returns, as for firms just being taken public, the market return R_{mt} itself could be used as a crude estimate of the subject firm's normal return. When there *are* data from a prior estimation period, the usual procedure is to estimate by ordinary least-squares parameters α and β in the linear model $R_t = \alpha + \beta R_{mt} + u_t$. Here u_t is an unobserved error, and β measures the average sensitivity of R_t to R_{mt}. Given estimates $\hat{\alpha}$ and $\hat{\beta}$ from the estimation period, the normal return on day t of the study period is taken to be $\hat{R}_t = \hat{\alpha} + \hat{\beta} R_{mt}$.

Whatever estimate of normal return is used, the deviation $e_t \equiv R_t - \hat{R}_t$ is referred to as the *residual* return. In the absence of specific information, one expects this to be zero *on average*. To assess this, the residual return is recorded day by day during an "event window"—a period beginning at some point after the end of the estimation period and extending beyond the event day. The hope is that the residual will show the extent and speed of the market's reaction to the announcement. Of course, the realized residual on any given day is subject to other influences as well, and this added noise may make it difficult to discern the real effect of the event in question. There are two ways of reducing this noise. If the focus of the study is not on a specific firm but on the general effect of events of the given type, it is common to average residuals across a cohort of firms, *all of whom* at some time experience the same event. In order to get a large enough sample to be effective, one has to stagger time periods. Thus, one averages the residuals of n firms on their own specific announcement day—$t = 0$, say—and on days $t \pm 1, t \pm 2, \ldots$. When event dates of different firms do differ, it is reasonable to think that the residuals for different firms on the same *relative* date t are nearly independent. In this case some of the idiosyncratic effects will average out and emphasize the common effect of the event being studied. Thus, the average residual across firms, $\bar{e}_t \equiv n^{-1} \sum_{j=1}^{n} e_{jt}$, should be more informative than the residual of a specific firm. Of course, the average of a finite number of independent, mean–zero variables will rarely be precisely zero. As a further way of emphasizing the common effect of the event, it is also common to *cumulate* these averages over time. That is, starting at the beginning of the event window—some day $-T$ prior to the announcement day, one records $\bar{e}_{-T}, \bar{e}_{-T} + \bar{e}_{-T+1}, \bar{e}_{-T} + \bar{e}_{-T+1} + \bar{e}_{-T+2}$, and so on, giving as cumulative average residual up through day t of the study $C_t = \sum_{\tau=-T}^{t} \bar{e}_\tau$. The plot of these residuals against time often gives clear indication of the effect and timing of the announcement on the market's perception of the firm's value. Cumulating in this way is even more important when the study pertains to one single firm.

15.1.2   A Sample Study

An early but still widely cited event study is that by Eugene Fama, Lawrence Fisher, Michael Jensen, and Richard Roll (1969) to gauge the effect of split announcements. A stock split merely replaces one's holding of s shares at price S_t by fs shares at price S_t/f, where f is the split factor. Since $sS_t \equiv fsS_t/f$, one would be justified in thinking stock splits irrelevant to the value of one's shares; nevertheless, there are potentially some indirect effects. A minor one is that the cost of selling the larger number of shares may well be greater, given that fees charged by some brokers increase with the size of the trade. A more compelling consideration is that the volatility—standard deviation of rates of return—tends to rise after splits. There is no conclusive

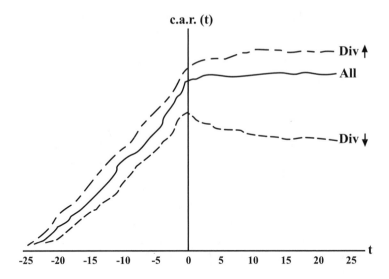

c.a.r. (t)

Div ↑

All

Div ↓

-25   -20   -15   -10   -5   0   5   10   15   20   25

t

Figure 15.1   Monthly cumulative average residuals of NYSE firms relative to announcement of stock split.

story as to why this happens, but there are a couple of possible sources. One is that bid-asked spreads are wider as proportions of post-split prices, so volatility can rise just from the effect of bid-asked bounce. Also, volatility tends to correlate with and perhaps be causally influenced by the volume of trading, and splits can bring in more small investors. One expects that the effects of transaction costs and volatility would depress a stock's price, but there may be positive effects also. It has been suggested that splits serve as credible signals of management's favorable beliefs about a firm's future prospects; that is, taking the definite action of splitting shares is supposed to be more informative than merely supplying analysts with favorable opinions. Also, split announcements are sometimes accompanied by other more tangible signals, such as announcements of changes in the dividend. The Fama et al. (1969) study was intended to see how these various considerations net out.

The researchers used *monthly* returns data for all the NYSE stocks with split factors f of at least $5 : 4$ (i.e., five new shares for every four old ones) during the period 1927–1959. There were a total of 940 such events involving 622 different firms. As described above, they first estimated normal returns each month from the regressions of each firm's monthly return on that of a market index. Then, for each month during an event window of about 2 years before and after the splits they averaged the residuals, $e_{jt} = R_{jt} - \left(\hat{\alpha}_j + \hat{\beta}_j R_{mt} \right)$, across the firms. Given that monthly data were used, the study was of little value in gauging the speed of the market's reaction, but their findings regarding the average effect of splits is very interesting. Their plots of cumulative average residuals looked much like those depicted in Fig. 15.1, which give a rough picture of how the average prices of the firms behaved before and after the split announcements. The noteworthy feature is that prices trended up strongly before the announcements (that is, return residuals were more often positive than negative) and then more or less leveled off. The cumulative average residuals for the entire cross section of 940 events show little change in the months following the split. Firms that announced increased dividends around the time of the split showed some continued "abnormal" growth thereafter, while those whose dividends decreased did worse than usual in relation to the market. The conclusions reached by Fama et al. were all about how anticipations of dividend changes influenced pre-split behavior of share prices. They attributed the pre-split rise in share prices to the market's anticipation that a split would occur *and be accompanied by* an increase in the dividend. Share prices of firms that did in fact deliver increases were sustained in the months that followed, while prices of the remaining firms tended to revert towards earlier levels. In retrospect, these researchers (and others) recognized that their findings really gave no basis for such a connection between splits and dividends. It is more likely that the firms that increased dividends after the split were those that experienced continued growth in earnings, whereas those that decreased dividends had less favorable prospects. Of course, there is an obvious explanation for the average increase in share prices *before* the splits. (How would *you* explain it?)

15.2 DYNAMIC TESTS

We now turn to tests of efficiency that deal with the persistent behavior of prices over time, rather than their response to isolated events. The history of such studies illustrates how scientific progress occurs in fits and starts, with many detours and wrong turns—or at least turns that *appear* to have been wrong given what we know now. After absorbing some uncomfortable facts about how prices actually behave in free markets, economists for a time equated market efficiency with the randomness of price changes. Only after the development of the dynamic equilibrium models of Chapter 14 was it understood that the two concepts were quite separate. We begin by describing some of the early statistical work to search for predictability, then look at how the dynamic pricing models raised doubts about the usefulness of simple tests of randomness. In the early 1980s the need to find different indicators of inefficiency led Robert Shiller (1981, 1982) and Stephen LeRoy and Richard Porter (1981) to propose tests based on *second* moments of price changes. We will examine their influential argument that prices are far too volatile to be consistent with "rational" beliefs, and then see that this could be consistent with cross-sectional evidence for a *momentum* effect in prices.

15.2.1 Early History

In his *General Theory of Employment, Interest, and Money*, J. M. Keynes remarked that financial markets are like a rather perverse beauty contest in which the goal is not to select the most beautiful contestant but to identify the one whom *others* regard as such. In other words, Keynes—a sophisticated investor himself, as well as the most influential economist of his day—thought that markets were driven more by psychology and whim than by considered judgment. Despite this early challenge, the serious effort to determine whether markets respond appropriately to information did not begin until the 1950s, when statisticians began to publish some sensational findings.[40] Applying various statistical tests to long time series of prices of commodities, individual stocks, and stock indexes, they concluded that changes in prices are simply unpredictable from their past values—that is, that they are purely...*random*. Not surprisingly, these assertions about the random character of stock prices sparked outrage and ridicule among economists. Is it not simply ludicrous that valuations placed on the engines of economic activity could be nothing more than randomly generated numbers? But was this really the proper conclusion to be drawn from what the statisticians found? After all, it was not that the valuations themselves are random; it was that the *increments* to values from one period to the next are unpredictable. Once it was understood in this way, many economists embraced the *random-walk hypothesis* with ideological fervor. The interpretation that led to this change of heart was as follows. If markets really do a good job of allocating scare capital among competing uses, then prices at any given moment must embody the information that is currently available to the market; that is, if today's best information indicates that some company will have strong earnings in the future, then that company should already be benefitting from a higher stock price and a lower cost of capital. Conversely, a company whose current prospects seem bleak should pay a penalty for devoting more resources to unproductive ends. But if current prices do fully reflect what is known, then increments to price must reflect *news*, and by definition news is information that is unpredictable from past knowledge. In short, prices change in response to news, and news is random, so price changes themselves must be random.

With this interpretation it seemed appropriate in evaluating market efficiency to continue the statistical tradition of testing whether prices are random walks. The formal model was $S_{t+\tau} = \nu\tau + S_t + u_\tau$, where drift ν allows for the evident long-run trends in prices and (over periods of the same length τ) $\{u_\tau\}$ are independent and identically distributed (i.i.d.) random variables. We can refer to this as the *additive* random-walk model. It implies that the changes in price over nonoverlapping intervals are statistically independent. Thus, in the additive random-walk model any sequence of price changes, although not necessarily having zero expected value, are still i.i.d. Since the past history of a stock's price changes is one component of current information, this model does capture some of the features that the intuitive view of an efficient market would lead us to expect. However, its implications are in one sense narrower, in that we know more at any given time than just the history of price changes; and yet in another sense they are broader, in that market efficiency might not necessarily require price changes to be identically distributed.

The first variation on this additive theme was to recast the model in terms of proportional changes, as $S_{t+\tau} = S_t \left(1 + \nu\tau + \varepsilon_\tau\right)$, with $\{\varepsilon_\tau\}$ still i.i.d. and $S_{t+\tau}$ being dividend-adjusted. This *multiplicative* random-walk model is roughly consistent with the observation that absolute price changes of high-priced stocks are typically larger than those for low-priced stocks. Dividing by S_t and subtracting unity give us a

---

[40]For a sampling of this early work, see Cootner (1964).

model in which the one-period simple rate of return follows an additive random walk: $R_{t,t+\tau} = \nu\tau + \varepsilon_\tau$. If the factors $\{1 + \nu\tau + \varepsilon_\tau\}$ are i.i.d. as lognormal, this is consistent with the model of trending geometric Brownian motion, wherein $dS_t = \mu S_t \cdot dt + \sigma S_t \cdot dW_t$ and $S_{t+\tau} = S_t \exp\left[\left(\mu - \sigma^2/2\right)\tau + \sigma\left(W_{t+\tau} - W_t\right)\right]$. Early tests of the multiplicative random walk model were like those described in Chapter 10 and carried out in Empirical Project 4, searching for "first moment" dependence by calculating autocorrelations of rates of return and regressing rates on a long string of their past values. As we have seen, such tests give little basis for challenging the random-walk hypothesis. Subsequently, researchers added other variables to the right side of the regressions, expanding the information set to include data from firms' financial statements, prices of other assets, and macroeconomic variables. In vector form with $\mathbf{X}_t = (X_{1t}, ..., X_{kt})'$ as a vector of k such explanatory variables—all of whose values are known at time t—and $\mathbf{R}_t$ as the rates of return for some number of earlier periods, this produced models of the form

$$R_{t,t+\tau} = \mu + \mathbf{X}_t\beta + \mathbf{R}_t\gamma + \varepsilon_\tau.$$

Still, tests of $H_0 : \beta = \mathbf{0}, \gamma = \mathbf{0}$ typically failed to reject the null hypothesis at conventional levels of significance. However, the limitless variety of variables to include in $\mathbf{X}_t$ made it inevitable that something would someday be found that was indeed correlated with the given history of rates of return.[41] One such "anomaly," discovered in the late 1970s, was that rates of return in the month of January were on average higher than in other months. Snooping for this amounted to including in $\mathbf{X}_t$ a dummy variable taking the value unity when t corresponds to January and zero otherwise. Various rationales were suggested, including price rebounds from tax-motivated selling at the end of the preceding year. But once the empirical regularity became known, it was not long before profit-seeking investors shifted the "January" effect to December,..., July, and so on.

15.2.2   Implications of the Dynamic Models

The evident opportunity for data mining eventually brought regression tests of market efficiency into disfavor. Moreover, developments in dynamic modeling like those discussed in Chapter 14 led economists to question whether these procedures were at all relevant, as it became understood that market efficiency by itself had no clear implication for whether future rates of return are to some extent predictable from current information. To see why, recall the implications for the dynamic behavior of prices in the Lucas model, where both the decisions and the beliefs of the representative economic agent were fully rational. In that model an asset's current equilibrium price satisfies

$$S_t = E_t\left[\delta_{t,t+1}\left(S_{t+1} + D_{t+1}\right)\right], \tag{15.1}$$

---

[41] Searching for such ex post correlations is a standard exercise in courses offered in some undergraduate business schools. One of my economics graduate students, then teaching a course in the local B school, occasionally would be asked to suggest new variables to include. One day, as a joke, he suggested that they include the regression *residual*, $e_\tau = R_{t,t+\tau} - \hat{\mu} + \mathbf{X}_t\hat{\beta} - \mathbf{R}_t\hat{\gamma}$. Discovering that this brought their R^2 statistics squarely up to 1.0000, his students proclaimed him a genuine financial guru.

where $\delta_{t,t+1} \equiv U'_{t+1}(c_{t+1})/U'_t(c_t) > 0$ represents the stochastic discount factor or pricing kernel. We saw in Chapter 14 that very special conditions were required within this model to make discounted price processes behave as martingales. For example, one condition that does the trick is for $\delta_{t,t+1}$ to be $\mathcal{F}_t$-measurable—that is, not stochastic at all, but this would require either that future consumption itself be known in advance or else that the representative agent be risk neutral. Alternatively, (as in Exercise 14.4) special joint conditions on utilities and the evolution of dividends would suffice. Except in such special circumstances, however, persistence in stochastic innovations to consumption would generally be transmitted to the discount factor and lead to some degree of first-moment dependence in prices and returns.

15.2.3 Excess Volatility

If current information can improve predictions of future returns relative to their long-run average values even in markets with fully informed and rational agents, what sort of evidence would be inconsistent with market inefficiency? Or is the hypothesis of market efficiency essentially irrefutable, and therefore not a genuine scientific hypothesis at all? It was with these questions in mind that Shiller (1981) and LeRoy and Porter (1981) began to look for evidence of inefficiency not in expected values of returns—first moments—but in *variances*. To understand their approach, we take (15.1) again as the starting point. Since this relation holds in each period, we also have $S_{t+1} = E_{t+1}[\delta_{t+1,t+2}(S_{t+2} + D_{t+2})]$ at $t+1$, and substituting this for S_{t+1} in (15.1) gives

$$S_t = E_t \{\delta_{t,t+1}D_{t+1} + E_{t+1}[\delta_{t,t+1}\delta_{t+1,t+2}(S_{t+2} + D_{t+2})]\}.$$

But

$$\delta_{t,t+1}\delta_{t+1,t+2} = \frac{U'_{t+1}(c_{t+1})}{U'_t(c_t)}\frac{U'_{t+2}(c_{t+2})}{U'_{t+1}(c_{t+1})} = \frac{U'_{t+2}(c_{t+2})}{U'_t(c_t)} \equiv \delta_{t,t+2},$$

which is the marginal rate of substitution of consumption at $t+2$ for consumption at t. Thus, by the tower property of conditional expectation, we obtain

$$S_t = E_t[\delta_{t,t+1}D_{t+1} + \delta_{t,t+2}(S_{t+2} + D_{t+2})],$$

and continuing this recursive process leads ultimately to

$$S_t = \sum_{j=1}^{\infty} E_t(\delta_{t,t+j}D_{t+j}). \tag{15.2}$$

Now let $S_t^* \equiv \sum_{j=1}^{\infty} \delta_{t,t+j}D_{t+j}$ be the *actual* realization of discounted future dividends, which, of course, is *not* known at time t. Then under the hypothesis of market efficiency S_t is a "rational" forecast of S_t^*; that is, it is the conditional expectation given all publicly available information $\mathcal{F}_t$. Therefore, as a tautology we can write $S_t^* = S_t + \varepsilon_t$, where $\varepsilon_t \equiv S_t^* - E_t S_t^* = S_t^* - S_t$ is the forecast *error*. Now it is a basic fact that rational forecast errors—deviations from conditional expectations—are uncorrelated with the forecasts themselves, for if they were not, the forecasts could be improved (in the sense of reducing mean-squared error). From this fact one can draw the obvious conclusion that the variance of S_t^* could be no smaller than the variance of S_t itself if the current price of the asset is an unbiased forecast of discounted cash flows. In other words, if S_t is a rational forecast, it must be true that $VS_t \leq VS_t^*$. It seems to follow, then, that if we track the course of price over time the process should be less volatile than the track of the discounted cash flows that were subsequently realized.

To test this, Shiller constructed an annual index of stock prices back to 1871 that was supposed to represent an extension of the S&P 500 (which actually began in the early 1950s with only a few dozen stocks). Using the *actual* dividend stream as observed from 1871 to the present, he constructed for each year t from 1871 forward a series of discounted subsequent dividends $\{S_t^*\}$, extending the series in time beyond the available data by assuming that dividends would continue to grow at the average rate since 1871. For this Shiller used several different versions of the discount factor: (1) a constant value corresponding to the geometric mean rate of return of the index over the entire history, (2) the prices of one-month default-free bonds, and (3) the marginal rate of substitution of future for current *aggregate* per-capita real consumption. Figure 15.2 is a free-hand sketch resembling the graph shown in Shiller's 2003 survey article. It depicts the actual S&P value and version (1) of the discounted future dividends. The obvious feature is that the stock price series bounces about wildly, whereas the discounted dividend process is relatively smooth, even during the great depression of the 1930s. Using the other versions of discount factors, the story is the qualitatively the same, although the discrepancy in volatility is less pronounced. Shiller's conclusion is that stock prices are *not* rational forecasts of discounted future dividends, but that they contain an irrational component. As he puts it, the market is subject to manic–depressive swings of mood, bouncing between extremes of pessimism and "irrational exuberance." Leroy and Porter (1981) reached much the same conclusion with a different experimental design. Wouldn't Keynes have smiled to see all this!

Not surprisingly, these findings and the explanations for them have been challenged repeatedly (and heatedly) since the early 1980s. The challenges have been from both economic and statistical directions. Shiller's initial findings were based on a constant discount factor, which, as we have seen, seems to require risk neutrality. High risk aversion, as with CRRA utilities $U_t(c_t) = c_t^{1-\gamma}/(1-\gamma)$ having large values of γ, can indeed induce much more variability in the discounted dividend series; but Shiller points out that unreasonably high values of γ would be needed to explain the observations. The statistical challenge has focused on his treatment of dividends and prices as *stationary* processes. If dividends themselves follow a random walk, as $D_{t+1} = \alpha + D_t + \varepsilon_{t+1}$, then the process really has no unconditional mean or variance; likewise for the price process. Shiller has countered by refining his theory in ways that are consistent with such nonstationary models, presenting variance bounds for increments to prices in terms of increments to the discounted dividend process. A more fundamental challenge can be made about theories based on a representative agent with additively separable utility. As we saw in the previous chapter, we now

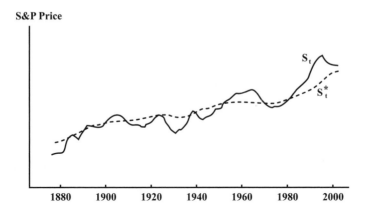

Figure 15.2   Actual S&P value S_t and discounted realized dividends S_t^*.

have theories based on habit persistence and loss aversion that do generate volatility that cannot be explained by fundamentals alone. We could not call these human traits "irrational," and they certainly do not arise from an inability to process information. But can such preference-driven factors be the whole explanation? One wonders why in markets with rational participants there are not enough well-capitalized players capable of recognizing that buying low and selling high generates substantial long-term gains, who could then act to smooth the market's cycles.

While the debate continues in some quarters, it is now widely acknowledged that there is considerable force in Shiller's argument that our vision of the future is sometimes blurrier than it need be. It *does* seem that markets often overreact to information. For example, there are often dramatic price moves following firms' announcements that earnings missed analysts' targets by minuscule amounts. Indeed, the entire market often lurches when a single large firm announces better or worse earnings than expected, then bounces back the next day. Why might such *apparent* overreactions occur? Perhaps we are often tempted to extrapolate from the very recent past to the future, disregarding or forgetting the previous long history of experience. Perhaps we forget that earnings processes of firms are, to a large although incomplete extent, *controlled* by managers. A sequence of bad earnings results typically prompts them to explore new product lines, engage in drastic cost cutting, and so on, in order to get the firms back on track. Conversely, a period of relative success can lead to overconfidence, overexpansion, and lax cost control. One who views things from a longer perspective is less apt to react so strongly to current events. There are many other instances in which investors seem to have just projected from past trends. For example, the inflow of money into mutual funds tends to lag price movements in the overall market. Supposedly sophisticated investors such as oil sheiks and Japanese billionaires were snapping up gold at more than $800/oz in the early 1980s, just before it began a 5-year downward spiral to the $300s, from which it did not recover even in nominal terms until well after Y2K. Then there are the more recent tech and real-estate "bubbles" and the subsequent financial melt-down of 2008.

These anecdotal observations suggest that there is a kind of feedback effect in prices. Perhaps a string of increases tends to spark investors' interest and push value beyond what is consistent with the "fundamentals." Perhaps this continues until some negative fundamental shock bursts the bubble and reverses the process. In his 2003 paper Shiller himself proposed a simple feedback model that illustrates the idea; specifically

$$S_t = c \int_{-\infty}^{t} e^{-\gamma(t-s)} \cdot dS_s + (1-c) \, F_t, \tag{15.3}$$

where $0 < c < 1$, $\gamma \geq 0$, and F_t is the "fundamental" source of value. Although current price S_t does depend on the fundamental, when $\gamma > 0$ it also has an extrapolative component that depends on past changes. (If $\gamma = 0$, on the other hand, we have $S_t = c \int_{-\infty}^{t} dS_s + F_t = cS_t + (1-c) \, F_t$, so that current price does equal the fundamental.) Expression (15.3) implies that the "instantaneous" change in S_t is

$$dS_t = c \cdot dS_t - c\gamma \int_{-\infty}^{t} e^{-\gamma(t-s)} \cdot dS_s + (1-c) \cdot dF_t$$

$$= \gamma \left(F_t - \frac{S_t}{1-c} \right) \cdot dt + dF_t. \tag{15.4}$$

Thus, the feedback model implies an error-correcting behavior, as the drift in $S_t - F_t$ is positive when $S_t < F_t(1-c)$ and negative when the inequality is reversed. Still, despite the tendency to self-correct, discrepancies between $\{S_t\}$ and $\{F_t\}$ can persist for many periods because of the erratic movements in $\{F_t\}$. Indeed, we can have perpetual overshooting back and forth if $\gamma > 1$, and in general higher values of γ imply higher volatility of dS_t relative to dF_t over a time sample. On the other hand, the feedback does not necessarily imply high autocorrelation in successive price moves over short periods, so the model is not intrinsically at odds with the near-random-walk behavior found in the early regression tests of efficiency.

We do see evidence of error correction in broad-based stock indexes at business-cycle frequencies, despite the very weak predictability of prices at short and moderate horizons. Stocks tend to reach their cyclic lows as we approach the troughs of business cycles but before the aggregate economic statistics signal an upturn. At such times the public mood is at its sourest and the supply of ready cash for investment is most limited. At the level of aggregation of typical portfolios there is some evidence of feedback, even though simple correlation and regression tests do not show it to be statistically significant in the individual stocks. Among money managers a well-known guide for stock picking is what is called *relative strength*. This expresses the idea that stocks that have performed well recently are apt to continue to do so; that is, that prices exhibit a certain degree of *momentum*. The paper by Narasimhan Jegadeesh and Sheridan Titman (1993) reports a careful statistical analysis to show that a market-neutral portfolio long recent "winners" and short recent "losers" does tend to produce excessive gains relative to a risk-based standard. The key here is to hold enough "names" to average out some of the idiosyncratic noise. More recently

Thomas George and Chuan-Yang Hwang (2004) reported that stocks whose prices are close to the 52-week high tend to outperform the market. In their view proximity to the high is a better predictor of future returns than is recent performance. Of course, there is clearly some association between the two, as most stocks whose prices are near their highs would have had strong recent gains. A couple of explanations have been conjectured, both in line with the efforts by behavioralists to attribute market phenomena to investors' psychological or perceptual limitations: (1) that the 52-week high serves as a mental "anchor" for investors, who consider that a stock near its high is apt to have topped out and therefore to be more sensitive to bad news than to good; and (2) that people have a tendency to take capital gains too quickly given the merits of the stock, while holding onto losers to postpone the pain of recognizing losses. Because of these irrational—or at least *non*rational—tendencies, the stocks remain undervalued relative to fundamentals even though at their highs, and so their prices usually continue to rise as the market gradually corrects the error. Although the story is certainly plausible, one should be a little skeptical of both explanations and the findings themselves. It is conceivable that the documented excess returns to these strategies merely provide the necessary compensation for greater risk. Prices of recent winners often take abrupt and terrifying plunges—perhaps because those who bought on the basis of simple extrapolation are inclined to bail out at the first sign of trouble.

EXERCISES

15.1 Show that pricing relation (15.1) would imply zero correlation between future rates of return $\{R_{t+1}\}$ and current observables $\{X_t\}$ if the representative agent's utility were of the form $U_t(c_t) = a_t + b_t c_t$, where $\{b_t\}_{t=0,1,\ldots}$ is any strictly positive, $\mathcal{F}_0$-measurable (known in advance) process .

15.2 Show that when $\delta_{t,t+1} = \delta$ in expression (15.2) we are back to the elementary dividend–discount model of Chapter 5, $S_t = \sum_{j=1}^{\infty} \delta^j E_t D_{t+j}$. How would the result change if the $\{\delta_{t,t+j}\}_{j=1}^{\infty}$ were not all the same but at least were all known in advance?

15.3 Given that $S_t = E_t S_t^*$ and $\varepsilon_t \equiv S_t^* - S_t$, show that $E S_t \varepsilon_t = 0$ and that $V S_t \leq V S_t^*$.

PART III

PARADIGMS FOR PRICING

CHAPTER 16

STATIC ARBITRAGE PRICING

We have now looked at two general approaches for modeling how assets' prices are formed in competitive, frictionless markets. The first was the CAPM, which is grounded on Markowitz' hypothesis that individuals choose portfolios so as to optimize a function of the mean and variance of return. The main implication is that assets' equilibrium expected rates of return depend on their covariances with the rate of return on the "market" portfolio. Specifically, the expected rate of return for each asset $j \in \{1, 2, ..., n\}$ is

$$E_t R_{j,t+1} = r_t + \lambda_t \beta_{jt}, \tag{16.1}$$

where (1) $R_{j,t+1}$ is the jth asset's simple rate of return between t and $t+1$ and r_t is the simple rate of return on a default-free, one-period, discount bond or riskless money-market fund; (2) $\beta_{jt} = C_t \left(R_{j,t+1}, R_{m,t+1} \right) / V_t R_{m,t+1}$; (3) $\lambda_t \equiv E_t R_{m,t+1} - r_t$ is the excess expected rate of return of the market portfolio; and (4) t subscripts on $E, C,$ and V indicate conditioning on what is known at t. Letting $S_{j,t+1} + D_{j,t+1}$ represent asset j's price plus dividend at $t+1$, the identity $1 + R_{j,t+1} = \left(S_{j,t+1} + D_{j,t+1} \right) / S_{jt}$ implies the following expression for the asset's current equilibrium price:

$$S_{jt} = \frac{E_t \left(S_{j,t+1} + D_{j,t+1} \right)}{1 + r_t + \lambda_t \beta_{jt}}.$$

Thus, in equilibrium an asset's price equals its expected per-share value next period *discounted* at a rate that depends on the extent to which its return tracks with that of the market as a whole.

The second modeling approach, discussed in Chapter 14, was based on the dynamic framework in which individuals maximize lifetime expected utility of consumption. The conclusion that comes from most models of this type is that the current price of asset j equals the expectation of the future value per share discounted by a *stochastic* factor that equals the "representative" investor's marginal rate of substitution of future for current consumption:

$$S_{jt} = E_t \left[\frac{U'_{t+1}(c_{t+1})}{U'_t(c_t)} (S_{j,t+1} + D_{j,t+1}) \right].$$

As we have seen, the two approaches can be reconciled approximately by appealing to a short-holding-period argument. Indeed, the correspondence is exact if derived in a dynamic, continuous-time framework under appropriate assumptions about processes $\{S_{jt}\}_{j=1}^n$. In any case the basic implication of the dynamic model is much the same as that of the CAPM—assets that pay off most handsomely in "good" states of the world, when aggregate consumption (or the return to the market portfolio) is high, and do worst in the "bad" states command relatively low current prices, while assets that provide *hedges* against aggregate shocks command premium prices.

While this implication of the optimizing models is compelling, we saw in Chapters 6 and 14 that the empirical evidence from the models is, at best, mixed. In terms of the cross-sectional implications, since the 1990s beta (as we have tried to measure it) has no longer seemed to account for much of the variation among stocks' average (over time) rates of return. Measures of the firms' sizes (market capitalizations) and book-to-market ratios were better predictors—at least during certain periods. While the jury is still out on what this might mean, one possible explanation is that size and book-to-market do (or did) proxy for other sources of risk than just variations in aggregate consumption. For example, firms with high book-to-market ratios are typically "distressed," having experienced low or negative earnings in the recent past. Such firms tend to be highly dependent on availability of credit, and so the fact that their expected rates of return are high (i.e., current prices *low*) may signify their high exposure to risk of a credit crunch.

16.1   PRICING PARADIGMS: OPTIMIZATION VERSUS ARBITRAGE

While it is possible to account for the effects of multiple risk sources within an optimizing framework, one of the guiding principles of intellectual enquiry suggests trying another approach. The principle, known as *Occam's razor*, is that one should choose the *simplest* among competing explanations of some phenomenon. One has had to swallow quite a few manifestly counterfactual assumptions in developing the optimizing models. Does it really make sense that people choose among risky prospects on the basis of the first two moments, as required in the Markowitz theory? Can we really ignore psychologists' experimental evidence that people do not obey the von Neumann–Morgenstern axioms of rational behavior? Do individuals really have "rational" or even common beliefs about the joint distributions of assets' future rates of return? Given the evident willingness of some people to take significant risks for tenuous or transitory gains—look at how they drive, smoke, and eat!—are we even sure that risk aversion is a primary influence on market prices? While the usefulness of the optimizing theories does not *require* affirmative answers to these questions, it would clearly be nice if testable implications could be obtained more simply; and it would be *really* nice if a theory could be found that was both simpler *and* better able to account for what we see in financial markets.

One approach with this potential is based not on optimizing behavior and not even on a concept of market equilibrium, but on the assumption that markets will not long offer opportunities for arbitrage. In the next three sections of this chapter we will look at relatively simple applications of this principle that involve static (buy-and-hold) portfolios, as opposed to dynamic portfolios that have to be constantly adjusted. The first application, Steve Ross' (1976) *arbitrage pricing theory* (APT), is intended to explain how prices of all "primary" assets—stocks, bonds, investment commodities—are related at a moment in time. In other words, it is an alternative to the CAPM. The second application is to default-free bonds, whose payout structures offer opportunities for arbitrage unless the bonds are correctly priced relative to each other. The final application is to forward contracts. The payoffs of these *derivative* assets are linked to the price of some "underlying" asset—a portfolio of stocks or bonds, gold, or other investment asset. Since the future payoffs of forward contracts can be expressed as linear functions of the future underlying price, they can be replicated by simple buy-and-hold, static portfolios of the underlying and default-free discount bonds. In Chapter 17 we shall take up models for other derivative securities, such as options, whose nonlinear payoff structures require *dynamic* trading strategies in order to price them by arbitrage. The remaining chapters of this part of the book develop further arbitrage-based tools and models for pricing options and other such nonlinear derivatives.

We have used arbitrage arguments informally thus far, understanding intuitively that arbitrages are deals that give something for nothing. It is now time to give a precise definition. Technically, there are two classes of arbitrages:

1. At a current cost of zero or less one secures a prospect that has *no* chance of future loss and *some* positive chance of positive future value.

2. At current *negative* cost, one secures a prospect that has no chance of future loss.

Thus, in type 1 arbitrages one bears no cost today but gets something of value, in the form of a *contingent* future gain. An example is a deal like that available briefly in 2000, when 3Com Corp. spun off its former subsidiary Palm Pilot. For a time before the spin-off and before Palm's shares were actually issued, the prospective shares traded "when issued" at prices that made Palm worth more than the inclusive parent. Anyone who was able sell Palm short and put the money into 3Com would clearly have made a very good deal. In type 2 arbitrages one gets something today for sure without having to pay for it in the future. An example would be selling a call option on a stock for more than it would cost to buy a portfolio of stocks and bonds that could be used to replicate the option's eventual payoff—whatever the payoff turned out to be. Another example would be buying a portfolio of bonds that replicated the payoffs of another bond (or portfolio of bonds) that could be sold for more. The guiding principle behind the "no arb" approach to pricing is the conviction that such opportunities for riskless gain would quickly be exploited, causing prices to adjust until the opportunities were eliminated. That they *would* be quickly eliminated just requires that people be aware of what is going on around them and that they always desire more of the things that money will buy. It matters not whether they be risk averse or whether they weave in and out of traffic with cigarette and cell phone in one hand and double cheeseburger in the other.

Now let us see how Steve Ross used the no-arb principle to arrive at implications for the cross section of expected returns—implications that can extend those of the optimizing models and possibly account for the empirical findings that are contrary to CAPM.

16.2 THE ARBITRAGE PRICING THEORY (APT)

In line with the no-arb principle, the fundamental assumption of the APT pertains not to preferences but to a general property of assets' rates of return. Specifically, Ross assumes that one-period rates of return of the $n + 1$ available assets have the following *linear factor structure*:

$$R_j = \mu_j + \beta_{j1}f_1 + \cdots + \beta_{jk}f_k + u_j, j = 0, 1, 2, ..., n. \qquad (16.2)$$

Here (1) R_j is the simple rate of return of asset j from t to $t+1$ (a period of unspecified length); (2) $\mu_j = E_t R_j$; (3) $\{f_i\}_{i=1}^{k}$ are possibly unobservable common "factors"— random variables with zero means; (4) the $\{\beta_{ji}\}$ are constants, sometimes called *factor loadings*; and (5) u_j is a mean–zero random variable. Time subscripts on R_j, the $\{f_i\}$, and u_j are omitted for brevity. The $\{f_i\}$ are called *common* factors because they are sources of risk affecting all risky assets to varying degree, while the loadings $\{\beta_{ji}\}_{i=1}^{k}$ indicate the sensitivity of asset j to these various risks. The factors are not necessarily observable variables, although one expects that they correspond to some risks that make economic sense. For example, f_1 might be a general "market" or aggregate consumption factor, in which case β_{j1} would have the same interpretation as the beta in the CAPM; f_2 might be a credit-risk factor; f_3 an inflation-risk factor; and so on. On the other hand, the realization of variable u_j in a given period represents developments whose effects are limited to asset j, such as management turnover, development of new proprietary technologies, or losses of assets due to fraud or natural disasters. Thus, u_j embodies what is called *idiosyncratic* risk. With this interpretation, it is reasonable to assume (and we do) that the $\{u_j\}_{j=1}^{n}$ are independent of each other and of the common factors. Of course, that $\mu_j = E_t R_j$ follows from the specification that u_j and the factors all have means equal to zero. (In essence, we simply *define* the f's as deviations from factor means.) As we shall see, it is crucial for the theory that k, the number of factors, be much smaller than n, the number of risky assets. Asset 0, which completes the collection of $n + 1$ assets, represents a riskless investment, such as a default-free US Treasury bill that matures at the end of the period. This has zero β's on all factors and also zero idiosyncratic risk, so $u_0 = 0$ and $R_0 = \mu_0 = r$, say, with probability one.

Understanding the APT requires a little use of linear algebra. As a first step, we need to represent (16.2) for all $n + 1$ assets at once using matrix notation. For this, let

$$\mathbf{R} = \begin{pmatrix} r_0 \\ R_1 \\ \vdots \\ R_n \end{pmatrix}, \mathbf{f} = \begin{pmatrix} f_1 \\ f_2 \\ \vdots \\ f_k \end{pmatrix}, \mathbf{B} = \begin{pmatrix} 0 & 0 & \cdots & 0 \\ \beta_{11} & \beta_{12} & \cdots & \beta_{1k} \\ \vdots & \vdots & \ddots & \vdots \\ \beta_{n1} & \beta_{n2} & \cdots & \beta_{nk} \end{pmatrix}, \mathbf{u} = \begin{pmatrix} 0 \\ u_1 \\ \vdots \\ u_n \end{pmatrix}.$$

Thus, $\mathbf{R}$ is an $(n + 1)$ vector of rates of return of the $n + 1$ assets, $\mathbf{f}$ is a k vector of the factors, $\mathbf{B}$ is an $(n + 1) \times k$ matrix of the factor loadings (its jth row contains asset j's sensitivities to the k factors), and $\mathbf{u}$ is an $(n + 1)$ vector of the idiosyncratic effects. The first row of $\mathbf{B}$ and the first element of $\mathbf{u}$ are zeros because asset 0 is itself riskless. With this notation (16.2) can be expressed compactly as

$$\mathbf{R} = \mu + \mathbf{Bf} + \mathbf{u}. \tag{16.3}$$

Now, let $\mathbf{p} = (p_0, p_1, ..., p_n)'$ represent a portfolio (where the prime denotes transposition, as usual) and $\mathbf{1}$ be a vector of units, with $\mathbf{p'1} = \mathbf{1'p} = \sum_{j=0}^{n} p_j$. Ordinarily we think of portfolios for which $\mathbf{p'1} = 1$, but we now consider portfolios such that (1) $\mathbf{p'p} > 0$ but (2) $\mathbf{p'1} = 0$. Point 1 means that not all the elements of $\mathbf{p}$ are zero (a totally uninteresting case), and this together with point 2 implies that some elements are positive and some are negative. Such are called *arbitrage portfolios*; it costs nothing to acquire them, because long positions in some assets are financed by short positions in others. Actually, it is helpful to think of such a $\mathbf{p}$ as the *change* in the composition of an ordinary investment portfolio—one with $\mathbf{p'1} = 1$—due to a reshuffling of assets; that is, rebalancing an ordinary investment portfolio is equivalent to *acquiring* an arbitrage portfolio. With this interpretation, negative p's can be achieved without the capacity to invest the proceeds of short sales.

Now let us try to choose a specific arbitrage portfolio $\mathbf{p}$ that satisfies the additional condition $\mathbf{p'B} = \mathbf{0}$. Written out, this is

$$(p_0, p_1, ..., p_n) \begin{pmatrix} 0 & 0 & \cdots & 0 \\ \beta_{11} & \beta_{12} & \cdots & \beta_{1k} \\ \vdots & \vdots & \ddots & \vdots \\ \beta_{n1} & \beta_{n2} & \cdots & \beta_{nk} \end{pmatrix} = \begin{pmatrix} \sum_{j=1}^{n} p_j \beta_{j1} \\ \sum_{j=1}^{n} p_j \beta_{j2} \\ \vdots \\ \sum_{j=1}^{n} p_j \beta_{jk} \end{pmatrix} = \begin{pmatrix} 0 \\ 0 \\ \vdots \\ 0 \end{pmatrix},$$

which represents k equations in the n unknowns, $p_1, p_2, ..., p_n$. (The riskless share p_0 does not enter, since the first row of $\mathbf{B}$ has only zeros.) A portfolio satisfying this requirement is said to be a *zero-beta portfolio*. Adding the requirement $\mathbf{p'1} = 0$ means that finding the zero-beta arbitrage portfolio requires solving $k + 1$ equations in $n + 1$

unknowns. There will always be the *trivial* solution $p_1 = p_2 = \cdots p_n = 0$ to such a homogeneous system, but we want something more. Now, we invoke the assumption that $n >> k$ (many more assets than factors). There will then be (infinitely) many vectors $\mathbf{p}$ that solve these equations, and hence infinitely many zero-beta arbitrage portfolios. Given this abundance of possibilities, we should be able pick one that satisfies the additional requirement of being *well diversified*, meaning that no one element of $\mathbf{p}$ dominates the others. Thus, each p_j should be on the order of n^{-1} in absolute value. For such a portfolio the product $\mathbf{p}'\mathbf{u} = \sum_{j=1}^{n} p_j u_j$ should be close to zero with high probability, since the idiosyncratic variables are independent and all have mean zero. Thus, we can write for the rate of return of our diversified, zero-beta, arbitrage portfolio

$$R_{\mathbf{p}} \equiv \mathbf{p}'\mathbf{R} = \mathbf{p}'(\mu + \mathbf{Bf} + \mathbf{u}) = \mathbf{p}'\mu + \mathbf{p}'\mathbf{Bf} + \mathbf{p}'\mathbf{u} \doteq \mathbf{p}'\mu.$$

Notice that by making the portfolio zero-beta we have eliminated all the factor risk, and by making it well diversified we have eliminated (almost all) of the idiosyncratic risk. In short, we have eliminated (almost) *all* risk and thus have a portfolio that costs nothing but earns rate $\mathbf{p}'\mu$ (almost) for sure.

This is where the no-arb assumption comes in. If the market offers no opportunity for arbitrage, the payoff of a portfolio that *costs* nothing must *be* nothing; that is, since its rate of return is certain, that certain value must be...*zero*. Were it positive, we would be getting *something* later for *nothing* now—a type 1 arbitrage. (Obviously, there is an easy way to profit from $\mathbf{p}'\mu < 0$ also.) Thus, any portfolio $\mathbf{p}$ with the properties $\mathbf{p}'\mathbf{1} = 0$ and $\mathbf{p}'\mathbf{B} = \mathbf{0}$ must also have the property $\mathbf{p}'\mu = 0$. Ross' brilliant insight was that this simple fact has profound implications for the relation among assets' *expected* rates of return. To see it requires a result from linear algebra. First, note the obvious fact that if there were a scalar constant δ and a vector $\lambda = (\lambda_1, ..., \lambda_k)'$ such that $\mu = \mathbf{1}\delta + \mathbf{B}\lambda$, then any vector $\mathbf{p}$ with $\mathbf{p}'\mathbf{1} = 0$ and $\mathbf{p}'\mathbf{B} = \mathbf{0}$ would make $\mathbf{p}'\mu = \mathbf{p}'\mathbf{1}\delta + \mathbf{p}'\mathbf{B}\lambda = 0$ also. The needed fact from linear algebra is the converse. Specifically, if $\mathbf{p}'\mu = 0$ for *all* vectors such that $\mathbf{p}'\mathbf{1} = 0$ and $\mathbf{p}'\mathbf{B} = \mathbf{0}$, then $\mathbf{1}$ and $\mathbf{B}$ "span" the $(n+1)$-dimensional linear space in which μ resides. In that case *there exist* scalar δ and vector λ such that $\mu = \mathbf{1}\delta + \mathbf{B}\lambda$. Written out, this means that

$$\begin{pmatrix} r \\ \mu_1 \\ \vdots \\ \mu_n \end{pmatrix} = \begin{pmatrix} \delta \\ \delta \\ \vdots \\ \delta \end{pmatrix} + \begin{pmatrix} 0 & 0 & \cdots & 0 \\ \beta_{11} & \beta_{12} & \cdots & \beta_{1k} \\ \vdots & \vdots & \ddots & \vdots \\ \beta_{n1} & \beta_{n2} & \cdots & \beta_{nk} \end{pmatrix} \begin{pmatrix} \lambda_1 \\ \lambda_2 \\ \vdots \\ \lambda_k \end{pmatrix}$$

or that $\mu_j = \delta + \lambda_1 \beta_{j1} + \cdots + \lambda_k \beta_{jk}$ for $j \in \{1, 2, ..., n\}$. Combined with the result $\mu_0 \equiv r = \delta$ for the riskless asset, we have as implications for expected returns and prices

$$ER_j \equiv \mu_j = r + \lambda_1 \beta_{j1} + \cdots + \lambda_k \beta_{jk} \qquad (16.4)$$

$$S_{jt} = \frac{E_t(S_{j,t+1} + D_{j,t+1})}{1 + r + \lambda_1 \beta_{j1} + \cdots + \lambda_k \beta_{jk}}.$$

Thus, in an arbitrage-free economy with many assets and a linear factor structure for rates of return, the expected rate of return of any asset is a linear function of its sensitivities to the factor risks. The λ values indicate the marginal effects on expected return of changes in factor sensitivities. Comparison with (16.1) shows that the case $k = 1$ with f_1 as the market factor corresponds precisely to the implication of the CAPM, without any assumptions about mean–variance preferences, short holding periods, or identical expectations. The result does not even require that the prices we observe be always consistent with an *equilibrium* in which everyone is satisfied with current holdings. What we do require is merely that there be a factor structure, that people know what the factor structure is, and that there be no arb opportunities at the moment we record prices. This is Occam's razor at its sharpest. Moreover, allowing for the possibility of multiple factors, such as credit risk, inflation risk, and term-structure risk, may explain some of the empirical anomalies found in testing the CAPM. Indeed, some empirical tests of the APT have found evidence for precisely these factors, in addition to the general procyclical "market" factor emphasized by the CAPM. Is all this too good to be true?

For a while the APT was hailed as the Rosetta stone of asset pricing, but it wasn't long before the critics began to point out some cracks. One unattractive feature is that there is no definitive way of identifying a unique factor structure, since if $\mathbf{R} = \mu + \mathbf{Bf} + \mathbf{u}$, then it is also true that $\mathbf{R} = \mu + \mathbf{B}^*\mathbf{f}^* + \mathbf{u}$, where $\mathbf{B}^* = \mathbf{BM}$, $\mathbf{f}^* = \mathbf{M}'\mathbf{f}$, and $\mathbf{M}$ is any of infinitely many orthogonal matrices—one such that $\mathbf{MM}' = \mathbf{I}$, the identity matrix. The best we can do is to find some putative candidates that span the relevant k-dimensional space in which the true factors reside. But although the model requires that investors themselves know the spanning factors, economists continue to argue about what they are—and even about how many there are. There is also an epistemological challenge to the APT. It stems from the fact that the central conclusion (16.4) holds only as an approximation, because no portfolio $\mathbf{p}$ with finitely many elements can diversify away all the idiosyncratic risk. That is, the critical condition $\mathbf{p}'\mathbf{u} = \mathbf{0}$ with probability one when the elements of $\mathbf{u}$ are independent and have mean zero can be approached only in the limit as the number of assets goes to infinity. Thus, if empirical findings with actual market data are not precisely in accord with (16.4)—and, of course, they never can be—then how can we decide whether the discrepancy is due to our misperceiving the economy's factor structure, or to errors in estimating μ and $\mathbf{B}$, or to the existence of arbitrage opportunities, or to the inadequacy of the theory? The criticism, in short, is that the APT is not, even in principle, refutable or falsifiable. As the critics argue, a model that cannot be refuted is essentially empty—like a statement that specifies how many angels can dance on the head of a pin—and if the APT does not specify precisely what to look for, then how could it be refuted, even given all the data in the world?

One can acknowledge these criticisms and yet retain considerable regard for the APT. We can be almost confident from the outset that any given theory provides only an imperfect and narrow description of reality. In other words, we can just accept that the APT is false, just as the CAPM is. These and all other models are merely provisional ways of bringing some order to what we see around us, reducing it to essentials that we can grasp and can use to make some predictions, and serving our needs until something better comes along. Whatever its shortcomings, the APT has unquestionably given us valuable insights about financial markets, and Ross' ingenious use of the no-arb principle in this context guides us toward other applications.

16.3   ARBITRAGING BONDS

A default-free, coupon-paying bond is nothing more than a *portfolio* of default-free *discount* bonds that mature on the dates at which the coupon bond makes its payments of coupons and principal. Thus, one could replicate the coupon-paying bond by holding such a portfolio of discount bonds of appropriate principal values. Moreover, if the coupon bond could be sold for more than it would cost to buy the portfolio—or the other way around—then there would be an opportunity for arbitrage. This much is obvious. Perhaps less obvious is the fact that any bond—whether it pays a coupon or not—can be replicated by *some* portfolio of other bonds that span the same payoff space. This implies that all their prices must be properly aligned if arbitrage is not to be possible, and so finding a misalignment means finding an opportunity for arbitrage. But how does one look for such a "misalignment"? To discover how, we must first understand what it means for the putative replicating bonds to "span the same payoff space," and this in turn requires us to characterize bonds in a special way.

Let us consider a collection of N_B default-free bonds of various maturities and coupon dates. The inventory of a dealer in US Treasury securities would be such a collection, and for each of the bonds in that collection there would be a pair of bid and asked prices at which the dealer stands ready to buy and sell. The data from Empirical Project 1 in the file BondQuotes.xls at the FTP site are a specific example. The file contains descriptions and actual price quotes as of mid-February 2008 for 81 bonds with maturities ranging from less than one month to about ten years. Now, let N_D be the number of distinct dates at which at least one bond in the collection is to make a cash payment. (There are 50 such dates for the 81 bonds in the file.) We can then associate with each bond a vector of length N_D whose elements indicate the dollar cash payments on the various dates, and we can represent the collection of bonds as an $N_B \times N_D$ matrix. As a more manageable example than that in BondQuotes.xls, the 14×4 matrix comprising the last four columns of Table 16.1 represents a collection of payments by 14 bonds selected from the 81 bonds in the file. The bid and asked columns show price quotations per \$100 principal value, and the columns headed by month/year are the bonds' payoffs per \$100 face. All the bonds mature on the 15th day of the indicated month and year.

Now let us think of a given bond's payoff vector as representing a point in Euclidean space of dimension N_D, the number of payment dates in our collection of bonds. Thus, the 14 bonds in the table correspond to 14 points in four-dimensional space $\Re_4$. Since the first three bonds make payments at only the first date, they can also be regarded as points in the one-dimensional subspace $\Re_1 \subset \Re_4$. Likewise, bonds 1–6 are points in subspace $\Re_2$, and all of bonds 1–10 are in subspace $\Re_3$. Now as we have already stated in connection with the APT, a collection of vectors *spans* a vector space if any vector (any point in the space) can be represented as a linear combination of those in the collection. Clearly, each of the first three bonds spans the one-dimensional

Table 16.1

Bond	Bid	Asked	8/08	2/09	8/09	2/10
1: 0.000%8/08	99.051	99.130	100.0000	0	0	0
2: 3.250%8/08	100.551	100.591	101.6250	0	0	0
3: 4.125%8/08	100.958	101.017	102.0625	0	0	0
4: 0.000%2/09	98.173	98.309	0	100.0000	0	0
5: 3.000%2/09	101.068	101.158	1.5000	101.5000	0	0
6: 4.500%2/09	102.572	102.609	2.2500	102.2500	0	0
7: 0.000%8/09	97.189	97.319	0	0	100.0000	0
8: 3.500%8/09	102.384	102.504	1.7500	1.7500	101.7500	0
9: 4.875%8/09	104.403	104.536	2.4375	2.4375	102.4375	0
10: 6.000%8/09	106.022	106.179	3.0000	3.0000	103.0000	0
11: 0.000%2/10	96.273	96.443	0	0	0	100.000
12: 3.500%2/10	103.018	103.177	1.7500	1.7500	1.7500	101.750
13: 6.500%2/10	108.749	108.966	3.2500	3.2500	3.2500	103.250
14: 4.750%2/10	107.380	107.593	2.3750	2.3750	2.3750	102.375

space corresponding to payoff date 8/08, in that each bond could be represented as a multiple of any other. Likewise, any two of bonds 4–6 span the two-dimensional space corresponding to 8/08 and 2/09, and thus any of bonds 1–6 could be represented uniquely as a linear combination of any two of bonds 4–6. (For examples, a portfolio that is short one unit of bond 4 and long three units of bond 5 replicates two units of bond 6, while a portfolio short 409 units of bond 5 and long 406 units of bond 6 replicates three units of bond 1.) On the other hand, although bonds 7–10 represent four distinct points in the subspace $\Re_3$ corresponding to the first three payment dates, the four vectors do not span that three-dimensional space, because any of one of them could be replicated with just two of the others. In other words, the 4×3 matrix that their payoffs occupy has rank 2. To see that directly, recall that the rank of a matrix is not changed if any row is multiplied by a nonzero constant or if a multiple of one row is added to or subtracted from another row. Subtracting the first row from each of rows 2–4 leaves the matrix

$$\begin{pmatrix} 0 & 0 & 100.0000 \\ 1.7500 & 1.7500 & 1.7500 \\ 2.4375 & 2.4375 & 2.4375 \\ 3.0000 & 3.0000 & 3.0000 \end{pmatrix},$$

whose second–fourth rows by themselves have rank 1. However, although bonds 7–10 alone do not suffice, we could span $\Re_3$ in many ways; with (1) any one of bonds 1–3, any one of bonds 4–6, and any one of bonds 7–10; (2) any two of bonds 4–6 and any one of bonds 7–10; (3) any one of 1–3 and any two of 7–10; and (4) any one of bonds 4–6 and any two of bonds 7–10. Thus, there are 84 ways of replicating any one of bonds 1–10, and there are many more ways still to replicate each bond in the entire set of 14. To check exhaustively that the price quotes in the table allow no arbitrages would require comparing the cost of buying each bond with the proceeds from selling each of its replicating portfolios, then comparing the proceeds from selling each bond with the cost of buying each of its replicating portfolios. This would clearly involve a lot of calculation. Is there any more practical way to spot mispricing?

Indeed there is, and the exercise in Empirical Project 1 points the way. Regressing the average of the bid and asked quotes for the 14 bonds on the four payoff columns (suppressing the intercept) gives four regression coefficients, .9900, .9818, .9728, .9625, which correspond to the implicit discount factors for sure cash receipts on the four dates. The regression also yields a vector of 14 residuals—the differences between the actual averages of the quotes and the fitted values from the regression. Table 16.2 presents these residuals in rank order, along with the numbers and descriptions of the corresponding bonds. From these residuals we might guess that coupon bond 13 is a little underpriced relative to the others and that discount (zero-coupon) bonds 1 and 11 are priced a little high. We can thus focus on these to find some possible arbitrage opportunities. As an example of how to do this, we will consider buying bond 13 and selling some portfolio that replicates it, and we will concentrate on replicating portfolios that contain bonds 1 and 11. Since bond 13 has payoffs on all four dates, replication will require four other bonds that span $\Re_4$. Clearly, the easiest spanning vectors to work with are zero-coupon bonds 1, 4, 7, and 11, which form the usual Euclidean basis in $\Re_4$. To find the replicating coefficients, we again use regression. With the payoff vector of bond 13 as the response variable and the payoffs of bonds 1, 4, 7, and 11 as predictors, and again suppressing the intercept, we get coefficients $c_{01} = c_{04} = c_{07} = 0.0325$ and $c_{11} = 1.0325$. The four residuals from the regression all equal zero, confirming the perfect fit and exact replication. The coefficients indicate that for every dollar of principal value of bond 13 we would want to hold $0.0325 in principal value of each of the first three bonds and $1.0325 principal value of bond 11. Buying two million dollars' face value of bond 13—that is, 2000 bonds each of $1000 principal value—at the quoted asked price would cost $2,179,320, and selling $65,000 face value of each of bonds 1, 4, and 7 plus $2,065,000 face value of bond 11 would bring in $65(990.51 + 981.73 + 971.89) + 2,065 (962.73) = \$2,179,405.90$, for a modest type 2 arbitrage gain of about $86. Since the residual for bond 7 makes it appear a bit underpriced, we might do better to replace it in the replicating portfolio by another that completes the span of $\Re_4$, such as bond 8 or 9. Using 8 yields an arbitrage profit of about $114, while using 9 gives about $110. Clearly, the dealer who issued these particular quotes was not exposed to great loss from other dealers or sharp, wealthy traders—and to none at all if unwilling to buy and sell the large amounts of bonds needed to work the arbs.

16.4 PRICING A SIMPLE DERIVATIVE ASSET

"Derivative" assets are so named because their values are in fact *derived* from those of the "underlying" assets to which their values are linked. The linkage is precisely specified by the contract that binds the two parties to the transaction. Typically,

Table 16.2   Bonds sorted by residuals from regression of quotes on payoffs

Bond	Residual
13: 6.500% 2/10	−.086
3: 4.125% 8/08	−.052
14: 4.750% 2/10	−.038
2: 3.250% 8/08	−.035
6: 4.500% 2/09	−.028
5: 3.000% 2/09	−.027
7: 0.000% 8/09	−.021
10: 6.000% 8/09	−.008
12: 3.500% 2/10	.015
8: 3.500% 8/09	.016
9: 4.875% 8/09	.017
4: 0.000% 2/09	.059
1: 0.000% 8/08	.093
11: 0.000% 2/10	.113

derivatives are contracts with finite lives, "expiring" at some future time T at which their values are determined by the contractually specified functions of the underlying prices. This connection between the prices of derivative and underlying often makes it possible to use the underlying asset and one or more other traded assets to replicate the payoff of the derivative at T. Moreover, the replication would be carried out in such a way that the portfolio neither requires nor generates additional funds during the life of the contract. In other words, the replicating portfolio is *self-financing*. In this case the value of the derivative at any $t < T$ must equal the value of the replicating portfolio at that date, or else there would be a type 2 arbitrage opportunity. Working the arb would amount to selling whichever of derivative or portfolio had the higher price and buying the other, pocketing the difference. At expiration date T the positions would then be "unwound" at zero net cash outlay.

A forward contract is really the simplest example of a derivative asset. This is a contract between two parties to exchange some underlying commodity or asset for an agreed-on amount of cash at some fixed future date T. Once the exchange is made, the contract terminates and all is done. As an example, a US importer of automobiles from the Republic of Korea (RK) faces an obligation to pay the exporter 100 million units of RK currency—the won—3 months from now. To avoid exchange-rate risk, the importer enters a forward contract with a bank to buy 100 million $= 10^8$ won in 3 months for $100,000 = \$10^5$. The quantity $f_0 = 10^5/10^8 = 10^{-3}$ represents the *forward price* of won in dollars, as determined by the agreement made at $t = 0$ (i.e., it is the 3-month forward \$/won exchange rate). The importer, who is to accept delivery of won, is said to be on the "long" side of the forward contract. At $t = T$ (3 months from now), the importer will pay the bank $f_0 \cdot 10^8 = 10^5$ dollars and receive (or have transferred to the exporter's bank) 10^8 won, whereupon all obligations terminate.

Abstracting from transaction costs and other market frictions, the forward price f_0 that is agreed on when the contract is initiated is such that the obligation has no net resale value to either party. However, as time progresses and the won/dollar exchange rate changes, the contract acquires positive value to one side and negative value to the other. Letting S_T be the time T spot exchange rate (\$/won), the per-unit value of the contract at expiration, when it has no remaining life, is $\mathfrak{F}(S_T, 0) \equiv S_T - f_0$ from the standpoint of the importer. The task is to determine the per-unit value $\mathfrak{F}(S_t, T - t)$ when the remaining life is $T - t \in [0, T]$.

For this we can use a replicating argument, as follows. To act now (time t) to secure one RK won at T, the importer could simply buy Korean currency at the current (spot) exchange rate and purchase a T-maturing RK unit bond, or, equivalently, deposit the appropriate amount in a Korean money-market fund. The cost of the bond or the required deposit would be $e^{-r_{RK}(T-t)}$ in RK currency, where r_{RK} is the prevailing continuously compounded spot rate. The dollar cost would be this amount multiplied by the dollar/won spot exchange rate at t, or $S_t e^{-r_{RK}(T-t)}$. Next, to arrange now for the cash outlay of f_0 at T (the cash flow of $-f_0$), the importer could simply sell f_0 dollar-denominated, T-expiring unit bonds, each worth $B(t, T) = e^{-r_{US}(T-t)}$, committing to redeem them at maturity. In other words, the importer could borrow $f_0 B(t, T)$ dollars, to be repaid at T with interest continuously compounded at rate r_{US}. The current dollar cash flow is $S_t e^{-r_{RK}(T-t)}$ out and $f_0 B(t, T) = f_0 e^{-r_{US}(T-t)}$ in, so the net cost of the replicating portfolio, and hence the arbitrage-free per-unit value of the contract that expires after $T - t$ units of time, is

$$\mathfrak{F}(S_t, T - t) = S_t e^{-r_{RK}(T-t)} - f_0 e^{-r_{US}(T-t)}. \tag{16.5}$$

Were someone to offer to buy and sell such existing contracts at a different price, there would be an arbitrage opportunity. To see what the initial forward price f_0 would have to be to make the contract's initial value zero, set $0 = S_0 e^{-r_{RK}T} - B(0, T)f_0$ and solve to get $f_0 = S_0 e^{-r_{RK}T}/B(0, T) = S_0 e^{(r_{US}-r_{RK})T}$.

More generally, if the underlying asset offered a continuously compounded yield at rate δ, then a forward contract initiated at time 0 and expiring at T would be worth

$$\mathfrak{F}(S_t, T - t) = S_t e^{-\delta(T-t)} - B(t, T)f_0 \tag{16.6}$$

at $t \in [0, T]$. Thus, for a contract that was actually initiated at t, the forward price that sets $\mathfrak{F}(S_t, T - t) = 0$ would be

$$f_t = \frac{S_t e^{-\delta(T-t)}}{B(t, T)}. \tag{16.7}$$

In the case that holding the underlying required a continuous net outlay for storing or insuring, then δ would be negative.

Exercise 16.4 shows a simple way to adjust formula (16.5) to allow for known future costs or dividends that are paid in lump sum. Note, however, that these simple forward pricing formulas apply only to financial assets or to commodities held primarily for investment purposes. For contracts on raw materials, agricultural goods, and the like, unpredictable fluctuations in production and consumption introduce additional complications that are beyond our scope.

There is also an additional complication in pricing futures contracts. While futures and forwards impose the same general obligations to make or take delivery of a commodity at specified price and date, futures commitments are traded on open markets rather than being negotiated directly between the parties. To make the markets sufficiently liquid, contracts are standardized, commodity by commodity, as to delivery dates, quantities, and delivery locations. One who takes a long position in a contract makes an initial margin deposit with a broker and then commits to pay the prevailing futures price per unit of commodity. Likewise, the counterparty who takes the corresponding short position posts initial margin and commits to delivering at the futures price. (The margin deposits are made merely to assure that the contract will be honored and can be in the form of interest-bearing securities.) Of course, the futures price for new commitments fluctuates from day to day, just as newly contracted forward prices would change in accord with (16.7). So far, apart from the required margin deposits, the arrangement looks just the same as for forwards, but now comes the difference. At the end of each day the futures contract is automatically renegotiated to commit the parties to trading at the *new* futures price then prevailing, and an amount of cash equal to the contract size times the one-day change in the futures price is added to or deducted from each party's margin balance. This process of *marking to market* gives the newly minted contract zero value but changes the value of the account with the broker. The effect is to apportion into daily increments whatever total gain or loss each party would have incurred in holding the contract to expiration. The resulting cash flow means that a futures commitment cannot be replicated precisely by a *static* buy-and-hold portfolio of underlying and bonds. Moreover, the interest rates at which the daily cash flows can be reinvested or financed also vary with market conditions, and in ways that cannot be foreseen when the futures contract is initiated; and this uncertainty in interest rates drives a small wedge between arbitrage-free forward and futures prices.[42] On the other hand, if there were no such uncertainty in interest rates, then futures and forward prices would coincide precisely in frictionless, arbitrage-free markets. For the proof, see Exercise 16.7.

EXERCISES

16.1 The economy is governed by a two-factor model that is consistent with the APT. Expected simple one-period rates of return for 11 assets and their corresponding factor loadings are shown below. Determine the constants λ_1 and λ_2 that relate

---

[42]To see just how they differ requires a model for the joint evolution of interest rates and the spot price of the underlying commodity. For examples and a detailed discussion, see Epps (2007, pp. 153–155 and 488–490).

the assets' mean rates of return to their factor sensitivities.

$$
\mu = \begin{pmatrix} 0.0300 \\ 0.0502 \\ 0.0794 \\ 0.0620 \\ 0.0747 \\ 0.0730 \\ 0.0609 \\ 0.0422 \\ 0.0907 \\ 0.0444 \\ 0.0910 \end{pmatrix}, \quad
\mathbf{B} = \begin{pmatrix} 0.0000 & 0.0000 \\ 0.2154 & 0.5306 \\ 1.4435 & 0.6846 \\ 1.0673 & 0.3559 \\ 1.3280 & 0.6061 \\ 1.3399 & 0.5388 \\ 0.5815 & 0.6413 \\ 0.5347 & 0.0488 \\ 1.9283 & 0.7388 \\ 0.0268 & 0.4614 \\ 1.9763 & 0.7172 \end{pmatrix}
$$

16.2 For the data in Exercise 16.1 find investment proportions p_0 in the riskless asset and p_1, p_2 in the first two risky assets such that

$$\mathbf{p} = (p_0, p_1, p_2, +.10, -.10, +.10, -.10, +.10, -.10, +.10, -.10)$$

is a zero-beta, arbitrage portfolio.

16.3 Using the data in Table 16.1, find some other opportunities for type 2 arbitrage.

16.4 A paranoiac investor foresees financial and political events that make it prudent to hold part of wealth in gold bullion. Not having enough cash to buy gold now, the investor plans to enter into a forward contract with a New York bank to take delivery of 1000 oz of gold in 3 months (i.e., at $T = .25$ years from now). The bank charges a monthly fee of $c = \$3$ per ounce to store and insure gold, to be paid at the end of each month. Spot gold is currently selling for $S_0 = \$800/\text{oz}$, and one can borrow and lend risklessly for periods up to 3 months at a continuously compounded, annualized rate $r = .04$. Determine the arbitrage-free 3-month forward price for gold f_0.

16.5 Show that (16.7) is consistent with relation $B_t(t', T) = B(t, T)/B(t, t')$ from (4.7) that gives the forward price at t for delivery at t' of a T-maturing discount bond.

16.6 The prices of riskless discount bonds paying one unit of US currency (\$) and one unit of European currency (€) on December 31, 2005 are tabulated below as of January 1, 2005 and May 1, 2005, along with the \$/€ exchange rate on those two dates:

	1/01	5/01
US bond price (\$)	0.932	0.949
Euro bond price (€)	0.914	0.935
\$/€ rate	0.937	0.893

On January 1 an American bank entered a forward contract to accept delivery of one million euros on December 31, 2005.

(a) Determine the forward price in $ per € that would have been established on January 1.

(b) Find the arbitrage-free value of this commitment (from the standpoint of the American bank) as of May 1.

16.7 If interest rates and bond prices were known in advance over the life of a forward contract, it would be possible to replicate its terminal value precisely by taking appropriate positions in futures, despite the practice of marking futures positions to market. To illustrate, consider a contract that expires in just 2 days. Measuring time in days for simplicity, let $B(0, 1)$, $B(0, 2)$, and $B(1, 2)$ represent the prices of 1- and 2-day unit discount bonds as of $t = 0$ and of a 1-day bond as of $t = 1$; let S_2 be the $t = 2$ spot price per unit of the underlying commodity; and let F_0, F_1 and f_0, f_1 be the per-unit futures and forward prices at $t = 0$ and $t = 1$ for delivery at $t = 2$. We assume (counterfactually) that $B(1, 2)$ is known as of $t = 0$, but of course the forward, futures, and spot prices after $t = 0$ are not known initially. At $t = 1$, there is no distinction at all between a forward contract and a futures contract, since both will expire in one day, so we must have $F_1 = f_1$. Of course, $f_2 = F_2 \equiv S_2$ since futures and forward prices for immediate delivery must coincide with the spot price in an arbitrage-free, frictionless market. Now the terminal, per-unit value of a forward contract initiated at $t = 0$ is $\mathfrak{F}(S_2, 0) = S_2 - f_0$. *Assuming* that $F_0 = f_0$ also, show that this terminal value could be replicated precisely—no matter what value S_2 might have—by taking appropriate positions in futures contracts at $t = 0$ and $t = 1$. Conclude that if future bond prices were known in advance, futures prices and forward prices would have to agree in arbitrage-free, frictionless markets.

CHAPTER 17

DYNAMIC ARBITRAGE PRICING

We have seen how Steve Ross used arbitrage arguments to develop a model for the cross section of expected rates of return of primary financial assets, such as stocks and bonds, whose values are not directly tied to those of other assets. We have also seen how the possibility of arbitrage restricts a dealer's bid and asked quotes for default-free bonds of various maturities. Finally, we have seen one example of how arbitrage arguments are used to price a derivative asset—one whose value *is* tied to the price of another. We saw that the payoff of a contract for forward delivery of an investment asset can, at least in frictionless markets, be exactly replicated by a static, buy-and-hold portfolio of the underlying asset and riskless bonds. It was possible to replicate the payoff of the forward because the contract's terminal value, $\mathfrak{F}(S_T, 0) = S_T - \mathsf{f}_0$, is a linear function of the values of these assets at terminal date T. Of course, values of portfolios are themselves linear functions of the values of the component assets, so replicating the payoff of the forward is just a matter of picking and holding the right portfolio. In this chapter we will take a first look at how arbitrage arguments are applied to determine values of assets whose payoffs are *nonlinear* functions of traded assets. In this situation, no buy-and-hold portfolio can precisely replicate the terminal payoff for all possible values of the underlying asset, so the portfolio has to be constantly rebalanced as time and underlying prices evolve. In other words, the replication has to be *dynamic*. Moreover, if we are to judge the value of the derivative by the value of the replicating portfolio, the portfolio must be self-financing, meaning that it does not require additional resources after once being put in place. Of course, buy-and-hold portfolios are automatically self-financing.

Determining the replicating portfolios for derivatives with nonlinear payoffs requires solving differential or partial differential equations (PDEs). One way of solving them is to use the martingale property of evolving conditional expectations, a process that is explained below. An alternative to pricing by solving PDEs is to apply what is now called the *fundamental theorem of asset pricing*. The theorem states that all assets and derivatives become martingales in *some* probability measure, once they are normalized by the price of some asset that serves as an appropriate numeraire. This fact allows arbitrage-free values of derivatives to be found without specifically working out the replicating portfolio or even writing down a PDE. We discuss martingale pricing in Chapter 19 and concentrate here on pricing by replication in the classic Black–Scholes–Merton setting in which the prices of underlying assets follow geometric Brownian motion.

17.1 DYNAMIC REPLICATION

Consider the problem of replicating a nonlinear payoff, such as that of a European-style call option on a share of common stock. A European call option expiring at some date T gives the holder the right (i.e., the *option*) to buy the underlying asset at a predetermined price at time T.[43] The predetermined price is called the *strike* or *exercise* price, and we use the symbol X to represent it. The contractual obligation imposed by the call differs from that of the forward contract, since the party on the long side of the forward *must* pay f_0 and take delivery of the underlying, whereas the holder of the option will do so only if it is advantageous. Thus, a European call is exercised if and only if $S_T - X > 0$ and is otherwise let to expire. This makes its terminal value to the holder (i.e., the value at T) not $S_T - X$ but $\max(S_T - X, 0)$ or, in more compact notation, $(S_T - X)^+$. A plot shows a function that equals zero for all values of S_T up to and including X, then increases dollar for dollar with S_T thereafter. Clearly, this kinked curve does not describe a *linear* function. To match the contract's market value for each value of $S_T \leq X$, a replicating portfolio would have to be worthless (e.g., containing no stock and no cash), while for each value of $S_T > X$ it would need to contain one share of stock and a maturing debt of $\$X$. Similarly, the value of a European-style put option, which gives the right to *sell* the underlying for X at T, is the nonlinear function $(X - S_T)^+$. To replicate this payoff the portfolio would have to be worthless for all $S_T \geq X$ and to be long X dollars' worth of T-maturing bonds and short the stock when $S_T < X$.

---

[43]"American" calls also give holders the right to buy at a predetermined price, but they can be exercised at any time up to and including T. By focusing on European options we avoid the difficult problem of valuing the right to early exercise, which is discussed briefly in Section 18.5. Despite the geographic names, both types of options are traded in markets around the world.

Replicating a derivative asset dynamically requires buying some initial portfolio of traded assets, then following a program for trading the assets in response to the passage of time and fluctuations in the underlying price (and any other "state" variables that influence the derivative's value). The trading must be such that (1) the portfolio's value at T matches that of the derivative asset and (2) the portfolio is self-financing. As we have seen, the latter requirement means that funds used to buy any asset must be financed by sales of others and that funds generated by sales must be used for new purchases. If the initial composition of the portfolio can be determined, and if an appropriate trading plan can be developed, then the initial value of this self-financing, replicating portfolio must match that of the derivative. Otherwise, by buying whichever is cheaper and shorting the other, one makes money now with no chance of losing in the future—a type 2 arbitrage.

17.2   MODELING PRICES OF THE ASSETS

The static policy that replicates the forward contract could be designed without any knowledge of how the price of the underlying commodity might change through time. This is so because one merely has to buy and hold with no intermediate trading. By contrast, designing a dynamic procedure does require having a model for prices of the assets used to replicate. For this introduction to the concept of dynamic replication we will adopt the model used by Fischer Black and Myron Scholes (1973) and Robert Merton (1973) in developing their famous formulas for prices of European-style options. They model the price process $\{S_t\}_{t \geq 0}$ of the underlying asset—which we henceforth refer to as a "stock"—as following geometric Brownian motion:

$$dS_t = \mu S_t \cdot dt + \sigma S_t \cdot dW_t.$$

Here μ and σ are constants, and Brownian motion $\{W_t\}_{t \geq 0}$ is adapted to the evolving information set (filtration) $\{\mathcal{F}_t\}_{t \geq 0}$. Thus, at t the current value and history of the stock's price are known. Recall from Chapter 11 that this model implies that sample paths of $\{S_t\}_{t \geq 0}$ are continuous and that the conditional distribution given $\mathcal{F}_t$ of the value at any future date is distributed as lognormal.

Under this model only one additional asset is needed to replicate the payoff of an option. We can use for this purpose discount, default-free bonds maturing at T or later. Alternatively, we can use the idealized money-market fund that holds and constantly rolls over bonds that are on the verge of maturing, thus earning the instantaneous spot rate of interest. In either case, for the basic Black–Scholes theory we assume that spot rate r_t is either constant (i.e., $r_t = r$ for each t) or, if time varying, is known in advance over the life of the option. This implies that future values of the bond and money fund are also known in advance, which means that we can work with just one *risky* asset—the underlying stock itself. To keep the notation simple, we will work initially with a money fund that earns a *constant* spot rate, having per-share value $M_t = M_0 e^{rt}$ at $t \geq 0$ and some arbitrary initial value M_0 (e.g., \$1). The price at t of a T-expiring discount bond paying \$1 at maturity is then $B(t, T) = e^{-r(T-t)} = M_t/M_T$. With this model we have $dM_t = r M_0 e^{rt} \cdot dt = r M_t \cdot dt$ and $dB(t, T) = r B(t, T) \cdot dt$

as the "instantaneous" changes in per-unit values of money fund and bond. For the time being we assume that the underlying stock pays no dividend and costs nothing to store. What happens when there are dividends and when short rates are time varying but known in advance is discussed at the end of the chapter.

Finally, the possibility of replicating precisely depends on some obviously counterfactual institutional assumptions about markets and the technology of trading. Specifically, we require that assets be perfectly divisible, that markets be always open and perfectly competitive, and that trading can take place instantly, continuously, and costlessly. While counterfactual, these assumptions let us focus on the principal market forces, in much the same way that we learn the principles of Newtonian mechanics by ignoring the effect of air on the acceleration of dense objects in Earth's gravitational field.

17.3   THE FUNDAMENTAL PARTIAL DIFFERENTIAL EQUATION (PDE)

To broaden our scope a bit, we shall consider how to price a *generic* T-expiring, European-style derivative asset. It could be a call option, a put option, or something else, like a "straddle" (combination of call and put). For that matter, it could still be a forward contract. Whatever it is, its value at expiration, $D(S_T, 0)$ say, is supposed to be a known function of the value of the underlying stock at T. Thus, for European calls and puts with strike price X we would have $D(S_T, 0) = C(S_T, 0) = (S_T - X)^+$ and $D(S_T, 0) = P(S_T, 0) = (X - S_T)^+$. The value at any $t \in [0, T]$ is represented explicitly as a function $D(S_t, T - t)$ of the two time-varying quantities S_t and $T - t$, the current underlying price and the time remaining until expiration. We shall assume that this function, whatever it turns out to be, is "smooth," in the sense of having at least one continuous time derivative and two continuous derivatives with respect to price. This will make it possible to apply Itô's formula. The goal in "pricing" the derivative asset is to specify precisely the function $D(S_t, T - t)$ for any $S_t \geq 0$ and any $t \in [0, T]$.

Let us now try to replicate the derivative's (as yet unknown) value at any such t using some number p_t shares of the stock and q_t units of the money fund. We want the value of this portfolio, $V_t \equiv p_t S_t + q_t M_t$, to equal $D(S_t, T - t)$ at each t, and, in particular, we want V_T to equal the known function $D(S_T, 0)$. Replication requires that the instantaneous changes in V_t and $D(S_t, T - t)$ be precisely the same with probability one, so we must have $dV_t = dD(S_t, T - t)$. Applying Itô's formula to V_t gives

$$dV_t = \left(\frac{\partial V_t}{\partial S_t} \cdot dS_t + \frac{\partial V_t}{\partial M_t} \cdot dM_t \right) + \left(\frac{\partial V_t}{\partial p_t} \cdot dp_t + \frac{\partial V_t}{\partial q_t} \cdot dq_t \right)$$
$$= (p_t \cdot dS_t + q_t \cdot dM_t) + (S_t \cdot dp_t + M_t \cdot dq_t).$$

Note that changes in p_t and q_t are under our control, so there is no contribution from derivative $\partial^2 V_t / \partial p_t \partial S_t$. Here the first term in parentheses represents the effects of changes in prices, while the second term results from changes in the composition of the portfolio. Although the composition will indeed change as we replicate dynamically, this second term has to be zero, since purchases of one asset must be financed by sales of the other. Thus

$$dV_t = p_t \cdot dS_t + q_t \cdot dM_t$$

$$= p_t S_t \cdot \frac{dS_t}{S_t} + (D - p_t S_t) \cdot \frac{dM_t}{M_t}$$

$$= p_t S_t \left(\mu \cdot dt + \sigma \cdot dW_t \right) + (D - p_t S_t)\, r \cdot dt$$

$$= [Dr + p_t S_t \left(\mu - r \right)] \cdot dt + p_t S_t \sigma \cdot dW_t, \qquad (17.1)$$

where D by itself denotes $D\left(S_t, T - t\right)$. Now we apply Itô's formula again to express the instantaneous change in D:

$$dD = -D_{T-t} \cdot dt + D_S \cdot dS_t + \frac{D_{SS}}{2} \cdot d\langle S \rangle_t$$

$$= -D_{T-t} \cdot dt + D_S S_t \mu \cdot dt + D_S \sigma S_t \cdot dW_t + \frac{D_{SS}}{2} S_t^2 \sigma^2 \cdot dt$$

$$= \left(-D_{T-t} + D_S S_t \mu + \frac{D_{SS}}{2} S_t^2 \sigma^2 \right) \cdot dt + D_S S_t \sigma \cdot dW_t. \qquad (17.2)$$

Here, subscripts on D indicate partial derivatives, as $D_{T-t} \equiv \partial D / \partial \left(T - t \right)$, and the arguments $\left(S_t, T - t \right)$ have been suppressed throughout. Note that

$$\frac{\partial D\left(S_t, T - t\right)}{\partial t} = \frac{\partial D\left(S_t, T - t\right)}{\partial \left(T - t\right)} \frac{\partial \left(T - t\right)}{\partial t}$$

$$= -D_{T-t},$$

which explains the sign of D_{T-t} in (17.2). The expressions for dV_t and dD show that both V_t and $D\left(S_t, T - t\right)$ are themselves Itô processes. Each has a trend or drift part—the dt term, and each has a stochastic part—the dW_t term. Requiring that $dV_t = dD$ with probability one means that the trend parts of the two expressions must be precisely the same, and likewise the stochastic parts. Equating the stochastic parts gives $p_t S_t \sigma = D_S S_t \sigma$, and hence $p_t = D_S$. Thus, the number of shares of the underlying stock held at t must equal the partial derivative of D with respect to S_t, which is called the *delta* of the derivative asset. Of course, at this point we do not yet know the value of D_S, since we do not yet know the function D. Be patient; we will find it.

Setting $p_t = D_S$ in the drift term of (17.2), subtracting the drift term in (17.1), and cancelling the $D_S S_t \mu$ terms give

$$0 = -D_{T-t} + D_S S_t r + \tfrac{1}{2} D_{SS} S_t^2 \sigma^2 - Dr. \qquad (17.3)$$

As a relation between the function D and its partial derivatives with respect to the two arguments $T - t$ and S_t, Eq. (17.3) constitutes a partial differential equation— the word "partial" distinguishing it from an "ordinary" differential equation for a function of just one variable. Loosely speaking, (17.3) represents the equation of motion of D as time progresses and the price of the underlying security changes.

Given the models for S_t and M_t, *any* derivative asset with S_t as underlying price—call option, put option, forward contract, or whatever—must follow precisely this PDE if there is to be no opportunity for arbitrage. For this reason, it is often called the *fundamental PDE* for derivatives pricing. The solutions to the PDE will depend on certain initial and boundary conditions that specify how D behaves at extreme values of the arguments. It is these extra conditions that define the particular derivative. As applied to European puts and calls, it happens that the solution requires only an initial condition.[44] That initial condition simply sets $D(S_T, 0)$ equal to whatever function of S_T is specified in the contract that defines the particular derivative asset: $(S_T - X)^+$ and $(X - S_T)^+$ for call and put with strike X, and $S_T - f_0$ for a forward contract. Noting similarities with the heat equation in physics, whose properties were already well known, Black and Scholes were able to give the following solutions for this PDE for European calls and puts:

$$C(S_t, \tau) = S_t \Phi \left[q^+ \left(\frac{S_t}{BX}, \tau \right) \right] - BX \Phi \left[q^- \left(\frac{S_t}{BX}, \tau \right) \right] \qquad (17.4)$$

$$P(S_t, \tau) = BX \Phi \left[q^+ \left(\frac{BX}{S_t}, \tau \right) \right] - S_t \Phi \left[q^- \left(\frac{BX}{S_t}, \tau \right) \right]. \qquad (17.5)$$

Here (1) $\tau \equiv T - t$ is the time to expiration, (2) $B \equiv B(t, T) = e^{-r\tau}$ is the price at t of a T-maturing unit discount bond, (3) Φ is the standard normal CDF, and (4)

$$q^{\pm}(s, \tau) \equiv \frac{\ln s \pm \sigma^2 \tau / 2}{\sigma \sqrt{\tau}}$$

for positive s and τ.

17.3.1 The Feynman–Kac Solution to the PDE

We shall study some of the properties of the Black–Scholes–Merton formulas in the next chapter, but to reduce some of the mystery in their development let us now look at an ingenious way to solve (17.3), attributed to physicist Richard Feynman (1948) and probabilist Mark Kac (1949). The method requires no knowledge of physics or of the myriad other techniques for solving PDEs, but we do need some understanding of probability theory and stochastic processes. Besides being useful of itself (and a thing of beauty!), the method also connects the Black–Scholes–Merton approach to pricing derivatives with the more modern martingale approach that we discuss in Chapter 19.

To see how this works, we need to recall the fundamental property of conditional expectation that was discussed in Section 3.4.2; namely, the tower property or the

---

[44]The solution subject to the condition that $D(S_T, 0)$ equal the contractually specified value at T also satisfies the conditions that must hold at the boundary values of S_t: $C(0, T - t) = 0$ and $C(+\infty, T - t) = +\infty$ for the call; $P(0, T - t) = B(t, T)X$ and $P(+\infty, T - t) = 0$ for a put; and $\mathcal{F}(0, T - t) = -B(t, T)f_0$ and $\mathcal{F}(+\infty, T - t) = +\infty$ for a forward contract. If we think of $D(S_t, T - t)$ as a function of $T - t$, then it makes sense to regard the specification of terminal value as an "initial" condition.

law of iterated expectations. To expand a bit on its meaning, let Y_T be the value of some $\mathcal{F}_T$-measurable random variable—one whose realization will be known at T; let $\mathcal{F}_t$ be the information available at $t \leq T$; and assume that $E |Y_T| < \infty$. Then the conditional expectation $E(Y_T \mid \mathcal{F}_t)$ exists and depends, in general, on aspects of the environment up through t. One can think of $E(Y_T \mid \mathcal{F}_t)$ as our best guess or forecast of Y_T given all that we know at t. Indeed, of the infinitely many $\mathcal{F}_t$-measurable quantities $\hat{Y}_t$ it is the one that minimizes mean squared error, $E\left[\left(Y_T - \hat{Y}_t\right)^2 \mid \mathcal{F}_t\right]$. Of course, as of any time $s < t$, we rarely know all the features of the time t environment on which our future forecast will be made, and so $E(Y_T \mid \mathcal{F}_t)$ is a random variable as of time s. The expectation of this random variable given $\mathcal{F}_s$ is $E[E(Y_T \mid \mathcal{F}_t) \mid \mathcal{F}_s]$. With the same interpretation as before, this is our best guess as of time s of what our best guess will be at time t. Now let us assume that $\mathcal{F}_s \subseteq \mathcal{F}_t$ for each $s < t$. This means that information can only grow or remain the same over time—that we will have never forgotten at t anything that was known at s. Under this condition the tower property proclaims that the best guess of the later best guess of Y_T is merely the *current* best guess of Y_T. In symbols, this is

$$E[E(Y_T \mid \mathcal{F}_t) \mid \mathcal{F}_s] = E(Y_T \mid \mathcal{F}_s).$$

Now let us think specifically of the information sets themselves as evolving through (continuous) time in such a way that $\mathcal{F}_s \subseteq \mathcal{F}_t$ for each $s < t$. As explained in Section 10.3.3, as t advances the collection or *filtration* $\{\mathcal{F}_t\}_{0 \leq t < T}$ catches more and more of the information that had previously slipped through and escaped our knowledge. As t advances, we can think of the expectations conditional on the time t element of the filtration as a stochastic process, $\{Q_t\}_{0 \leq t \leq T}$, with $Q_t \equiv E(Y_T \mid \mathcal{F}_t)$ depending, potentially, on the entire history of the environment up to t. Clearly, $Q_T \equiv E(Y_T \mid \mathcal{F}_T)$ is just Y_T itself, since the value of Y_T is included in information $\mathcal{F}_T$. More generally, the entire process $\{Q_t\}_{0 \leq t \leq T}$ is adapted to $\{\mathcal{F}_t\}_{0 \leq t \leq T}$, since for any t the value of Q_t and its past are known once $\mathcal{F}_t$ becomes available. Now here is the critical feature of this conditional-expectations process:

> The process $\{Q_t\}_{0 \leq t \leq T} \equiv \{E(Y_T \mid \mathcal{F}_t)\}_{0 \leq t \leq T}$ is a martingale adapted to $\{\mathcal{F}_t\}_{0 \leq t \leq T}$.

The proof of this is simple. First, by regarding initial value Y_0 as deterministic, we can take the unconditional expectation $E(\cdot)$ to mean $E(\cdot \mid \mathcal{F}_0)$. There are two things to prove: (1) integrability—that $E|Q_t| < \infty$ for each t—and (2) the fair-game property, that $E(Q_t \mid \mathcal{F}_s) = Q_s$ for $0 \leq s \leq t \leq T$. Here is the first:

$$E|Q_t| = E[|E(Y_T \mid \mathcal{F}_t)|] \leq E[E(|Y_T| \mid \mathcal{F}_t)] = E|Y_T| < \infty.$$

The fair-game part is left as an exercise. We need two more basic facts: (1) an $\{\mathcal{F}_t\}$-martingale $\{Q_t\}$ has the property that

$$E(Q_t - Q_s \mid \mathcal{F}_s) = E(Q_t \mid \mathcal{F}_s) - Q_s = Q_s - Q_s = 0$$

for $s \leq t$—thus, increments to a martingale have conditional expectation zero; and (2) subject to good behavior on the part of $\{\sigma_t\}_{0 \leq t \leq T}$, the Itô process $\{X_t\}_{0 \leq t \leq T}$

that evolves as $dX_t = \mu_t \cdot dt + \sigma_t \cdot dW_t$ is a martingale if and only if drift process $\{\mu_t\}_{t \geq 0}$ is identically zero. This is easily remembered by reflecting that the expected value of the "instantaneous" increment, $E(dX_t \mid \mathcal{F}_t) = \mu_t \cdot dt + \sigma_t \cdot E(dW_t \mid \mathcal{F}_t) = \mu_t \cdot dt$, must be zero for martingales. A sufficient condition for the good behavior of $\{\sigma_t\}_{0 \leq t \leq T}$ is that $E \int_0^T \sigma_t^2 \cdot dt < \infty$. If $\{X_t\}_{t \geq 0}$ is a geometric Brownian motion with $dX_t = \mu X_t \cdot dt + \sigma X_t \cdot dW_t$, then $\sigma_t = \sigma \bar{X}_t$. Exercise 17.2 shows that such a process meets the sufficient condition and is a martingale when $\mu = 0$.

Now let us apply all this to solving the fundamental PDE when our stock's price evolves as $dS_t = \mu S_t \cdot dt + \sigma S_t \cdot dW_t$. As a first try—which won't quite get us where we want to go but will point the right way—let us define $Q_t = E_t D(S_T, 0)$, where we now use $E_t(\cdot)$ for $E(\cdot \mid \mathcal{F}_t)$. Notice—crucial point!—that Q_T is just $D(S_T, 0)$, the *known* function of S_T that specifies the derivative's payoff at expiration. We assume, of course, that $E|D(S_T, 0)| < \infty$. Now for $t < T$ we can express Q_t as a function $Q_t = D(S_t, T - t)$, since under geometric Brownian motion the conditional distribution of S_T given $\mathcal{F}_t$ depends only on *current* value S_t and time. (When the dust settles and we have got the right version of Q_t, $D(S_t, T - t)$ will emerge as the time t value of the derivative, but for now the symbol just represents some undetermined function of price and time.) Then, assuming that D is sufficiently smooth, Itô's formula gives

$$dQ_t = -D_{T-t} \cdot dt + D_S \cdot dS_t + \frac{D_{SS}}{2} \cdot d\langle S \rangle_t$$

$$= \left(-D_{T-t} + D_S \mu S_t + \frac{D_{SS}}{2} \sigma^2 S_t^2 \right) \cdot dt + D_S \sigma S_t \cdot dW_t.$$

But the drift term here must be zero, since $\{Q_t\}$ is an $\{\mathcal{F}_t\}$-martingale. Thus, our function D obeys the PDE

$$0 = -D_{T-t} + D_S \mu S_t + \frac{D_{SS}}{2} \sigma^2 S_t^2. \tag{17.6}$$

By construction, this is a PDE whose solution is known (or at least determinable); that is,

$$D(S_t, T - t) = Q_t = E_t Q_T = E_t D(S_T, 0) = \int D(s, 0) f(s \mid \mathcal{F}_t) \cdot ds,$$

where $f(s \mid \mathcal{F}_t)$ is the conditional density. In other words, we can find $D(S_t, T - t)$ for $t < T$ just by working out the integral. However—and this is why our first try does not work—(17.6) is not quite the same as the PDE (17.3) that describes our derivative asset. That PDE has r instead of μ in the second term, and it contains the additional term $-rD$. So, what to do? The obvious thing to do—well, it was obvious for Richard Feynman—is to try to *modify* the definition of Q_t so as to produce the correct PDE.

To this end, construct another martingale as the conditional expectation of $D(S_T, 0)$ discounted at the riskless rate:

$$Q_t' \equiv E_t e^{-rT} D(S_T, 0) = e^{-rT} E_t D(S_T, 0).$$

Now, defining a *new* function D implicitly for $t < T$ by putting $e^{-rt}D(S_t, T - t)$ equal to Q_t' (happily, the new D will turn out to be the function we want!) and applying Itô's formula give

$$dQ_t' = e^{-rt}\left(-D_{T-t} + D_S\mu S_t + \frac{D_{SS}}{2}\sigma^2 S_t^2 - rD\right) \cdot dt + D_S\sigma S_t \cdot dW_t.$$

Since $\{Q_t'\}$ is still an $\{\mathcal{F}_t\}$ martingale, we thus have a new PDE:

$$0 = -D_{T-t} + D_S\mu S_t + \frac{D_{SS}}{2}\sigma^2 S_t^2 - rD.$$

This does have the $-rD$ term missing from (17.6), but we still need somehow to replace μ with r in the second term. So let us just *pretend* that the model for dS_t has mean drift equal to the riskless rate, as $dS_t = rS_t \cdot dt + \sigma S_t \cdot dW_t$. With this sleight of hand we have indeed constructed a function that satisfies the fundamental PDE and whose solution is known in general terms. Thus, $e^{-rt}D(S_t, T - t) = \hat{E}_t\left[e^{-rT}D(S_T, 0)\right]$ implies that

$$D(S_t, T - t) = e^{-r(T-t)}\hat{E}_t D(S_T, 0) = B(t, T)\hat{E}_t D(S_T, 0). \tag{17.7}$$

Here, the "hat" on E indicates that the expectation is calculated under the fictional assertion that the stock's drift term is r instead of the actual μ. Now all that remains is to work out the expectation of $D(S_T, 0)$ under this pseudo model for $\{S_t\}$.

Of course, the switch from μ to r does seem a little bold if not downright shady, but we will see in Chapter 19 that there is indeed a justification that makes sense—and not only to theoretical physicists but even to probabilists and economists.

17.3.2 Working out the Expectation

We have come a long way, but finding $\hat{E}_t D(S_T, 0)$ can still require a bit of work, depending on the nature of the function. We will start by using (17.7) to price a simple forward contract, whose value we have already found by static replication, then turn to the more challenging problem of pricing a European put. In both cases we continue to assume that the underlying asset pays no dividend and that the interest rate is a constant, but we will see very soon how to generalize.

For the forward contract we have $D(S_T, 0) = \mathfrak{F}(S_T, 0) = S_T - \mathfrak{f}_0$, where $\mathfrak{f}_0$ is the forward price that was negotiated at $t = 0$. Now recall that when $\{S_t\}_{t \geq 0}$ is a geometric Brownian motion the terminal value S_T is distributed as lognormal; specifically, with $S_t \in \mathcal{F}_t$ we have

$$S_T = S_t \exp\left[\left(\mu - \frac{\sigma^2}{2}\right)(T - t) + \sigma(W_T - W_t)\right]$$

$$\sim S_t \exp\left[\left(\mu - \frac{\sigma^2}{2}\right)(T - t) + \sigma\sqrt{T - t} \cdot Z\right],$$

where $Z \sim N(0,1)$. Changing μ to r and taking the expectation give $\hat{E}_t S_T = S_t e^{r(T-t)}$, so the value of the forward contract at $t \in [0, T]$ is

$$\mathfrak{F}(S_t, T-t) = B(t, T)\left[S_t e^{r(T-t)} - \mathrm{f}_0\right] = S_t - B(t, T)\,\mathrm{f}_0. \tag{17.8}$$

This corresponds to (16.6) when the dividend or cost rate δ in that formula equals zero. Here, the integration problem involved in finding $\hat{E}_t \mathfrak{F}(S_T, 0)$ was trivial because $\mathfrak{F}(S_T, 0)$ depends *linearly* on S_T.

Now let us deal with the put option, whose terminal value is decidedly *nonlinear* in S_T. To simplify notation, we will value a T-expiring European put with strike price X as of $t = 0$. Again changing μ to r and now using terminal value function $D(S_T, 0) = P(S_T, 0) = (X - S_T)^+$, we have

$$\hat{E}(X - S_T)^+ = E\left[X - S_0 e^{(r-\sigma^2/2)T + \sigma\sqrt{T}\cdot Z}\right]^+$$
$$= \int_{-\infty}^{\infty}\left[X - S_0 e^{(r-\sigma^2/2)T + \sigma\sqrt{T}\cdot z}\right]^+ \frac{1}{\sqrt{2\pi}}e^{-z^2/2} \cdot dz.$$

Now $S_0 e^{(r-\sigma^2/2)T + \sigma\sqrt{T}\cdot z} < X$ (the condition for the put to be in the money) if and only if z is less than

$$\frac{\ln(X/S_0) - rT + \sigma^2 T/2}{\sigma\sqrt{T}} = \frac{\ln(BX/S_0) + \sigma^2 T/2}{\sigma\sqrt{T}} \equiv q^+\left(\frac{BX}{S_0}, T\right),$$

where $B \equiv B(0, T) = e^{-rT}$. Limiting the range of integration to such values of z removes the annoying $[\cdot]^+$ from the integrand, giving

$$\hat{E}(X - S_T)^+ = X\Phi(q^+) - S_0 e^{rT}\int_{-\infty}^{q^+}\frac{1}{\sqrt{2\pi}}e^{-\sigma^2 T/2 + \sigma\sqrt{T}\cdot z - z^2/2} \cdot dz.$$

Changing variables in the integral as $y = z - \sigma\sqrt{T}$ yields

$$\int_{-\infty}^{q^+ - \sigma\sqrt{T}}\frac{1}{\sqrt{2\pi}}e^{-y^2/2} \cdot dy = \Phi\left(q^+ - \sigma\sqrt{T}\right) \equiv \Phi(q^-)$$

and gives for the initial value of the T-expiring put

$$P(S_0, T) = B(0, T)\,\hat{E}(X - S_T)^+$$
$$= BX\Phi\left[q^+\left(\frac{BX}{S_0}, T\right)\right] - S_0\Phi\left[q^-\left(\frac{BX}{S_0}, T\right)\right],$$

where

$$q^\pm\left(\frac{BX}{S_0}, T\right) \equiv \frac{\ln(BX/S_0) \pm \sigma^2 T/2}{\sigma\sqrt{T}}.$$

This corresponds to the Black–Scholes expression for $P(S_t, \tau)$ in (17.5) when $t = 0$ and $\tau \equiv T - t = T$.

17.4 ALLOWING DIVIDENDS AND TIME-VARYING RATES

Many underlying assets on which options trade produce some positive cash flow in the form of dividends or interest. For example, many common stocks pay quarterly cash dividends, the values of which—although clearly not known far in advance—are highly predictable during the company's fiscal year. Since companies' payment schedules vary, dividends of component stocks in diversified stock indexes like the S&P 500 dribble out more or less continuously. In principle, these could be used as they are received to buy more of the portfolio itself. Likewise, holdings of foreign currencies, like the RK (Republic of Korea) won in the example of forward contracts in the previous chapter, usually earn interest at the going foreign rate. In effect, this is automatically reinvested, causing the value of the underlying position to grow more or less continuously. Whatever the form in which they occur, the presence of such payments to holders of the underlying asset affects the composition of replicating portfolios and therefore alters the pricing formulas for derivative assets. Although there is no perfectly correct, simple adjustment to the Black–Scholes–Merton formulas to allow for lump-sum cash dividends such as those paid by individual common stocks, it is not difficult to handle *continuous* payments that are instantaneously reinvested. If cash payments are made at a known, continuously compounded rate δ and reinvested in the stock as they are received, then a share purchased at time 0 would be worth $S_t^+ \equiv S_t e^{\delta t}$ at t. To construct a self-financing portfolio that dynamically replicates the derivative, we would then have to work with this *cum*-dividend position. Exercise 17.3 shows that the appropriate change in formulas (17.4) and (17.5) is to replace S_t by $S_t e^{-\delta(T-t)}$. The same change applies for formula (17.8), which makes it correspond to the general formula (16.6) worked out in Section 16.4. If $\delta > 0$, the factor $e^{-\delta(T-t)} < 1$ is the number of shares that one would have to hold at t in order to wind up with one share at T if cash flows are continuously reinvested. Of course, if $\delta < 0$, these are *outflows* that have to be financed by selling the underlying, in which case one would need to start out with more than one share. Recalling expression (16.7) in the discussion of forward contracts, notice that $S_t e^{-\delta(T-t)}/B(t,T) = f_t$ would be the *forward* price for time T delivery of the underlying on a contract *initiated* at t. With this substitution and again putting $\tau \equiv T - t$ formulas (17.4) and (17.5) become

$$C(S_t, \tau) = Bf_t \Phi\left[q^+\left(\frac{f_t}{X}, \tau\right)\right] - BX\Phi\left[q^-\left(\frac{f_t}{X}, \tau\right)\right] \qquad (17.9)$$

$$P(S_t, \tau) = BX\Phi\left[q^+\left(\frac{X}{f_t}, \tau\right)\right] - Bf_t \Phi\left[q^-\left(\frac{X}{f_t}, \tau\right)\right], \qquad (17.10)$$

where, for positive s and τ,

$$q^\pm(s, \tau) \equiv \frac{\ln s \pm \sigma^2 \tau/2}{\sigma\sqrt{\tau}}.$$

We have simplified the derivation of the pricing formulas by regarding the short rate of interest as constant over the life of the derivative. But while short rates are typically far less volatile than prices of stocks, they do vary over time, and their future course is by no means fully predictable. To allow for stochastic rates requires a model for how they and the price of the underlying asset jointly evolve. How stock options and other derivatives can be priced under these conditions is briefly outlined in Section 19.6. However, to allow for *variation* in short rates, so long as they are considered known in advance, involves no substantive changes in the Black–Scholes formulas. In this case instantaneous changes in the value of the money fund are still predictable, so we just have $dM_t/M_t = r_t \cdot dt$ instead of $dM_t/M_t = r \cdot dt$, and the development from relations (17.1)–(17.3) remains the same. In the Feynman–Kac solution we replace $Q'_t \equiv e^{-rT} E_t D\left(S_T, 0\right)$ by $Q'_t \equiv \exp\left(-\int_0^T r_u \cdot du\right) E_t D\left(S_T, 0\right)$ and define $D\left(S_t, T-t\right)$ for $t < T$ implicitly by $\exp\left(-\int_0^t r_u \cdot du\right) D\left(S_t, T-t\right) \equiv Q'_t$, so that $D(t, T-t) = \exp\left(-\int_t^T r_u \cdot du\right) \hat{E}_t D\left(S_T, 0\right)$. Now, recalling the discussion in Section 4.3, if short rates are known in advance, then they must be the same as the forward rates implicit in prices of discount bonds, so $\{r_u\}_{t \leq u \leq T} = \{r_t(u)\}_{t \leq u \leq T}$. Thus, assuming that short rates are known amounts simply to assuming that they coincide with forward rates. Since current prices of discount bonds are representable explicitly in terms of instantaneous forward rates, as $B\left(t, T\right) = \exp\left(-\int_t^T r_t\left(u\right) \cdot du\right)$, we also have the representation $B(t, T) = \exp\left(-\int_t^T r_u \cdot du\right)$ when spot and forward rates coincide. Thus, formulas (17.9) and (17.10) continue to hold under these conditions.

EXERCISES

17.1 Prove that conditional expectations process $\{Q_t \equiv E\left(Y_T \mid \mathcal{F}_t\right)\}_{0 \leq t \leq T}$ obeys the fair-game property of martingales when Y_T is an $\mathcal{F}_T$-measurable random variable with $E|Y_T| < \infty$.

17.2 Prove that the process $\{X_t\}_{0 \leq t \leq T}$ is a martingale when $dX_t = \sigma X_t \cdot dW_t$.

17.3 If the underlying asset for a derivative pays a dividend at a continuously compounded rate δ, then a replicating portfolio containing the underlying and discount bonds would generate a continuous flow of cash. This could be reinvested continuously in one or both assets—and *would* be reinvested, since cash is sterile. Suppose that it is reinvested solely in the underlying asset. In this case the value of $e^{-\delta T}$ shares of the underlying, purchased at $t = 0$ for $S_0 e^{-\delta T}$, would be worth $S_t^+ \equiv S_t e^{-\delta(T-t)}$ after t units of time, and a replicating portfolio containing p_t units of the augmented, *cum*-dividend position and q_t units of the money fund would then be worth $V_t^+ = p_t S_t^+ + q_t M_t$. On the other hand, the derivative's value $D\left(S_t, T-t\right)$ at time t depends just on the actual market price of one share S_t, since one who holds the derivative has no claim to the dividend. Following steps like those used to derive (17.3), show that replication using $\{V_t^+\}$ instead of $\{V_t\}$ leads to the modified fundamental PDE:

$$0 = -D_{T-t} + D_S S_t(r - \delta) + \tfrac{1}{2} D_{SS} S_t^2 \sigma^2 - Dr. \tag{17.11}$$

17.4 Explain why the Feynman–Kac solution of (17.11) subject to the initial conditions for calls and puts leads to the formulas (17.9) and (17.10) when the underlying pays dividends continuously at rate δ.

17.5 Show that the expression $\mathfrak{F}(S_t, T - t) = S_t e^{-\delta(T-t)} - f_0 e^{-r(T-t)}$ for the value of a forward contract satisfies PDE (17.11) when $\{S_t\}_{0 \le t \le T}$ is a geometric Brownian motion.

17.6 $\{S_t\}_{t \ge 0}$ is a geometric Brownian motion, with $dS_t = \mu S_t \cdot dt + \sigma S_t \cdot dW_t$, and $M_t = M_0 e^{rt}$ is the value of the money fund at t. Under what condition is normalized process $\{S_t^* \equiv S_t/M_t\}$ a martingale?

CHAPTER 18

PROPERTIES OF OPTION PRICES

In the previous chapter we worked out the Black–Scholes–Merton formulas for prices of European options as an exercise in dynamic replication. We can now step back and consider some general properties that prices of options must have—regardless of the dynamics of underlying price—if we are to rule out simple arbitrages. After verifying that the Black–Scholes formulas satisfy these basic conditions, we will proceed to examine some of their other features. In particular, we will see that the formulas actually provide the recipes for the replicating portfolios of stock and money fund. Running such a replicating portfolio is an example of what is called *delta hedging*. We will explain how delta hedging has been used as an attempt to "insure" an arbitrary stock portfolio. Finally, we will look briefly at the problem of pricing American-style options—those that can be exercised at any time up to expiration.

18.1 BOUNDS ON PRICES OF EUROPEAN OPTIONS

It is easy to see that prices of European options must fall within certain limits if there is to be no opportunity for arbitrage. We assume, as usual, that the underlying price is bounded below by zero and that there are no transaction costs.

Starting with a lower bound for values of call options, notice that the obvious inequality $(S_T - X)^+ \geq S_T - X$ implies that the call's terminal value is at least as great as that of a forward contract with forward price X. Assuming that the underlying pays a continuous dividend at rate $\delta \geq 0$, the time t value of such a forward would be $\mathfrak{F}(S_t, T - t) = S_t e^{-\delta(T-t)} - B(t, T) X$. Although this can well be negative, the value of a call option cannot; for if $C(S_t, T - t) < 0$ we could buy the call for a negative sum, getting cash up front, then simply hold it for possible additional value at expiration. This would be a type 2 arbitrage. Moreover, even when the forward contract has positive value, the call must always be worth at least as much; for if $0 < C(S_t, T - t) < S_t e^{-\delta(T-t)} - B(t, T) X$, one could short $e^{-\delta(T-t)}$ shares of stock and buy the call and X units of T-maturing discount bonds, pocketing $S_t e^{-\delta(T-t)} - B(t, T) X - C(S_t, T - t) > 0$ in cash. At T the bonds would generate $\$X$ in cash, but one would need S_T in cash to replace the borrowed stock and the dividends foregone by the lender. If the call expired worthless ($S_T \leq X$), this would leave $X - S_T \geq 0$. If the call were in the money ($S_T > X$), then one would use the $\$X$ in cash to exercise it and claim the stock with which to repay the lender, for zero net cash flow. Either way we have a type-2 arbitrage. Thus, we have as arbitrage-free lower bound

$$C(S_t, T - t) \geq \max[\mathfrak{F}(S_t, T - t), 0] = \left[S_t e^{-\delta(T-t)} - B(t, T) X\right]^+.$$

The least upper bound for $C(S_t, T - t)$ turns out to be $S_t e^{-\delta(T-t)}$. Were this violated and $C(S_t, T - t) > S_t e^{-\delta(T-t)}$, an arbitrageur could sell the call and buy $e^{-\delta(T-t)}$ shares of stock, reinvesting the dividends as received and thereby winding up with one share at T. There are then two relevant possibilities for time T. If $S_T \leq X$ the buyer of the call would not exercise it, so the arbitrageur would retain the one share of stock worth $S_T \geq 0$. If $S_T > X$ the call would be exercised and the share of stock would be given up in return for $X > 0$. Either way, the arbitrageur gets a positive cash flow at t and has nothing to lose at T—another type-2 arbitrage. Thus, arbitrage-free bounds on the values of European calls are

$$\left[S_t e^{-\delta(T-t)} - B(t, T) X\right]^+ \leq C(S_t, T - t) \leq S_t e^{-\delta(T-t)}. \tag{18.1}$$

Exercise 18.1 shows that arbitrage-free prices of European puts are bounded as

$$\left[B(t, T) X - S_t e^{-\delta(T-t)}\right]^+ \leq P(S_t, T - t) \leq B(t, T) X. \tag{18.2}$$

Except in special cases, such as when $S_t \doteq 0$, these bounds are too wide to be of much value in pricing options. For example, with $S_t = X = 100$, $\delta = 0$, and $B(t, T) = .95$ the value of a call could be anywhere between $\$5$ and $\$100$. Still, without specific models for the evolution of $\{S_t\}$ and $\{B(t, T)\}$, this is simply the best we can do. On the other hand, in a frictionless, arbitrage-free economy there is an exact expression for the *difference* between the values of European calls and puts that have the same strike prices and expirations. The relation, known as *European put–call parity*, is essential to a basic understanding of options. Establishing it is a trivial exercise in *static* replication. Observe that the difference between the time T values of the call and put is $(S_T - X)^+ - (X - S_T)^+$. A moment's thought (or drawing a picture) shows that this equals $S_T - X$, regardless of the relation between S_T and X. But, as we know, $S_T - X$ is the terminal per-unit value (to the long side) of a forward contract with forward price X. Since none of call, put, or forward produces any intermediate cash flows, the per-unit value of a position that is long the call and short the put must at *all* times equal the per-unit value of a forward contract, $\mathfrak{F}(S_t, T - t)$. Thus, we have the following lockstep parity between values of European call and put having the same X and T:

$$C(S_t, T - t) - P(S_t, T - t) = S_t e^{-\delta(T-t)} - B(t, T) X. \qquad (18.3)$$

An obvious—and important!—implication is that, having priced one of these options using an appropriate model, the arbitrage-free price of the other can be deduced immediately by simple arithmetic.

18.2 PROPERTIES OF BLACK–SCHOLES PRICES

If the Black–Scholes formulas are to be useful, they must certainly obey these basic no-arbitrage conditions that apply regardless of the behavior of the underlying and bonds. Let us see whether they do. For reference, here are the formulas again:

$$C(S_t, \tau) = B f_t \Phi \left[q^+ \left(\frac{f_t}{X}, \tau \right) \right] - B X \Phi \left[q^- \left(\frac{f_t}{X}, \tau \right) \right] \qquad (18.4)$$

$$P(S_t, \tau) = B X \Phi \left[q^+ \left(\frac{X}{f_t}, \tau \right) \right] - B f_t \Phi \left[q^- \left(\frac{X}{f_t}, \tau \right) \right], \qquad (18.5)$$

where $\tau \equiv T - t$, $B \equiv B(t, T)$, $f_t \equiv S_t e^{-\delta(T-t)} / B(t, T)$, and

$$q^{\pm}(s, \tau) \equiv \frac{\ln s \pm \sigma^2 \tau / 2}{\sigma \sqrt{\tau}}, s > 0, \tau > 0. \qquad (18.6)$$

First, let us verify that these are consistent with the initial conditions of the fundamental PDE—the ones that specify the options' terminal values. Noting that $f_t \to S_T$ as $t \to T$, suppose first that $S_T > X$, in which case the call is in the money and the put is out. We then have $\ln(f_T / X) = -\ln(X / f_T) > 0$, so as $t \to T$ the arguments of $\Phi(\cdot)$ in both terms of (18.4) approach $+\infty$, while the arguments in (18.5) approach $-\infty$. But $\Phi(+\infty) = 1$ and $\Phi(-\infty) = 0$ imply that $C(S_T, 0) = S_T - X$ and $P(S_T, 0) = 0$ when $S_T > X$. It is easy to see in the same way that $C(S_T, 0) = 0$ and $P(S_T, 0) = X - S_T$ when $S_T < X$. The case $S_T = X$ is handled a little differently, as shown in Exercise 18.2.

To see that the Black–Scholes formulas are consistent with put–call parity, note from definition (18.6) that $q^+(s^{-1}, \tau) = -q^-(s, \tau)$ and $q^-(s^{-1}, \tau) = -q^+(s, \tau)$. Thus, putting $q^{\pm} \equiv q^{\pm}(f_t / X, \tau)$ for brevity, the difference between the prices of call and put is

$$
\begin{aligned}
C\left(S_t, \tau\right) - P\left(S_t, \tau\right) &= B\left[f_t \Phi\left(q^+\right) - X\Phi\left(q^-\right)\right] - B\left[X\Phi\left(-q^-\right) - f_t\Phi\left(-q^+\right)\right] \\
&= Bf_t\left[\Phi\left(q^+\right) + \Phi\left(-q^+\right)\right] - BX\left[\Phi\left(q^-\right) + \Phi\left(-q^-\right)\right] \\
&= Bf_t - BX \text{ (since } \Phi\left(z\right) + \Phi\left(-z\right) = 1 \text{ for all } z) \\
&= e^{-\delta\tau}S_t - B\left(t, T\right)X.
\end{aligned}
$$

One interesting feature of the Black–Scholes formulas is that they do not involve the stock's mean drift rate μ. This parameter mysteriously vanished in the process of constructing the replicating portfolio and developing the fundamental PDE. Since the PDE itself does not involve μ, its solution clearly could not depend on μ either. Indeed, to use Feynman–Kac to solve the PDE, we had to *pretend* that $\mu = r$ in working out the expectation of terminal value. The irrelevance of μ for the arbitrage-free value of an option is indeed surprising at first, because the mean drift of the underlying price clearly influences the *objective* probability that the option will wind up in the money. Of course, the cold logic of the derivation provides the ultimate explanation, and we will get another take on it in Chapter 19 by changing the way we measure probability. Still, it is possible to supply a little intuition now, as follows. In the Black–Scholes framework options' payoffs can be replicated with portfolios of traded assets, and so any option position can be precisely hedged, eliminating all risk. We will see precisely how to do this in the next section. Then, since holding an option does not *necessarily* subject one to risk, an option should be priced the same way *regardless* of peoples' attitudes toward risk. In particular, the price of an option in the real world should be the same as that in a make-believe world in which everyone was risk neutral. In that case, the option's current value should equal the expectation of its value at expiration, discounted at the riskless rate. Of course, in this fictional risk-neutral world the underlying itself would be priced so that its expected total return equaled that of a riskless bond or money fund. But for $E_t S_T / S_t = E_t \exp\left[\left(\mu - \sigma^2/2\right)\left(T - t\right) + \sigma\left(W_T - W_t\right)\right] = e^{\mu(T-t)}$ to equal $M_T / M_t = B(t, T)^{-1} = e^{r(T-t)}$ requires $\mu = r$.

Given the irrelevance of μ, the only parameter from the underlying geometric BM model for $\{S_t\}$—and the only thing in the formulas that is not readily observable—is the volatility, σ. It is important to see how the Black–Scholes prices depend on σ. Differentiating (18.4) and (18.5) with respect to σ shows that prices of call and put alike increase monotonically with volatility. This makes sense, because options are really bets that the stock price will move in one direction or the other—up for calls and down for puts. Of course, the values are protected against moves in the wrong direction since out-of-money options do not have to be exercised. What happens for extreme values of σ is particularly interesting. Exercise 18.3 shows that

$$
\lim_{\sigma \downarrow 0} C\left(S_t, T - t\right) = \max\left[0, S_t e^{-\delta(t-t)} - B(t, T)X\right] \tag{18.7}
$$

$$
\lim_{\sigma \downarrow 0} P\left(S_t, T - t\right) = \max\left[B(t, T)X - S_t e^{-\delta(t-t)}, 0\right] \tag{18.8}
$$

and

$$\lim_{\sigma \uparrow +\infty} C(S_t, T - t) = S_t e^{-\delta(t-t)} \tag{18.9}$$

$$\lim_{\sigma \uparrow +\infty} P(S_t, T - t) = B(t, T) X. \tag{18.10}$$

Note that these limits correspond to the general no-arbitrage bounds, (18.1) and (18.2). The facts that the Black–Scholes values range between the no-arb bounds and that they are strictly increasing for $0 < \sigma < \infty$ have an important implication; namely, there is some single value of volatility that equates the Black–Scholes price to any observed *market* price that is itself within the arbitrage-fee bounds. To see this, write the call and put formulas momentarily just as functions of σ, as $C^{\text{B-S}}(\sigma)$ and $P^{\text{B-S}}(\sigma)$. Solving the equation $C^{\text{B-S}}(\sigma) = C^M$ (the market price) for σ produces what is called the *implicit volatility* of the call option, as determined by the market price. Since function $C^{\text{B-S}}(\sigma)$ is one to one, it has a single-valued inverse, so we can *represent* implicit volatility as $\hat{\sigma} = (C^{\text{B-S}})^{-1}(C^M)$, although finding an explicit solution requires a numerical procedure. Likewise, $\hat{\sigma} = (P^{\text{B-S}})^{-1}(P^M)$ represents the implicit volatility of the put. Exercise 18.4 shows that implicit volatilities from call and put with the same expiration and strike must agree if the market prices obey put–call parity.

Since volatility is the only truly unobservable parameter that drives options in the Black–Scholes model, it is the only thing about which traders who rely on the formulas can profoundly disagree. Among this group, a superior ability to *forecast* volatility equates to the ability to make trading gains.

Now consider what the Black–Scholes formulas imply about the effects of instantaneous changes in the underlying price. Differentiating $C(S_t, T - t)$ and $P(S_t, T - t)$ partially with respect to S_t takes a bit of work, because S_t appears (via f_t) in the arguments of Φ as well as outside. However, working it all out will show that the result is precisely the same as if one ignored the presence of f_t in the CDFs. Thus

$$C_S \equiv \frac{\partial C(S_t, \tau)}{\partial S_t} = e^{-\delta \tau} \Phi \left[q^+ \left(\frac{f_t}{X}, \tau \right) \right] \tag{18.11}$$

$$P_S \equiv \frac{\partial P(S_t, \tau)}{\partial S_t} = -e^{-\delta \tau} \Phi \left[q^- \left(\frac{X}{f_t} \right) \right]. \tag{18.12}$$

These so-called *deltas* of the options (i.e., $\Delta C / \Delta S$, $\Delta P / \Delta S$) are of great importance. Note that $C_S \geq 0$ and $P_S < 0$ for $t < T$.[45] Thus, in a comparative-static sense, the call's value increases with increases in the underlying price, while the put's value decreases. This makes sense, given the rights these options confer on the holder to buy the stock and to sell the stock at fixed prices, respectively. It is not difficult to show that the second derivatives, C_{SS} and P_{SS}, called the *gammas*, are both positive, so that both $C(S_t, \tau)$ and $P(S_t, \tau)$ are convex functions of S_t.

---

[45]The call's delta is strictly positive for $t < T$ except at $S_t = 0$.

One can also work out the effects of changes in the strike price to show that

$$C_X = -B(t,T)\,\Phi\left[q^-\left(\frac{f_t}{X},\tau\right)\right] \tag{18.13}$$

$$P_X = B(t,T)\,\Phi\left[q^+\left(\frac{X}{f_t},\tau\right)\right]. \tag{18.14}$$

18.3 DELTA HEDGING

The formulas for the partial derivatives with respect to S_t and X show that the Black–Scholes call and put prices can be written simply as

$$C(S_t, T-t) = C_S S_t + C_X X \tag{18.15}$$

$$P(S_t, T-t) = P_S S_t + P_X X. \tag{18.16}$$

Now recall the replicating argument in Section 17.3 that was used to derive the formulas. We constructed a portfolio of a no-dividend stock and a money-market fund (or equivalently of the underlying and riskless bonds) with the same value as that of a generic derivative asset, $D(S_t, T-t) = p_t S_t + q_t M_t$, and we saw that $p_t = D_S$ and $q_t M_t = D(S_t, T-t) - D_S S_t$.[46] Thus, the Black–Scholes formulas do give the precise recipes for constructing the replicating portfolio at any $t \in [0, T]$. For example, for the call we hold C_S units of the underlying and $C_X X = C(S_t, T-t) - C_S S_t$ worth of bonds or money fund. As t advances and S_t fluctuates, the partial derivatives C_S and C_X evolve to maintain the equality between values of replicating portfolio and option. Moreover, the trades that are required to do this are self-financing. For a proof see Exercise 18.5.

    Figure 18.1 shows that the composition of the replicating portfolio can be read off a plot of $C(S_t, T-t)$ versus S_t. There the kinked bold line depicts the terminal value function of a call with strike price X. The bold convex curve depicts the Black–Scholes call function at some $t < T$ as a function of S_t. Its slope approaches that of the terminal value function above X (i.e., unity) as S_t increases. At any given S_t the call is worth $C(S_t) \equiv C(S_t, T-t)$, and the slope of the tangent line at the point $(S_t, C(S_t))$, namely $C_S = C_S(S_t, T-t)$, represents the number of units of the underlying in the replicating portfolio. The vertical dashed line extending down from $(S_t, C(S_t))$ has length $C_S S_t$ and is therefore the value of the position in the underlying. This exceeds the value of the option itself, and the (negative) difference, $C(S_t) - C_S S_t$, is the value of the short position in the money fund (or bonds) that completes the replicating portfolio. At higher underlying price S_t', the number of units $C_S(S_t')$ in the underlying is closer to unity and a larger short position in the money fund is required.

---

[46]Recalling Exercise 17.3, when the underlying pays a dividend continuously at rate δ we replicate a generic T-expiring derivative worth $D(S_t, T-t)$ at t with $p_t = D_S e^{\delta(T-t)}$ units of a *cum*-dividend position. That position results from starting with $e^{-\delta T}$ shares at time 0 and reinvesting the dividends as they are received. Each such *cum*-dividend unit is worth $S_t^+ = S_t e^{-\delta(T-t)}$ at t, so that $p_t S_t^+ = D_S S_t$. Thus, the corresponding number of ordinary *ex*-dividend shares is still just D_S.

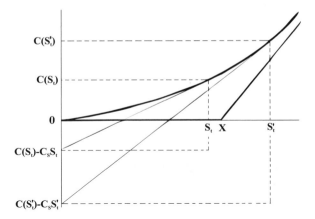

Figure 18.1   Finding the replicating portfolio from graph of call function.

Since the Black–Scholes formulas are recipes for replication, one can in principle use them to create *synthetically* options not actually traded in the market. As a very relevant example, suppose that one holds a nice, Markowitz-approved, diversified portfolio of stocks and bonds. Its current value is $V_t = \sum_{j=1}^{n} p_j S_{jt}$, where p_j is the number of units (e.g., shares) of asset j and S_{jt} is its current price. *Assuming* that the process $\{V_t\}$ follows geometric BM, as $dV_t = \mu_\mathbf{p} V_t \cdot dt + \sigma_\mathbf{p} V_t \cdot dW_t$, one could use the Black–Scholes formula to determine the arbitrage-free value $P\left(V_t, T - t\right)$ of a put that gives the right to sell the portfolio at T for some arbitrary amount $X_\mathbf{p}$. If one owned such an option along with the portfolio itself, the net value of assets could not fall below $X_\mathbf{p}$. Why? Because if $V_T < X_\mathbf{p}$, one could exercise the option, "putting" the portfolio to some counterparty in exchange for $X_\mathbf{p}$ dollars in cash. Holding such an option would thus provide temporary insurance against declines in the value of the diversified portfolio. Of course, puts on some such arbitrary portfolio are not apt to be sold on an exchange. However, the Black–Scholes formula shows precisely how to manufacture such an instrument on one's own. The replicating portfolio for the option would contain $P_V < 0$ units of the underlying and have $P\left(V_t, T - t\right) - P_V V_t$ dollars invested in a money-market fund earning the instantaneous spot rate.   To acquire this replicating portfolio one would simply sell off the fraction $|P_V|$ of the investment portfolio for $|P_V| V_t$ dollars, then buy $P\left(V_t, T - t\right) - P_V V_t$ dollars' worth of riskless bonds or money fund. Should the value of the investment portfolio decline, one would need to sell more of it in order to keep up the hedge.   This is because $P_{VV} = \partial P_V / \partial V_t > 0$ and $V_t \downarrow$ implies $P_V \downarrow$, approaching -1 as $V_t \to 0$ (assuming $\delta = 0$). In this limiting case all of the investment portfolio will ultimately have been sold off in order to provide a money-fund balance of X_p at T. Conversely, should the value of the investment portfolio increase from the initial level, one would need to buy additional units.

This scheme for portfolio insurance was widely marketed and adopted by investment funds during the 1980s. Unfortunately, while the principle often did work well enough on a small scale, its widespread application proved disastrous. The contemporaneous actions of traders to dump large blocks of stock (and surrogate index futures) as prices fell and their haste to buy when prices rose produced some very choppy markets, culminating in the meltdown on October 19, 1987. Ironically, despite the adoption of institutional changes designed to prevent its repetition, that October event significantly reduced the subsequent predictive value of the Black–Scholes formulas, having made people's perceptions of investment risks incompatible with the continuous, lognormal price processes on which the formulas depend.

18.4 DOES BLACK–SCHOLES STILL WORK?

The year 1973 marked not only the publication of the Black–Scholes–Merton formulas but also the opening of the Chicago Board Options Exchange and the first opportunity to trade options in a central marketplace. Before the CBOE was launched options were traded only over the counter through put-and-call brokers, and only by the most sophisticated investors. The narrow market and the difficulty of matching price quotations with contemporaneous underlying prices made it difficult to test any model of option pricing, but tests of Black–Scholes were soon carried out with the data that became available through the CBOE. Given that most option traders actually based their subjective valuations on the Black–Scholes formulas—merely inserting their own volatility parameters—it is not surprising that Mark Rubinstein (1985) found in early data only minor inconsistencies with market prices. As alluded to above, however, things changed drastically after the 1987 crash.

A common way to characterize the problem is by comparing the implicit volatilities corresponding to options with the same expiration date but different strikes. If $C^M(X)$ and $C^{\text{B-S}}(\sigma, X)$ represent market and Black–Scholes valuations of a call with strike X, the implicit volatility $\hat{\sigma}(X)$ then satisfies $C^{\text{B-S}}(\hat{\sigma}(X), X) = C^M(X)$. By virtue of put–call parity, the same value $\hat{\sigma}(X)$ should equate Black–Scholes and market prices of puts as well. Now actual volatility is a property of the underlying price process and not of the options traded on it. Thus, implicit volatilities from options with different strikes should all be the same if the pricing formulas are correct. However, since 1987 plots of $\hat{\sigma}(X)$ versus X have not been flat at all, but have shown a pronounced slope and curvature. Empirical Project 6 (at the end of this chapter) shows that the pictures one gets with options on the S&P 500 index typically resemble that in Fig. 18.2, which plots implicit volatilities versus strikes of February-expiring European-style options on the S&P as of late January 2008, at a moment when the index level itself was at about 1302. Plots of this sort are referred to as *smile curves*, although the lopsided shape one usually sees with S&P options is a little more like a "smirk." The smile reflects the high implicit volatilities of options whose strikes are far away from the current underlying price—that is, options that are deep in or far out of the money. Because the functions $C^{\text{B-S}}$ and $P^{\text{B-S}}$ are strictly increasing in σ, this indicates that prices of such options are in some sense too high relative to those near the money. In effect, since options benefit from big moves in underlying prices, the market seems to regard such moves as more likely than is consistent

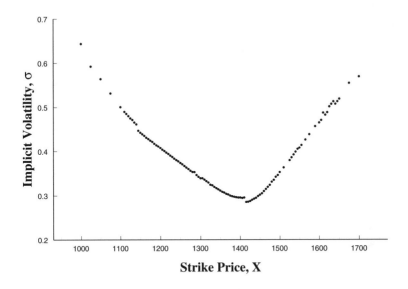

Figure 18.2   Implicit volatilities versus strike prices for S&P 500 options.

with the continuous, relatively tame sample paths of geometric Brownian motion. The lopsidedness or smirk effect indicates that out-of-money puts command an extra premium in the market, as if investors place particular value on their ability to "insure" portfolios against market crashes.

Another way to characterize the current failing of Black–Scholes is through what is called the *term structure* of implicit volatilities. Implicit volatilities of a sample of options with the same strike but different expirations $\{T_j\}_{j=1}^m$ are usually quite different. Smile curves tend to flatten out as T increases. Also, plots of $\hat{\sigma}$ against T (holding X constant) typically decline as T increases when markets have been especially volatile in the recent past, and the plots usually increase with T when markets have been especially calm. It is as if volatility is expected to return to some normal level over time—a feature that we shall see in Chapter 20 is incorporated in the current models of time-varying volatility. Even more troubling, for given T the shape of the locus $(T, \hat{\sigma}(T))$ varies as market conditions change.

The smiles, smirks, and term-structure patterns in options continue to motivate the search for better models than geometric Brownian motion. We shall look at several of these in Chapters 20 and 21, then in the final chapters see how to use them to price options.

18.5   AMERICAN-STYLE OPTIONS

A European-style option can be exercised only at expiration date T. Is there ever a situation in which the holder would *want* to exercise it early? In other words, for sufficiently large S_t, is it ever possible that $C(S_t, T - t) < S_t - X$; and for sufficiently small S_t, is it possible that $P(S_t, T - t) < X - S_t$? Take another look at the call and put formulas (18.4) and (18.5). Consider first the call. If $\delta = 0$ (no dividend on the underlying), then $C(S_t, T - t) - (S_t - X) > X(1 - B) > 0$; that is, in the absence of a dividend and assuming that interest rates are positive, the call prior to expiration is always worth more alive than dead. However, if $\delta > 0$ the slope of the call function, $C_S = e^{-\delta(T-t)}\Phi(q^+)$, approaches $e^{-\delta(T-t)} < 1$, so for sufficiently high S_t, function $C(S_t, T - t)$ will cut the $(S_t - X)$ line, and early exercise would be optimal. The intuition is that when S_t is large enough, the gain from exercising and starting to receive dividends exceeds what is gained by postponing the cash payment of X. For puts we need to consider what happens as S_t becomes small. It is easy to see from (17.10) that $\lim_{S_t \to 0} P(S_t, T - t) - (X - S_t) = BX - X < 0$. This holds regardless of whether there is a dividend. The intuition is that when S_t has fallen sufficiently, getting the cash flow $X - S_t$ immediately outweighs the potential gain from holding the put until T in hopes that S_t will decline further. For example, if the underlying were a common stock of a firm that had dissolved in bankruptcy with $S_t = 0$, one would certainly want to get the cash X immediately rather than wait.

These considerations show that American calls are subject to (rational) early exercise only when the underlying asset is to pay a dividend during the life of the option. Otherwise, American calls should be priced the same as European calls. However, the early-exercise potential of American puts always has positive value, making them worth more than their European counterparts. All exchange-traded US equity options are in fact of American style, while some stock index options are American (e.g., the S&P 100s) and some are European (S&P 500s). There are not many examples of exchange-traded options that come in *both* flavors and thus permit a direct comparison of market values. One example is the options on foreign exchange ("ForEx" or "FX" in street jargon) that are traded on the Philadelphia exchange. Here the underlying (deposits held at foreign banks) does earn interest, so American options on currencies are indeed subject to early exercise.[47]

For puts and calls alike, the threshold or boundary value of S_t that prompts early exercise depends on the time remaining until expiration. Considering just American puts for illustration, the critical boundary is a function $\{\mathfrak{B}_t\}_{0 \leq t \leq T}$ at which the "hold" value of the option just equals the exercise value: $P^A(\mathfrak{B}_t, T - t) = X - \mathfrak{B}_t$. In other words, when $S_t < \mathfrak{B}_t$ it is advantageous to exercise; otherwise, the option's live value is at least as great and one should hold on. A few qualitative features about the boundary are easily determined. One is that $\mathfrak{B}_T = X$, since at expiration one would always want to exercise if and only if $S_T < X$. Another is that $\mathfrak{B}_t$ is an increasing function of time. The intuition is that the live option has less chance of getting farther in the money the shorter the time to expiration, $T - t$. On the other

---

[47]Options are also traded on futures contracts, such as contracts on the S&P 500 index. On exercise, an in-the-money futures call pays in cash the difference between the futures price and the strike. It also gives the holder a freshly initiated position in the futures contract, which, as does a forward contract, has no initial value. Although futures contracts provide no cash flow to the holder, it turns out that American-style futures options are subject to early exercise. For an explanation, see Epps (2007, p. 273).

hand, one cannot determine precisely what the boundary is at any t without knowing the option's live value at that point, and one cannot determine the option's live value at t without knowing at what price to exercise at future dates. Thus, while there are many approximations to the values of American-style options, there are no exact formulas, even in the simple geometric BM/constant-r setting of Black and Scholes.[48]

EXERCISES

18.1 Show that when $S_T \geq 0$ the inequality $X - S_T \leq (X - S_T)^+ \leq X$ implies the no-arb bounds (18.2) on values of European puts.

18.2 Show that formulas (18.4) and (18.5) satisfy $C(S_T, 0) = P(S_T, 0) = 0$ when $S_T = X$.

18.3 Verify the limiting expressions in (18.7)–(18.10).

18.4 Given a put–call pair with the same X and T whose market prices obey put–call parity, show that the corresponding implicit volatilities, $\hat{\sigma}_C$ and $\hat{\sigma}_P$, must be equal.

18.5 Restricting for simplicity to the no-dividend case ($\delta = 0$), use relation (18.15) and fundamental PDE (17.3) to show that the call-replicating portfolio of C_S units of stock and $C_X X / M_t$ units of the money fund is self-financing.

EMPIRICAL PROJECT 6

The purpose of this project is to determine whether the Black–Scholes model still explains the current structure of prices of certain European options. File S&Poptions.prn in folder Project 6 at the FTP site contains bid and asked quotes as of $t = $ January 23, 2008 on pairs of S&P call and put options expiring at $T_1 = $ February 15, 2008, $T_2 = $ June 20, 2008, and $T_3 = $ December 19, 2008. There are $K_1 = 116$ strikes for February, $K_2 = 50$ strikes for June, and $K_3 = 50$ strikes for December. At the time of the quotes the value of the underlying S&P 500 index was $S_t = 1301.55$. You will use these data to calculate the options' implicit volatilities and to plot smile curves like that in Fig. 18.2. Here are the specific steps to follow.

1. The various options expire on the third Friday of the expiration month. Dating from $t = $ January 23, the February options expire in 23 days, the June options in 149 days, and the December options in 331 days. Express these as fractions of a 365-day year to determine the corresponding times to expiration, $\{T_j - t\}_{j=1}^3$.

---

[48]The *binomial* pricing model is often used to price American-style derivatives when the underlying is assumed to follow geometric BM. This involves dividing the option's remaining life $T - t$ into n intervals and approximating the conditional distribution of $\ln S_T$ given S_t as a binomial that converges to normal as $n \to \infty$. For an extensive discussion and applications, see Chapter 5 of Epps (2007).

2. We will regard the S&P portfolio as paying dividends at continuous average rates $\{\delta_j\}_{j=1}^3$ for the three maturity dates. Although the riskless interest rate is, in principle, observable, it is not clear exactly what specific rate applies for the marginal trader. Thus, the interest rate and the dividend rate will be deduced from the options' prices themselves, in conformity with European put–call parity; that is, letting $\tau_j \equiv T_j - t$,

$$C\left(S_t, \tau_j; X_k\right) - P\left(S_t, \tau_j; X_k\right) = S_t e^{-\delta_j \tau_j} - B\left(t, T_j\right) X_k, \qquad (18.17)$$

for options with maturity dates $\{T_j\}_{j=1}^3$ and strikes $\{X_k\}_{k=1}^{K_j}$. For this, first calculate for the call at each strike X_k the average of the respective bid and asked quotes. Do the same for the put. For the moment, take these to represent the market prices of the various options. Now calculate the difference between the prices of calls and puts, as per the left side of (18.17).

3. Working separately with the options for each expiration date T_j, regress the difference between the prices of call and put on underlying price S_t and on the *negative* of the strikes, $\{-X_k\}$. Since S_t is the same for all options, its value takes the place of the intercept term in the regression. You can either (a) create a column of K_j values all equal to S_t as a predictor variable and *suppress* the intercept that is normally supplied in regression routines, or (b) allow an intercept, regress the difference in prices just on the strikes, then divide the intercept by S_t. Now, with $\hat{a}_j$ and $\hat{b}_j$ as the estimated coefficients on S_t and the strikes for expiration T_j, we will have $\delta_j = -\ln\left(\hat{a}_j\right)/\tau_j$ and $r_j = -\ln(\hat{b}_j)/\tau_j$ for the average dividend rate and average spot rate during $[t, T_j]$.

4. For values of $X_k \leq S_t$ you will find the implicit volatilities from the *asked* quotes of the (in-the-money) *calls*, while for values of $X_k > S_t$ you will find them from the asked quotes of the (in-the-money) *puts*. Letting $O^{\text{B-S}}\left(S_t, \tau_j; X_k, r_j, \delta_j, \sigma\right)$ and $O^{\text{M}}\left(S_t, \tau_j; X_k\right)$ represent Black–Scholes prices and asked quotes of *generic* T_j-expiring options with strike X_k (that is, $O = C$ for calls and $O = P$ for puts), the implicit volatility $\hat{\sigma}$ is the (unique) root that satisfies

$$O^{\text{B-S}}\left(S_t, \tau_j; X_k, r_j, \delta_j, \hat{\sigma}\right) - O^{\text{M}}\left(S_t, \tau_j; X_k\right) = 0. \qquad (18.18)$$

An approximate value can be found by evaluating the left side of (18.18) on a fine grid of $\hat{\sigma}$ values, but a faster and more accurate method is to use a routine for finding roots of nonlinear equations. (A routine based on the simple bisection algorithm works perfectly well.) **Note:** Before searching for a root at a given τ_j and X_k, be sure that the asked quotes do not violate the arbitrage-free lower bounds

$$C\left(S_t, \tau_j; X_k\right) \geq S_t e^{-\delta_j \tau_j} - B\left(t, T_j\right) X_k$$
$$P\left(S_t, \tau_j; X_k\right) \geq B\left(t, T_j\right) X_k - S_t e^{-\delta_j \tau_j}.$$

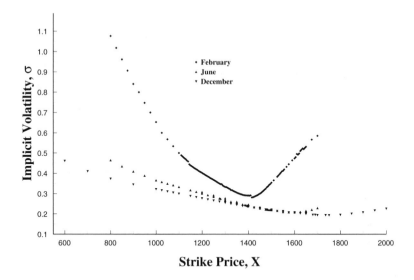

Figure 18.3   January 2008 smile curves for S&P options expiring in February, June, and December.

There will be no solution for any option that does violate one or the other of these bounds, so just exclude it from the study.[49]

5. Once the implicit volatilities are found for each T_j, make a scatter plot to represent the smile curve. Figure 18.3 gives an idea of what to expect. You can see that there is both substantial curvature and substantial asymmetry and that these diminish as time to expiration increases.

---

[49]Violations of such bounds—and of put–call parity—can occur in the data for a number of reasons. Some of the quotes may be "stale"—not updated quickly enough to conform to the current underlying price. Also, quotes are expressed in minimum increments of $.05, and for options far out of the money this can represent a significant fraction of the arbitrage-free price.

CHAPTER 19

MARTINGALE PRICING

This chapter presents another approach to pricing derivatives that does not directly involve replicating arguments. The approach is based on the fact that suitably normalized prices of assets can be turned into martingales when markets allow no opportunities for arbitrage. In the Black–Scholes environment the actual method of solution is formally the same as that in the Feynman–Kac solution of the fundamental PDE; that is, one still calculates the expected value of the derivative's terminal payoff using a modified distribution for the underlying in which its mean drift μ is replaced by r. However, we now come at this directly, without having to derive or solve a PDE or to imagine ourselves in some make-believe risk-neutral world or to restrict ourselves to models based on geometric Brownian motions. Martingale pricing is now the approach generally favored by most theorists and practitioners in the derivatives field.[50] The method relies on what is now called the *fundamental theorem of asset pricing*. Some background is needed before that can be made intelligible.

---

[50]Exceptions are the "financial engineers," who are typically more comfortable with differential equations than with probability theory.

Quantitative Finance. By T.W. Epps
Copyright © 2009 John Wiley & Sons, Inc.

19.1 SOME PREPARATION

To understand martingale pricing one needs to be familiar with the concept of changes of measure, so at this point one should review Sections 2.4 and 3.5. In particular, one needs to understand the concepts of absolute continuity of measures and the special meaning attached to the term "equivalent measures." While measures are the fundamental concept and we generally think of probability mass and density functions as being derived from them, it may be simpler at first to connect from the other direction. Thus, if X is a discrete random variable with countable support $\mathbb{X}$ on a probability space $(\Omega, \mathcal{F}, \mathbb{P})$, and if f_X is the probability mass function, then X induces a probability measure $\mathbb{P}_X$ on $(\Re, \mathcal{B})$ (the real line and the Borel subsets thereof) as $\mathbb{P}_X(B) = \sum_{x \in B \cap \mathbb{X}} f_X(x) = \sum_{x \cap \mathbb{X}} \mathbf{1}_B(x) f_X(x)$, where B is any member of $\mathcal{B}$. Thus, measure $\mathbb{P}_X$ is a mapping from $\mathcal{B}$ to $[0, 1]$. Likewise, if random variable X is continuous with PDF f_X, then $\mathbb{P}_X(B) = \int_B f_X(x) \cdot dx = \int \mathbf{1}_B(x) f_X(x) \cdot dx$. Using Stieltjes integrals the two cases can be handled simultaneously as $\mathbb{P}_X(B) = \int \mathbf{1}_B(x) \cdot dF_X(x)$, where F_X is the CDF. Induced measure $\mathbb{P}_X$ connects with the overall measure $\mathbb{P}$ on $(\Omega, \mathcal{F})$ as $\mathbb{P}_X(B) = \mathbb{P}(\{\omega : X(\omega) \in B\}) \equiv \mathbb{P}[X^{-1}(B)]$.

Now let $\hat{F}$ be a different CDF that assigns a different measure $\hat{\mathbb{P}}_X$ to at least some set of real numbers. The critical question for our purposes is whether there is any set $\mathcal{S}$ that is null in one measure but not in the other—a "null" event being one having probability zero. The measures $\mathbb{P}_X$ and $\hat{\mathbb{P}}_X$ are *equivalent* if and only if they have the same null sets; i.e., $\mathbb{P}_X\{\mathcal{S}\} = 0 \Leftrightarrow \hat{\mathbb{P}}_X\{\mathcal{S}\} = 0$. Obviously, if F is a continuous CDF (the CDF of a continuous random variable) and $\hat{F}$ is a step function (the CDF of a discrete random variable), then the associated measures can never be equivalent. That is because all the unit probability of the discrete distribution is concentrated at isolated points, yet the entire set of such points is null under the continuous measure. As a slight oversimplification that is nonetheless a useful guide, we can say the following: If the two random variable are either both discrete or both continuous, then their probability measures are equivalent if and only if the PDFs (for continuous random variables) or probability mass functions (for discrete random variables) are positive on precisely the same sets.[51] Here are some examples:

1. Measures associated with the following pairs of continuous distributions are equivalent: (a) $N(0, 1)$ and $N(5, 10)$, (b) $N(0, 1)$ and Student's t, (c) $f(x) = 1$ for $0 \leq x \leq 1$ and $\hat{f}(x) = 3x^2$ for $0 \leq x \leq 1$. However, neither of the measures in (c) is equivalent to any of those in (a) or (b).

---

[51] This oversimplifies in the sense that the PDF of any continuous random variable can be altered arbitrarily at any *countable* number of points without altering the implied measure. This is why we refer to various *versions* of the PDF. Any such version f_X really represents an *equivalence class* $\{f_X\}$ of functions that agree "almost everywhere" with respect to Lebesgue measure. Thus, to put it correctly we would say that measures $\mathbb{P}_X$ and $\hat{\mathbb{P}}_X$ are equivalent if and only if there is a set $\mathbb{S}$ with $\mathbb{P}_X(\mathbb{S}) = \hat{\mathbb{P}}_X(\mathbb{S}) = 1$ such that equivalence classes $\{f_X\}$ and $\{\hat{f}_X\}$ each contain *some* member—some version—that is positive on $\mathbb{S}$ and zero elsewhere.

2. Measures associated with the following pairs of discrete distributions are equivalent: (a) $f(x) = 2^x e^{-2}/x!$ for $x \in \{0, 1, 2, ...\}$ and $\hat{f}(x) = 2^{-(x+1)}$ for $x \in \{0, 1, 2, ...\}$; (b) $f(x) = \binom{10}{x}(.2)^x(.8)^{10-x}$ for $x \in \{0, 1, ..., 10\}$ and $\hat{f}(x) = \frac{1}{11}$ for $x \in \{0, 1, ..., 10\}$. However, neither of the measures in (b) is equivalent to either of those in (a).

Our interest in financial applications will be in the behavior of stochastic processes that evolve in time—specifically, price processes. If $\{S_t\}_{0 \leq t \leq T}$ is such a process that is adapted to a filtration $\{\mathcal{F}_t\}_{0 \leq t \leq T}$ (so that the value of S_t and its history are known when information $\mathcal{F}_t$ becomes available), then there is a probability measure $\mathbb{P}$ that governs this process—and all other such adapted processes of interest. Here the $\mathcal{F}$ of our overall probability space $(\Omega, \mathcal{F}, \mathbb{P})$ is a σ field that contains each member of the filtration; in particular, $\mathcal{F}_T \subset \mathcal{F}$. (The additional condition $\mathcal{F} \subset \mathcal{F}_T$ is also allowed, in which case $\mathcal{F} = \mathcal{F}_T$.) $\mathbb{P}$ "governs" the processes in the sense that for each process $\{S_t\}_{0 \leq t \leq T}$ it tell us the probability that $S_t \in B$ for every Borel set B and each time t. In principle, many alternative measures could govern our price processes, and these different measures would assign different probabilities to some events. Nevertheless, any two such measures $\mathbb{P}$ and $\hat{\mathbb{P}}$ would be equivalent so long as $\mathbb{P}\{S_t \in \mathcal{S}\} = 0 \Leftrightarrow \hat{\mathbb{P}}\{S_t \in \mathcal{S}\} = 0$ for each t, each (measurable) set $\mathcal{S}$, and each process $\{S_t\}_{0 \leq t \leq T}$ that is defined on our (filtered) probability space.

19.2 FUNDAMENTAL THEOREM OF ASSET PRICING

Having gained (or regained) some theoretical perspective, we can now explain the famous theorem that characterizes prices of financial assets. We have implicitly in the background a probability space consisting of a set of outcomes Ω, a filtration of information sets $\{\mathcal{F}_t\}_{0 \leq t \leq T}$ that evolve on some finite time interval $[0, T]$, an overall σ field $\mathcal{F}$ with $\mathcal{F}_T \subset \mathcal{F}$, and a probability measure $\mathbb{P}$ on $(\Omega, \mathcal{F})$ that determines distributions of prices of all assets for all $t \in [0, T]$. We call $\mathbb{P}$ the *natural* measure, since it describes things as they are in "nature." We will take it for granted that prices of all traded assets in our frictionless economy are nonnegative with probability one. We also assume that there is at least one asset whose price cannot equal zero at any t under measure $\mathbb{P}$. The price of such an asset can then serve as a *numeraire*. Because it is strictly positive, we can divide other prices by it and thus express them relative to the numeraire. Typical choices for the numeraire are the T-maturing default-free discount bond with price process $\{B(t, T)\}_{0 \leq t \leq T}$ and the money-market fund, with price process $\{M_t\}_{0 \leq t \leq T}$. We do not use the dollar (or other currency) as numeraire, because sterile cash serves for transacting only and is not a viable *asset* in an arbitrage-free economy. For simplicity we will maintain for now the Black–Scholes assumption that short rate of interest r is constant and positive. This guarantees that bond and money-fund processes are themselves strictly positive and grow deterministically at the same rate r as time advances; that is, $M_t = M_0 e^{rt}$ and $B(t, T) = e^{-r(T-t)} = M_t/M_T = e^{-rT} e^{rt} = B(0, T) e^{rt}$. Under this condition everything that follows works in exactly the same way regardless of which numeraire we choose, but to make the notation more compact, we shall use $\{M_t\}_{0 \leq t \leq T}$. Then, if $\{S_t\}_{0 \leq t \leq T}$ is the price process of some arbitrary asset, we will say that $\{S_t^* \equiv S_t/M_t\}_{0 \leq t \leq T}$ is its *normalized* version. Section 19.6 describes how the analysis changes when the short rate is stochastic and when other numeraires than $\{M_t\}$ or $\{B(t,T)\}$ are used.

As we shall see, the fundamental theorem will specify conditions under which normalized price processes are martingales. In order to grasp the significance of this, it is important to understand that neither actual prices of traded assets nor their normalized versions are ordinarily martingales in reality (i.e., under natural measure $\mathbb{P}$). We know that the current price of any asset must be such that the expected future dividend-adjusted price compensates *at least* for the opportunity cost of tying up cash and postponing consumption. Thus, we would always have $M_t < E_t M_T = M_T$ for a money fund growing at a known short rate and $B(t,T) < B(T,T) = 1$ for a discount unit bond. However, suppose that we consider price processes that are normalized by $\{M_t\}$. Now any process that is constant over time is *trivially* a martingale, since $X_t = c$ for all $t \geq 0$ implies that $X_t = E_t X_T = E_t c = c$. Therefore, since $M_t^* \equiv M_t/M_t = 1$, the money fund process normalized by itself is clearly a martingale under $\mathbb{P}$ or any equivalent measure. Likewise, $\{B(t,T)^* = B(t,T)/M_t\}_{0 \leq t \leq T}$ is a martingale under $\mathbb{P}$ when there is no uncertainty in interest rates, since

$$\frac{B(t,T)}{M_t} = \frac{e^{-r(T-t)}}{M_0 e^{rt}} = \frac{1}{M_0 e^{rT}} = \frac{B(T,T)}{M_T}.$$

On the other hand, if short-rate process $\{r_u\}_{t \leq u \leq T}$ was not presumed to be known during $[t,T]$, then M_T would also be uncertain at t and we would *not* generally have

$$B(t,T)M_t^{-1} = B(T,T) E_t M_T^{-1} = M_0^{-1} E_t \exp\left(-\int_t^T r_u \cdot du\right).$$

Nor would we expect any other normalized *risky* price process $\{S_t^* \equiv S_t/M_t\}$ to be a martingale under $\mathbb{P}$, even if $\{M_t\}$ were deterministic. That is because the required fair-game condition $E_t(S_T/M_T) = S_t/M_t$ is equivalent to $E_t(S_T/S_t) = M_T/M_t = B(t,T)^{-1}$, and the latter condition would hold for all risky assets only if their prices did not reflect risk. In other words, prices normalized by $\{M_t\}_{0 \leq t \leq T}$ would *all* be martingales only in markets with risk-neutral investors. Although our thought experiments in the last chapter led us to consider such make-believe worlds, we do not expect people in *real* markets to be risk neutral. Fortunately, the following *fundamental theorem* gives us our martingales without a trip to Oz.

Theorem: If markets offer no opportunities for arbitrage, then there exists a measure $\hat{\mathbb{P}}$ equivalent to natural measure $\mathbb{P}$ under which the normalized (integrable) price process $\{S_t^* \equiv S_t/M_t\}_{0 \leq t \leq T}$ of any traded asset is a martingale adapted to information process $\{\mathcal{F}_t\}_{0 \leq t \leq T}$. Conversely, if such an equivalent martingale measure does exist, then markets offer no opportunities for arbitrage.

The integrability requirement here is just a technical one to ensure the existence of conditional expectations, and there are also technical reasons to restrict processes to an arbitrary but *finite* interval of time. What really matters is the theorem's assurance that by tweaking our calculations of probabilities we can give normalized price processes the crucial fair-game property. We shall see very shortly why this is so useful and how it can be done. While the proof of the theorem in a continuous-time environment is very deep and requires special technical conditions on prices and filtrations, the basic idea behind it is surprisingly simple and intuitive.

To see the intuition, we can just fix two arbitrary times $t < T$ and show that the absence of arbitrage is necessary and sufficient for there to be an equivalent measure $\hat{\mathbb{P}}$ such that $\hat{E}_t S_T^* = S_t^*$. Of course, the "hat" on E indicates that expectation is taken under $\hat{\mathbb{P}}$. First, let us see that the presence of an arbitrage opportunity is inconsistent with the existence of an equivalent martingale measure (EMM); that is, we will show that $A \implies$ not $\hat{\mathbb{P}}$. By the principle of the contrapositive we will have thus shown that $\hat{\mathbb{P}} \implies$ not A, which is to say that the existence of an EMM implies the absence of arbitrage. Now, to work an arbitrage between times t and T, we would need to be able to sell a dollar's worth of one asset (stock or money fund), buy a dollar's worth of the other, and thereby obtain a positive chance of gain at T with no chance of loss. (Exchanging cash for a default-free bond or a share of the money fund is just such an arbitrage, which is why we cannot regard cash as an asset if we regard markets as arbitrage-free.) If such a risk-free chance of gain could be obtained by selling the stock and buying the money fund, then we would have[52]

$$\mathbb{P}_t \left\{ \frac{M_T}{M_t} \geq \frac{S_T}{S_t} \right\} = \mathbb{P}_t \left\{ \frac{S_T}{M_T} \leq \frac{S_t}{M_t} \right\} \equiv \mathbb{P}_t \left\{ S_T^* \leq S_t^* \right\} = 1$$

$$\text{and} \tag{19.1}$$

$$\mathbb{P}_t \left\{ \frac{M_T}{M_t} > \frac{S_T}{S_t} \right\} = \mathbb{P}_t \left\{ S_T^* < S_t^* \right\} > 0.$$

If it could be done instead by the opposite trade, we would have

$$\mathbb{P}_t \left\{ \frac{M_T}{M_t} \leq \frac{S_T}{S_t} \right\} = \mathbb{P}_t \left\{ \frac{S_T}{M_T} \geq \frac{S_t}{M_t} \right\} \equiv \mathbb{P}_t \left\{ S_T^* \geq S_t^* \right\} = 1$$

$$\text{and} \tag{19.2}$$

$$\mathbb{P}_t \left\{ \frac{M_T}{M_t} < \frac{S_T}{S_t} \right\} = \mathbb{P}_t \left\{ S_T^* > S_t^* \right\} > 0.$$

One or the other of these would have to hold if there were an opportunity for arbitrage. But condition (19.1) implies that $\mathbb{P}_t \left\{ S_T^* > S_t^* \right\} = 0$, while (19.2) implies that $\mathbb{P}_t \left\{ S_T^* < S_t^* \right\} = 0$. As these are $\mathbb{P}$-null events, the probabilities under any equivalent measure $\hat{\mathbb{P}}$ would have to be zero as well. In either case, then, all the probability mass of S_T^* must lie *at or on one side of* S_t^*, and so it is impossible to have $\hat{E}_t S_T^* = S_t^*$.

Next, let us see that not $A \implies \hat{\mathbb{P}}$, which is to say that an EMM *does* exist in an economy without arbitrages. This is easy, for "not A" means one of two things: either (1) that returns of the two assets will be the same with probability one, or (2)

---

[52]To simplify notation, we write $\mathbb{P}_t \left\{ M_T/M_t \geq S_T/S_t \right\}$, and so on, rather than $\mathbb{P} \left(\left\{ \omega : M_T/M_t \geq S_T (\omega) /S_t \right\} \mid \mathcal{F}_t \right)$.

that switching a dollar's worth of either asset into the other *may* produce a loss. In the first case, we have $\mathbb{P}_t\{M_T/M_t = S_T/S_t\} \equiv \mathbb{P}_t\{S_T^* = S_t^*\} = 1$. In that case the normalized process is constant and is therefore trivially a martingale under $\mathbb{P}$ and under any equivalent measure. In the second case we have $\mathbb{P}_t\{M_T/M_t > S_T/S_t\} = \mathbb{P}_t\{S_T^* < S_t^*\} > 0$ and $\mathbb{P}_t\{M_T/M_t < S_T/S_t\} = \mathbb{P}_t\{S_T^* > S_t^*\} > 0$, so that the distribution of S_T^* places probability mass on both sides of S_t^*. In that case, by shifting some probability mass from one side to the other—but not creating or filling any holes (null sets)—we could always create an equivalent measure for which $\hat{E}_t S_T^* = S_t^*$.

19.3 IMPLICATIONS FOR PRICING DERIVATIVES

We are now ready to see how the fundamental theorem is used to value European-style derivative assets. In principle, we can consider any kind of derivative so long as its expiration value at T is a known and integrable function of the price process of one or more traded assets, $\{S_{jt}\}$. However, to simplify we shall just consider derivatives that depend on a single price at T, as $D(S_T, 0)$.

First, we must figure out how to change our model for $\{S_t\}_{0 \leq t \leq T}$ so that normalized version $\{S_t^* = S_t/M_t\}_{0 \leq t \leq T}$ becomes a martingale. In this chapter we restrict discussion to the dynamics for $\{\bar{S}_t\}$ and $\{M_t\}$ that were specified originally by Black and Scholes and Merton, except that we allow the underlying to pay dividends continuously at rate δ. In that case Itô's formula shows that normalized *cum*-dividend process $\{S_t^{+*} = S_t^+/M_t \equiv S_t e^{\delta(t-T)}/M_t\}$ that we would use for replication evolves as

$$
\begin{aligned}
dS_t^{+*} &= \frac{1}{M_t} \cdot dS_t^+ - \frac{S_t^+}{M_t^2} \cdot dM_t \\
&= \frac{S_t^+}{M_t}\frac{dS_t^+}{S_t^+} - \frac{S_t^+}{M_t}\frac{dM_t}{M_t} \\
&= \frac{S_t^+}{M_t}\left[\frac{S_t e^{\delta(t-T)} \cdot dS_t/S_t + S_t e^{\delta(t-T)}\delta \cdot dt}{S_t^+} - r \cdot dt\right] \\
&= (\mu + \delta - r)\, S_t^{+*} \cdot dt + \sigma S_t^{+*} \cdot dW_t.
\end{aligned}
$$

To make $\{S_t^{+*}\}_{0 \leq t \leq T}$ a martingale, we would need to replace μ with $r - \delta$, as was done in Chapter 17 in applying the Feynman–Kac technique to solve the fundamental PDE. However, μ is an extrinsic constant that, in accordance with our model, governs the way the price process evolves; that is, in our real, measure-$\mathbb{P}$ world μ contributes the critical average trend in $\{S_t\}$ that compensates investors for the asset's risk and for the immediate consumption that they forego in holding it. Very simply, μ is what it is, and we are not free to change it. What to do?

Suppose that we just create a new Brownian process with time t value $\hat{W}_t \equiv W_t + (\mu + \delta - r)\, t/\sigma$. Then $W_t = \hat{W}_t - (\mu + \delta - r)\, t/\sigma$, and our equation of motion for $\left\{S_t^{+*}\right\}$ becomes

$$dS_t^{+*} = (\mu + \delta - r)\, S_t^{+*} \cdot dt + \sigma S_t^{+*} \cdot d\left[\hat{W}_t - \frac{(\mu + \delta - r)\, t}{\sigma}\right] \tag{19.3}$$

$$= \sigma S_t^{+*} \cdot d\hat{W}_t.$$

Having no explicit drift term, this does *look* like a martingale process, but of course it is not. The drift has just been moved into the manufactured $\{\hat{W}_t\}$, which is therefore not a *standard* Brownian motion under natural measure $\mathbb{P}$. Clearly, averaging over sample paths in $\mathbb{P}$ gives $E\hat{W}_t = (\mu + \delta - r)\, t/\sigma$, which need not equal zero. Still, we know from the fundamental theorem that there is *some* equivalent measure under which $\left\{S_t^{+*}\right\}$ is a martingale, so there must be a $\hat{\mathbb{P}}$ under which this trending Brownian motion $\{\hat{W}_t\}$ *does* behave as a standard Brownian motion. In fact, a proposition known as *Girsanov's theorem* shows rigorously that such a measure does exist—and actually tells us how to construct it. The details of this are rather involved, but the basic idea is simple. In effect, the switch to measure $\hat{\mathbb{P}}$ changes the probabilities associated with bundles of sample paths of $\left\{\hat{W}_t\right\}_{0 \leq t \leq T}$ in such a way that $\hat{E}\hat{W}_t = 0$, yet it preserves all the other features of a Brownian motion. For example, if the average trend were positive, then $\hat{\mathbb{P}}$ would make paths that wind up above $(\mu + \delta - r)\, T/\sigma$ less likely than under $\mathbb{P}$ and would increase the odds for those that wind up below.[53]

Now we are ready for the final step. If our derivative is to present no opportunity for arbitrage, its normalized price $\{D(S_t, T - t)/M_t\}_{0 \leq t \leq T}$ must also be a martingale under $\hat{\mathbb{P}}$. This means that $D(S_t, T - t)/M_t = \hat{E}_t D\left(S_T, 0\right)/M_T$, and so

$$D(S_t, T - t) = \frac{M_t}{M_T} \hat{E}_t D\left(S_T, 0\right)$$

$$= e^{-r(T-t)} \hat{E}_t D\left(S_T, 0\right)$$

$$= B\left(t, T\right) \hat{E}_t D\left(S_T, 0\right). \tag{19.4}$$

Calculating this expectation under new measure $\hat{\mathbb{P}}$ thus gives us the current arbitrage-free value of the derivative—and all without solving or even writing down a PDE!

Unfortunately, there remains one annoying question. What if there is some other measure $\tilde{\mathbb{P}}$ under which $\left\{S_t^{+*}\right\}$ is a martingale? In principle there could be many of these, since the martingale property just restricts the first moment—and since different measures would generally produce different values for expectations of nonlinear functions like $D\left(S_T, 0\right)$, how can we be sure that we have the measure that gives the right value for $D(S_t, T - t)$? We will see in the final section that the key to uniqueness is whether our economy makes it possible to replicate the payoff of the derivative using traded assets—that is, whether assets' prices "span" the relevant risk space. We have already seen in Section 17.3 that derivatives whose payoffs are smooth functions of underlying price can be replicated with the underlying and money fund when

---

[53]For the full statement and proof of Girsanov, see Shreve (2004, pp. 212–214) or Epps (2007, pp. 118–121).

$\{S_t\}_{0 \le t \le T}$ is a geometric Brownian motion and $\{M_t\}$ is deterministic. Thus, we will defer the question of uniqueness and review the steps for pricing a European-style derivative under these conditions. Here they are:

1. Change the drift of $\{S_t\}$ from μ to $r - \delta$, where δ is the continuous dividend rate.

2. Work out the conditional expectation of $D(S_T, 0)$ when (with $\tau \equiv T - t$)

$$S_T = S_t e^{(r-\delta-\sigma^2/2)\tau+\sigma(\hat{W}_T-\hat{W}_t)} \sim S_t e^{(r-\delta-\sigma^2/2)\tau+\sigma\sqrt{\tau}\cdot Z} \qquad (19.5)$$

and $Z \sim N(0,1)$.

3. Discount the result by multiplying by $B(t,T) = e^{-r(T-t)}$, the current value of a riskless, T-maturing, discount unit bond.

After applying these steps to some specific derivatives, we will see (1) how martingale pricing relates to the dynamic equilibrium pricing of Chapter 14, (2) how to proceed when $\{M_t\}_{0 \le t \le T}$ is not deterministic, (3) how to work with other numeraires, and (4) how to verify uniqueness.

19.4 APPLICATIONS

1. **Forward contracts.** A forward contract to exchange cash for some asset or investment commodity at date T is then worth $\mathfrak{F}(S_T, 0) \equiv S_T - f_0$ per unit to the party who receives the asset, where S_T and f_0 are the spot price at T and the forward price. If underlying price process $\{S_t\}$ follows geometric BM with drift μ and the asset generates a continuous dividend at rate δ, then changing the drift to $r - \delta$ and working out the expectation give for the value of the contract with remaining life $\tau \equiv T - t$

$$\begin{aligned}
\mathfrak{F}(S_t, \tau) &= B(t,T)\left(\hat{E}_t S_T - f_0\right) \\
&= B(t,T)\left[\hat{E}_t S_t e^{(r-\delta-\sigma^2/2)\tau+\sigma(\hat{W}_T-\hat{W}_t)} - f_0\right] \\
&= B(t,T)\left(S_t e^{(r-\delta)\tau} - f_0\right) \\
&= e^{-\delta\tau} S_t - B(t,T) f_0.
\end{aligned}$$

Of course, this is precisely the result we obtained in Chapter 17 by using the Feynman–Kac technique to solve the fundamental PDE. It is also the result obtained in Chapter 16 by a static replication argument that required no specific model for the underlying dynamics of $\{S_t\}$. Indeed, since $\mathfrak{F}(S_T, 0) = S_T - f_0$ is linear in S_T, we need no such model for martingale pricing either. All we need is that $\{S_t^+/M_t = S_t e^{-\delta\tau}/M_t\}$ is a martingale under our EMM $\hat{\mathbb{P}}$, for then (recognizing that $S_T^+ \equiv S_T$ since $\tau = 0$ at $t = T$)

$$\frac{\mathfrak{F}(S_t, \tau)}{M_t} = \hat{E}_t \frac{\mathfrak{F}(S_T, 0)}{M_T} = \hat{E}_t \frac{S_T - f_0}{M_T} = \frac{S_t^+}{M_t} - \frac{f_0}{M_T},$$

so that

$$\mathfrak{F}\left(S_t, \tau\right) = S_t^+ - \frac{M_t}{M_T} f_0 = e^{-\delta\tau} S_t - B(t, T) f_0.$$

2. **European puts**. The value of a T-expiring European put with strike price X and remaining life $\tau \equiv T - t$ is

$$
\begin{aligned}
P\left(S_t, \tau; X\right) &= B(t, T)\hat{E}_t \left(X - S_T\right)^+ \\
&= B(t, T)\hat{E}_t \left[X - S_t e^{\left(r - \delta - \sigma^2/2\right)\tau + \sigma\left(\hat{W}_T - \hat{W}_t\right)}\right]^+ \\
&= B(t, T) \int_{-\infty}^{\infty} \left[X - S_t e^{\left(r - \delta - \sigma^2/2\right)\tau + \sigma\sqrt{\tau}\cdot z}\right]^+ f(z) \cdot dz \\
&= B(t, T) \int_{-\infty}^{q^+(X/f_t)} \left[X - S_t e^{\left(r - \delta - \sigma^2/2\right)\tau + \sigma\sqrt{\tau}\cdot z}\right] f(z) \cdot dz,
\end{aligned}
$$

where $f(z) = (2\pi)^{-1/2} e^{-z^2/2}$,

$$q^\pm(s, \tau) \equiv \frac{\ln s \pm \sigma^2\tau/2}{\sigma\sqrt{\tau}},$$

and $f_t = S_t e^{(r-\delta)\tau}$. Carrying out the integration gives formula (17.10). The Feynman–Kac trick to solve the Black–Scholes PDE involved exactly the same calculations. Clearly, for martingale pricing we do still need a specific model for the underlying price, because $\hat{E}_t P\left(S_T, 0; X\right) = \hat{E}_t \left(X - S_T\right)^+$ cannot be expressed in terms of $\hat{E}_t S_T$ alone.

3. **European calls**. The value of a T-expiring European call with strike price X and remaining life $\tau \equiv T - t$ is

$$
\begin{aligned}
C\left(S_t, \tau; X\right) &= B(t, T)\hat{E}_t \left(S_T - X\right)^+ \\
&= B(t, T)\hat{E}_t \left[S_t e^{\left(r - \delta - \sigma^2/2\right)\tau + \sigma\left(\hat{W}_T - \hat{W}_t\right)} - X\right]^+ \\
&= B(t, T) \int_{q^+(f_t/X)}^{\infty} \left[S_t e^{\left(r - \delta - \sigma^2/2\right)\tau + \sigma\sqrt{\tau}\cdot z} - X\right] f(z) \cdot dz.
\end{aligned}
$$

Working out the integral gives formula (17.9). Alternatively, having already priced the comparable put option, the solution could be found from put–call parity:

$$
\begin{aligned}
C\left(S_t, \tau; X\right) &= P\left(S_t, \tau; X\right) + S_t e^{-\delta\tau} - e^{-r\tau} X \\
&= P\left(S_t, \tau; X\right) + B(t, T)\left(f_t - X\right).
\end{aligned}
$$

4. **"Digital" options**. A T-expiring European-style digital option is worth some fixed amount X if S_T is in some specified range. For a specific case, consider

a digital call that is worth X at time T provided $S_T > K$ for some $K > 0$. The terminal value is $C^D(S_T, 0; X, K) = X\mathbf{1}_{(K,\infty)}(S_T)$. Thus

$$
\begin{aligned}
C^D(S_t, \tau; X, K) &= B(t,T)\hat{E}_t\left[X\mathbf{1}_{(K,\infty)}(S_T)\right] \\
&= B(t,T)X\int_{q^+(f_t/K)}^{\infty}\frac{1}{\sqrt{2\pi}}e^{-z^2/2}\cdot dz \\
&= B(t,T)X\Phi\left[-q^+(f_t/K)\right].
\end{aligned}
$$

5. **"Threshold" calls.** A T-expiring European-style threshold call allows the holder to buy the underlying at X provided the terminal underlying price is above a still higher threshold, $S_T > K > X$. The terminal value is therefore

$$
C^{\text{Th}}(S_T, 0; X, K) = (S_T - X)\mathbf{1}_{(K,\infty)}(S_T).
$$

Writing this as

$$
C^{\text{Th}}(S_T, 0; X, K) = (S_T - K)\mathbf{1}_{(K,\infty)}(S_T) + (K - X)\mathbf{1}_{(K,\infty)}(S_T)
$$

shows that $C_{\text{Th}}(S_T, 0; X, K)$ is simply the sum of the values of a "plain vanilla" call with strike K and a digital that pays $K - X$ if $S_T > K$. Thus

$$
C^{\text{Th}}(S_t, \tau; X, K) = C(S_t, \tau; K) - C^D(S_t, \tau; K - X, K).
$$

It is often possible in this way to decompose contingent claims into others that can easily be priced.

19.5 MARTINGALE VERSUS EQUILIBRIUM PRICING

Now that we understand the mechanics of martingale pricing, we can step back a pace for a broader perspective. The first task is to see how martingale pricing relates to the dynamic equilibrium model of Chapter 14. Recall that in the representative-agent, equilibrium framework of Section 14.3 the price of any asset—whether primary or derivative—satisfies the relation

$$
S_t = E_t\left[\frac{U'_{t+1}(c_{t+1})}{U'_t(c_t)}\right](S_{t+1} + D_{t+1}) \equiv E_t\delta_{t,t+1}(S_{t+1} + D_{t+1}). \tag{19.6}
$$

Here, U'_t and U'_{t+1} are the marginal utilities for consumption at dates t and $t+1$ (from the representative agent's additively separable utility function of lifetime consumption), and their ratio, $\delta_{t,t+1}$, is the marginal rate of substitution of future consumption for current consumption. Clearly, $\delta_{t,t+1}$ serves to discount the asset's expected future dividend-corrected value back to its present value, but since future consumption is not $\mathcal{F}_t$-measurable—not known at t—it is a *stochastic* discount factor. We are going to see that (19.6) corresponds to the expression

$$
S_t = \hat{E}_t\left[\frac{M_t}{M_{t+1}}(S_{t+1} + D_{t+1})\right] \tag{19.7}
$$

that we get by shifting from natural measure $\mathbb{P}$ to martingale measure $\hat{\mathbb{P}}$. Here, as usual, $M_t = M_0 \exp\left(\int_0^t r_s \cdot ds\right)$ is the evolving price of a unit of our money-market fund, but now we can be general enough to allow short rate $\{r_t\}$ to be both time varying and unpredictable. Now since (19.6) must hold for each asset in the economy, it must also hold for the money fund. Thus, $M_t = E_t \delta_{t,t+1} M_{t+1}$, and so, putting $\rho_{t,t+1} \equiv \delta_{t,t+1} M_{t+1}/M_t$, we have (1) $\rho_{t,t+1} > 0$, (2) $E_t \rho_{t,t+1} = 1$, and (3)

$$
S_t = E_t \left[\rho_{t,t+1} \frac{M_t}{M_{t+1}} (S_{t+1} + D_{t+1}) \right]. \tag{19.8}
$$

Now, to see how this amounts to expectation under a new measure, put $Y_{t+1} \equiv (S_{t+1} + D_{t+1}) M_t / M_{t+1}$ and let $f_t(\cdot)$ be the conditional density of Y_{t+1} under natural measure $\mathbb{P}$. The expected value under $\mathbb{P}$ is then $E_t Y_{t+1} = \int y f_t(y) \cdot dy = \int_S y f_t(y) \cdot dy$, where S is the set of real numbers y on which $f_t(y) > 0$. (Typically, in our models $S = (0, \infty)$ or $[0, \infty)$.) Also, if B is any Borel set of real numbers, we can represent $\mathbb{P}(Y_{t+1} \in B \mid \mathcal{F}_t)$ as $\int_B f_t(y) \cdot dy = \int_{B \cap S} f_t(y) \cdot dy$, where $\mathcal{F}_t$ is the information available at t. Now suppose that $\hat{f}_t$ is another density that takes positive values on precisely the same set of real numbers S that f_t does, and let $\hat{\mathbb{P}}$ be the corresponding measure. $\mathbb{P}$ and $\hat{\mathbb{P}}$ are thus *equivalent* measures. Then

$$
\hat{\mathbb{P}}(Y_{t+1} \in B \mid \mathcal{F}_t) = \int_{B \cap S} \hat{f}_t(y) \cdot dy = \int_{B \cap S} \frac{\hat{f}_t(y)}{f_t(y)} f_t(y) \cdot dy,
$$

so that the *change* of measure from $\mathbb{P}$ to $\hat{\mathbb{P}}$ can be achieved by multiplying the PDF f_t by the factor $\hat{f}_t / f_t$. Of course, since it is the ratio of densities that are positive on the same sets, this factor is always positive on S. (The integral equals zero if $B \cap S = \emptyset$.) Moreover, the ratio has the property that

$$
E_t \frac{\hat{f}_t(Y_{t+1})}{f_t(Y_{t+1})} = \int_S \frac{\hat{f}_t(y)}{f_t(y)} f_t(y) \cdot dy = \int_S \hat{f}_t(y) \cdot dy = 1.
$$

But these are precisely the properties of the weighted stochastic discount factor $\rho_{t,t+1}$, which can therefore be regarded as a ratio of densities. Thus, the expectation under $\mathbb{P}$ in (19.8) corresponds to expectation under a new but equivalent measure $\hat{\mathbb{P}}$, as (19.7). From this comes our martingale pricing formula

$$
\frac{S_t}{M_t} = \hat{E}_t \left(\frac{S_{t+1} + D_{t+1}}{M_{t+1}} \right),
$$

as applied to the dividend-corrected value of the stock (or whatever the asset is). We have thus seen that in the equilibrium, representative-agent framework there is a measure such that normalized, dividend-adjusted prices are martingales. Of course, this is consistent with the fundamental theorem of asset pricing; and that theorem must indeed apply here, since there could be no *equilibrium* in a market with self-interested, informed agents and opportunities for arbitrage.

19.6 NUMERAIRES, SHORT RATES, AND EQUIVALENT MARTINGALE MEASURES

Throughout our examples thus far we have used as numeraire the price $\{M_t\}$ of the idealized money-market fund that invests continuously at the "short" (instantaneous spot) rate of interest. The fair-game property of normalized prices under measure $\hat{\mathbb{P}}$ then gives

$$M_t^{-1} D\left(S_t, T - t\right) = \hat{E}_t\left[M_T^{-1} D\left(S_T, 0\right)\right]. \tag{19.9}$$

Now, if the short rate at which $\{M_t\}$ continuously grows is a known constant r, as we have often assumed, then $M_T = M_t e^{r(T-t)}$ is also known at t. In that case M_T^{-1} can be removed from the expectation, and the arbitrage-free value of the derivative becomes

$$D\left(S_t, T - t\right) = \frac{M_t}{M_T} \hat{E}_t D\left(S_T, 0\right). \tag{19.10}$$

Moreover, since $\{M_t\}$ is deterministic under this condition, total returns from investments in the money fund and riskless bonds must coincide—otherwise, there is a clear opportunity for arbitrage. Thus, $M_t / M_T = B\left(t, T\right) / B(T, T) = B(t, T)$, and (19.10) can be expressed equivalently as

$$D\left(S_t, T - t\right) = B(t, T)\hat{E}_t D\left(S_T, 0\right), \tag{19.11}$$

just as we had worked out previously in (19.4). This implies that under $\hat{\mathbb{P}}$ the expected total return from holding the derivative is precisely the same as the certain return from holding a riskless discount bond:

$$\frac{\hat{E}_t D\left(S_T, 0\right)}{D\left(S_t, T - t\right)} = \frac{1}{B(t, T)} \equiv \frac{B\left(T, T\right)}{B(t, T)}. \tag{19.12}$$

Moreover, taking $D\left(S_T, 0\right) = S_T$ (making the "derivative" just the underlying asset itself, equivalent to a call option with zero strike price) shows that $\hat{E}_t S_T / S_t = 1/B(t, T)$ also. Thus, when we assign probabilities to sample paths in accordance with $\hat{\mathbb{P}}$ we get the same result as we would under natural measure $\mathbb{P}$ if investors were risk neutral, requiring no added compensation for risk. For that reason, the equivalent martingale measure (EMM) $\hat{\mathbb{P}}$ is often called the *risk-neutral measure*. Be assured, however, that this does *not* mean that we are back in the world of make believe. It just happens that derivatives would be priced the same in that world as in the real one, assuming no opportunities for arbitrage.

Of course, in reality the short rate is not constant but wanders about in continuous time, just as prices of stocks do, although usually with much lower volatility. How would the pricing formulas change if the short rate were regarded as time varying but still known in advance? Actually, as we saw in Section 17.4, there would be no change at all apart from replacing $r\tau \equiv r(T - t)$ by $\int_t^T r_u \cdot du$ in the models for the underlying and the money fund. If path $\{r_t\}_{0 \leq t \leq T}$ is $\mathcal{F}_0$-measurable—and therefore known at each $t \in [0, T]$—then so is $M_T = M_t \exp\left(\int_t^T r_u \cdot du\right)$, and relation

$B(t, T) = M_t/M_T$ and formula (19.11) still apply to justify risk-neutral pricing. Notice in particular that when the short rate is considered to be known in advance it makes no difference whether we normalize by bond price process $\{B(t,T)\}_{0 \leq t \leq T}$ or by $\{M_t\}_{0 \leq t \leq T}$, for either choice of numeraire would lead to the same result. Specifically, since $M_t/M_T = B(t,T)/B(T,T) = B(t,T)$ when $\{M_t\}_{0 \leq t \leq T}$ is deterministic, we have

$$\frac{D(S_t, T-t)}{B(t,T)} = \frac{\hat{E}_t D(S_T, 0)}{B(T,T)} = \hat{E}_t D(S_T, 0),$$

as per (19.11).

Although allowing short rates to vary predictably has no consequence for how we price European-style derivatives, things do change significantly when we account for the fact that future short rates are also imperfectly predictable. In that case $M_T = M_t \exp\left(\int_t^T r_u \cdot du\right)$ is also unpredictable, and while formula (19.9) remains valid, formulas (19.10) and (19.11) do not. To find the expected value in (19.9), we must have an explicit model for the joint evolution of the short rate and the underlying price. There are various models for the short rate. Two of the best known ones are the mean-reverting processes proposed by by Oldrich Vasicek (1977),

$$dr_t = (a - br_t) \cdot dt + \sigma \cdot d\hat{W}'_t, \tag{19.13}$$

and John Cox, Jonathan Ingersoll, and Steve Ross (1985),

$$dr_t = (a - br_t) \cdot dt + \sigma\sqrt{r_t} \cdot d\hat{W}'_t.$$

Here $\{\hat{W}'_t\}$ represents a standard Brownian motion under martingale measure $\hat{\mathbb{P}}$, which is typically distinct from but correlated with the process $\{\hat{W}_t\}$ that drives the derivative's underlying price $\{S_t\}$. Given any such $\{r_t\}$ process that is adapted to our evolving information $\{\mathcal{F}_t\}$, the current prices of discount bonds are determined via (19.9) as

$$B(t,T) = M_t \hat{E}_t M_T^{-1} B(T,T) = \hat{E}_t M_t M_T^{-1} = \hat{E}_t e^{-\int_t^T r_u \cdot du}. \tag{19.14}$$

As an example, solving for r_t in the Vasicek model yields

$$r_t = e^{-bt} r_0 + a\beta(t) + \sigma \int_0^t e^{-b(t-s)} \cdot d\hat{W}'_s, \tag{19.15}$$

where $\beta(t) \equiv b^{-1}\left(1 - e^{-bt}\right)$ for $t \geq 0$. Then working out the expectation in (19.14) gives[54]

$$B(t,T) = \exp\left\{-\int_t^T \beta(T-u)\left[a - \frac{\sigma^2}{2}\beta(T-u)\right] du\right\} e^{-\beta(T-t)r_t}. \tag{19.16}$$

---

[54]See Epps (2007, pp. 463–472), for the derivation and the corresponding expression under the model of Cox et al.

Since formulas (19.10) and (19.11) no longer follow from (19.9) when M_T is uncertain, relation (19.12) no longer holds either under measure $\hat{\mathbb{P}}$. Thus, if we take the bond process $\{B(t, T)\}_{0 \le t \le T}$ as numeraire, $\hat{\mathbb{P}}$ is no longer the appropriate measure for pricing. For this reason we often refer to $\hat{\mathbb{P}}$ as the "spot" EMM rather than the "risk-neutral" EMM when the instantaneous spot rate is regarded as uncertain. On the other hand, since bond prices cannot (we assume) be negative, they are perfectly satisfactory choices of numeraire; it is just that the change of numeraire requires a different measure. Specifically, the relation $D(S_t, T - t) = B(t, T)E^T D(S_T, 0)$ that corresponds to (19.11) holds under what is called the *T-forward measure*, $\mathbb{P}^T$. For example, in the Vasicek model the change of measure from $\hat{\mathbb{P}}$ to $\mathbb{P}^T$ makes $\hat{W}_t - \sigma \int_0^t \beta(T - s) \cdot ds$ a Brownian motion. In general, corresponding to a specific numeraire there will be a specific measure that makes normalized prices martingales and allows us to price our derivatives. Exercise 19.2 finds the EMM that applies when underlying price $\{S_t\}_{0 \le t \le T}$ itself is used as numeraire, as it can be so long as it is positive with probability one in the applicable model.

For derivatives on fixed-income products, such as bond options and interest rate caps, it is obviously essential to recognize the uncertainty in interest rates. There is an extensive literature on arbitrage-free models for interest rates, much of which is quite advanced and beyond the scope of a survey course in quantitative finance.[55] In any case, for options of moderate term on common stocks, stock indexes, and foreign currencies the modeling of interest-rate risk is not so critical. Much more is to be gained by using better models than geometric Brownian motion for the underlying. We will begin to study some of these better models in Chapter 20, then put them to work in pricing options in Chapters 22 and 23.

19.7 REPLICATION AND UNIQUENESS OF THE EMM

As we have just seen, different EMMs generally correspond to different numeraires. It is time to address the issue of whether different EMMs might correspond to any one numeraire. In that case, focusing on numeraire $\{M_t\}$ for definiteness, if there were two equivalent measures $\hat{\mathbb{P}}$ and $\tilde{\mathbb{P}}$ that made $\{S_t/M_t\}_{0 \le t \le T}$ a martingale, then the calculations of $D(S_t, T - t)$ as $M_t \hat{E}_t M_T^{-1} D(S_T, 0)$ and as $M_t \tilde{E}_t M_T^{-1} D(S_T, 0)$ might not agree unless terminal value $D(S_T, 0)$ were of the form $a + bS_T$. We will now see that the EMM for a given numeraire is unique if and only if markets are *complete*, in the sense that there are enough securities to enable any financial risk to be hedged away by holding an appropriate self-financing portfolio. In other words, in complete markets it is possible to replicate the payoff of any given contingent claim by holding such a self-financing portfolio.

Before we try to see the connection between uniqueness and replication, we should reassure ourselves that the ability to replicate the payoff of a particular derivative

---

[55]Damiano Brigo and Fabio Mercurio (2001) and Marek Musiela and Marek Rutkowski (2005) give comprehensive treatments of the subject. Mark Joshi (2003), Steven Shreve (2004), and Epps (2007) give briefer surveys.

could even potentially solve our problem. Indeed, taking the uniqueness theorem at face value, it appears that it could *not* do so. Why? Because no matter how well developed our markets are we can always invent financial risks that cannot be hedged. For example, two people might make a bet on the cumulative rainfall at city hall in Peoria, Illinois, from midnight on January 1 through noon on July 4. How could either of them replicate that payoff by holding traded securities? While markets have indeed become much more nearly complete in response to widespread desires to hedge specific risks, they could never eliminate every conceivable financial peril.[56] So if markets are not and can never be complete, how can martingale pricing ever deliver unique arbitrage-free prices for a specific derivative? The answer is that if traded assets do make it possible to replicate the payoff of that specific derivative, then we can narrow down the EMM to some member of an "equivalence class." Any member of that class will assign the same probabilities to the derivative's various potential payoffs and thus price the derivative just the same, even though any two members of the class might well differ on the prospects for rain in Peoria.

Thus reassured that it is worth the effort, let us develop some sense of why replication is both necessary and sufficient for pricing a specific derivative. First, let us see why replication assures uniqueness, or, in symbols, $R \Longrightarrow U$. We can add to the generality of the result by considering a derivative whose value $D\left(\mathbf{S}_T, M_T, 0\right)$ at T depends on a k vector of prices, one of which is the numeraire. Since, by hypothesis, we can replicate all such functions with a self-financing portfolio, it follows that we can replicate a digital option that pays M_T if $\left(\mathbf{S}_T, M_T\right)$ takes a value in some k-dimensional (measurable) set $A \in \Re_k$—and nothing otherwise. Let the time t value of the replicating portfolio be $V_t \equiv \mathbf{p}_t' \mathbf{S}_t + q_t M_t$, and represent the derivative's payoff as $\mathbf{1}_A\left(\mathbf{S}_T, M_T\right) M_T$, where

$$\mathbf{1}_A\left(\mathbf{S}_T, M_T\right) = \left\{ \begin{array}{ll} 1, & \left(\mathbf{S}_T, M_T\right) \in A \\ 0, & \left(\mathbf{S}_T, M_T\right) \notin A \end{array} \right. .$$

Then by the fundamental theorem of pricing there is at least one measure $\hat{\mathbb{P}}$ in our arbitrage-free economy such that

$$
\begin{aligned}
V_t &= M_t \hat{E}_t M_T^{-1} D\left(\mathbf{S}_T, M_T, 0\right) \qquad\qquad (19.17)\\
&= M_t \hat{E}_t \mathbf{1}_A\left(\mathbf{S}_T, M_T\right) \\
&= M_t \hat{\mathbb{P}}\left\{\left(\mathbf{S}_T, M_T\right) \in A \mid \mathcal{F}_t\right\}.
\end{aligned}
$$

The value V_t therefore pins down the specific value of $\hat{\mathbb{P}}\left\{\left(\mathbf{S}_T, M_T\right) \in A \mid \mathcal{F}_t\right\}$, as $M_t^{-1} V_t$. But set A being arbitrary and the relation holding for all t means that the pricing relation determines the evolving conditional probability that is assigned to *every* (measurable) set. It therefore determines the measure itself up to an equivalence class; that is, if we could trade with someone who used an EMM $\tilde{\mathbb{P}}$ such that $\tilde{\mathbb{P}}\left\{\left(\mathbf{S}_T, M_T\right) \in A \mid \mathcal{F}_t\right\} \neq \hat{\mathbb{P}}\left\{\left(\mathbf{S}_T, M_T\right) \in A \mid \mathcal{F}_t\right\}$ for some A, then we could work an arbitrage on the hapless individual. On the other hand, if $\tilde{\mathbb{P}}\left\{\left(\mathbf{S}_T, M_T\right) \in A \mid \mathcal{F}_t\right\}$

---

[56]Indeed, "weather" derivatives now let us hedge bets on certain specific meteorological events.

and $\hat{\mathbb{P}}\{(\mathbf{S}_T, M_T) \in A \mid \mathcal{F}_t\}$ did coincide for all A but assigned different values to outcomes of Peoria rainfall, then $\tilde{\mathbb{P}}$ and $\hat{\mathbb{P}}$ would be in the same equivalence class and would work equally well for pricing derivatives with payoffs depending only on $\{\mathbf{S}_t, M_t\}_{0 \le t \le T}$.

Now let us see if we can also understand why uniqueness of the EMM implies that we can replicate, that is, why $U \implies R$. It helps to think about this in reverse and try to see that *not* $R \implies$ *not* U. That is easy to see, for without replication there is simply no relation such as (19.17) to pin down $\hat{\mathbb{P}}\{(\mathbf{S}_T, M_T) \in A \mid \mathcal{F}_t\}$ and no reason why $M_t\tilde{\mathbb{P}}\{(\mathbf{S}_T, M_T) \in A \mid \mathcal{F}_t\}$ could not represent the derivative's value as well as $M_t\hat{\mathbb{P}}\{(\mathbf{S}_T, M_T) \in A \mid \mathcal{F}_t\}$. Or, to look at it another way, suppose that we somehow knew the current value of a derivative $D(\mathbf{S}_t, M_t, T - t)$ and set up a self-financing portfolio $\mathbf{p}_t'\mathbf{S}_t + q_t M_t$ with the same current value in the effort to replicate. However, since by assumption replication is not possible, it must be that, no matter what portfolio $(\mathbf{p}_t', q_t)$ we chose, there would be some time t at which $dD(\mathbf{S}_t, M_t, T - t) - \mathbf{p}_t' \cdot d\mathbf{S}_t - q_t \cdot dM_t = \Delta_t \ne 0$. But then the discrepancy Δ_t must respond to some risk or risks that $\{\mathbf{S}_t, M_t\}$ do not "span"; and in that case, even though two measures $\hat{\mathbb{P}}$ and $\tilde{\mathbb{P}}$ both make $\left\{M_t^{-1}\mathbf{S}_t\right\}_{0 < t \le T}$ a (vector) martingale, they could assign different probabilities to these other influences and therefore different values to $D(\mathbf{S}_t, M_t, T - t)$.

What does all this mean in practice? So long as we stay in the safe world of Black and Scholes, where prices of assets evolve as continuous geometric BMs and where values of derivatives on those underlying assets are smooth enough to apply Itô's formula, then we know that replication is possible and that our martingale solutions are unique. That is comforting. However, when we progress to certain more descriptive models than geometric BMs, as we do in Chapters 20 and 21, we will encounter situations in which replication is not possible. As we shall see in the final chapters, this does not mean that things are hopeless, but it does mean that we will need to find other ways to pin down the relevant equivalence class of martingale measures.

EXERCISES

19.1 The price of one share of a non-dividend-paying stock is initially $S_0 = 1$, and it evolves under natural measure $\mathbb{P}$ as $dS_t = \mu S_t \cdot dt + \sigma S_t \cdot dW_t$, where $\mu = 0.1$, $\sigma = 0.2$. The price of one unit of the money fund is initially $M_0 = 1$ and evolves as $dM_t = rM_t \cdot dt$, where $r = .05$. In each case below use martingale pricing to find the arbitrage-free value at $t = 0$ of a European-style derivative expiring at $T = 1$ and having the given terminal value $D(S_1, 0)$.

    (a) $D(S_1, 0) = 1$
    (b) $D(S_1, 0) = 1 - \mathbf{1}_{(0,1)}(S_1)$
    (c) $D(S_1, 0) = (S_1 - 1)^+$
    (d) $D(S_1, 0) = (S_1 - 1)^+ + (1 - S_1)^+$
    (e) $D(S_1, 0) = (S_1 - 1)^2$

19.2 $\{M_t\}$ evolves as $dM_t/M_t = r_t \cdot dt$, where short rate process $\{r_t\}_{t\geq0}$ is deterministic, and a stock's price evolves as $dS_t = \mu S_t \cdot dt + \sigma S_t \cdot d\bar{W}_t$ under natural measure $\mathbb{P}$. Since $\mathbb{P}(S_t > 0) = 1$ for each t under this model, $\{S_t\}$ itself would qualify as a numeraire. Find a process $\{\alpha_t\}_{t\geq0}$ such that normalized process $\{M_t^{**} \equiv M_t/S_t\}_{0\leq t}$ is a martingale under the measure $\mathbb{P}^{**}$ that makes $\{W_t^{**} = W_t - \alpha_t\}$ a standard Brownian motion.

19.3 Show that (19.15) satisfies the SDE (19.13).

19.4 Find $\partial \ln B(t,T)/\partial t$ when $B(t,T)$ takes the form (19.16), and evaluate the derivative at $T = t$.

19.5 $\{M_t\}$ evolves as $dM_t/M_t = r_t \cdot dt$; short rate process $\{r_t\}_{t\geq0}$ evolves under spot EMM $\hat{\mathbb{P}}$ as per (19.13); and the price $B(t,T)$ of a T-expiring discount bond is given by (19.16). Find a process $\{\alpha_t\}_{t\geq0}$ such that normalized process $\{M_t^T \equiv M_t/B(t,T)\}_{0\leq t<T}$ is a martingale under the T-forward measure $\mathbb{P}^T$ that makes $\left\{W_t^T = \hat{W}_t - \alpha_t\right\}$ a standard Brownian motion.

19.6 A stock's price evolves as $dS_t = \mu_t S_t \cdot dt + \sigma_t S_t \cdot dW_t$, where drift and volatility processes $\{\mu_t\}$ and $\{\sigma_t\}$ are time varying but $\mathcal{F}_0$-measurable. The stock pays no dividend. Likewise, short-rate process $\{r_t\}$ is also $\mathcal{F}_0$-measurable, and the normalized process $\{S_t^*\}_{0\leq t\leq T}$ is given by $S_t^* \equiv S_t/M_t$ with $M_t = M_0 \exp\left(\int_0^t r_s \cdot ds\right)$. Now, corresponding to (19.3), we have $dS_t^* = \sigma_t S_t^* \cdot d\hat{W}_t$, where $\hat{W}_t = W_t + \int_0^t (\mu_s - r_s)/\sigma_s \cdot ds$. Invoking Girsanov's theorem, we can still find a measure $\hat{\mathbb{P}}$ such that $\left\{\hat{W}_t\right\}_{0\leq t\leq T}$ is a Brownian motion and $\{S_t^*\}_{0\leq t\leq T}$ is a martingale. **(a)** Find the random variable S_t^* that solves the SDE $dS_t^* = \sigma_t S_t^* \cdot d\hat{W}_t$ subject to initial condition $S_t^* = S_0^*$. **(b)** Determine the distributions of S_T^* and of S_T under $\hat{\mathbb{P}}$ as of time $t \in [0,T]$. (*Hint:* Recall from Section 11.1 the properties of Itô integrals of deterministic functions with respect to BM.) **(c)** Explain how, if at all, the answers in (b) would change if drift process $\{\mu_t\}$ were not $\mathcal{F}_0$-measurable. What would happen if $\{r_t\}$ were also stochastic?

19.7 In the error-correcting model (15.3) suppose that the fundamental that drives the stock's price evolves as $dF_t = \alpha F_t \cdot dt + \beta F_t \cdot dW_t$, so that $\{F_t\}$ is a geometric BM under natural measure $\mathbb{P}$. Assume for simplicity that the stock pays no dividend. **(a)** Using the relation (15.4), find expressions for drift and volatility processes $\{\mu_t\}$ and $\{\sigma_t\}$ such that $dS_t = \mu_t S_t \cdot dt + \sigma_t S_t \cdot dW_t$ under $\mathbb{P}$. Are these processes deterministic or stochastic? **(b)** Introducing a money-market fund that evolves as $dM_t = r_t M_t \cdot dt$, what would be required of a measure $\hat{\mathbb{P}}$ if it is to make $\{S_t^* = S_t/M_t\}_{0\leq t\leq T}$ a martingale? **(c)** Under such a $\hat{\mathbb{P}}$, and assuming an $\mathcal{F}_0$-measurable short-rate process $\{r_t\}$, how could one price at $t \in [0,T]$ a European-style derivative on the stock that is worth $D(S_T, 0)$ at expiration? In particular, if $D(S_T, 0) = (S_T - X)^+$, would a formula like that of Black and Scholes still apply?

CHAPTER 20

MODELING VOLATILITY

Despite the fact that dynamic rational expectations models assure us that price changes in markets with rational, informed agents are not necessarily uncorrelated with their past, we have seen that there is little statistical evidence of such first-moment dependence, at least at short and moderate horizons. Indeed, this was evident from the correlation and regression tests carried out in Empirical Project 4. However, in that project the search for predictability in the *second* moment was substantially more productive. Evidence for this was found by regressing stocks' squared, one-period rates of return on past squared values and on the inverse of the initial price level, as

$$R_t^2 = \alpha + \beta_1 R_{t-1}^2 + \cdots + \beta_5 R_{t-5}^2 + \gamma S_{t-1}^{-1} + u_t, \qquad (20.1)$$

where $R_t = \ln(S_t/S_{t-1})$ with S_t dividend corrected. With long samples of high-frequency data we almost always find positive and statistically significant estimates of the β's at the first few lags, indicating that levels of volatility—high or low—tend to persist for a while. The statistical evidence for the inverse relation between volatility and initial price level is weaker, but in time series of stock returns that extend over many years one does often find positive and statistically significant estimates of γ

Quantitative Finance. By T.W. Epps
Copyright © 2009 John Wiley & Sons, Inc.

in specifications such as (20.1).[57] This phenomenon is referred to as the *leverage effect*, since some have attributed it to the link between a firm's share price and its debt/equity ratio; however, more is certainly involved. James Ohlson and Stephen Penman (1985) find that there are increases in volatility even after stock splits, despite the fact that such events have no direct bearing on a firm's capital structure. Moreover, to be justified in attributing the effect to leverage in the usual sense would require tracking not just the value of a firm's equity but the level of its debt as well.

This chapter presents some of the purely statistical models that have been put forth to capture volatility persistence and price-level effects in continuous and discrete time. These are "statistical" in that they were developed just to describe the data and are not developed from economic theory, such as the dynamic equilibrium models for market prices. In the next chapter we look at models that allow price processes to have abrupt jumps. Together, the volatility and jump models set the stage for some of the advanced theories of derivatives prices that will be covered in Chapters 22 and 23. These extend the Black–Scholes–Merton theory to more realistic underlying price processes than geometric Brownian motion and provide much better fits to market prices of options.

20.1 MODELS WITH PRICE-DEPENDENT VOLATILITY

Both price-level effects and volatility persistence can in principle be explained in a model similar to geometric Brownian motion but with a volatility coefficient that depends on current and/or past prices, as

$$\frac{dS_t}{S_t} = \mu \cdot dt + \sigma \left(\{S_s\}_{0 \leq s \leq t} \right) \cdot dW_t.$$

These models have the technical advantage that just one risk source, the Brownian motion $\{W_t\}_{t \geq 0}$, drives everything in the model. As will be explained later, this means (given an appropriate specification of $\sigma(\{S_s\}_{0 \leq s \leq t})$) that it is still possible to replicate the payoff of a derivative asset with the underlying and riskless bonds. We look first at a model in which volatility depends only on *current* price S_t.

20.1.1 The Constant-Elasticity-of-Variance Model

In an early attempt to capture price-level effects on volatility, John Cox and Steve Ross (1976) proposed what is called the constant-elasticity-of-variance (CEV) model. This is represented by the stochastic differential equation (SDE)

$$\frac{dS_t}{S_t} = \mu \cdot dt + \sigma_0 S_t^{\gamma-1} \cdot dW_t,$$

where $\{W_t\}$ is a Brownian motion under natural measure $\mathbb{P}$, σ_0 is a positive constant, and $0 < \gamma \leq 1$. When $\gamma = 1$ this is just a geometric Brownian motion,

---

[57]For the evidence see Andrew Christie (1982) and Epps (1996).

but when $\gamma < 1$ the volatility of the proportional price change moves counter to the price level, in line with the empirical finding of leverage effects. The label "CEV" applies because the *elasticity* of volatility, $d\ln(\sigma_0 S_t^{\gamma-1})/d\ln S_t = \gamma - 1$, is constant. While many other functional forms would have the same qualitative effect, this has the attraction that there is a known solution to the SDE above. In the case $\gamma = 1$ (geometric BM) we know by applying Itô's formula to $d\ln S_t$ that $S_T = S_t \exp\left[(\mu - \sigma^2/2)(T-t) + \sigma(W_T - W_t)\right]$, from which we infer that the conditional distribution of S_T given time t information is lognormal. The solution when $\gamma < 1$ is much more complicated, and here we shall just comment on some of its features. The distribution function $F_t(s) = \Pr(S_t \leq s)$ equals zero for $s < 0$ and is continuous and strictly increasing for $s > 0$, but its value at the origin, $F_t(0)$, is positive. This is because the model implies that the origin is an "absorbing" state; that is, price can hit zero, and if it does it remains there in perpetuity—an event that corresponds to a firm's terminal bankruptcy. Since the conditional distribution of S_T is known, it is possible to work out expressions for values of European options via martingale pricing. Making $\{S_t^* \equiv S_t/M_t\}_{0 \leq t \leq T}$ a martingale just requires removing the drift from $dS_t^*/S_t^* = (\mu - r_t) \cdot dt + \sigma_0 S_t^{\gamma-1} \cdot dW_t$.[58] Girsanov's theorem again assures us that there is a measure $\hat{\mathbb{P}}$ under which $\hat{W}_t = W_t + \sigma_0^{-1} \int_0^t S_s^{1-\gamma}(\mu - r_s) \cdot ds$ is a Brownian motion and under which $dS_t^*/S_t^* = \sigma_0 S_t^{\gamma-1} \cdot d\hat{W}_t$. Following the recipe for martingale pricing, one then finds the time t value of a European-style derivative worth $D(S_T, 0)$ at expiration as

$$D(S_t, T-t) = M_t \hat{E}_t M_T^{-1} D(S_T, 0).$$

Abstracting from uncertainty in short-rate process $\{r_t\}$, this simplifies to

$$D(S_t, T-t) = B(t,T) \int D(s,0) \cdot d\hat{F}_t(s),$$

where $\hat{F}_t$ is the conditional CDF of S_T when $\mu(T-t)$ is replaced by $\int_t^T r_s \cdot ds$. The CEV model does eliminate some of the volatility smile or smirk associated with the Black–Scholes formulas, whereby options with strikes below the current underlying price have higher implicit volatilities than do options near the money. However, the model does little to capture the phenomenon of volatility persistence, since price shocks have to be large enough to change price substantially before there is much effect on volatility.

20.1.2 The Hobson–Rogers Model

A formulation that does lead to volatility persistence was proposed by David Hobson and L. C. G. Rogers (1998). The idea is to let volatility depend on the displacement

---

[58]If there is a continuous dividend at instantaneous rate δ_t, then we would need to make $\left\{S_t^{+*} = S_t \exp\left(-\int_t^T \delta_s \cdot ds\right)/M_t\right\}$ a martingale to price a T-expiring derivative. We can just ignore dividends for now to simplify the notation. To allow for them, one can always just replace r_t with $r_t - \delta_t$ in the martingale pricing formulas.

of current price from some "normal" level, as measured by a geometrically declining average of past values. Specifically, they model the logarithm of the normalized stock price, $s_t \equiv \ln(S_t/M_t)$, as

$$ds_t = \mu(\Delta_t) \cdot dt + \sigma(\Delta_t) \cdot dW_t.$$

Here $\Delta_t = s_t - \lambda \int_{-\infty}^{t} e^{-\lambda(t-u)} s_u \cdot du$ is the measure of price displacement, with $\lambda > 0$ determining the rate of decay of the influence of past price levels on the normal price. Many specifications of $\sigma(\Delta_t)$ would have the effect of increasing volatility when price has moved substantially off track. Perhaps the simplest example is $\sigma(\Delta_t) = \sigma_0 \Delta_t^2$. Unfortunately, no reasonable specification, including this one, produces a model for which the conditional distribution of S_T is thus far known. Pricing options off this model thus requires either solving a PDE numerically or using simulation to estimate expected payoffs under the martingale measure.[59]

20.2 AUTOREGRESSIVE CONDITIONAL HETEROSKEDASTICITY MODELS

The common way to capture volatility persistence in discrete-time economic data is through a class of models known as *autoregressive conditional heteroskedasticity* (ARCH) processes and their extensions. To understand these it is helpful first to know something about more conventional models of time series.

In the 1960s and 1970s statisticians George Box and Gwilym Jenkins popularized a family of models for stationary time series known as ARMA models, standing for "autoregressive, moving average." These are in fact hybrids of two other types. A process described by

$$X_t = \mu + u_t - \theta_1 u_{t-1} - \cdots - \theta_q u_{t-q}$$

is called a *moving-average* process of order q, abbreviated MA(q). Here μ and the θs are constants, and the $\{u_t\}$ are i.i.d. shocks with mean zero and some variance σ^2. The shocks are often called *innovations*. Since the innovations in different periods are independent and since X_t depends only on q past values, it follows that X_t and X_{t-j} are uncorrelated for $j > q$. That is, the autocorrelation function, $\{\rho_j = C(X_t, X_{t-j})/VX_t\}_{j=1,2,\ldots}$, dies out after q lags. Modeling a process in which correlations linger for many periods thus requires many parameters. Another way of modeling persistent processes is with an *autoregressive* (AR) model. An AR(p) model has the form

$$X_t = \mu + \phi_1 X_{t-1} + \cdots + \phi_p X_{t-p} + u_t,$$

where again μ and the ϕs are constants and the $\{u_t\}$ are the innovations. This model appears to relate X_t just to its most recent p realizations, and indeed it is true that

---

[59]For an extensive empirical analysis of the Hobson–Rogers model, including its ability to fit market prices of options, see Olesia Verchenko (2008).

$E_{t-1}X_t = \mu + \phi_1 X_{t-1} + \cdots + \phi_p X_{t-p}$. However, since each variable on the right depends also on p lags farther back, the model implies that autocorrelations $\{\rho_j\}$ never die out. In other words, past innovations continue to have some effect on future values. Economic time series are often well represented by such models, since many economic variables—unemployment rates, money growth rates, gross domestic product, interest rates—are highly persistent. Combining the AR and MA principles leads to an ARMA(p, q) model:

$$X_t = \mu + \phi_1 X_{t-1} + \cdots + \phi_p X_{t-p} + u_t - \theta_1 u_{t-1} - \cdots - \theta_q u_{t-q}.$$

This hybrid form with lagged realizations of both the variable and the shock can achieve autocorrelation profiles of very flexible shapes, even with small values of p and q. This makes it possible to mimic the behavior of data with just a few parameters. For example, the ARMA(1, 1) model $X_t = \mu + \phi_1 X_{t-1} + u_t - \theta_1 u_{t-1}$ can represent many monthly economic series reasonably well. Note that the constant-expected-returns model that was the basis for early efficient markets tests, $R_t = \mu + u_t$, can itself be considered a degenerate ARMA process with just two parameters, but of course this special case implies zero autocorrelations at all lags.

All models of the ARMA class are said to be *homoskedastic*, meaning that the conditional variance of X_t given what is known at $t - 1$ is the same for all t and all past realizations of the process. But we know that this feature does not characterize assets' returns, since volatility (conditional standard deviation) is to some extent predictable from its past. Having been accustomed to using and studying ARMA models, it is natural that the econometricians who first tried to model this conditional *hetero*skedasticity would use the same sort of setup. The first attempt was through ARCH models, in which the variance of the shock u_t is made a linear function of past squared values:

$$\sigma_t^2 = \delta + \gamma_1 u_{t-1}^2 + \cdots + \gamma_q u_{t-q}^2.$$

Thus, in this ARCH(q) form the simple constant-expected-return model would be $R_t = \mu + \sigma_t \varepsilon_t$ with $\varepsilon_t = u_t/\sigma_t$. Clearly, as in the MA models of Box and Jenkins, many parameters could be needed to describe a process having "long memory" in volatility. The generalized ARCH (GARCH) was a way to extend memory and achieve flexibility by adding terms in lagged values of σ_t^2 itself. The general GARCH(p, q) specification is

$$\sigma_t^2 = \delta + \alpha_1 \sigma_{t-1}^2 + \cdots + \alpha_p \sigma_{t-p}^2 + \gamma_1 u_{t-1}^2 + \cdots + \gamma_q u_{t-q}^2.$$

As for the ARMA class of models, it turns out that small values of p and q afford enough flexibility to capture the most prominent features of the data. For example, the GARCH(1,1) process

$$\sigma_t^2 = \delta + \alpha \sigma_{t-1}^2 + \gamma u_{t-1}^2 \tag{20.2}$$

is often used to capture volatility persistence in returns. To see why it allows for persistence at long lags, just substitute recursively for lagged values of σ_t^2 on the right side:

$$\sigma_t^2 = \delta + \alpha \left(\delta + \alpha \sigma_{t-2}^2 + \gamma u_{t-2}^2 \right) + \gamma u_{t-1}^2$$

$$= \cdots$$

$$= \delta(1 + \alpha + \alpha^2 + \cdots) + \gamma \left(u_{t-1}^2 + \alpha u_{t-2}^2 + \cdots \right)$$

While ARCH/GARCH models and further variations have been very successful in modeling discrete-time data, they do have limitations. Indeed, some have referred to these not as "models" but as "representations." A representation that fits at observation intervals of one day will not work for weekly data, monthly data, and so forth, as new values of p, q, and the coefficients will be required. The only apparent solution for this frequency dependence is to model price and volatility processes in *continuous* time.

20.3 STOCHASTIC VOLATILITY

The CEV and Hobson–Rogers models do present volatility as stochastic processes in continuous time, but they are driven just by the single Brownian motion that drives price itself. This has a significant advantage for martingale pricing of derivatives, because it allows their payoffs to be replicated and therefore pins down the specific martingale measure that corresponds to a given numeraire. However, relying on a single risk source does limit the richness of effects that can be described. There have been many attempts to allow for other driving forces that are independent of the fundamental BM or at least not *fully* dependent on it. Since the additional risk source is typically unobservable, not corresponding to the price of any traded asset, volatility is referred to as "stochastic" in models of this sort. One such model, due to Steve Heston (1993), has become an industry benchmark because of its generality and tractability.

Heston's model starts out with a process for price itself that looks deceptively like geometric BM:

$$dS_t = \mu_t S_t \cdot dt + \sigma_t S_t \cdot dW_t. \tag{20.3}$$

The important difference here is that mean drift and volatility processes $\{\mu_t\}$ and $\{\sigma_t\}$ are now not merely time varying, but stochastic. As we now know, in using martingale methods to price derivatives, we wind up replacing $\{\mu_t\}$ by short-rate process $\{r_t\}$, and so the original drift process is really irrelevant in that application. The focus, then, is on the all-important volatility process. Heston models *squared* volatility as another Itô process:

$$d\sigma_t^2 = \left(\alpha - \beta \sigma_t^2 \right) \cdot dt + \gamma \sigma_t \cdot dW_t'. \tag{20.4}$$

Several features of this specification are noteworthy. First, the parameterization of the drift term, with $\alpha > 0$ and $\beta > 0$, causes the process to be mean reverting; that is, the process has, on average, an upward trend whenever $\sigma_t^2 < \alpha/\beta$ and a downward trend when $\sigma_t^2 > \alpha/\beta$, so α/β is a base to which squared volatility tends to be attracted. This is a nice feature, since we would not think that a stock's volatility could wander off into the stratosphere, as can the price level itself. Also, notice that diffusion coefficient $\gamma\sigma_t$ depends on volatility itself, rather than on its square. The purpose is not so much to achieve realism as to make the model easy to use in pricing derivatives. How it does this is explained in Chapter 23. Finally, notice that the energy source $\{W_t'\}$ that derives volatility is a BM different from that driving price itself. This means that volatility and price do not necessarily move in lockstep; instead, positive price shocks can sometimes be accompanied by positive changes in volatility and sometimes by negative changes. However, as we will see later, explaining the shapes of volatility smiles in options does require that these two driving forces be correlated, and *negatively*. That is, when price moves up, volatility usually (but not always) declines, whereas volatility tends to increase following negative price shocks. Indeed, even a cursory look at price charts reveals such asymmetry, since bear markets tend to be very choppy and bull markets relatively tame. In any case, the Heston model does allow the correlation between W_t and W_t' to take any arbitrary value $\rho \in (-1, 1)$, the specific value being something for the data to tell us. Exercise 20.2 shows how to construct dependent standard BMs $\{W_t\}$ and $\{W_t'\}$ from an independent pair.

20.4 IS REPLICATION POSSIBLE?

In order to price derivatives when the underlying price follows any of these models for volatility, we will need to change from natural measure $\mathbb{P}$ to an EMM—specifically, given the usual choice of numeraire, to the equivalent measure $\hat{\mathbb{P}}$ that makes $\{S_t^* = S_t/M_t\}_{0 \le t \le T}$ a martingale. In the CEV and Hobson–Rogers models the single Brownian motion that drives price also drives volatility, so it is possible to replicate derivatives on the underlying with just the underlying asset and money fund. As we saw in Section 19.7, this means that for those models the required EMM $\hat{\mathbb{P}}$ is unique. Thus, once we have adjusted the model so as to make $\{S_t^*\}$ a martingale, we need only work out an expected value to find the arbitrage-free price of a European-style derivative, as $D(S_t, T - t) = M_t\hat{E}_t M_T^{-1} D(S_T, 0)$.

Things are not so simple in the Heston framework. The value of any nonlinear derivative on $\{S_t\}$ will depend on the process $\{\sigma_t\}$, and unless there is some portfolio of traded assets with which $\{\sigma_t\}$ is instantaneously perfectly correlated, there is no way to hedge volatility risk. As we saw in Section 19.7, this tells us that there is now a multiplicity of relevant EMMs and a multiplicity of possible arbitrage-free valuations of a derivative. When we change the measure to an EMM $\hat{\mathbb{P}}$, the model changes in some way that we can only conjecture, apart from the fact that the mean drift in $\{S_t\}$ becomes the riskless rate. Let us simply take as primitive assumption that $\{S_t\}$ evolves under $\hat{\mathbb{P}}$ as specified in (20.3) and (20.4), except with r_t in place of μ_t in the drift; that is, as

$$dS_t = r_t S_t \cdot dt + \sigma_t S_t \cdot d\hat{W}_t$$
$$d\sigma_t^2 = (\alpha - \beta\sigma_t^2) \cdot dt + \gamma\sigma_t \cdot d\hat{W}_t',$$

where $\{\hat{W}_t\}$ and $\{\hat{W}_t'\}$ are standard Brownian motions under $\hat{\mathbb{P}}$ with $E\hat{W}_t\hat{W}_t' = \rho t$ for each $t \geq 0$. By regarding this as a primitive assumption there is no need to try to relate the models under $\mathbb{P}$ and $\hat{\mathbb{P}}$ if our interest is just in pricing derivatives on the underlying $\{S_t\}$.

As we shall see, this indeterminacy of EMM $\hat{\mathbb{P}}$ is a common problem in models that attempt to overcome some of the inadequacies of geometric BM. The standard approach is to use market prices of certain traded derivatives, such as exchange-traded options, to validate the model and to *infer* the values of the relevant parameters. Once these have been found, they can be used to estimate arbitrage-free prices of other derivatives that are not actively traded.

Here, briefly, is how the parameters can be determined. Once we see how to price options in the Heston framework, as we shall do in Chapter 23, we will be able to express the time t arbitrage-free price of a European-style option in terms of (1) the observable S_t, (2) the (assumed known) future course of short rates $\{r_s\}_{t\leq s\leq T}$, (3) the unobservable parameters α, β, γ, ρ, and (4) the unobserved initial volatility σ_t, which can just be regarded as a fifth parameter. From a sample of prices of traded options one can then find the values of these parameters that produce the best agreement between prices implied by the model and those observed in the market. Although various indicators of "agreement" between model and market are available, a common choice is the sum of squared price discrepancies; that is, if using a sample of m call options with market prices, expirations, and strikes $\{C_j^M, T_j, X_j\}_{j=1}^m$, one would find the parameter values that solve

$$\min_{\alpha,\beta,\gamma,\rho,\sigma_t} \sum_{j=1}^m \left[C_j^M - C^H\left(S_t, T_j - t; X_j; \alpha, \beta, \gamma, \rho, \sigma_t\right)\right]^2, \qquad (20.5)$$

where $C^H\left(\cdot\right)$ represents the value implied by the Heston model at the given set of arguments. Another approach is to minimize the sum of squared differences between Black–Scholes implicit volatilities of market and model prices. That is, one finds the $\hat{\sigma}_j^M$ that equates each B-S price $C^{\text{B-S}}\left(S_t, T_j - t; X_j; \hat{\sigma}_j^M\right)$ to each C_j^M, and likewise the $\hat{\sigma}_j^H$ that equates $C^{\text{B-S}}$ to the corresponding C^H. This $\hat{\sigma}_j^H$ will, of course, depend on the particular parameter values at which C^H is evaluated. The estimates $\hat{\alpha}, \hat{\beta}, \hat{\gamma}, \hat{\rho}, \hat{\sigma}_t$ are then the values that solve $\min_{\alpha,\beta,\gamma,\rho,\sigma_t} \sum_{j=1}^m \left(\hat{\sigma}_j^M - \hat{\sigma}_j^H\right)^2$. Even though in this framework B-S is no longer the appropriate model, using it to convert to implicit volatilities reduces the undue influence of options that have high prices just because they are deep in the money or have long times to expiration.

EXERCISES

20.1 Letting L be the lag operator, with the property that $L^k c = c$ for any constant and $L^k u_t = u_{t-k}$ for a discrete-time process $\{u_t\}$, show that the GARCH(1,1) specification (20.2) implies that $\sigma_t^2 = (1 - \alpha L)^{-1}(\delta + \gamma u_{t-1}^2) = (1-\alpha)^{-1}\delta + \gamma\sum_{j=0}^\infty \alpha^j u_{t-1-j}^2$ when $|\alpha| < 1$.

20.2 Let $\{W_t\}_{t\geq 0}$ and $\{W_t^*\}_{t\geq 0}$ be independent, standard BMs, so that $EW_t = EW_t^* = 0$, $VW_t = VW_t^* = t$, and $EW_tW_t^* = 0$. Show how to construct from these another standard BM $\{W_t'\}$ with the property that $EW_tW_t' = \rho t$ for any $\rho \in (-1, 1)$.

20.3 With squared volatility evolving as in (20.4), what would be the instantaneous change in volatility itself?

20.4 If $\gamma = 0$ in the Heston model, then volatility process $\{\sigma_t\}$ is deterministic ($\mathcal{F}_0$-measurable), although still time varying.

    (a) Assuming that short-rate process $\{r_t\}_{0\leq t\leq T}$ is also deterministic, show that S_T is distributed as lognormal conditional on the information available at $t \in [0, T]$; specifically, show that

$$\ln S_T \sim N\left[\ln S_t + \int_t^T (r_u - \sigma_u^2/2)\cdot du, \int_t^T \sigma_u^2 \cdot du\right]$$

    conditional on $\mathcal{F}_t$.

    (b) Let $\{S_t\}_{0\leq t\leq T}$ represent the price process for a stock that pays no dividend. Under the conditions of part (a), show that $C\left(S_0, T; \{\sigma_u^2\}_{0\leq u\leq T}\right)$, the arbitrage-free value at time 0 of a T-expiring European call with strike X, is

$$C\left(S_0, T; \{\sigma_u^2\}_{0\leq u\leq T}\right) = S_0\Phi\left[q^+\left(\frac{S_0}{BX}, T\right)\right]$$
$$-BX\Phi\left[q^-\left(\frac{S_0}{BX}, T\right)\right], \quad (20.6)$$

    where $B \equiv B(0, T) = \exp\left(-\int_0^T r_u \cdot du\right)$, and

$$q^\pm(s, T) \equiv \frac{\ln s \pm \left(\int_0^T \sigma_u^2 \cdot du\right)/2}{\sqrt{\int_0^T \sigma_u^2 \cdot du}}.$$

20.5 If $\gamma \neq 0$ in the Heston model, then volatility is stochastic. But suppose $\rho = 0$, so that BMs $\{\hat{W}_t\}_{0\leq t\leq T}$ and $\{\hat{W}_t'\}_{0\leq t\leq T}$ in measure $\hat{\mathbb{P}}$ are uncorrelated and therefore independent. Show under this condition that the arbitrage-free value at $t = 0$ of a T-expiring European call with strike X is $EC\left(S_0, T; \{\sigma_u^2\}_{0\leq u\leq T}\right)$, where $C(\cdot, \cdot; \cdot)$ is as given by (20.6).

CHAPTER 21

DISCONTINUOUS PRICE PROCESSES

Empirical Project 4 showed that marginal distributions of securities' rates of return typically have thicker tails than do normal distributions of the same mean and variance. The effect is most pronounced in high-frequency data, when price is recorded at daily or intraday intervals, but it is to some extent evident even in monthly returns. Chapter 9 described some "mixture" models for marginal distributions of returns that can capture to some extent this excessive frequency of outliers. The discrete- and continuous-time models for volatility that we have just examined also generate price processes whose increments are mixtures. When volatility of price changes over time—as a function either of price itself or some other stochastic influence—changes in log price in different periods are, in effect, random draws from distributions having different variances. Thus, the models with price-dependent and stochastic volatility can explain the general features found in the data. On the other hand, these volatility-based models fail in one significant respect. The effect of variable volatility should be most pronounced over longer periods, since this gives more time for volatility to change substantially, yet it is in high-frequency data that we see the strongest evidence of outliers.

The evidence for this comes not only from the prices of primary securities but also from prices of options on those securities. One of the steps in Empirical Project

6 was to calculate the implicit volatility of the S&P index from index options of different strikes. Of course, under the Black–Scholes assumption that price follows geometric Brownian motion with volatility σ, there should be no systematic variation in the implicit volatilities backed out from prices of different options on the same underlying asset. By contrast, as we saw in Fig. 18.2, plots of implicit volatility versus strike show marked curvature and asymmetry—"smiles" and "smirks," with higher values for low strikes than for high strikes. Moreover, studies have shown that the degree of curvature varies erratically over time and with political and economic conditions. While smiles and smirks in implicit volatilities could be accounted for just by the behavior of the stochastic discount factor and the strong desire to hedge against declines in consumption, the rapid changes in their shapes over time suggest that changes in expectations are also involved. Specifically, the way the market prices options suggests that traders perceive large moves in underlying prices—particularly large declines—to be far more likely than they would be if log-price processes were continuous with normal increments. And the fact that smiles and smirks are far more pronounced in prices of short-term options suggests that underlying prices are susceptible to abrupt and precipitous changes.

These empirical observations motivate efforts to model prices as discontinuous processes, subject to abrupt movements of random and sometimes large magnitude. Additional motivation comes from the belief that prices *should* react rapidly to new information in efficiently functioning markets. Thus motivated, we will look at several of the best-known models that allow for discontinuities. The first of these, proposed by Robert Merton (1976), relies on the Poisson process to trigger randomly spaced jumps of random sizes in what would otherwise be a continuous geometric Brownian motion. Next comes a class of *subordinated* processes that began with a paper by Dilip Madan and Eugene Seneta (1990) and was further developed by Madan and other coauthors. In these models price moves in an erratic fashion with infinitely many jumps in any positive interval of time. Finally, we shall look at a model in which price consists of value units that branch or reproduce at random intervals in continuous time. The resulting *branching process* allows price to remain constant for stretches of time then jump instantaneously to new levels, just as they do when assets trade in real markets.

21.1 MERTON'S JUMP–DIFFUSION MODEL

We saw in Section 10.3 that a Poisson process $\{N_t\}_{t \geq 0}$ is an integer-valued, nondecreasing, discontinuous process that evolves in continuous time. Setting $N_0 = 0$, we can think of N_t as representing the number of events (of some sort) that have taken place during the interval $(0, t]$. More generally, during any interval $(s, t]$ the process increases by a integer-valued amount $N_t - N_s$, which is distributed as Poisson with parameter $\lambda (t - s)$. Note that this allows $N_t - N_s$ to be zero with positive probability. The positive constant λ, called the *intensity parameter*, equals the expected change in the process per unit time. Thus, the mean number of jumps during any interval of time is proportional to the length of the interval; and since the mean and variance of any Poisson variate are equal, it is also true that the variance of $N_t - N_s$ is proportional to $t - s$. The other defining property of the process is that increments over nonoverlapping intervals of time are independent; that is, $N_u - N_t$ is independent of $N_s - N_q$ for any $q < s \leq t < u$. In the application to prices of assets, we usually think of $N_t - N_s$ as counting the number of value-relevant news events during $(s, t]$.

Since the Poisson process has a positive chance of remaining constant during any interval, a plot of $\{N_t\}$ versus t looks a bit like a poorly designed staircase that has tread sizes of random length but with steps of uniform, unit heights. That is, putting $\Delta N_t \equiv N_{t+\Delta t} - N_t$, the relations $\Pr(\Delta N_t = 1) = \lambda \Delta t + o(\Delta t) = 1 - \Pr(\Delta N_t = 0)$ and $\Pr(\Delta N_t = n) = o(\Delta t)$ for all $n > 1$ indicate that the change dN_t during the *infinitesimal* interval $(t - dt, t]$ can take only the values zero and unity. Thus, in the plot of $\{N_t\}$ versus t the randomly timed jumps are always by one unit, so $N_t - N_{t-} \equiv \lim_{h \to 0}(N_t - N_{t-h})$ (left-hand limit) equals either zero or unity. ($\{N_t\}$ is *right*-continuous, so $N_{t+} - N_t = \lim_{h \to 0}(N_{t+h} - N_t) = 0$ always.)

Merton's idea was to model a security's price process as having two components. One of these would be the nicely behaved, continuous geometric Brownian motion of Black and Scholes, while the other part would be a series of shocks of random sizes, triggered by a Poisson process. The intuition is that price evolves slowly most of the time because of "liquidity trading" as people rebalance portfolios, contribute from current saving, and withdraw funds for consumption purposes on their own independent schedules, but that information shocks occasionally produce discontinuities. The size of the jump in price, when it does occur, is essentially unpredictable, but we can model it as having some continuous distribution. Thus, in differential form Merton's model is as follows:

$$dS_t = \mu S_{t-} \cdot dt + \sigma S_{t-} \cdot dW_t + S_{t-} U \cdot dN_t. \qquad (21.1)$$

Here μ and σ are the usual mean drift and volatility of the Brownian component, U is the shock associated with the event marked by a possible jump in $\{N_t\}$, and $\{W_t\}$ and $\{N_t\}$ are a Brownian motion and Poisson process, respectively, under natural measure $\mathbb{P}$. All of $\{W_t\}, \{N_t\}$, and U are supposed to be independent. Notice that on the right side S is evaluated at $t-$, which represents the instant "just before" the possible jump at t.

To solve this stochastic differential equation we need an extension of Itô's formula that allows for jumps. For this, write (21.1) as $dS_t = dY_t + S_{t-} U \cdot dN_t$, where $\{Y_t\}$ represents the continuous part of $\{S_t\}$. Now introduce a function $f(t, S_t)$ that is continuously differentiable with respect to t and twice-continuously differentiable with respect to S_t. The extended Itô formula for the "instantaneous" change in the function's value is then

$$df(t, S_t) = f_t \cdot dt + f_S \cdot dY_t + \frac{f_{SS}}{2} \cdot d\langle Y \rangle_t + f(t, S_t) - f(t, S_{t-}).$$

Here the first three terms are the familiar ones from continuous stochastic calculus, while the last two represent the sudden effect of any jump that might occur at t. Writing out the dY_t and $d\langle Y \rangle_t$ components from the continuous part in the usual way gives

$$df\left(t, S_t\right) = \left(f_t + f_S \mu S_{t-} + \frac{f_{SS}}{2} \sigma^2 S_{t-}^2\right) \cdot dt + \sigma S_{t-} \cdot dW_t$$
$$+ f\left(t, S_t\right) - f\left(t, S_{t-}\right).$$

To solve the SDE (21.1) we apply the Itô formula to $f\left(t, S_t\right) = \ln S_t$ (which does not depend explicitly on t). Evaluating the partial derivatives at S_{t-}, we have $f_S = S_{t-}^{-1}, f_{SS} = S_{t-}^{-2}$, and

$$d \ln S_t = \left(\mu - \frac{\sigma^2}{2}\right) \cdot dt + \sigma \cdot dW_t + \ln S_t - \ln S_{t-}.$$

But since $Y_t - Y_{t-} = 0$ for the continuous part, the increment in log price from $t-$ to t is

$$\ln S_t - \ln S_{t-} = \ln\left(S_{t-} + S_{t-}U \cdot dN_t\right) - \ln S_{t-}$$
$$= \ln\left(1 + U \cdot dN_t\right)$$
$$= \ln\left(1 + U\right) \cdot dN_t.$$

(To justify the last step, see Exercise 21.2.) Clearly, this makes sense only if $\Pr(1 + U > 0) = 1$. Combining the continuous and jump parts gives

$$d \ln S_t = \left(\mu - \frac{\sigma^2}{2}\right) \cdot dt + \sigma \cdot dW_t + \ln\left(1 + U\right) \cdot dN_t,$$

and putting this in integral form gives the solution to (21.1):

$$\ln S_t - \ln S_0 = \left(\mu - \frac{\sigma^2}{2}\right) t + \sigma W_t + \sum_{j=0}^{N_t} \ln\left(1 + U_j\right). \qquad (21.2)$$

Here, the last term registers the effect of the $N_t \geq 0$ shocks that take place during $(0, t]$, and we define $U_0 \equiv 0$ to handle the case $N_t = 0$. To determine the actual distribution of $\ln S_t$ requires a model for the distribution of the random variables $\{1 + U_j\}_{j=1}^{\infty}$. The simplest idea is to model them as i.i.d. lognormal, in which case the change in log price is seen to be the sum of $N_t + 1$ independent normals. In discrete time this is essentially James Press' (1967) "compound-events model," which was described in Chapter 9. The continuous-time version is often referred to as the *jump–diffusion model*, since it combines a discontinuous process with a continuous "diffusion" process—the geometric Brownian motion.

Figure 21.1 gives a visual comparison of the jump–diffusion model (JD) with geometric BM (GBM) and Heston's stochastic volatility (SV) model. The upper part of the figure shows simulated sample paths of $\{S_t/S_0\}_{0 \leq t \leq 1}$ for each model. These

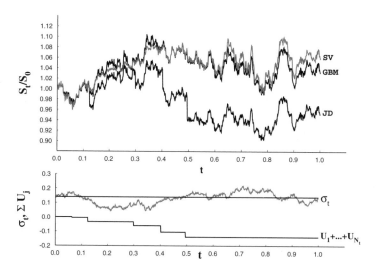

Figure 21.1   Simulated sample paths of geometric BM (GBM), JD, and SV processes.

were generated at intervals $\Delta t = 10^{-3}$ as[60]

$$\left(\frac{\Delta S_t}{S_t}\right)^{\text{GBM.}} = \mu_0 \cdot \Delta t + \sigma_0 \cdot \Delta W_t$$

$$\left(\frac{\Delta S_t}{S_{t-}}\right)^{\text{JD.}} = \mu_1 \cdot \Delta t + \sigma_1 \cdot \Delta W_t + U \cdot \Delta N_t$$

$$\left(\frac{\Delta S_t}{S_t}\right)^{\text{SV}} = \mu_0 \cdot \Delta t + \sigma_t \cdot \Delta W_t$$

$$\Delta \sigma_t^2 = \left(\alpha - \beta \sigma_t^2\right) \cdot \Delta t + \gamma \sigma_t (\rho \cdot \Delta W_t + \bar{\rho} \cdot \Delta W_t').$$

The lower figure tracks the level of stochastic volatility σ_t as it wanders about initial value σ_0, and also the accumulated jumps, $\sum_{j=0}^{N_t} U_j$. A careful inspection of the upper figure shows the SV path (in gray) to be smoother than that of GBM when $\sigma_t < \sigma_0$ and more erratic when $\sigma_t > \sigma_0$. This is qualitatively consistent with the prolonged periods of high and low volatility that we see in actual prices.

---

[60] ΔW_t and $\Delta W_t'$ were generated as $\sqrt{\Delta t}$ times independent standard normals; μ_1 and σ_1 were set to equate the mean and standard deviation of $(\Delta S_t / S_{t-})^{\text{JD.}}$ to $\mu_0 = 0.04$ and $\sigma_0 = 0.14$, compensating for the effect of jumps; β was set to $\alpha/\sigma_0^2 = 0.1/\sigma_0^2$ to attract σ_t back to σ_0.; ΔN_t was set equal to unity if a random draw from uniform $(0, 1)$ produced a value less than $\theta \cdot \Delta t = 2 \cdot \Delta t$, and to zero otherwise; jump sizes $\{U_j\}$ were generated independently as $\ln(1 + U_j) \sim N(-0.03, .01^2)$; and $\gamma = 0.3$ and $\bar{\rho} \equiv \sqrt{1 - \rho^2} = \sqrt{1 - (-0.7)^2}$.

21.2   THE VARIANCE–GAMMA MODEL

The other discrete-time mixture model considered in Chapter 9 was Student's t. The t distribution can, in fact, be generated as a mixture of normals with variances equal to inverses of gamma variates. Recall that a gamma variate Y is a continuous random variable with PDF $f(y) = \Gamma(\alpha)^{-1} \beta^{-\alpha} y^{\alpha-1} e^{-y/\beta}$ for $y > 0$, where parameters $\alpha > 0$ and $\beta > 0$ govern the distribution's shape and scale, respectively. Recall, too, that the mean and variance are $EY = \alpha\beta$ and $VY = \alpha\beta^2$, and that the moment-generating function is $\mathcal{M}_Y(\zeta) \equiv Ee^{\zeta Y} = (1 - \beta\zeta)^{-\alpha}$ for $\zeta < \beta^{-1}$. The mixing process involves first obtaining a realization $Y = y$ from the gamma distribution, then drawing from a normal distribution with mean zero and variance $\sigma^2(y) = y^{-1}$. With $\alpha = \frac{1}{2}$ and $\beta = 2/\nu$ for some $\nu \geq 1$, this produces a thick-tailed Student variate with ν degrees of freedom. This has mean zero and variance $\nu/(\nu - 2)$ when $\nu > 2$, so we have to rescale and add an appropriate constant to get values that fit particular data.

Now an alternative way to create a random variable with a thick-tailed distribution would be to set the normal variance equal to the gamma variate itself instead of its inverse. Unlike Student's t, however, the resulting marginal distribution,

$$ f(s) = \int_0^\infty \frac{1}{\sqrt{2\pi y}} e^{-s^2/(2y)} \Gamma(\alpha)^{-1} \beta^{-\alpha} y^{\alpha-1} e^{-y/\beta} \cdot dy, $$

cannot be expressed in closed form. This makes it harder to estimate the model's key parameters by maximum likelihood and accounts for the preeminence of the t distribution to represent returns in discrete time.

On the other hand, the variance–gamma (VG) model arises in a very elegant way in continuous time, and in that environment it is easier to use than the Student model for pricing derivatives. The VG arises through a beautiful concept known as a *subordination*, which we now explain. Suppose that $\{X_t\}_{t\geq 0}$ is some arbitrary stochastic process evolving in continuous time. Let $\{\mathfrak{T}_t\}_{t\geq 0}$ be another process having the special properties $\mathfrak{T}_0 = 0$, $\mathfrak{T}_t \geq \mathfrak{T}_s$ for $t > s$, and increments $\mathfrak{T}_t - \mathfrak{T}_s$ that are i.i.d. on periods of equal length. Thus, $\{\mathfrak{T}_t\}$ is a nondecreasing process, as is time itself. In fact, we will think of $\{\mathfrak{T}_t\}$ as representing *operational* time. In effect, it allows our clock to speed up at certain times (as in the segments of the curve in Fig. 21.2 that have slopes greater than unity) and slow down at others (as in the segments with $d\mathfrak{T}_t/dt < 1$). In this context we call $\{\mathfrak{T}_t\}$ a *directing process*. The idea is to let the directing process control the speed at which $\{X_t\}$ evolves in calendar time. This is done just by defining a new process $\{X_{\mathfrak{T}_t}\}_{t\geq 0}$, which is $\{X_t\}$ *subordinated* to $\{\mathfrak{T}_t\}$. In intervals of calendar time during which the directing process increases rapidly, the subordinated process evolves rapidly, whereas in slow periods it moves but little.

The VG model works by subordinating a Brownian motion to a gamma *process*. A gamma process $\{\mathfrak{T}_t\}_{t\geq 0}$ starts with $\mathfrak{T}_0 = 0$ and moves upward in discrete jumps, the *expected* number of which in any interval of time is infinite. Although the process moves in discrete jumps, the increment $\mathfrak{T}_t - \mathfrak{T}_s$ over a time span $(s, t]$ has a gamma distribution, so that the changes can take any nonnegative values. Since gamma

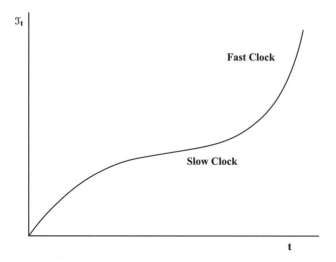

Figure 21.2   Directing process running the operational clock.

variates are (like all "proper" random variables) finite with probability one, it is evident that "most" of the jumps during any interval of time must be vanishingly small.

To model the logarithm of the price of a security, we would propose a VG process of the form

$$\ln S_t = \ln S_0 + \mu t + \gamma \mathfrak{T}_t + \sigma W_{\mathfrak{T}_t}, \tag{21.3}$$

where $\{W_t\}_{t \geq 0}$ and $\{\mathfrak{T}_t\}_{t \geq 0}$ are a Brownian motion and a gamma process under natural measure $\mathbb{P}$. Given the initial information $\mathcal{F}_0$, $\mathfrak{T}_t$ is distributed as gamma with shape parameter t/ν and scale parameter ν for some $\nu > 0$, with PDF

$$f_{\mathfrak{T}_t}(\tau) = \Gamma\left(\frac{t}{\nu}\right)^{-1} \nu^{-t/\nu} \tau^{t/\nu - 1} e^{-\tau/\nu}, \tau > 0.$$

With this setup the expected change in operational time during $[0, t]$ is $(t/\nu)\,\nu = t$ and the variance is $(t/\nu)\,\nu^2 = t\nu$. Thus, operational time moves at the same speed as calendar time on average, but there is variation about the mean; and it is this variation that gives rise to the thick tails in the marginal distribution of $\ln S_t$. Notice that the discontinuous movement in $\{\mathfrak{T}_t\}$ confers the same discontinuity to the log-price process also, since during the positive increment of operational time from $\mathfrak{T}_{t-}$ to $\mathfrak{T}_t$ the Brownian motion changes by $W_{\mathfrak{T}_t} - W_{\mathfrak{T}_{t-}}$. The inclusion of the term $\gamma \mathfrak{T}_t$ in (21.3) allows the distribution to have some skewness, the sign of which is governed by parameter γ.

21.3 STOCK PRICES AS BRANCHING PROCESSES

The idea of running a BM by a randomly advancing clock is very elegant and pleasing mathematically. However, the VG implementation of the idea does have some unattractive features as a model for share prices. One in particular is that price takes unboundedly many jumps in each arbitrarily short interval of time. At the level of market microstructure, this is no better characterization of how things actually work than is the continuous geometric Brownian motion. We know that in actual markets prices move by discrete amounts equal to integral multiples of the institutionally established minimum "tick." Until the late 1990s most US stocks on NASDAQ, NYSE, and the other exchanges traded in multiples of one-eighth of a dollar, with ticks in 16ths and 32ds of a dollar for stocks selling at very low prices. Since then, the tick size for all stocks has been reduced in phases to the present value of $.01. Because they are stated as multiples of the minimum tick and remain constant until changed, the bid and asked quotes by dealers and specialists are themselves discontinuous processes in continuous time that move up or down by multiples of $.01. Trade prices, on the other hand, are recorded only when trades occur and thus cannot properly be regarded as continuous-time processes. A nice way to model the quotes is by generalizing to continuous time the well known discrete-time, integer-valued *Bienayme–Galton–Watson* (B-G-W) process.

The B-G-W process was first proposed as a model for the number of males with a particular family name. Here, briefly, is the idea, using the name "Jones" as an example. Let $n \in \{0, 1, 2, ...\}$ count generations of Joneses and for each n let $\{Y_{n,j}\}_{j=1}^{\infty}$ be i.i.d., nonnegative, integer-valued random variables. Variable $Y_{n,j}$ will represent the number of male offspring of individual j in generation n that survive to reproductive age. For example, the $\{Y_{n,j}\}$ might be modeled under natural measure $\mathbb{P}$ as binomial or as Poisson—both of which assign positive probability to the event $Y_{n,j} = 0$. We start the process at generation $n = 0$ with $S_0 \geq 1$ males having the name Jones. At $n = 1$ the jth of these individuals would have produced some number of male progeny $Y_{0,j}$ to carry on the name. The collection of all these offspring at generation 0 now constitutes the next generation of male Joneses, which number $S_1 = Y_{0,1} + Y_{0,2} + \cdots + Y_{0,S_0}$. If $S_1 > 0$, then at time $n = 2$ each of these S_1 individuals will have produced some further random number of progeny, giving rise to the next generation of $S_2 = Y_{1,1} + Y_{1,2} + \cdots + Y_{1,S_1}$ members; and so on with $S_n = Y_{n-1,1} + Y_{n-1,2} + \cdots + Y_{n-1,S_{n-1}}$ members at generation n. Figure 21.3 illustrates the idea starting with $S_0 = 1$, a lone individual who has three offspring. One of these has no son, but the other two generate a total of five male progeny, so we have $S_2 = 5$ at generation 2.

A B-G-W process may continue indefinitely, but it may also eventually die out. The Jones line would terminate at generation n if and only if $S_n > 0$ and $Y_{n,1} = Y_{n,2} = \cdots = Y_{n,S_n} = 0$. If p is the probability that any given male leaves behind no male offspring (i.e., $p = \Pr(Y_{n,j} = 0)$ for each n and each j), then the probability of extinction at generation n is p^{S_n}. It is easy to see that extinction will eventually occur if μ, the expected number of each male's progeny, is less than unity. Note that

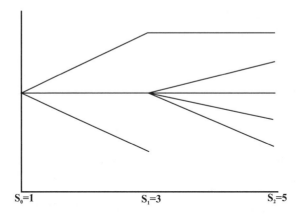

$S_0=1$ $S_1=3$ $S_2=5$

Figure 21.3   Three generations of a branching process.

the expected number of Joneses at $n = 1$ is simply $ES_1 = E\left(Y_{0,1} + \cdots + Y_{0,S_0}\right) = \mu S_0$. Likewise $ES_2 = E\left[E\left(S_2 \mid S_1\right)\right] = E\left(\mu S_1\right) = \mu^2 S_0$, and so on, with $ES_n = \mu^n S_0$. One can see from this that ES_n approaches zero as $n \to \infty$ if $\mu < 1$, that $ES_n \to \infty$ if $\mu > 1$, and that $\{S_n\}_{n=0}^{\infty}$ is a discrete-time martingale if $\mu = 1$.

To adapt this model to continuous time, introduce a continuous-time, integer-valued, nondecreasing process that counts the number of generations in intervals of calendar time. A ready example is the Poisson process $\{N_t\}_{t \geq 0}$. We assume that this is independent of the $\{Y_{n,j}\}$. For the application to stock prices we switch from counting Joneses to counting the number of ticks in the quoted price, and we now think of N_t as the number of events up to t that could potentially lead to a change in the quote. Letting S_n represent the price in ticks after n such events, the price at time t is then S_{N_t}. This is simply the total number of "value units" generated after the N_t events that have occurred during $(0, t]$. Notice that the model is really just another subordination scheme, but now we have subordinated a discrete-time B-G-W process $\{S_n\}_{n=0,1,2,\ldots}$ to a continuous-time, integer-valued directing process $\{N_t\}_{t \geq 0}$. Note, too, that although it does evolve in continuous time, $\{S_{N_t}\}_{t \geq 0}$ is still an integer-valued process, representing the share price at t in *cents*. In the application to stock prices "extinction" corresponds to a firm's terminal bankruptcy, from which point its stock is worthless.

Much of the mathematics of the model remains straightforward. For example, the stock's expected price at t is

$$ES_{N_t} = E\left\{E\left[S_{N_t} \mid N_t\right]\right\} = S_0 E\mu^{N_t}.$$

Recalling the discussion of probability-generating functions (PGFs) in Section 3.4.3), this is

$$ES_{N_t} = S_0 e^{\lambda t(\mu - 1)} \tag{21.4}$$

when $N_t \sim \text{Poisson}(\lambda t)$. From this it is evident that, as in the discrete-time formulation, (1) ES_{N_t} explodes as $t \to \infty$ if $\mu > 1$, (2) $ES_{N_t} \to 0$ if $\mu < 1$, and (3) $\{S_{N_t}\}_{t \geq 0}$ is a martingale if $\mu = 1$. In the last case it is possible to show that the limiting distribution of a normalization of S_{N_t} as $S_0 \to \infty$ is that of the compound-events model.[61]

21.4   IS REPLICATION POSSIBLE?

We saw in Section 20.4 that the extra source of risk in Heston's stochastic volatility model makes it impossible to replicate a derivative using traded primary assets—and thus impossible to identify directly the relevant equivalent measure for martingale pricing. How do our discontinuous models fare on this score? Unfortunately, when underlying prices are subject to random jumps it is also true that markets are "incomplete"—in the sense that not all risks can be hedged. The reason is that there is no way to set up a self-financing portfolio that always replicates a nonlinear derivative when underlying price can take jumps of unpredictable size. This is not difficult to see. Suppose at t the underlying takes a jump of $\Delta S_t \equiv S_t - S_{t-}$. In that case the value $V_{t-} = p_{t-}S_{t-} + q_{t-}M_t$ of our replicating portfolio jumps by $\Delta V_t = p_{t-}\Delta S_t$,[62] while the value of the derivative moves by $\Delta D_t \equiv D\left(S_{t-} + \Delta S_t, T - t\right) - D\left(S_{t-}, T - t\right)$. To replicate we would thus need to have $p_{t-} = \Delta D_t / \Delta S_t$ shares of the underlying just prior to the jump, and this equality would have to hold for *all* potential values of ΔS_t. Now, if our derivative happened to be a forward contract, with

$$D\left(S_t, T - t\right) \equiv \mathfrak{F}\left(S_t, T - t\right) = S_t e^{-\delta(T-t)} - B(t, T)\mathfrak{f}_0,$$

then $\Delta \mathfrak{F}_t / \Delta S_t = e^{-\delta(T-t)}$ for all $|\Delta S_t| > 0$, so setting $p_{t-} = e^{-\delta(T-t)}$ would indeed reproduce the effect of any such abrupt change. However, for an option or other derivative whose value depends *nonlinearly* on underlying price, the value of $\Delta D_t / \Delta S_t$ would depend on the unpredictable size of ΔS_t.

As for the stochastic volatility model, the inability to replicate means that there are in general a multiplicity of relevant EMMs, and thus a multiplicity of possible arbitrage-free solutions for $D\left(S_t, T - t\right)$. As was true in Heston's SV framework, we will again just assume for each discontinuous model that it applies under EMM $\hat{\mathbb{P}}$ and infer its parameters by "fitting" them to observed prices of traded options. Of course, the parameters must be such that $\{S_t^* = S_t/M_t\}$ (or $\{S_t^{+*}/M_t = S_t e^{-\delta(T-t)}/M_t\}$ if there is a dividend at rate δ) is a martingale under $\hat{\mathbb{P}}$. In the next chapter we begin to see how to price nonlinear derivatives under these more realistic models.

---

[61] For details, see Epps (1996).

[62] Note that $M_t = M_{t-}$ since $\left\{M_t = M_0 \exp\left(\int_0^t r_s \cdot ds\right)\right\}$ is a continuous process.

EXERCISES

21.1 Given that $\Pr(\Delta N_t = n) = (\lambda \Delta t)^n e^{-\lambda \Delta t}/n!$ for $n \in \{0, 1, 2, ...\}$, show that
(a) $\Pr(\Delta N_t = 0) = 1 - \lambda \Delta t + o(\Delta t)$, **(b)** $\Pr(\Delta N_t = 1) = \lambda \Delta t + o(\Delta t)$,
and **(c)** $\Pr(\Delta N_t = n) = o(\Delta t)$ for all $n > 1$.

21.2 Assuming that $\{N_t\}$ is a Poisson process and that U is any random variable such
that $\Pr(1 + U > 0) = 1$, show that $\Pr[\ln(1 + U \cdot dN_t) = \ln(1 + U) \cdot dN_t] = 1$.

21.3 Calculate the mean and variance of $\ln S_t$ given that $\{S_t\}$ evolves under natural
measure $\mathbb{P}$ as in (21.2) and the $\{U_j\}$ are i.i.d. with $\ln(1 + U_j) \sim N(\theta, \xi^2)$.

21.4 Assume that $\{S_t\}$ evolves under natural measure $\mathbb{P}$ as in model (21.2) and that
a money-market fund evolves as $dM_t = r M_t \cdot dt$. What would be required of
an equivalent measure $\hat{\mathbb{P}}$ if it is to make $\{S_t^* = S_t/M_t\}_{0 \le t \le T}$ a martingale in
the context of this model?

21.5 Calculate the mean and variance of $\ln S_t$ given that $\{S_t\}$ evolves as in model
(21.3).

21.6 Assume that $\{S_t\}$ evolves under natural measure $\mathbb{P}$ as in model (21.3) and that
a money-market fund evolves as $dM_t = r M_t \cdot dt$. What would be required of
an equivalent measure $\hat{\mathbb{P}}$ if it is to make $\{S_t^* = S_t/M_t\}_{0 \le t \le T}$ a martingale in
the context of this model?

21.7 Assume that an asset's price under natural measure $\mathbb{P}$ evolves as branching
process $\{S_{N_t}\}_{t \ge 0}$ with $E S_{N_t}$ as in (21.4). Let the numbers of progeny $\{Y_{n,j}\}$
of all individuals j in all generations n be i.i.d. under natural measure $\mathbb{P}$ as

$$\mathbb{P}(Y = y) = \binom{k}{y} \theta^y (1 - \theta)^{k-y}, \; y \in \{0, 1, 2, ..., k\},$$

where $k \ge 2$ is an integer and $\theta \in (0, 1)$. If a money-market fund evolves as
$dM_t = r M_t \cdot dt$, what would be required of an equivalent measure $\hat{\mathbb{P}}$ if it is to
make $\{S_t^* = S_t/M_t\}_{0 \le t \le T}$ a martingale in the context of this model?

CHAPTER 22

OPTIONS ON JUMP PROCESSES

In this and the final chapter we will see how to value European-style derivative assets when price processes for underlying assets are modeled more realistically than as geometric Brownian motion. We begin with Merton's (1976) jump–diffusion model, for which elementary methods deliver explicit formulas for values of European options. For the other discontinuous models of the previous chapter—variance–gamma and the branching process—there are indirect methods of pricing European options that rely on some more advanced tools. These methods require numerically integrating certain complex-valued functions that are derived from integral transforms of probability distributions known as *characteristic functions*. Section 22.3 provides a brief introduction to characteristic functions and explains a simple technique for using them to price European options. The final section shows how the technique is applied, using all three discontinuous models of Chapter 21 as examples. Chapter 23 then applies the same methods to the more complicated issue of pricing options under Heston's SV model and under a further extension, due to David Bates (1996), that combines stochastic volatility in the diffusion part with Poisson-driven jumps.

The main goal of these final chapters is to show how to carry out martingale pricing in more advanced settings, and so it may help to have a quick review of the general procedure. Everything we do relies on the conclusion of the fundamental theorem of asset pricing. The theorem states that in arbitrage-free markets there exists an EMM—a probability measure $\hat{\mathbb{P}}$ that is equivalent to *natural* measure $\mathbb{P}$—under which appropriately normalized prices are martingales. As should now be familiar, to "normalize" the prices we express them in terms of a numeraire; that is, we divide them by some other price that remains strictly positive. The ordinary unit of cash will not work as numeraire, because cash is not a viable asset when there are riskless alternatives that provide a positive return. Common choices of numeraire for pricing T-expiring derivatives are (1) the evolving price $\{B(t,T)\}_{0\le t\le T}$ of an idealized default-free, discount bond paying \$1 at maturity and (2) the price $\{M_t\}_{0\le t\le T}$ of an idealized money-market fund that holds and continuously rolls over discount bonds on the verge of maturing. As explained in Section 19.6, different EMMs are usually associated with different numeraires, but it happens that the EMMs for $\{B(t,T)\}$ and $\{M_t\}$ coincide under the simplifying assumption that short rates are known in advance. In what follows we continue to normalize with $\{M_t\}$, and we do assume that future instantaneous spot rates of interest are known and constant over the lives of the options. This eliminates having to model an additional stochastic process, and it introduces little error in pricing short- or medium-term options on common stocks and stock indexes. When the short rate is a constant r out to the option's expiration at T, the time t values of a T-maturing discount bond and the money fund are $B(t,T) = e^{-r(T-t)}$ and $M_t = M_0 e^{rt}$, respectively, with $B(T,T) = 1$ and $M_T = M_0 e^{rT}$, both known in advance. Thus, if $D(S_t, T-t)$ is the value at $t \in [0,T]$ of some generic, European-style derivative with terminal value $D(S_T, 0)$, the martingale property of the normalized process $\{D(S_t, T-t)/M_t\}_{0\le t\le T}$ under an EMM $\hat{\mathbb{P}}$ and the assumption that M_T is known at t imply that

$$M_t^{-1} D(S_t, T-t) = \hat{E}_t\left[M_T^{-1} D(S_T, 0)\right] = M_T^{-1}\hat{E}_t D(S_T, 0).$$

This leads at once to the equivalent pricing formulas

$$D(S_t, T-t) = \frac{M_t}{M_T}\hat{E}_t D(S_T, 0)$$
$$= B(t,T)\hat{E}_t D(S_T, 0).$$

Of course, since replication of nonlinear derivatives is not possible in any model with jumps of unpredictable size, there are many potential EMMs, and we must somehow determine which one the market uses. Once we find the appropriate $\hat{\mathbb{P}}$ and adjust the dynamics so as to make $\{S_t^* \equiv S_t/M_t\}_{0\le t\le T}$ a $\hat{\mathbb{P}}$ martingale (or $\{S_t^{+*} \equiv S_t e^{-\delta(T-t)}/M_t\}_{0\le t\le T}$ if there is a continuous dividend at rate δ), we have merely to evaluate the expectation and obtain a solution. Doing this is relatively easy in Merton's jump–diffusion model, and so we begin there.

22.1   OPTIONS UNDER JUMP–DIFFUSIONS

Recall that Merton's model characterizes the price of the underlying asset through the following stochastic differential equation:

$$dS_t = \mu S_{t-} \cdot dt + \sigma S_{t-} \cdot dW_t + S_{t-} U \cdot dN_t.$$

Here $\{N_t\}_{t\geq 0}$ is a Poisson process with intensity $\lambda > 0$, $\{W_t\}_{t\geq 0}$ is an independent standard Brownian motion, and U is a random variable independent of $\{N_t\}$ and $\{W_t\}$ with $\Pr(1+U>0) = 1$. The solution to this SDE (subject to initial value $S_0 > 0$) was found in Chapter 21 to be

$$\ln S_t = \left[\ln S_0 + \left(\mu - \frac{\sigma^2}{2}\right)t + \sigma W_t\right] + \sum_{j=0}^{N_t} \ln(1 + U_j). \qquad (22.1)$$

To complete the specification, we put $U_0 \equiv 0$ (to handle the case $N_t = 0$) and model $\{\ln(1+U_j)\}_{j=1}^{\infty}$ as i.i.d. normal. This makes $\ln S_t$ the sum of a normal random variable with mean $\ln S_0 + (\mu - \sigma^2/2)t$ and variance $\sigma^2 t$ (the term in brackets) plus a Poisson-distributed number of additional independent normals. Parameterizing the common distribution of $\{\ln(1+U_j)\}_{j=1}^{\infty}$ a little strangely, as normal with mean $\ln(1+\eta) - \xi^2/2$ and variance ξ^2, will simplify expressions obtained later on. Now, since $\{N_t\}$ is independent of everything else, we can condition on the event $N_t = n$, where n is any nonnegative integer, without affecting the behavior of $\{W_t\}$ or the $\{U_j\}$. This makes $\ln S_t$ the sum of $n + 1$ independent normals, and so $\ln S_t$ itself normally distributed with mean equal to the sum of the means, $\ln S_0 + (\mu - \sigma^2/2)t + n[\ln(1+\eta) - \xi^2/2]$, and variance equal to the sum of the variances, $\sigma^2 t + n\xi^2$.

Thus, just as random variable S_t is lognormal when *process* $\{S_t\}_{t\geq 0}$ is a geometric Brownian motion, so, too, S_t is *conditionally* lognormal in the jump model. This fact makes it possible to derive Black–Scholes-like formulas for prices of European-style derivatives in four steps. To explain by example, let us take the current time to be $t = 0$ and consider the derivative to be a T-expiring European put with strike price X. To simplify, we assume that the underlying asset pays no dividend. Here are the steps in outline form, details to be filled in later:

- First, change the values of relevant parameters so as to make $\{S_t^* = S_t/M_t\}$ a martingale. In applications we will have to find the set that corresponds to the particular EMM $\hat{\mathbb{P}}$ that the market uses.

- Then, find $\hat{E}\left[(X - S_T)^+ \mid N_T = n\right]$, which is the conditional expectation under $\hat{\mathbb{P}}$ of terminal value $P(S_T, 0) = (X - S_T)^+$ *given* that price will have had $N_T = n$ jumps. Since the conditional distribution of S_T is lognormal, the result will resemble the Black–Scholes put formula without the discount factor e^{-rT}, except that drift and volatility parameters will depend on n.

- Next, replace realization n of random variable N_T with N_T itself and take the expected value of $\hat{E}\left[(X - S_T)^+ \mid N_T\right]$ using the model $N_T \sim$ Poisson (λT); that is, apply the tower property of conditional expectation and calculate $\hat{E}(X - S_T)^+ = \hat{E}\left\{\hat{E}\left[(X - S_T)^+ \mid N_T\right]\right\}$. Since the Poisson distribution is supported on the nonnegative integers, the result will be the sum of infinitely many terms with probability weights that ultimately fall off rapidly toward zero.

■ Finally, discount the put's expected terminal value back to $t = 0$ using the risk-free rate. This just amounts to multiplying by $M_0/M_T = B(0,T)$, the current price of a default-free, discount, unit bond.

Now here are the details:

1. If the process $\{S_t/M_t\}_{0 \le t \le T}$ is a martingale in some measure $\hat{\mathbb{P}}$, then it will be true that $\hat{E}(S_T/M_T) = S_0/M_0$. We will evaluate $\hat{E}(S_T/M_T)$ and see what change is needed to equate the result to S_0/M_0. Setting $t = T$ in (22.1) and exponentiating give

$$S_T = S_0 e^{(\mu - \sigma^2/2)T + \sigma W_T} \prod_{j=0}^{N_T} (1 + U_j).$$

First take expectations conditional on N_T. Independence of $\{W_t\}$ and the $\{U_j\}$ lets us pass the expectation operator through the factors, as

$$\hat{E}(S_T \mid N_T) = \hat{E}\left[S_0 e^{(\mu - \sigma^2/2)T + \sigma W_T} \mid N_T\right] \prod_{j=0}^{N_T} \left[\hat{E}(1 + U_j) \mid N_T\right].$$

Now, because of independence, all the conditional expectations are the same as the *unconditional* expectations, so

$$\hat{E}(S_T \mid N_T) = \left[S_0 e^{(\mu - \sigma^2/2)T} \hat{E} e^{\sigma W_T}\right] \prod_{j=0}^{N_T} \hat{E}(1 + U_j).$$

The factor in brackets reduces to $S_0 e^{\mu T}$ because $\hat{E} e^{\sigma W_T} = e^{\sigma^2 T/2}$, and since each of the $\{1 + U_j\}$ is lognormal with parameters $\ln(1 + \eta) - \xi^2/2$ and ξ^2, it follows (Exercise 22.1) that $\hat{E}(1 + U_j) = e^{\ln(1+\eta)} = 1 + \eta$. Leaving this in exponential form for the moment, we thus have $\hat{E}(S_T \mid N_T) = S_0 e^{\mu T} e^{\ln(1+\eta)N_T}$. Now, (Exercise 22.2) if $Y \sim$ Poisson (θ) for some $\theta > 0$, then $\hat{E} e^{\zeta Y} = e^{\theta(e^{\zeta} - 1)}$ for any real number ζ. (This is the probability-generating function of Y evaluated at ζ.) Putting $\theta = \lambda T$ and $\zeta = \ln(1 + \eta)$ gives $\hat{E} e^{\ln(1+\eta)N_T} = e^{\lambda T \eta}$ and an exceedingly simple expression for expected future price: $\hat{E} S_T = S_0 e^{(\mu + \lambda \eta)T}$. Now for $\{S_t/M_t\}_{0 \le t \le T}$ to be a martingale requires that

$$M_0^{-1} S_0 = \hat{E} M_T^{-1} S_T = \frac{S_0 e^{(\mu + \lambda \eta)T}}{M_0 e^{rT}} = M_0^{-1} S_0 e^{(\mu - r + \lambda \eta)T}.$$

To equate the first and last expressions, we must change the trend parameter of the geometric Brownian motion from μ to $r - \lambda \eta$ (or to $r - \delta - \lambda \eta$ if there is a continuously paid dividend at rate δ). Formally, this involves applying Girsanov's theorem and changing to a measure $\hat{\mathbb{P}}$ in which

$\left\{\hat{W}_t \equiv W_t - (\mu - r + \lambda\eta)\, t/\sigma\right\}$ is a standard BM. As explained in Section 20.4, we will need to "reverse-engineer" all of λ, η, ξ, and σ by inferring their values in any given application from prices of traded derivatives. This can be done once we finish the task of expressing the arbitrage-free price as a function of these parameters.

2. Now we must find the conditional expectation of terminal value $(X - S_T)^+$ in measure $\hat{\mathbb{P}}$. Making the substitution $\mu = r - \lambda\eta$, we have that $\ln S_T$ is distributed as normal with mean

$$\hat{E}\ln S_T = \ln S_0 + \left(r - \lambda\eta - \frac{\sigma^2}{2}\right)T + n\left[\ln\left(1+\eta\right) - \frac{\xi^2}{2}\right]$$

and variance $\hat{V}\ln S_T = \sigma^2 T + n\xi^2$ conditional on the event $N_T = n$. Reordering the terms in the mean as

$$\ln S_0 + (r - \lambda\eta)\,T + n\ln\left(1+\eta\right) - \left(\frac{\sigma^2 T + n\xi^2}{2}\right)$$

and setting $r_n \equiv (r - \lambda\eta) + n\ln\left(1+\eta\right)/T$ and $\sigma_n^2 \equiv \left(\sigma^2 + n\xi^2/T\right)$, we have under $\hat{\mathbb{P}}$ that $\ln\left(S_T/S_0\right)$ is distributed as $N\left[\left(r_n - \sigma_n^2/2\right)T, \sigma_n^2 T\right]$ conditional on $N_T = n$. All this notation is designed to create obvious parallels with the Black–Scholes case. In that model $\ln\left(S_T/S_0\right) \sim N\left[\left(r - \sigma^2/2\right)T, \sigma^2 T\right]$ under $\hat{\mathbb{P}}$, and the value of the put at $t = 0$ was given by

$$P^{\text{B-S}}\left(S_0, T; r, \sigma\right) = e^{-rT}\hat{E}\left(X - S_T\right)^+$$
$$= e^{-rT}X\Phi\left[q^+\left(\frac{e^{-rT}X}{S_0}\right), T\right]$$
$$- S_0\Phi\left[q^-\left(\frac{e^{-rT}X}{S_0}\right), T\right]$$

where

$$q^{\pm}\left(s, \tau\right) \equiv \frac{\ln s \pm \sigma^2\tau/2}{\sigma\sqrt{\tau}}.$$

Multiplication by e^{rT} gives an expression for the expectation of the put's terminal value under $\hat{\mathbb{P}}$: $\hat{E}\left(X - S_T\right)^+ = e^{rT}P^{\text{B-S}}\left(S_0, T; r, \sigma\right)$. Now, by analogy, on substituting r_n for r and σ_n^2 for σ^2 and letting $q_n^{\pm}$ be the corresponding version of $q^{\pm}$, we have in the present case that

$$\hat{E}\left[\left(X - S_T\right)^+ \mid N_T = n\right] = X\Phi\left[q_n^+\left(\frac{e^{-rT}X}{S_0}\right)\right]$$
$$- S_0 e^{r_n T}\Phi\left[q_n^-\left(\frac{e^{-rT}X}{S_0}\right)\right]. \quad (22.2)$$

If we somehow knew that there would be precisely n information events during $(0, T]$, then discounting this expression back to $t = 0$ at the riskless rate would give us the initial value of the option.

3. Since we do not know the number of information events, we must in fact average the conditional valuations over the various integer values that n could have, weighting each by its Poisson probability under $\hat{\mathbb{P}}$. In other words, we must "iterate" the expectation by writing (22.2) as a function of N_T and finding the expected value. Letting $p(n; \lambda T) \equiv (\lambda T)^n e^{-\lambda T}/n!$ for $n \in \{0, 1, 2, ...\}$ be the PMF of N_T under the Poisson model, this gives us the put's unconditional expected terminal value under $\hat{\mathbb{P}}$:

$$
\hat{E}(X - S_T)^+ = \sum_{n=0}^{\infty} \left\{ X\Phi\left[q_n^+\left(\frac{e^{-rT}X}{S_0}\right)\right] - S_0 e^{r_n T}\Phi \right.
$$
$$
\left. \left[q_n^-\left(\frac{e^{-rT}X}{S_0}\right)\right]\right\} p(n; \lambda T).
$$

4. Finally, we discount the expected terminal value at the *actual* riskless rate to get the arbitrage-free value of the put under the Merton model:

$$
P^M(S_0, T) = e^{-rT}\hat{E}(X - S_T)^+.
$$

For computational purposes it helps to express P^M in a different form. Suppose that we have programmed a function, macro, or module to calculate the Black–Scholes put price in terms of the initial underlying price, time to expiration, strike price, interest rate, and volatility, as $P^{\text{B-S}}(S_0, T; X, r, \sigma)$. Replacing r, σ by r_n, σ_n and "undoing" the implicit discounting at rate r_n by multiplying by $e^{r_n T}$, we have the following computationally efficient formula for $P^M \equiv P^M(S_0, T; X, r, \sigma, \lambda, \eta, \xi)$

$$
P^M = e^{-rT}\sum_{n=0}^{\infty} e^{r_n T} P^{\text{B-S}}(S_0, T; X, r_n, \sigma_n) p(n; \lambda T). \tag{22.3}
$$

To speed things up even more, one can calculate the Poisson probabilities recursively as the sum progresses, starting with $p(0; \lambda T) = e^{-\lambda T}$ and using $p(n; \lambda T) = \lambda T p(n - 1; \lambda T)/n$ for $n \geq 1$. Of course, one has to truncate the sum after some finite number of terms. To see how far to carry it in order to achieve a given degree of precision, one can bound the put as $P^{\text{B-S}}(S_0, T; X, r_n, \sigma_n) \leq e^{-r_n T}X$ and use one of several available formulas for upper bounds on $\Pr(N_T > n)$ when $N_T \sim$ Poisson (λT); e.g., $\Pr(N_T > n) \leq 1 - \exp[-\lambda T/(n + 1)]$.

With appropriate estimates of σ, λ, η, and ξ the Merton model can give a reasonable fit to market prices of options having short times to expiration. Suppose that we apply (22.3) with such an appropriate set of parameters and calculate Merton prices for each of various strikes $\{X_j\}_{j=1}^{m}$, then for each strike back out the corresponding Black–Scholes implicit volatility by solving for $\hat{\sigma}_j$ in

$$
P^M(S_0, T; X_j, r, \sigma, \lambda, \eta, \xi) = P^{\text{B-S}}(S_0, T; X_j, r, \hat{\sigma}_j).
$$

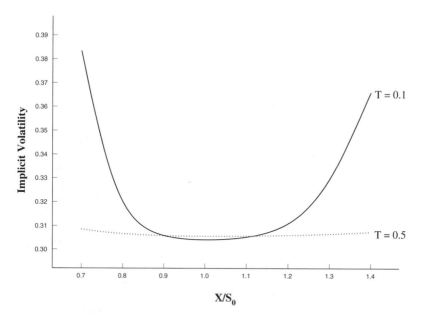

Figure 22.1   Implicit volatilties versus "moneyness" (X/S_0) for short- and long-term options under jump–diffusion dynamics.

By plotting the $\{\hat{\sigma}_j\}$ versus the $\{X_j\}$, one can see that Merton prices can indeed generate pronounced volatility smiles. However, if the model is parameterized so as to fit prices of short-term options, it typically does *not* fit prices of options having longer term. The reason is that increments to log price over nonoverlapping periods of equal length are i.i.d., because increments to the Poisson and BM processes and the jump sizes all have this property. Also, given the way we have modeled the jump sizes, the increments have finite variance. It thus follows from the central-limit theorem that increments over *long* periods—which are sums of i.i.d., finite-variance increments over many *sub*periods—must converge in distribution to the normal as the length of the period increases. In other words, the noise produced by the Poisson shocks just averages out in the long run, and we are back to the lognormal model for terminal price. The result, shown formally in Exercise 22.4, is that Merton prices converge to Black–Scholes prices as $T \to \infty$, and for parameters that fit short-term options reasonably well the convergence is simply too rapid to give good fits for large T. The two plots of volatility smiles in Fig. 22.1 (based on a model with $\lambda = .1$, $\eta = 0$, $\xi = .2$, $\sigma = .3$, and $r = .05$) illustrate the problem.

It thus appears that Poisson-counted jumps alone are not the complete answer to the volatility smile, but before we can consider more realistic models, we shall need some additional tools.

22.2 A PRIMER ON CHARACTERISTIC FUNCTIONS

We have made use of the moment-generating function (MGF) repeatedly in previous chapters—the function $\mathcal{M}_Y(\zeta) = Ee^{\zeta Y}$ for random variable Y. We now need a related function that overcomes some significant limitations of the MGF. What are the limitations? The first is that MGFs do not exist for all distributions; that is, the calculation $Ee^{\zeta Y} = \int_{-\infty}^{\infty} e^{\zeta y} f(y) \cdot dy$ (in the continuous case) or $Ee^{\zeta Y} = \sum_{y \in \mathbb{Y}} e^{\zeta y} f(y)$ (in the discrete case) or $Ee^{\zeta Y} = \int_{-\infty}^{\infty} e^{\zeta y} \cdot dF(y)$ (in the general case) does not always converge to a finite number. The reason is that when $\zeta > 0$ the exponential explodes rapidly as $y \to \infty$, and likewise as $y \to -\infty$ when $\zeta < 0$. The MGF is said to exist only if the integral converges in some neighborhood of the origin—that is, for all $|\zeta|$ less than some positive ε. For this it is necessary that *all* the moments $\left\{ E|Y|^k \right\}_{k=1,2,\ldots}$ are finite. Thus, many random variables (e.g., Student's t) do not have MGFs.

This is a great annoyance, because when they do exist MGFs can deliver powerful results about distributions of functions of random variables. For one thing, $\mathcal{M}_Y$ *characterizes* the distribution of Y in the same way that CDF F does, since no two different distributions have the same MGF. For another, we can often deduce the distributions of a function of random variables by calculating its MGF. For example, if $\{Y_j\}_{j=1}^n$ are i.i.d. with MGF $\mathcal{M}_Y$ and $S_n \equiv \sum_{j=1}^n Y_j$, then $\mathcal{M}_{S_n}(\zeta) = \mathcal{M}_{Y_1}(\zeta)\mathcal{M}_{Y_2}(\zeta) \cdots \cdots \mathcal{M}_{Y_n}(\zeta) = \mathcal{M}_Y(\zeta)^n$. If we can recognize the form of the resulting MGF as belonging to a particular distribution, then we will know what that distribution is. However—and here is the other significant limitation—if we cannot recognize the form of such an MGF, then it is difficult to invert it analytically and recover the underlying distribution.[63]

Fortunately, there is another "integral transform" with all the MGF's useful properties, which does exist for all distributions, and which is more easily inverted. To introduce it, recall Euler's formula: $e^{iy} = \cos y + i \sin y$, where $i = \sqrt{-1}$ is the "imaginary" unit and y is a real number. The transformation e^{iy} takes the real number y from the real line to a point on the perimeter of the unit circle in the complex plane. This point lies at angle y with respect to the positive real axis, as shown in Fig. 22.2 for a value of $y \in (0, \pi/2)$. Thus, if y is the realization of random variable Y and ζ is an arbitrary real number, $e^{i\zeta y}$ takes ζy to a point on the perimeter at angle ζy from the positive real axis. If ζY has some distribution f_ζ, then the distribution of $e^{i\zeta Y}$ can be visualized by simply wrapping f_ζ around the unit circle as in Fig. 22.3, matching up the point $\zeta y = 0$ on the line with the point $(1, 0)$ on the circle and extending counterclockwise for positive ζy and clockwise for negative ζy. Wrapping the distribution in this way merely attaches the density or probability associated with ζy to the point $e^{i\zeta y}$ on the edge of the circle. The value of the *characteristic function* (CF) of Y at ζ is just the center of mass of this wrapped-around distribution. Representing the CF at ζ as $\phi(\zeta)$ (or as $\phi_Y(\zeta)$ if it is not otherwise understood to

---

[63]Inversion of MGFs is certainly *possible*. It is just that the required tools of complex integration are not in everyone's toolkit.

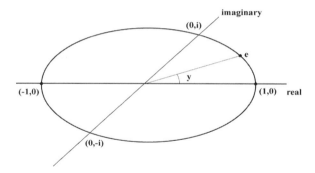

Figure 22.2 The point e^{iy} in the complex plane for some $y \in (0, \pi/2)$.

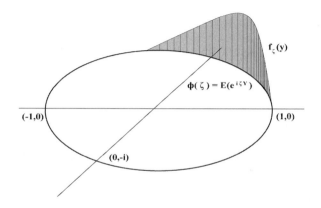

Figure 22.3 The characteristic function of Y at some $\zeta > 0$ as center of mass of a wrapped-around PDF of ζY.

apply to Y), we have $\phi(\zeta) = Ee^{i\zeta Y} = \int e^{i\zeta y} \cdot dF(y)$. The *function* ϕ thus assigns to each real number ζ a complex number somewhere on or within the unit circle in the complex plane. ϕ is in fact the Fourier (or Fourier–Stieltjes) transform of the distribution F. In Fig. 22.3 the value of ϕ at a particular ζ is represented by the dot within the unit circle. As ζ is varied, then of course the distribution of ζY changes, with f_ζ spreading out as $|\zeta|$ increases and contracting as $|\zeta|$ declines. As f_ζ changes in this way, its center of mass correspondingly moves about within the complex plane. Clearly, at $\zeta = 0$ all the (unit) probability mass of ζY lies at $(1,0)$, and so $\phi(0) = 1$, a real number. In general, however, ϕ is a complex-valued function, which can be visualized as a continuous curve snaking around within an infinitely long cylinder, as in Fig. 22.4.

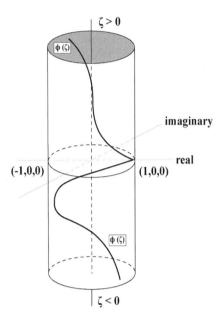

Figure 22.4   Characteristic function as a continuous curve in a cylinder whose cross section at $\zeta \in \Re$ is a circle in the complex plane.

When the MGF does exist as a function $\mathcal{M}(\zeta)$ for ζ in some neighborhood of the origin, then the CF is obtained merely by replacing ζ with $i\zeta$. For example, MGFs and CFs of several distributions we have already encountered are shown in Table 22.1. All the forms for the CF apply to all real ζ.

As for the MGF when it exists, there is a one-to-one relation between CFs. and distributions. There are several inversion formulas for CFs that allow one to recover the PDF or CDF rather simply by integrating a multiple of the CF with respect to ζ. For our purposes the most useful formula is the following:

$$F(y) = \frac{1}{2} - \lim_{c \to \infty} \int_{-c}^{c} \frac{e^{-i\zeta y}}{2\pi i\zeta} \phi(\zeta) \cdot d\zeta. \tag{22.4}$$

Table 22.1   MGFs and CFs of common distributions

Distribution	MGF, $\mathcal{M}(\zeta)$	CF, $\phi(\zeta)$
Normal (μ, σ^2)	$e^{\zeta\mu + \zeta^2\sigma^2/2}$	$e^{i\zeta\mu - \zeta^2\sigma^2/2}$
Gamma (α, β)	$(1 - \zeta\beta)^{-\alpha}, \zeta < 1/\beta$	$(1 - i\zeta\beta)^{-\alpha}$
Poisson (λ)	$\exp\left[\lambda\left(e^{\zeta} - 1\right)\right]$	$\exp\left[\lambda\left(e^{i\zeta} - 1\right)\right]$
Binomial (n, θ)	$\left[1 + \theta\left(e^{\zeta} - 1\right)\right]^{n}$	$\left[1 + \theta\left(e^{i\zeta} - 1\right)\right]^{n}$
Cauchy (δ, γ)	Does not exist	$e^{i\zeta\delta - |\gamma\zeta|}$

This applies to all continuous distributions at any real y and to all other distributions when y is not a point at which there is positive probability mass.[64]

22.3 USING FOURIER METHODS TO PRICE OPTIONS

Suppose that our goal is to price a T-expiring European-style option—either a call or a put. By the put–call parity relation developed in Section 18.1, the difference in prices of call and put with the same strike X and expiration T equals the arbitrage-free value of a T-expiring forward contract with forward price X. Since the value of the forward contract is known independently of the underlying dynamics, working out the arbitrage-free value of either option gives us the other by simple arithmetic. Thus, we can focus on pricing European puts. Now, given our model for the dynamics of underlying price $\{S_t\}$, suppose that we can somehow get a computable expression for the conditional CF of time T price S_T under equivalent martingale measure $\hat{\mathbb{P}}$, based on the information $\mathcal{F}_t$ that is available at any $t \in [0, T]$. We represent the value of the conditional CF at ζ as $\hat{\phi}_{S_T}(\zeta \mid \mathcal{F}_t)$, the "$\hat{}$" reminding us of the change of measure. Then with this function and relation (22.4) we can evaluate the conditional CDF of S_T under $\hat{\mathbb{P}}$ at any real number y as

$$\hat{F}_{S_T}(y \mid \mathcal{F}_t) = \frac{1}{2} - \lim_{c \to \infty} \int_{-c}^{c} \frac{e^{-i\zeta y}}{2\pi i \zeta} \hat{\phi}_{S_T}(\zeta \mid \mathcal{F}_t) \cdot d\zeta. \tag{22.5}$$

We will be able to use this representation for the conditional CDF of S_T to calculate $\hat{E}_t (X - S_T)^+$, the time t expectation of the terminal payoff of a T-expiring put with strike price X, and we can do this with a single numerical integration.

The trick in doing this is based on an identity that we now develop. First, note that all our models imply that $\hat{F}_{S_T}(y \mid \mathcal{F}_t) = 0$ for all $y < 0$, in line with the limited-liability feature of asset prices. This implies that $\int_{-\infty}^{X} \hat{F}_{S_T}(y \mid \mathcal{F}_t) \cdot dy = \int_{0}^{X} \hat{F}_{S_T}(y \mid \mathcal{F}_t) \cdot dy$. Next, evaluating the integral by parts, we have

$$\int_{0}^{X} \hat{F}_{S_T}(y \mid \mathcal{F}_t) \cdot dy = y \hat{F}_{S_T}(y \mid \mathcal{F}_t) \Big|_{0}^{X} - \int_{0}^{X} y \cdot d\hat{F}_{S_T}(y \mid \mathcal{F}_t)$$

$$= X \hat{F}_{S_T}(X) \mid \mathcal{F}_t - \int_{0}^{X} y \cdot d\hat{F}_{S_T}(y \mid \mathcal{F}_t). \tag{22.6}$$

---

[64]The notation $\lim_{c \to \infty} \int_{-c}^{c}$ indicates Cauchy's construction of the improper integral, which is commonly used in complex analysis. It is evaluated as $\lim_{c \to \infty} \left(\int_{0}^{c} + \int_{-c}^{0} \right)$ and is used here because the integral $\int_{-\infty}^{\infty} (2\pi i \zeta)^{-1} e^{-i\zeta y} \phi(\zeta) \cdot d\zeta$ need not exist in the usual (Lebesgue) construction.

Expressing the first term as $X \int_0^X d\hat{F}_{S_T}(y \mid \mathcal{F}_t)$ and combining with the second term give

$$
\int_0^X \hat{F}_{S_T}(y \mid \mathcal{F}_t) \cdot dy = \int_0^X (X - y) \cdot d\hat{F}_{S_T}(y \mid \mathcal{F}_t) \tag{22.7}
$$
$$
= \int_0^\infty (X - y)^+ \cdot d\hat{F}_{S_T}(y \mid \mathcal{F}_t)
$$
$$
= \hat{E}_t (X - S_T)^+.
$$

Thus, we can find the expected terminal payoff of the put merely by integrating from 0 to X the conditional CDF of S_T, and we can get the conditional CDF from the CF via (22.5). Although it appears that this will involve double numerical integration, we shall see that this is not the case.

Now, with most of the models we consider, it is easier to represent the conditional CF of $s_T \equiv \ln S_T$ than the CF of S_T itself, and so we will need somehow to work with that. Fortunately, a simple change of variable saves the day. Note that

$$
\hat{F}_{S_T}(y \mid \mathcal{F}_t) \equiv \hat{\mathbb{P}}(S_T \leq y \mid \mathcal{F}_t)
$$
$$
= \hat{\mathbb{P}}(s_T \leq \ln y \mid \mathcal{F}_t)
$$
$$
= \hat{F}_{s_T}(s \mid \mathcal{F}_t),
$$

where $s \equiv \ln y$. Therefore, we can set $y = e^s$ in (22.7) and change the limits of integration accordingly to get

$$
\hat{E}_t (X - S_T)^+ = \int_0^X \hat{F}_{S_T}(y \mid \mathcal{F}_t) \cdot dy
$$
$$
= \int_{-\infty}^{\ln X} \hat{F}_{s_T}(s \mid \mathcal{F}_t) e^s \cdot ds.
$$

Now we can use (22.5), replacing $\hat{F}_{s_T}$ in the preceding expression by its representation in terms of the conditional CF of s_T, to get

$$
\hat{E}_t (X - S_T)^+ = \frac{X}{2} - \frac{1}{2} \int_{-\infty}^{\ln X} \left[\lim_{c \to \infty} \int_{-c}^{c} \frac{e^{-i\zeta s}}{2\pi i \zeta} \hat{\phi}_{s_T}(\zeta \mid \mathcal{F}_t) \cdot d\zeta \right] e^s \cdot ds.
$$

Switching the order of integration, integrating with respect to s, simplifying, and then discounting by the time t value of a discount unit bond, finally lead to the following formula for the value of the put:

$$
P(S_t, T - t) = B(t, T) \frac{X}{2} \left[1 - \frac{1}{\pi} \lim_{c \to \infty} \int_{-c}^{c} \frac{X^{-i\zeta}}{i\zeta + \zeta^2} \hat{\phi}_{s_T}(\zeta \mid \mathcal{F}_t) \cdot d\zeta \right]. \tag{22.8}
$$

Evaluating this requires just one numerical integration with respect to ζ, and this can be done in a fraction of a second on even a Pentium II desktop computer.[65] The value of the integral is necessarily a real number, and so only the real part of the integrand must be integrated.

Having priced the put, we can get the arbitrage-free price of a call with the same expiration and strike from put–call parity as

$$
\begin{aligned}
C\left(S_t, T-t\right) &= P\left(S_t, T-t\right) + \mathfrak{F}\left(S_t, T-t\right) \\
&= P\left(S_t, T-t\right) + S_t e^{-\delta(T-t)} - B\left(t, T\right) X,
\end{aligned}
$$

where δ is the continuous dividend rate paid by the underlying asset.

22.4 APPLICATIONS TO JUMP MODELS

We will now apply formula (22.8) to price put options under the three discontinuous models of Chapter 21. For brevity of notation, we assume in each case that we are standing at $t = 0$ with underlying price S_0 and discount bond price $B\left(0, T\right)$ known and that the underlying asset pays no dividend (i.e., $\delta = 0$). Since we are at $t = 0$, conditioning on initial information $\mathcal{F}_0$ will be understood, with $E\left(\cdot\right)$ meaning the same thing as $E(\cdot \mid \mathcal{F}_0)$. We begin with Merton's jump–diffusion in order to compare the method with the direct calculation already completed. In each case all we need is to develop the conditional CF of S_T under martingale measure $\hat{\mathbb{P}}$, which would be used with formula (22.8) along with a computer routine for numerical integration.

1. **Merton's jump–diffusion.** Recall from Section 22.1 that, conditional on the occurrence of $N_T = n$ price shocks, $s_T \equiv \ln S_T$ is distributed as normal under $\hat{\mathbb{P}}$ with mean $s_0 + \left(r - \lambda\eta - \sigma^2/2\right)T + n\left[\ln\left(1+\eta\right) - \xi^2/2\right]$ and variance $\sigma^2 T + n\xi^2$, and that N_T is distributed as Poisson $\left(\lambda T\right)$. For brevity we can temporarily write the mean as $s_0 + \psi T + n\theta$, where $\psi \equiv r - \lambda\eta - \sigma^2/2$ and $\theta \equiv \ln\left(1+\eta\right) - \xi^2/2$. The CF of $\ln S_T$ under $\hat{\mathbb{P}}$ then comes from a simple application of the tower property of conditional expectation. Thus,

$$
\hat{\phi}_{s_T}\left(\zeta\right) = \hat{E}e^{i\zeta s_T} = \hat{E}\left[\hat{E}\left(e^{i\zeta s_T} \mid N_T\right)\right].
$$

From the first row of Table 22.1 we find that the conditional CF in the brackets is

$$
\begin{aligned}
\hat{E}\left(e^{i\zeta s_T} \mid N_T\right) &= e^{i\zeta\left(s_0 + \psi T + N_T\theta\right) - \zeta^2\left(\sigma^2 T + N_T\xi^2\right)/2} \\
&= S_0^{i\zeta} e^{\left(i\zeta\psi - \zeta^2\sigma^2/2\right)T} \cdot e^{\left(i\zeta\theta - \zeta^2\xi^2/2\right)N_T}.
\end{aligned}
$$

---

[65]There are many schemes for evaluating numerically a definite integral of the form $\int_a^b g\left(x\right) \cdot dx$; e.g., the trapezoidal method, Simpson's rule, Romberg integration, and Gaussian quadrature. These are implemented in many of the standard computational software packages. Even the simple trapezoidal rule gives very fast and accurate results in this application. Whatever method is used, the finite limits $\pm c$ need to expand as the time to expiration diminishes. This is because the CF decays more slowly as $|\zeta| \to \infty$ the more concentrated is the distribution of s_T. Also, it is clear that one must avoid evaluating the integrand at $\zeta = 0$.

Here the only random variable is N_T, and so iterating the expectation gives

$$\hat{E}e^{i\zeta s_T} = S_0^{i\zeta} e^{\left(i\zeta\psi - \zeta^2\sigma^2/2\right)T} \cdot \hat{E}e^{\left(i\zeta\theta - \zeta^2\xi^2/2\right)N_T}$$

$$= S_0^{i\zeta} e^{\left(i\zeta\psi - \zeta^2\sigma^2/2\right)T} \hat{\mathcal{M}}_{N_T}\left(i\zeta\theta - \frac{\zeta^2\xi^2}{2}\right).$$

Finally, using the formula for the Poisson MGF in Table 22.1, we express $\hat{\mathcal{M}}_{N_T}(\cdot)$ by replacing ζ with $i\zeta\theta - \zeta^2\xi^2/2$ and λ with λT to obtain the following expression for the CF of $s_T \equiv \ln S_T$ under $\hat{\mathbb{P}}$:

$$\hat{\phi}_{s_T}(\zeta) = S_0^{i\zeta} e^{\left(i\zeta\psi - \zeta^2\sigma^2/2\right)T} \exp\left[\lambda T\left(e^{i\zeta\theta - \zeta^2\xi^2/2} - 1\right)\right]. \qquad (22.9)$$

Using this in (22.8) to carry out a numerical integration actually involves less computational effort, holding accuracy constant, than does evaluating the sum in the direct expression (22.3).

2. **Variance–gamma.** Recall that in the VG model the log of price evolves as a Brownian motion that operates at a speed that is different from calendar time and represented by a process $\{\mathfrak{T}_t\}$, whose increments $\mathfrak{T}_t - \mathfrak{T}_s$ for $t > s$ have a gamma distribution. Specifically, we had

$$\ln S_t = \ln S_0 + \mu t + \gamma \mathfrak{T}_t + \sigma W_{\mathfrak{T}_t}, \qquad (22.10)$$

where $\mathfrak{T}_t$ is distributed as gamma with shape parameter t/ν and scale parameter ν for some $\nu > 0$. The PDF is

$$f_{\mathfrak{T}_t}(\tau) = \Gamma\left(\frac{t}{\nu}\right)^{-1} \nu^{-t/\nu} \tau^{t/\nu - 1} e^{-\tau/\nu}, \tau > 0.$$

Thus, from the properties of Brownian motion, we have that $s_T \equiv \ln S_T$ is distributed as $N\left(s_0 + \mu T + \gamma \mathfrak{T}_T, \sigma^2 \mathfrak{T}_T\right)$, conditional on $\mathfrak{T}_T$. From Table 22.1 the conditional CF is thus

$$\phi_{s_T|\mathfrak{T}_T}(\zeta) = e^{i\zeta(s_0 + \mu T + \gamma\mathfrak{T}_T) - \zeta^2\sigma^2\mathfrak{T}_T/2}$$

$$= S_0^{i\zeta} e^{i\zeta\mu T} \cdot e^{\left(i\zeta\gamma - \zeta^2\sigma^2/2\right)\mathfrak{T}_T}.$$

Again, the tower property delivers the CF conditional on the information available at $t = 0$:

$$\phi_{s_T}(\zeta) = S_0^{i\zeta} e^{i\zeta\mu T} E e^{\left(i\zeta\gamma - \zeta^2\sigma^2/2\right)\mathfrak{T}_T} = S_0^{i\zeta} e^{i\zeta\mu T} \mathcal{M}_{\mathfrak{T}_T}\left(i\zeta\gamma - \frac{\zeta^2\sigma^2}{2}\right).$$

Referring to the gamma MGF in Table 22.1, we replace ζ with $i\zeta\gamma - \zeta^2\sigma^2/2$, β with ν, and α with T/ν to obtain the following expression for the c.f. under natural measure $\mathbb{P}$:

$$\phi_{s_T}(\zeta) = S_0^{i\zeta} e^{i\zeta\mu T}\left[1 - \left(i\zeta\gamma - \frac{\zeta^2\sigma^2}{2}\right)\nu\right]^{-T/\nu}. \qquad (22.11)$$

Finally, we convert to an EMM $\hat{\mathbb{P}}$ by setting $\mu = r + \theta/\nu$, where $\theta \equiv \ln\left[1 - \nu\left(\gamma + \sigma^2/2\right)\right]$:

$$\hat{\phi}_{s_T}(\zeta) = S_0^{i\zeta} e^{i\zeta(r+\theta/\nu)T} \left[1 - \left(i\zeta\gamma - \frac{\zeta^2\sigma^2}{2}\right)\nu\right]^{-T/\nu}. \qquad (22.12)$$

Exercise 22.7 shows that $\hat{\phi}_{s_T}(\zeta) \to S_0^{i\zeta} \exp\left[i\zeta\left(r - \sigma^2/2\right)T - \zeta^2\sigma^2 T/2\right]$ as $\nu \to 0$, which corresponds to the CF of $s_T = \ln S_T$ under geometric Brownian motion.

3. **Branching process.** In the branching model the stock price at time T when there are N_T branches during the period $[0, T]$ is expressed as $S_{N_T} = Y_1 + Y_2 + \cdots + Y_{S_{N_T-1}}$, where the $\{Y_j\}$ are i.i.d. random variables taking nonnegative integer values and S_{N_T-1} is the price just before the last of the N_T branches. For this model it is easier to derive the CF of S_{N_T} itself, rather than of its logarithm. This is done indirectly by first deriving the probability-generating function (PGF), $\Pi_{S_{N_T}}(\varphi) \equiv E\varphi^{S_{N_T}}$ for an arbitrary complex number φ, and then setting $\varphi = e^{i\zeta}$. The PGF is found recursively, as follows. Conditioning on there being $N_T = n$ branches during $[0, T]$ and on the price S_{n-1} after $n - 1$ branches, we have

$$\begin{aligned}
\Pi_{S_{N_T}|N_T=n,S_{n-1}}(\varphi) &= E\varphi^{Y_1+Y_2+\cdots+Y_{S_{n-1}}} \\
&= E\varphi^{Y_1} \cdot E\varphi^{Y_2} \cdot \cdots \cdot E\varphi^{Y_{S_{n-1}}} \\
&= [\Pi_Y(\varphi)]^{S_{n-1}},
\end{aligned}$$

where Π_Y is the common PGF governing the number of "offspring" from each value element of S_{n-1}. Working backward by conditioning now on S_{n-2} gives

$$\begin{aligned}
\Pi_{S_{N_T}|N_T=n,S_{n-2}}(\varphi) &= E\left[\Pi_Y(\varphi)\right]^{S_{n-1}} \\
&= \{\Pi_Y\left[\Pi_Y(\varphi)\right]\}^{S_{n-2}} \equiv \left[\Pi_Y^{[2]}(\varphi)\right]^{S_{n-2}},
\end{aligned}$$

where the superscript "[2]" denotes the "iterated" function $\Pi_Y(\Pi_Y)$ in which Π_Y operates on itself. Continuing to work backward in this fashion to known initial price S_0 gives the PGF of S_{N_T} conditional just on $N_T = n$, as

$$\Pi_{S_{N_T}|N_T=n}(\varphi) = \Pi_Y^{[n]}(\varphi)^{S_0}. \qquad (22.13)$$

Finally, the CF of S_{N_T} comes by replacing φ by $e^{i\zeta}$ and n by N_T, then taking the expectation. Modeling N_T as Poisson (λT), this gives

$$\phi_{S_{N_T}}(\zeta) = \sum_{n=0}^{\infty} \left[\Pi_Y^{[n]}\left(e^{i\zeta}\right)\right]^{S_0} \frac{(\lambda T)^n e^{-\lambda T}}{n!}.$$

We now have to find a measure $\hat{\mathbb{P}}$ equivalent to natural measure $\mathbb{P}$ that makes $\{S_{N_t}/M_t\}_{0 \le t \le T}$ a martingale. Expression (21.4) implies that $ES_{N_T} = S_0 e^{\lambda T(\mu - 1)}$, so

$$EM_T^{-1} S_{N_T} = M_0^{-1} e^{-rT} ES_{N_T} = M_0^{-1} S_0 e^{\lambda T(\mu - 1 - r/\lambda)}$$

when $M_T = M_0 e^{rT}$ is deterministic. An EMM $\hat{\mathbb{P}}$ must equate $\hat{E} M_T^{-1} S_{N_T}$ to $M_0^{-1} S_0$ in accord with the fair-game property of martingales. To make the switch, we simply model $\{N_t\}$ and Y (the number of offspring per value unit) in any way that (1) preserves equivalence (keeps the supports the same) and (2) puts $\hat{E} Y = \mu = 1 + r/\lambda$. A good strategy is to start with a model for Y under $\mathbb{P}$ in which the iterates of the PGF have closed form, since this greatly facilitates the calculation of $\phi_{S_{N_T}}$ and of option values.[66]

EXERCISES

22.1 Show that $EU_j = \eta$ when $1 + U_j$ is distributed as lognormal with parameters $\ln(1 + \eta) - \xi^2/2$ and ξ^2.

22.2 Show that $Ee^{\zeta Y} = e^{\theta(e^\zeta - 1)}$ for any real number ζ when $Y \sim \text{Poisson}(\theta)$.

22.3 Given that random variable Y has mean μ, variance σ^2, and CF $\phi(\zeta) = Ee^{i\zeta Y}$, express the CF of the standardized version, $(Y - \mu)/\sigma$.

22.4 Using the notation $\psi \equiv r - \lambda\eta - \sigma^2/2$ and $\theta \equiv \ln(1 + \eta) - \xi^2/2$ as in (22.9), the mean and variance of $s_T \equiv \ln S_T$ under EMM $\hat{\mathbb{P}}$ are

$$\hat{E} s_T = s_0 + (\psi + \theta\lambda) T$$
$$\hat{V} s_T = \left[\sigma^2 + \left(\xi^2 + \theta^2\right)\lambda\right] T.$$

Setting $Z_T \equiv \left(s_T - \hat{E} s_T\right)/\sqrt{\hat{V} s_T}$, (a) express $\ln \hat{\phi}_{Z_T}(\zeta)$, the natural logarithm of the CF of Z_T under $\hat{\mathbb{P}}$; (b) determine $\lim_{T \to \infty} \hat{\phi}_{Z_T}(\zeta)$; and (c) state what this limit implies for the distribution of Z_T as $T \to \infty$.

22.5 We have seen that a drawback of the jump–diffusion model is its implication that the (standardized) $\ln S_T$ converges in distribution to the normal as $T \to \infty$, and that this causes implicit-volatility curves of options based on the model to flatten out too quickly to fit the market data. The convergence is a simple consequence of the central-limit theorem, given that increments to $\ln S_t$ are i.i.d. with finite variance. Section 9.3 described a discrete-time model for the continuously

---

[66]For such a model, see Theodore Harris (1989, p. 9). Section 9.4.4 of Epps (2007) gives details on pricing options under the branching model.

compounded rate of return in which the variance of $\ln S_{t+1} - \ln S_t$ is *infinite*. Suppose in the jump–diffusion framework that we were to model $\ln (1 + U)$ not as normal but as an infinite-variance stable variate. The *symmetric* stable distributions have very simple CFs involving three parameters that correspond to median, scale, and tail thickness. Representing these as η, ξ, and α, the CF is $\phi (\zeta) = \exp (i\zeta\eta - \xi |\zeta|^{\alpha})$. When $\alpha = 1$, this is the Cauchy distribution, which has neither mean nor variance; when $\alpha = 2$, it is the normal distribution, which has all its moments; but when $1 < \alpha < 2$, the distribution has a mean but variance is infinite. Thus, with $\ln (1 + U) \sim \text{stable}(\eta, \xi, \alpha)$, the problem of convergence to normality would be overcome, and it would be a simple matter to express the conditional CF of $\ln S_T$ under $\mathbb{P}$. Nevertheless, martingale pricing is not possible in this model. Why?

22.6 Under an EMM $\hat{\mathbb{P}}$ the mean and variance of $s_T \equiv \ln S_T$ in model (21.3) are

$$\hat{E}s_T = s_0 + \left(r + \frac{\theta}{\nu} + \gamma\right) T$$
$$\hat{V}s_T = \left(\sigma^2 + \gamma^2\nu\right) T,$$

where $\theta \equiv \ln \left[1 - \nu \left(\gamma + \sigma^2/2\right)\right]$; and $\ln \hat{\phi}_{s_T} (\zeta)$, the logarithm of the CF, is given by

$$i\zeta s_0 + i\zeta \left(r + \frac{\theta}{\nu}\right) T - \frac{1}{\nu} \ln \left[1 - \left(i\zeta\gamma - \frac{\zeta^2\sigma^2}{2}\right)\nu\right] T. \qquad (22.14)$$

Setting $Z_T \equiv \left(s_T - \hat{E}s_T\right) / \sqrt{\hat{V}s_T}$, (a) express $\ln \hat{\phi}_{Z_T} (\zeta)$, the natural logarithm of the CF of Z_T under $\hat{\mathbb{P}}$; (b) determine $\lim_{T\to\infty} \hat{\phi}_{Z_T} (\zeta)$; and (c) state what this limit implies for the distribution of Z_T as $T \to \infty$.

22.7 Working from expression (22.14), show that

$$\lim_{\nu\to 0} \ln \hat{\phi}_{s_T} (\zeta) = i\zeta \left[s_0 + \left(r - \frac{\sigma^2}{2}\right) T\right] - \frac{\zeta^2\sigma^2 T}{2},$$

and conclude that the CFs of $\ln S_T$ under the VG and geometric BM models coincide in the limit.

22.8 In the branching process suppose that we model the number of offspring Y as Poisson(θ) for some $\theta > 0$. Find the iterates $\Pi_Y^{[n]} (\varphi)$ of the PGF for $n = 1, 2, 3$.

22.9 Let Y be a discrete random variable supported on $\{0, 1, 2, ...\}$ with PGF $\Pi_Y (\varphi)$. (a) Show that $\Pi_Y (0) = \mathbb{P} (Y = 0)$, that $\Pi_Y (1) = 1$, and that $EY = \Pi_Y' (\varphi)|_{\varphi=1}$. (b) In the branching model suppose that we start at initial value S_0 and attain the value S_n after n branches. Show that $\mathbb{P} (S_n = 0) = \Pi_Y^{[n]} (0)^{S_0}$. (c) In the branching model show that $\lim_{n\to\infty} \mathbb{P} (S_n = 0) = 1$ if and only if

$EY \leq 1$. (Hint: Thinking of the definition of $\Pi_Y (\varphi)$ as $E\varphi^Y$, sketch a plot of a generic PGF for $\varphi \in [0, 1]$ and consider how the plots would differ for cases $EY \leq 1$ and $EY > 1$. Then think how plots of the iterates $\Pi_Y^{[n]} (\varphi)$ would behave as n increases and the function Π_Y is applied repeatedly to itself.) (**d**) Conclude that $\lim_{T \to \infty} \mathbb{P} (S_{N_T} = 0) = 1$ if $EY \leq 1$ when $\{N_t\}_{t \geq 0}$ is a Poisson process. (**e**) Assuming that $\hat{\mathbb{P}}$ is an EMM and that $EY \leq 1$ (the expectation still under natural measure $\mathbb{P}$), determine $\lim_{T \to \infty} \hat{\mathbb{P}} (S_{N_T} = 0)$.

CHAPTER 23

OPTIONS ON STOCHASTIC VOLATILITY PROCESSES

We have seen that Merton's jump–diffusion model fails to explain the "term structure" of option prices. In particular, plotting implicit volatility curves from Merton prices for a series of increasing times to expiration shows that the volatility smile flattens out much too quickly. The same limitation applies to the variance–gamma model, and for the same reason—as increments to log-price over equal intervals of time are i.i.d. and of finite variance, they quickly converge to normal as time to expiration increases. In Chapter 20 we saw that models that allowed volatility to change had just the opposite problem, in that they did not show sufficiently pronounced volatility smiles at *short* horizons. The obvious idea is to combine the two features, stochastic volatility and jumps, and we will look at a model that does this in the final section of the chapter. But to take things in smaller bites, we will begin first to implement Heston's SV model alone, which requires some new and nontrivial concepts. Once the ideas are mastered in this setting, it is not difficult to allow for Merton-style discontinuities in price paths.

Recall that the Heston model characterizes the underlying price process under martingale measure $\hat{\mathbb{P}}$ through the following stochastic differential equations:

$$dS_t = rS_t \cdot dt + \sigma_t S_t \cdot d\hat{W}_t \tag{23.1}$$

$$d\sigma_t^2 = (\alpha - \beta\sigma_t^2) \cdot dt + \gamma\sigma_t \cdot d\hat{W}_t', \tag{23.2}$$

where α, β, γ are positive constants and $\left\{\hat{W}_t, \hat{W}_t'\right\}$ are standard Brownian motions under $\hat{\mathbb{P}}$. Recall that the specification of the volatility equation allows for mean-reverting behavior of volatility and keeps σ_t^2 from becoming negative. The fact that volatility is subject to separate stochastic influences beyond the shocks driving price itself obviously greatly complicates the task of deducing the distribution of terminal price S_T in the equivalent martingale measure. The problem is much easier when the two Brownian motions are independent, and since they are Gaussian processes, this just requires them to be uncorrelated. However, it turns out that the model does a much better job of fitting the data when correlation is allowed. Specifically, explaining the "smirk" in implicit volatilities—the extreme high values at low strikes—requires negative correlation between increments to log price and to volatility. Our main effort will be directed toward this more general model; however, it is instructive to see how and why one can proceed more directly in the case of independent shocks, and we begin with that case. This also has the virtue of following the chronological development of the theory.

23.1   INDEPENDENT PRICE/VOLATILITY SHOCKS

To see how to proceed with martingale pricing when $E\hat{W}_t\hat{W}_t' = 0$ for each $t \geq 0$, let us consider first a simple generalization of the Black–Scholes theory when volatility is time varying but *deterministic,* so that its course over the life of an option is known in advance. The relevance of this situation to the one at hand will become apparent right away.

Recall that in the Black–Scholes framework of geometric Brownian motion, $dS_t = rS_t \cdot dt + \sigma S_t \cdot d\hat{W}_t$, the solution for price at any $t \geq 0$ given initial price S_0 has the familiar lognormal specification, $S_t = S_0 e^{(r-\sigma^2/2)t+\sigma\hat{W}_t}$ and $\ln S_t = \ln S_0 + (r - \sigma^2/2) t + \sigma\hat{W}_t$. Now let us break the option's life into subintervals $(0, t_1]$ and $(t_1, T]$ and allow volatility to differ in the two segments, as $\sigma_t = \sigma_1$ for $t \in (0, t_1]$ and $\sigma_t = \sigma_2$ for $t \in (t_1, T]$. The change in log price over $(0, T]$ is then the sum of two independent parts, $\ln S_{t_1} - \ln S_0$ and $\ln S_T - \ln S_{t_1}$:

$$\ln S_{t_1} - \ln S_0 = \left(r - \frac{\sigma_1^2}{2}\right)(t_1 - 0) + \sigma_1\left(\hat{W}_{t_1} - \hat{W}_0\right)$$

$$\ln S_T - \ln S_{t_1} = \left(r - \frac{\sigma_2^2}{2}\right)(T - t_1) + \sigma_2\left(\hat{W}_T - \hat{W}_{t_1}\right),$$

Here, initial values $\hat{W}_0 = 0$ and $t = 0$ are included in the first line just to make the two expressions look more alike. Adding the parts gives

$$\ln S_T - \ln S_0 = rT - \left[\frac{\sigma_1^2 t_1 + \sigma_2^2 (T - t_1)}{2} \right] + \sigma_1 (\hat{W}_{t_1} - \hat{W}_0) + \sigma_2 \left(\hat{W}_T - \hat{W}_{t_1} \right).$$

The numerator of the term in brackets can be written as $\bar{\sigma}_{[0,T]}^2 T$, where

$$\bar{\sigma}_{[0,T]}^2 = \sigma_1^2 \cdot \left(\frac{t_1}{T} \right) + \sigma_2^2 \cdot \left(\frac{T - t_1}{T} \right)$$

is the time-weighted average of the squared volatilities. The sum of the two weighted (independent) increments to Brownian motion is distributed as $N \left(0, \bar{\sigma}_{[0,T]}^2 T \right)$, which is to say as $\bar{\sigma}_{[0,T]} \sqrt{T} \cdot Z$ where $Z \sim N(0, 1)$. Thus, we have that

$$\ln S_T - \ln S_0 \sim \left(r - \frac{\bar{\sigma}_{[0,T]}^2}{2} \right) T + \bar{\sigma}_{[0,T]} \sqrt{T} \cdot Z, \qquad (23.3)$$

showing that S_T is still lognormally distributed, with variance proportional to the *weighted average* of squared volatilities over the two subintervals.

Generalizing, partition $(0, T]$ into subintervals $\{(t_{j-1}, t_j]\}_{j=1}^n$, where $0 = t_0 < t_1 < \cdots < t_{n-1} < t_n = T$, and let the volatility on the jth interval be σ_j. Then (23.3) still applies, but with a new time-weighted average of squared volatilities:

$$\bar{\sigma}_{[0,T]}^2 = \sum_{j=1}^n \left(\frac{t_j - t_{j-1}}{T} \right) \sigma_j^2.$$

Finally, letting $n \to \infty$ and $\max_{j \in \{1,2,\ldots,n\}} |t_j - t_{j-1}| \to 0$, and assuming Riemann integrability, we can allow for *continuous* changes in volatility as

$$T^{-1} \lim_{n \to \infty} \sum_{j=1}^n (t_j - t_{j-1}) \sigma_j^2 = T^{-1} \int_0^T \sigma_t^2 \cdot dt = \bar{\sigma}_{[0,T]}^2.$$

So long as the volatility process $\{\sigma_t^2\}_{t \geq 0}$ is *deterministic* lognormality still holds, and since the Black–Scholes formulas rely just on the lognormality of S_T, those formulas still apply with one trivial change. All that is required is to replace the constant σ with $\bar{\sigma}_{[0,T]}$, the square root of the time-weighted average of squared volatilities. For example, allowing for continuously paid dividends on the underlying at rate δ, the formula for the value of a T-expiring put at $t = 0$ is

$$P \left(S_0, T; \bar{\sigma}_{[0,T]}^2 \right) = B(0, T) \left\{ X \Phi \left[q^+ \left(\frac{X}{f_0}, T \right) \right] - f_0 \Phi \left[q^- \left(\frac{X}{f_0}, T \right) \right] \right\},$$
$$(23.4)$$

where $f_0 = S_0 e^{-\delta T} / B(0, T)$ and, for positive s and T,

$$q^{\pm}(s, T) = \frac{\ln s \pm \bar{\sigma}_{[0,T]}^2 T / 2}{\bar{\sigma}_{[0,T]} \sqrt{T}}.$$

Thus, generalizing Black–Scholes to allow for time-varying, deterministic volatility is straightforward.

It is obviously crucial for this result that the volatility process be deterministic, but of course it is completely unrealistic to suppose that we could in fact know the future course of volatility. Still, the result is useful because it leads to a way to price options when volatility is stochastic but independent of the shocks to price. As usual, it is the process of conditioning that simplifies an otherwise difficult task. Given independence, the *conditional* distribution of terminal price given average volatility is still lognormal, and so the discounted conditional expectation of the terminal value of a put or call under the martingale measure is still given by (23.4). The next step is to iterate the expectation and express the unconditional value of the put as

$$P\left(S_0, T\right) = B\left(0, T\right) \hat{E}\left\{ \hat{E}\left[\left(X - S_T\right)^+ \mid \bar{\sigma}^2_{[0,T]} \right] \right\}$$
$$= \hat{E}P\left(S_0, T; \bar{\sigma}^2_{[0,T]}\right).$$

Although working out the expected value analytically seems to be out of the question, the expectation can be approximated in a number of ways. The most straightforward way is by simulation. For this one would break up the option's lifetime into small subintervals of length $t_j - t_{j-1} = T/n$, say, and approximate the volatility over the jth interval as

$$\sigma^2_{t_j} = \sigma^2_{t_{j-1}} + \left(\alpha - \beta\sigma^2_{t_{j-1}}\right)\frac{T}{n} + \gamma\sigma_{t_{j-1}}Z_j\sqrt{\frac{T}{n}},$$

where $\{Z_j\}_{j=1}^n$ are i.i.d. as $N\left(0, 1\right)$. Starting at some initial value $\sigma^2_{t_0}$, a sample path of volatility could be constructed recursively by taking n draws from the standard normal distribution, and from this the average volatility $\bar{\sigma}^2_{[0,T]}$ could be approximated as $n^{-1}\sum_{j=1}^n \sigma^2_{t_{j-1}}$. This one realization of $\bar{\sigma}^2_{[0,T]}$ would then be plugged into the Black–Scholes formula to obtain one realization of $P\left(S_0, T; \bar{\sigma}^2_{[0,T]}\right)$. The entire process would be repeated $m - 1$ more times and the m realizations of the option's value averaged to estimate $\hat{E}P\left(S_0, T; \bar{\sigma}^2_{[0,T]}\right)$.

While this is straightforward, it will turn out that a much faster solution is possible as a special case of the more general model to which we now turn, which allows dependence between shocks to price and to volatility. This is where we will again see characteristic functions in action.

23.2   DEPENDENT PRICE/VOLATILITY SHOCKS

The conditioning approach that we have just described fails when the Brownian motions $\{\hat{W}_t, \hat{W}'_t\}$ in (23.1) and (23.2) are correlated, because the random paths taken by volatility and price are then loosely connected. Specifically, we can no longer just generate a time path of volatility over the life of the option and then assume that this value drives all the price moves during $[0, T]$.[67] Nevertheless, there is a way to price options in the case $E\hat{W}_t\hat{W}'_t = \rho t$ for any $\rho \in (-1, 1)$, and it is much faster than simulation—holding precision constant. Heston (1993) was the first to show that this could be done by applying Fourier methods in a way that required two numerical integrations. Fortunately, once the CF of $\ln S_T$ is determined under EMM $\hat{\mathbb{P}}$, we now know how to price the option in one step using the method described in Section 22.3, and this cuts the computational time roughly in half. However, in the SV model finding the CF is a much more involved process than for the discontinuous models treated in Chapter 22.

To get started, it will make things easier to change notation as $v_t \equiv \sigma_t^2$ and $s_t \equiv \ln S_t$. With this change our SV model under $\hat{\mathbb{P}}$ is

$$ds_t = \left(r - \frac{v_t}{2}\right) \cdot dt + \sqrt{v_t} \cdot d\hat{W}_t$$

$$dv_t = (\alpha - \beta v_t) \cdot dt + \gamma\sqrt{v_t} \cdot d\hat{W}'_t. \tag{23.5}$$

Here the first expression comes from applying Itô's formula to $f(S_t) = \ln S_t = s_t$. For simplicity we continue to assume that the short rate of interest has the known value r over the life of the option and that the underlying asset pays no dividend. (Allowing for a continuous, proportional dividend at rate δ involves just replacing r with $r - \delta$ in the expressions to follow, and allowing both of these to vary deterministically over time just amounts to replacing $(r - \delta)(T - t)$ by $\int_t^T (r_s - \delta_s) \cdot ds$.)

To proceed, let $\hat{\phi}(\zeta; s_t, v_t, T - t) = \hat{E}_t e^{i\zeta s_T} \equiv \hat{E}\left(e^{i\zeta s_T} \mid \mathcal{F}_t\right)$ be the conditional CF of $s_T \equiv \ln S_T$ under $\hat{\mathbb{P}}$ given information $\mathcal{F}_t$, where ζ is any real number. We will use the fact that conditional expectations *process*

$$\left\{\hat{E}\left(e^{i\zeta s_T} \mid \mathcal{F}_t\right) = \hat{\phi}(\zeta; s_t, v_t, T - t)\right\}_{0 \leq t \leq T}$$

is a (complex-valued) martingale to find a PDE that the CF must satisfy. It is a martingale because of the tower property of conditional expectation. Thus, for $0 \leq u \leq t$

---

[67]This is not to say that the option's value cannot be estimated by simulation—merely that it is more difficult and slower. To do it, we would start at $\hat{W}_0 = \hat{W}'_0 = 0$ and simulate step by step a correlated bivariate series $\left\{\hat{W}_{t_j}, \hat{W}'_{t_j}\right\}_{j=1}^n$ as $\hat{W}_{t_j} = \hat{W}_{t_{j-1}} + Z_j\sqrt{T/n}$ and $\hat{W}'_{t_j} = \hat{W}'_{t_{j-1}} + \rho Z_j\sqrt{T/n} + \bar{\rho}Z'_j\sqrt{T/n}$, with $\bar{\rho} \equiv \sqrt{1 - \rho^2}$ and $\left\{Z_j, Z'_j\right\}_{j=1}^n$ as independent draws from $N(0, 1)$. Then starting with given σ_0^2, S_0 we would generate $\left\{\sigma_{t_j}^2, S_{t_j}\right\}_{j=1}^n$ progressively as

$$\sigma_{t_j}^2 = \sigma_{t_{j-1}}^2 + \left(\alpha - \beta\sigma_{t_{j-1}}^2\right)\frac{T}{n} + \gamma\sigma_{t_{j-1}}\left(\hat{W}'_{t_j} - \hat{W}'_{t_{j-1}}\right)$$

$$\ln S_{t_j} = \ln S_{t_{j-1}} + \left(r - \frac{\sigma_{t_{j-1}}^2}{2}\right)\frac{T}{n} + \sigma_{t_{j-1}}\left(\hat{W}_{t_j} - \hat{W}_{t_{j-1}}\right).$$

Plugging terminal value $S_{t_n} \equiv S_T$ into $(X - S_T)^+$ gives one pseudorealization of the put's value at expiration, and averaging these over replications and discounting give an estimate of the put's value at $t = 0$.

we have

$$\hat{E}\left[\hat{\phi}\left(\zeta;s_t,v_t,T-t\right)\mid\mathcal{F}_u\right]\equiv\hat{E}\left[\hat{E}\left(e^{i\zeta s_T}\mid\mathcal{F}_t\right)\mid\mathcal{F}_u\right]$$
$$=\hat{E}\left(e^{i\zeta s_T}\mid\mathcal{F}_u\right)$$
$$=\hat{\phi}\left(\zeta;s_u,v_u,T-u\right).$$

Now, setting $\tau\equiv T-t$ for brevity and using Itô's formula, we have

$$d\hat{\phi}\left(\zeta;s_t,v_t,\tau\right)=-\hat{\phi}_\tau\cdot dt+\hat{\phi}_s\cdot ds_t+\frac{\hat{\phi}_{ss}}{2}\cdot d\left\langle s\right\rangle_t+\hat{\phi}_v\cdot dv_t+\frac{\hat{\phi}_{vv}}{2}\cdot d\left\langle v\right\rangle_t+\hat{\phi}_{sv}\cdot d\left\langle s,v\right\rangle_t,$$

where $d\left\langle s\right\rangle_t=v_t\cdot dt$, $d\left\langle v\right\rangle_t=\gamma^2 v_t\cdot dt$, and $d\left\langle s,v\right\rangle_t=\rho\gamma v_t\cdot dt$ are, respectively, the quadratic variations of s_t and v_t and the quadratic *covariation*. The covariation is proportional to ρ, which is the correlation between the Brownian motions. Using the expressions for ds_t, dv_t, and the increments to the quadratics, we have, on collecting terms, the following expression for $d\hat{\phi}\left(\zeta;s_t,v_t,\tau\right)$:

$$\left[-\hat{\phi}_\tau+\hat{\phi}_s\left(r-\frac{v_t}{2}\right)+\frac{\hat{\phi}_{ss}}{2}v_t+\hat{\phi}_v\left(\alpha-\beta v_t\right)+\frac{\hat{\phi}_{vv}}{2}\gamma^2 v_t+\hat{\phi}_{sv}\rho\gamma v_t\right]\cdot dt$$
$$+\hat{\phi}_s\sqrt{v_t}\cdot d\hat{W}_{1t}+\hat{\phi}_v\gamma\sqrt{v_t}\cdot d\hat{W}_{2t}.$$

Now, since the expected change in a martingale is zero, the drift term in this expression must be zero also, and this condition gives the following PDE that must be satisfied by the conditional CF:

$$0=-\hat{\phi}_\tau+\hat{\phi}_s\left(r-\frac{v_t}{2}\right)+\frac{\hat{\phi}_{ss}}{2}v_t+\hat{\phi}_v\left(\alpha-\beta v_t\right)+\frac{\hat{\phi}_{vv}}{2}\gamma^2 v_t+\hat{\phi}_{sv}\rho\gamma v_t.\quad(23.6)$$

The solution must also satisfy the initial condition, corresponding to $\tau=0$ (i.e., $t=T$), that sets

$$\hat{\phi}\left(\zeta;s_T,v_T,0\right)=\hat{E}_T e^{i\zeta s_T}=e^{i\zeta s_T}.$$

Of course, we must also have $\hat{\phi}\left(0;s_t,v_t,\tau\right)=1$ for all s_t,v_t,τ, since any CF must have this property.

The relatively simple structure of the PDE, in which v_t enters linearly throughout, makes it possible to find a solution of the form

$$\hat{\phi}\left(\zeta;s_t,v_t,\tau\right)=\exp\left[g\left(\tau;\zeta\right)+h\left(\tau;\zeta\right)v_t+i\zeta s_t\right].\quad(23.7)$$

Here, we must have $g\left(0;\zeta\right)=h\left(0;\zeta\right)=0$ for all ζ to satisfy the initial condition and $g\left(\tau;0\right)=h\left(\tau;0\right)=0$ for all $\tau\in[0,T]$ to satisfy $\hat{\phi}\left(0;s_t,v_t,\tau\right)\equiv1$. To see that (23.7) is a solution, calculate the partial derivatives and plug into (23.6). The partial derivatives are

$$\hat{\phi}_\tau = \hat{\phi} \cdot (g' + h'v_t)$$
$$\hat{\phi}_s = \hat{\phi} \cdot i\zeta$$
$$\hat{\phi}_{ss} = -\hat{\phi} \cdot \zeta^2$$
$$\hat{\phi}_v = \hat{\phi} \cdot h$$
$$\hat{\phi}_{vv} = \hat{\phi} \cdot h^2$$
$$\hat{\phi}_{sv} = \hat{\phi} \cdot i\zeta h,$$

where g' and h' are derivatives with respect to τ. Inserting into (23.6) gives the condition

$$0 = [-g' + i\zeta r + \alpha h] + \left[-h' - \frac{i\zeta + \zeta^2}{2} + (i\rho\gamma\zeta - \beta) h + \frac{\gamma^2 h^2}{2} \right] v_t.$$

If this is to hold for all values of v_t, it is necessary that each bracketed quantity be zero. This gives rise to the following *ordinary* differential equations in τ:

$$\frac{dg(\tau; \zeta)}{d\tau} = i\zeta r + \alpha h(\tau; \zeta) \qquad (23.8)$$

$$\frac{dh(\tau; \zeta)}{d\tau} = -\left(\frac{i\zeta + \zeta^2}{2} \right) + (i\rho\gamma\zeta - \beta) h(\tau; \zeta) + \frac{\gamma^2 h^2(\tau; \zeta)}{2}. \qquad (23.9)$$

When $\gamma \neq 0$ these have the following solutions subject to initial condition $g(0; \zeta) = h(0; \zeta) = 0$:

$$h(\tau; \zeta) = \frac{B - D}{\gamma^2} \frac{e^{D\tau} - 1}{1 - Qe^{D\tau}} \qquad (23.10)$$

$$g(\tau; \zeta) = i\zeta r \tau + \frac{\alpha}{\gamma^2} \left[(D - B)\tau - 2\ln\left(\frac{1 - Qe^{D\tau}}{1 - Q} \right) \right], \qquad (23.11)$$

where

$$D \equiv \sqrt{B^2 - 2A\gamma^2}$$
$$A \equiv -\frac{i\zeta + \zeta^2}{2}$$
$$B \equiv i\zeta\rho\gamma - \beta$$
$$Q \equiv \frac{B - D}{B + D}.$$

One can easily see that these also satisfy the necessary conditions $g(\tau; 0) = h(\tau; 0) = 0$.

Plugging the expressions for g and h into (23.7) gives the CF of $\ln S_T$ under $\hat{\mathbb{P}}$. While its form is not *beautiful*, it is highly *computable*, and with this computable expression in hand, pricing the European put proceeds just as it did for the applications in Chapter 22.

23.3   STOCHASTIC VOLATILITY WITH JUMPS IN PRICE

Having seen how to price European options under SV dynamics by Fourier methods, we will now adapt the method to a model with Merton-style, Poisson-driven jumps in the price paths. Our "SV–jump" model (often called the "Bates" model after David Bates (1996)) under martingale measure $\hat{\mathbb{P}}$ is now

$$dS_t = (r - \lambda\eta) \, S_{t-} \cdot dt + \sqrt{v_t} S_{t-} \cdot d\hat{W}_t + U S_{t-} \cdot dN_t$$
$$dv_t = (\alpha - \beta v_t) \cdot dt + \gamma\sqrt{v_t} \cdot d\hat{W}_t'.$$

Here $v_t = \sigma_t^2$ is squared volatility, $\{N_t\}_{t\geq 0}$ is a Poisson process with intensity λ, and jump sizes $\{U\}_{j=1}^{\infty}$ are i.i.d. as $\ln(1 + U) \sim N[\ln(1+\eta) - \xi^2/2, \xi^2]$. Brownian motions $\left\{\hat{W}_t, \hat{W}_t'\right\}_{t\geq 0}$ may be correlated with each other but are independent of $\{N_t\}_{t\geq 0}$ and the jump sizes, and $\{N_t\}_{t\geq 0}$ and $\{U_j\}_{j=1}^{\infty}$ are themselves mutually independent. With this parameterization we have the simple result that $EU = \eta$. Now Itô's formula lets us write $d\ln S_t$ as

$$ds_t = \left(r - \lambda\eta - \frac{v_t}{2}\right) \cdot dt + \sqrt{v_t} \cdot d\hat{W}_t + \ln(1 + U) \cdot dN_t.$$

In the constant-volatility case we were able to express S_t itself under $\hat{\mathbb{P}}$ as

$$S_t = S_0 \exp\left[\left(r - \lambda\eta - \frac{\sigma^2}{2}\right)t + \sigma\hat{W}_t\right] \prod_{j=0}^{N_t} (1 + U_j), \qquad (23.12)$$

where $U_0 \equiv 0$. In the present case the corresponding expression is

$$S_t = S_0 \exp\left[(r - \lambda\eta) t - \frac{1}{2}\int_0^t v_s \cdot ds + \int_0^t \sqrt{v_s} \cdot d\hat{W}_s\right] \prod_{j=0}^{N_t} (1 + U_j),$$

which clearly specializes to (23.12) when $v_t = \sigma^2$ for all t.

As with SV dynamics alone, pricing a T-expiring put option at time t with remaining life $\tau \equiv T - t$ involves three steps: (1) find $\hat{\phi}(\zeta; s_t, v_t, \tau)$, the conditional CF of $s_T = \ln S_T$ under EMM $\hat{\mathbb{P}}$; (2) express the conditional CDF of s_T, $\hat{F}_{s_T}(s; s_t, v_t, \tau) = \hat{\mathbb{P}}\{s_T \leq s \mid \mathcal{F}_t\}$, in terms of the CF via the inversion formula as an integral with respect to ζ; (3) translate from the CDF of s_T to that of S_T by a change of variable; and (4) evaluate the put by integrating $\hat{F}_{S_T}(S; s_t, v_t, \tau)$ from $S = 0$ to $S = X$ (the strike price). The only thing that is new is to see how step (1) must be changed to accommodate jumps.

Again, we will use the fact that conditional expectations processes are martingales to find the conditional CF. Since the conditional CF given information $\mathcal{F}_t$ is just the conditional expectation of $e^{i\zeta s_T}$, it follows that the evolving process

$$\left\{\hat{\phi}(\zeta; s_t, v_t, \tau)\right\}_{0 \leq t \leq T} = \left\{\hat{E}\left(e^{i\zeta s_T} \mid \mathcal{F}_t\right)\right\}_{0 \leq t \leq T}$$

is a martingale adapted to filtration $\{\mathcal{F}_t\}_{0 \le t \le T}$. This in turn implies that the expected change in the process is zero. By the extension of Itô's formula that allows for jumps, we have

$$
\begin{aligned}
d\hat{\phi} = &-\hat{\phi}_\tau \cdot dt + \hat{\phi}_s \left(r - \lambda\eta - \frac{v_t}{2} \right) \cdot dt + \hat{\phi}_s \sqrt{v_t} \cdot d\hat{W}_t + \hat{\phi}_{ss} \frac{v_t}{2} \cdot dt \\
&+ \hat{\phi}_v \left(\alpha - \beta v_t \right) \cdot dt + \hat{\phi}_v \gamma \sqrt{v_t} \cdot d\hat{W}'_t + \hat{\phi}_{vv} \gamma^2 \frac{v_t}{2} \cdot dt + \hat{\phi}_{sv} \rho\gamma v_t \cdot dt \\
&+ \left[\hat{\phi}\left(\zeta; s_t, v_t, \tau \right) - \hat{\phi}\left(\zeta; s_{t-}, v_t, \tau \right) \right] \cdot dN_t,
\end{aligned}
$$

where $s_t = s_{t-} + (N_t - N_{t-}) \ln (1 + U)$ and s_{t-} and N_{t-} are the left-hand limits. Taking expectations conditional on $\mathcal{F}_{t-}$ eliminates the terms in $d\hat{W}_t$ and $d\hat{W}'_t$ but not the term in dN_t, which is

$$
\hat{E}\left[\hat{\phi}\left(\zeta; s_t, v_t, \tau \right) - \hat{\phi}\left(\zeta; s_{t-}, v_t, \tau \right) \mid \mathcal{F}_{t-} \right] \lambda \cdot dt.
$$

Equating the result to zero leaves us with

$$
\begin{aligned}
0 = &-\hat{\phi}_\tau + \hat{\phi}_s \left(r - \lambda\eta - \frac{v_t}{2} \right) + \hat{\phi}_{ss} \frac{v_t}{2} + \hat{\phi}_v \left(\alpha - \beta v_t \right) + \hat{\phi}_{vv} \gamma^2 \frac{v_t}{2} \qquad (23.13) \\
&+ \hat{\phi}_{sv} \rho\gamma v_t + \hat{E}\left[\hat{\phi}\left(\zeta; s_t, v_t, \tau \right) - \hat{\phi}\left(\zeta; s_{t-}, v_t, \tau \right) \mid \mathcal{F}_{t-} \right] \lambda.
\end{aligned}
$$

This is a partial integral/differential equation that must be solved subject to initial condition $\hat{\phi}\left(\zeta; s_T, v_T, 0 \right) = e^{i\zeta s_T}$ and $\hat{\phi}\left(0; s_t, v_t, \tau \right) = 1$. Again, the structure of the model makes it possible to find a solution of the form

$$
\hat{\phi}\left(\zeta; s_t, v_t, \tau \right) = \exp\left[g\left(\tau; \zeta \right) + h\left(\tau; \zeta \right) v_t + i\zeta s_t + K\lambda\tau \right] \qquad (23.14)
$$

for certain complex-valued functions g and h and complex constant K. The constant K is the only change that is produced by adding Poisson-driven jumps to the SV process. Again, we have the initial conditions $g\left(0; \zeta \right) = h\left(0; \zeta \right) = 0$ and necessary conditions $g\left(\tau; 0 \right) = h\left(\tau; 0 \right) = 0$. To verify that the solution is of this form, calculate the derivatives of $\hat{\phi}$,

$$
\begin{aligned}
\hat{\phi}_\tau &= \hat{\phi}\left(g' + h'v_t + K\lambda \right) \\
\hat{\phi}_s &= \hat{\phi}i\zeta \\
\hat{\phi}_{ss} &= -\hat{\phi}\zeta^2 \\
\hat{\phi}_v &= \hat{\phi}h \\
\hat{\phi}_{vv} &= \hat{\phi}h^2 \\
\hat{\phi}_{sv} &= \hat{\phi}i\zeta h,
\end{aligned}
$$

and the expectation in (23.13):

$$
\begin{aligned}
\hat{E}\left[\hat{\phi}\left(\zeta; s_t, v_t, \tau \right) - \hat{\phi}\left(\zeta; s_{t-}, v_t, \tau \right) \mid \mathcal{F}_{t-} \right] &= e^{g + hv_t + i\zeta s_{t-} + K\lambda\tau} \left[\hat{E} e^{i\zeta \ln(1+U)} - 1 \right] \\
&= \hat{\phi} \cdot \left\{ e^{\left[i\zeta \ln(1+\eta) - (i\zeta + \zeta^2)\xi^2/2 \right]} - 1 \right\}.
\end{aligned}
$$

Here, g' and h' are derivatives with respect to τ, and (as above) we have suppressed the arguments of $\hat{\phi}$ for brevity. Note that all the derivatives and the expectation are proportional to $\hat{\phi}$, so if (23.14) is to solve (23.13) for all ζ and values of the state variables, we must have

$$
0 = -(g' + h'v_t + K\lambda) + i\zeta \left(r - \lambda\eta - \frac{v_t}{2} \right) - \zeta^2 \frac{v_t}{2} + h(\alpha - \beta v_t)
$$
$$
+ h^2 \gamma^2 \frac{v_t}{2} + i\zeta h\rho\gamma v_t + \left\{ e^{\left[i\zeta \ln(1+\eta) - (i\zeta + \zeta^2)\xi^2/2 \right]} - 1 \right\} \lambda.
$$

Setting K equal to the last term cancels it out, whereupon separating the remaining terms proportional to v_t from those that are not yields the condition

$$
0 = [-g' + i\zeta(r - \lambda\eta) + \alpha h] + \left[-h' - \frac{i\zeta + \zeta^2}{2} + (i\rho\gamma\zeta - \beta)h + \gamma^2 \frac{h^2}{2} \right] v_t.
$$

For this to hold for *all* realizations of v_t requires

$$
\frac{dg(\tau;\zeta)}{d\tau} = i\zeta(r - \lambda\eta) + \alpha h(\tau;\zeta) \tag{23.15}
$$
$$
\frac{dh(\tau;\zeta)}{d\tau} = -\frac{i\zeta + \zeta^2}{2} + (i\rho\gamma\zeta - \beta)h(\tau;\zeta) + \frac{\gamma^2 h(\tau;\zeta)^2}{2}. \tag{23.16}
$$

Solving (23.16) first for h and then (23.15) for g subject to the initial conditions gives, when $\gamma \neq 0$,

$$
h(\tau;\zeta) = \frac{B - D}{\gamma^2} \frac{e^{D\tau} - 1}{1 - Qe^{D\tau}}
$$
$$
g(\tau;\zeta) = [i\zeta(r - \lambda\eta)]\tau + \frac{\alpha}{\gamma^2} \left[(-B + D)\tau - 2\ln \left(\frac{1 - Qe^{D\tau}}{1 - Q} \right) \right],
$$

where D, A, B, and Q are just the same as in the Heston model.

Having verified the solution, we can express $\hat{\phi} \equiv \hat{\phi}(\zeta; s_t, v_t, \tau)$, the conditional CF under $\hat{\mathbb{P}}$, as

$$
\hat{\phi} = \exp \left\{ g(\tau;\zeta) + h(\tau;\zeta)v_t + i\zeta s_t + \left[(1+\eta)^{i\zeta} e^{A\xi^2} - 1 \right] \lambda\tau \right\}. \tag{23.17}
$$

We then proceed to steps 2–4 to price the put option.

23.4 FURTHER ADVANCES

The continuing effort to explain the smile curve in prices of options has produced many other models, each of which has its own advantages and weaknesses. Darrell Duffie et al. (2000), Bjorn Eraker et al. (2003), and Eraker (2004) have argued that stochastic volatility should also be modeled as a discontinuous process. (See Exercise 23.4 below.) While this does improve the fit, it does so at the cost of adding still more parameters that must somehow be estimated or found by calibration. Hua Fang (2002) has developed a more parsimonious model in which volatility is constant but the jump intensity is itself a mean-reverting stochastic process, $\{\lambda_t\}_{t\geq 0}$. Ernst Eberlein and Ulrich Keller (1995) and Eberlein et al. (1998) have proposed an alternative subordinated process to the VG model, in which operational time is driven by an inverse Gaussian distribution. To pick up the thick tails in marginal distributions while preserving the finite-moment feature of martingales, Peter Carr and Liuren Wu (2003) have developed a model based on an asymmetric stable distribution. Another promising alternative, proposed by Craig Edwards (2005), is to allow the stochastic process for the log of underlying price to switch randomly among alternative forms, reflecting the apparent tendency of markets to move among "regimes" of high and low volatility, fast and slow growth. As yet, however, no model passes the test of providing good fits over time with a single, stable set of parameters. This is the challenge for future researchers.

EXERCISES

23.1 When the underlying price follows geometric Brownian motion with constant volatility σ, the conditional distribution of $s_T = \ln S_T$ under $\hat{\mathbb{P}}$ given information $\mathcal{F}_t$ is $N\left[(r - \sigma^2/2)\tau, \sigma^2\tau\right]$, where $\tau \equiv T - t$. The corresponding CF is

$$\hat{\phi}(\zeta; s_t, \tau) = \exp\left[i\zeta(r - \frac{\sigma^2}{2})\tau - \frac{\zeta^2\sigma^2\tau}{2} + i\zeta s_t\right]. \tag{23.18}$$

By setting $\alpha = \beta = \gamma = 0$ in the Heston model, we are back to the geometric BM model with (in the new notation) $\sigma = \sqrt{v_t}$. Verify that (23.7) reduces to (23.18) in this case.

23.2 When $\gamma = 0$ in the Heston model but α and β are positive, volatility is time varying but deterministic, and solutions (23.10) and (23.11) for $g(\tau; \zeta)$ and $h(\tau; \zeta)$ in (23.8) and (23.9) no longer apply. **(a)** Find the appropriate solutions for this case. **(b)** How can we express volatility itself as function of time when $\gamma = 0$?

23.3 Under the conditions of Exercise 23.2 find the solution for the conditional CF of $\ln S_T$, $\phi(\zeta; s_t, v_t, \tau) = e^{g(\tau;\zeta) + h(\tau;\zeta)v_t + i\zeta s_t}$, and determine the implied conditional distribution of $\ln S_T$ under $\hat{\mathbb{P}}$.

23.4 To further embellish the SV–jump model, we could allow discontinuities in volatility as well as in price. For example, we could model price and (squared) volatility under $\hat{\mathbb{P}}$ as

$$ds_t = \left(r - \lambda\eta - \frac{v_t}{2}\right) \cdot dt + \sqrt{v_t} \cdot d\hat{W}_t + \ln(1 + U_s) \cdot dN_t$$

$$dv_t = (\alpha - \beta v_t) \cdot dt + \gamma\sqrt{v_t} \cdot d\hat{W}_t' + \ln(1 + U_v) \cdot dN_t.$$

In this setup the same independent Poisson process $\{N_t\}_{t \geq 0}$ drives the shocks to both. The volatility shocks would have to be such as to keep volatility positive; for example, $1 + U_v$ could have an exponential or a gamma distribution. Since volatility seems empirically to pick up when price declines abruptly, we would want the price and volatility shocks to be negatively correlated. This could be done by making $\ln(1 + U_s)$ normal *conditional* on U_v, with conditional mean $\ln(1 + \eta) - \xi^2/2 - \kappa \ln(1 + U_v)$ for some $\kappa > 0$, and conditional variance ξ^2. Following steps like those leading up to (23.13), find the partial differential/integral equation that would be satisfied by the conditional CF for $\ln S_t$, $\hat{\phi}(\zeta; s_t, v_t, \tau)$.

EMPIRICAL PROJECT 7

In this project you will use the Fourier methods of Section 22.3 to determine the prices for a sample of options that are implied by the variance–gamma (VG) and SV–jump models. Specifically, you will calculate the value at $t = 0$ of a T-expiring European put with strike X as

$$P(S_0, T) = B(0, T) \frac{X}{2} \left[1 - \frac{1}{\pi} \lim_{c \to \infty} \int_{-c}^{c} \frac{X^{-i\zeta}}{i\zeta + \zeta^2} \hat{\phi}_{s_T}(\zeta) \cdot d\zeta \right], \qquad (23.19)$$

where $\hat{\phi}_{s_T}(\zeta)$ is the CF of $s_T \equiv \ln S_T$ under a particular model. Parity relation $C(S_0, T) = P(S_0, T) + S_0 e^{-\delta T} - B(0, T) X$ gives the value of a call with the same X and T. The integral in (23.19) must be evaluated numerically, using a sufficiently large value of c to ensure convergence to penny accuracy. You will use the expressions for the VG and SV–jump CFs in (22.12) and (23.17), respectively. The options to be priced are the February, June, and December 2008 options on the S&P 500 whose implicit volatilities as of January 23, 2008 were found in Empirical Project 6. The strikes and implicit volatilities are in files FebIV, JunIV, and DecIV in folder Project 7 at the FTP site. The underlying price, time to expiration (in years), and average annualized interest and dividend rates are given below.

Expiration	S_0	T	r	δ
Feb.	1301.55	23/365	.0521	.0444
June	1301.55	149/365	.0319	.0246
Dec.	1301.55	331/365	.0277	.0221

The following parameters were found by minimizing the sum of squared differences between implicit volatilities of market and model prices at each expiration date. You will use these values to price the options.

VG Model

Expiration	γ	σ	ν
Feb.	-.37039	.37803	.28599
June	-.33684	.23707	.39077
Dec.	-.22657	.22667	.88127

SV–jump Model

Expiration	α	β	γ	ρ	σ_0	η	θ	ξ
Feb.	1.5760	10.030	7.600	-.453	.420	-.121	.0118	.00100
June	.428	3.525	1.424	-.694	.222	-.133	.440	.00050
Dec.	.322	3.943	2.776	-.422	.273	-.149	.573	.00001

Once values of puts for all strikes at each expiration have been calculated, convert those with $X \leq S_0$ to calls via put–call parity. Then determine the implicit volatility that corresponds to each price and plot along with the implicit volatilities from market prices. Figures 23.1–23.6 show what to expect.

Here are some comments on the methods and results:

1. Implicit volatilities were used as the criterion for fitting the models so as to lessen the effects of cross-sectional variation in "moneyness"—the price–strike relation. Minimizing the sum of squared differences in *prices* would give undue weight to options deep in the money, while minimizing the sum of squared *proportionate* differences would give extra influence to options *out* of money. Using implicit volatilities limits these distortions, although the smile phenomenon still puts more somewhat weight on options away from the money. This accounts to some degree for the models' especially bad fits to February options at strikes near S_0.

2. It would be possible to fit options of all terms simultaneously, but treating them separately reveals the instability of the best-fitting parameter values. Despite the separate treatment for each T, neither model does a good job of matching the highly curved and asymmetric smile for February.[68]

3. The three-parameter VG model fits these data almost as well as the more elaborate eight-parameter SV–jump. Moreover, the best-fitting VG parameters vary less with the options' terms.

---

[68] In fairness to the models, the day on which the options' prices were quoted (January 23, 2008) was one of unusual volatility, as the markets were in the early stages of the 2008 financial meltdown caused by the housing bubble and the subprime lending debacle.

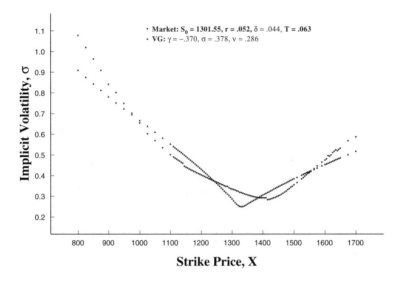

Figure 23.1   Implicit volatilities: VG versus February S&P options.

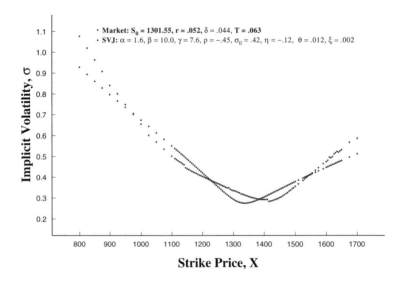

Figure 23.2   Implicit volatilities: SV–jump (SVJ) versus February S&P options.

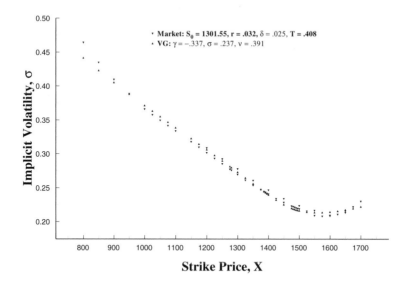

Figure 23.3   Implicit volatilities: VG versus June S&P options.

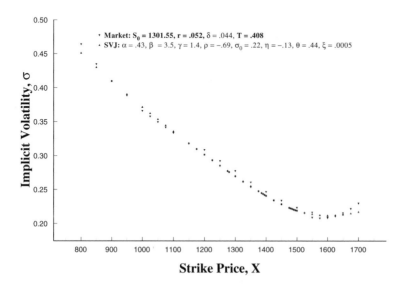

Figure 23.4   Implicit volatilities: SV–jump (SVJ) versus June S&P options.

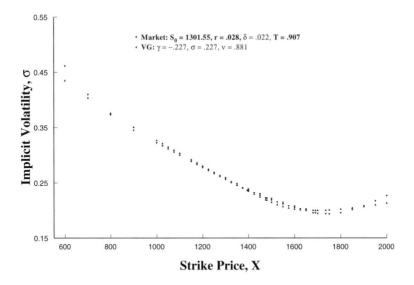

Figure 23.5   Implicit volatilities: VG versus December S&P options.

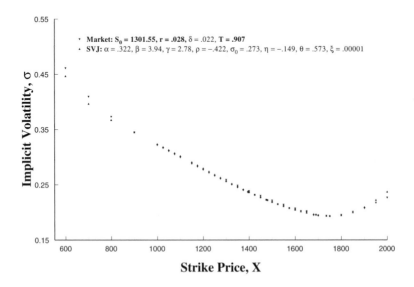

Figure 23.6   Implicit volatilities: SV–jump (SVJ) versus December S&P options.

Solutions to Selected Exercises

4.1

Compounding	Value
Annual	$1.05^{10} = 1.6289$
Semiannual	$1.025^{20} = 1.6386$
Quarterly	$1.0125^{40} = 1.6436$
Continuous	$e^{10(.05)} = 1.6487$

4.3 As inferred from the prices of strips, the bond with $15 coupon is worth

$$15\,(.97531 + .95123 + .92774 + .90484) + 1000\,(.90484) = 961.23.$$

This is priced correctly. However, the $30 bond is underpriced, since to replicate its payoffs with strips would cost

$$30\,(.97531 + .95123 + .92774 + .90484) + 1000\,(.90484) = 1017.61.$$

A firm with an inventory of strips could make a $500 gain at no risk by (1) buying 100 of the $30-coupon bonds, (2) selling three strips that mature on the first three coupon dates, and (3) selling 103 strips of 2-year maturity.

4.5 $r_t = -\partial \ln B(t,T)/\partial T|_{T=t} \doteq -\frac{1}{\Delta t} \ln B(t, t+\Delta t)$ for small Δt, so for $\Delta t = .002$ we have $r_t \doteq -500 \ln B(t, t+.002) \doteq .05$.

4.7 The bond's price, yield to maturity, and coupon are related as

$$B(0, c; .5, 1.0, 1.5, 2.0) = c \sum_{j=1}^{4} (1.05)^{-j/2} + 1000 (1.05)^{-2},$$

so that $c = \left[1019.97 - 1000 (1.05)^{-2}\right] / \sum_{j=1}^{4} (1.05)^{-j/2} = 30$.

5.1 Since $\sigma_{MX} = 0$, the variance of the daily return is $\sigma_p^2 = p^2\sigma_M^2 + (1-p)^2 \sigma_X^2$.

(**a**) Differentiating with respect to p and equating to zero give

$$\frac{d\sigma_p^2}{dp} = 2p\sigma_M^2 - 2(1-p)\sigma_X^2 = 0$$

$$p_* = \frac{\sigma_X^2}{\sigma_M^2 + \sigma_X^2}.$$

The second derivative is proportional to the positive quantity $\sigma_M^2 + \sigma_X^2$, so the solution does represent a minimum of σ_p^2, and it clearly satisfies the required constraint. The estimated values yield $p_* = \frac{14}{31}$.

(**b**) The mean and standard deviation of the minimum-variance portfolio's rate of return are

$$\mu_* = \frac{14}{31}(.00057) + \frac{17}{31}(.00123) \doteq .00093$$

$$\sigma_* = \sqrt{\left(\frac{14}{31}\right)^2 (.00017) + \left(\frac{17}{31}\right)^2 (.00014)} \doteq .0088.$$

(**c**) The highest mean return subject to the constraint $0 \le p \le 1$ would be at $p = 0$ (all the funds in XOM), where we would have $\mu_X = .00123$ and $\sigma_X = \sqrt{.00014} \doteq .01183$.

5.2 The plot of the feasible set is shown in the figure below as the solid hyperbolic curve. It can be produced in a spreadsheet by treating (σ_p, μ_p) as a parametric curve. For this, create a column of equally spaced values of p between zero and one, enter the formulas for σ_p and μ_p in the adjacent columns, and then plot μ_p against σ_p. The efficient set is the upper branch of the hyperbola.

5.3 The optimal attainable point (σ_0, μ_0) is where the feasible set is tangent to an indifference curve. To find the optimal investment share p_0, equate the expression for the slope of an indifference curve with the expression for the slope of the feasible set and solve for p_0. For this it is easier to work in the (σ_p^2, μ_p) plane where the indifference curves are linear. On the indifference

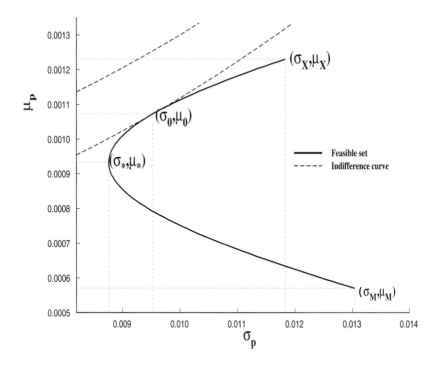

curve corresponding to the highest attainable utility U_0^* we have $\mu_p = U_0^* + \rho\sigma_p^2/2$ and $d\mu_p/d\sigma_p^2 = \rho/2$. The slope of the feasible set is

$$\frac{d\mu_p}{d\sigma_p^2} = \frac{d\mu_p/dp}{d\sigma_p^2/dp} = \frac{1}{2}\frac{\mu_M - \mu_X}{p(\sigma_M^2 + \sigma_X^2) - \sigma_X^2}.$$

Equality of the slopes implies

$$\rho = \frac{\mu_M - \mu_X}{p(\sigma_M^2 + \sigma_X^2) - \sigma_X^2},$$

and from this the solutions for the optimal investment shares for M and X (i.e., MSFT and XOM) are found to be

$$p_0 = \frac{\rho^{-1}(\mu_M - \mu_X) + \sigma_X^2}{\sigma_M^2 + \sigma_X^2}, \quad 1 - p_0 = \frac{\rho^{-1}(\mu_X - \mu_M) + \sigma_M^2}{\sigma_M^2 + \sigma_X^2}.$$

The constraint $0 \le p_0 \le 1$ is satisfied so long as $\rho^{-1}(\mu_X - \mu_M) < \sigma_X^2$ and $\rho^{-1}(\mu_M - \mu_X) < \sigma_M^2$. If $\mu_X \ge \mu_M$ this just requires $0 < \rho \le (\mu_X - \mu_M)/\sigma_X^2$.

Plugging the expression for p_0 into those for μ_p and σ_p^2 and simplifying give

$$\mu_0 = \frac{\rho^{-1}\left(\mu_X - \mu_M\right)^2 + \mu_X \sigma_M^2 + \mu_M \sigma_X^2}{\sigma_M^2 + \sigma_X^2}$$

$$\sigma_0^2 = \frac{\rho^{-2}\left(\mu_X - \mu_M\right)^2 + \sigma_X^2 \sigma_M^2}{\sigma_M^2 + \sigma_X^2}.$$

With $\rho = 10$ the data for Exercise 5.1 give $\mu_0 \doteq .00107$ and $\sigma_0 \doteq .00953$. With these values $U^*\left(\mu_0, \sigma_0^2\right) \equiv U_0^* \doteq .000618$, so the indifference curve passing through (σ_0, μ_0) has the equation $\mu = .000618 + 5\sigma^2$. This shows up in the figure for Exercise 5.2 as a dashed line. For reference, there is also a higher (and unattainable) indifference curve satisfying $\mu = .000800 + 5\sigma^2$.

6.1 Multiplying and summing give

$$\sum_{j=1}^{n} p_{mj}\mu_j = r\sum_{j=1}^{n} p_{mj} + \left(\mu_m - r\right)\sum_{j=1}^{n} p_{mj}\beta_j$$

$$= r + \left(\mu_m - r\right)\sum_{j=1}^{n} p_{mj}\beta_j,$$

since $\sum_{j=1}^{n} p_{mj} = 1$. The sum on the left equals μ_m, and so we must have $\sum_{j=1}^{n} p_{mj}\beta_j = 1$ if (6.5) is to be correct. But (6.6) implies

$$\sum_{j=1}^{n} p_{mj}\beta_j = \sum_{j=1}^{n} p_{mj}\frac{\sigma_{jm}}{\sigma_m^2}$$

$$= \frac{1}{\sigma_m^2}\sum_{j=1}^{n} p_{mj}E\left(R_j - \mu_j\right)\left(R_m - \mu_m\right)$$

$$= \frac{1}{\sigma_m^2}E\left[\sum_{j=1}^{n} p_{mj}\left(R_j - \mu_j\right)\right]\left(R_m - \mu_m\right)$$

$$= \frac{1}{\sigma_m^2}E\left(R_m - \mu_m\right)^2 = 1.$$

6.3 For movements along an indifference curve in the (V, E) plane we have $0 = dU^i\left(E, V\right) = U_E^i \cdot dE + U_V^i \cdot dV$. Thus, the quantity $\gamma_i \equiv -U_E^i/(2U_V^i) > 0$ is proportional to dV/dE, the *inverse* of the slope of the indifference curve. The slope itself, which is proportional to dE/dV, represents the increase in expected rate of return that is needed to just compensate for a marginal increase in risk; therefore, γ_i^{-1} can be interpreted as a measure of risk aversion, and γ_i itself as a measure of risk *tolerance*. If all the $\{\gamma_i\}$ changed by the factor α, then so would $\gamma_m \equiv \sum_{i=1}^{I}\gamma_i$, and by expression (6.2) the market's required compensation for risk, $\left(\mu_m - r\right)/\sigma_m^2$, would change by the factor α^{-1}. For example, a change by a factor $\alpha < 1$ would correspond to an increase in risk aversion and an increase in the slope of the SML. In principle, this could be detected by plotting estimates of assets' expected rates of return against estimates of their betas.

7.1 $\mathbb{P}_A \sim \mathbb{P}_B$ implies that $E_A u(W_0 + X) = 1 \cdot u(13) = E_B u(W_0 + X) = .80 \cdot u(14) + .20 \cdot u(10) = .80$. In this case $E_C u(W_0 + X) = .20 = E_D u(W_0 + X) = .25 \cdot .80 = .20$, so the preference for $\mathbb{P}_C$ over $\mathbb{P}_D$ is not consistent with $\mathbb{P}_A \sim \mathbb{P}_B$.

7.3 As per the hint, $\mathbb{P}_A$ is equivalent to a compound prospect that offers .90 chance of playing the game $\{.50, .00, .50\}$ and a .10 chance of playing $\{.00, .00, 1.0\}$ (i.e., getting nothing for sure), as represented below. This is because the compound prospect gives $.90 \cdot .50 = .45$ chance of 6000 and $.90 \cdot .50 + .10 \cdot 1 = .55$ chance of nothing, just as does $\mathbb{P}_A$. Similarly, $\mathbb{P}_B = \{.00, .90, .10\}$ is equivalent to a .90 chance at $\{.00, 1.0, .00\}$ and a .10 chance of playing $\{.00, .00, 1.0\}$. But $\mathbb{P}_C$ and $\mathbb{P}_D$ can be resolved into compound prospects that deliver the same games as do $\mathbb{P}_A$ and $\mathbb{P}_B$, respectively, although with different probabilities. By the substitution axiom $\mathbb{P}_B \succ \mathbb{P}_A$ implies that $\{.00, 1.0, .00\} \succ \{.50, .00, .50\}$, but in this case we should also have $\mathbb{P}_D \succ \mathbb{P}_C$.

$$\mathbb{P}_A = \begin{array}{c} {\scriptstyle .90} \nearrow \{.50, .00, .50\} \\ {\scriptstyle .10} \searrow \{.00, .00, 1.0\} \end{array} \quad , \quad \mathbb{P}_B = \begin{array}{c} {\scriptstyle .90} \nearrow \{.00, 1.0, .00\} \\ {\scriptstyle .10} \searrow \{.00, .00, 1.0\} \end{array}$$

$$\mathbb{P}_C = \begin{array}{c} {\scriptstyle .002} \nearrow \{.50, .00, .50\} \\ {\scriptstyle .998} \searrow \{.00, .00, 1.0\} \end{array} \quad , \quad \mathbb{P}_D = \begin{array}{c} {\scriptstyle .002} \nearrow \{.00, 1.0, .00\} \\ {\scriptstyle .998} \searrow \{.00, .00, 1.0\} \end{array}$$

7.5 We find that $E_A X = .20 \cdot 99 = 19.8$, $E_B X = .01 \cdot 999 + .99 \cdot 9 = 18.9$, $V_A X = .20 \cdot 99^2 - 19.8^2 = 1568.16$, $V_B X = .01 \cdot 999^2 + .99 \cdot 9^2 - 18.9^2 = 9702.99$. Thus, the Markowitz investor would choose $\mathbb{P}_A$. However, an EU maximizer with $W_0 = 1$ and $u(W_1) = \ln W_1 = \ln(1 + X)$ would choose $\mathbb{P}_B$, since $E_B \ln(1 + X) = .01 \cdot \ln 1000 + .99 \cdot \ln 10 \doteq 2.349 > E_A \ln(1 + X) = .2 \cdot \ln 100 \doteq .921$.

7.7 Putting $X \equiv S_1/S_0 - 1/B$ (the *excess* return of the stock over the bond), writing optimality condition (7.3) as $Eu'(W_1) X = 0$, and differentiating implicitly with respect to B with $dX/dB = 0$ give

$$0 = -\frac{W_0}{B^2} Eu''(W_1) X + \frac{\partial s^u}{\partial B} S_0 Eu''(W_1) X^2$$

and

$$\frac{\partial s^u}{\partial B} = \frac{W_0 Eu''(W_1) X}{B^2 S_0 Eu''(W_1) X^2} = -\frac{W_0}{B} \frac{\partial s^u}{\partial W_0},$$

which verifies (7.6). Now differentiate $Eu'(W_1)X = 0$ with $dX/dB = 1/B^2$ and solve to get

$$B\frac{\partial s^u}{\partial B} = \frac{W_0 Eu''(W_1)X}{BS_0 Eu''(W_1)X^2} - \frac{s^u S_0 Eu''(W_1)X + Eu'(W_1)}{BS_0 Eu''(W_1)X^2}$$

$$= -\frac{\partial s^u}{\partial \ln W_0} - \frac{\partial s^u}{\partial \ln S_0}$$

and hence (7.7).

8.1 Certainty equivalent c_A satisfies

$$E_A u(W_1) = .50\left(e^{-1} - 1\right) = u(W_0 + c_A)$$
$$= e^{c_A} - 1,$$

which yields $c_A \doteq -.3799$. Similarly, $c_B = +.3799$ and $c_C = 0$. Function $u(W_1)$ is convex for values $W_1 < W_0$ and concave for $W_1 > W_0$, implying risk aversion for gains and risk taking for losses. This is consistent with the calculated certainty equivalents; for the fact that $c_A > E_A X = -.5$ (the expected gain for $\mathbb{P}_A$) indicates that the *cost* of playing the unfavorable game is less than that of its expected payoff, while $c_B < E_B X$ indicates that the favorable $\mathbb{P}_B$ is worth less than its "fair" value. On the other hand, unlike Kahneman and Tversky's (1979) value function, this function implies risk neutrality toward prospects offering symmetric gains and losses, since $u(W_0 + x) = -u(W_0 - x)$ for each x implies that $c_C = E_C X = 0$ for $\mathbb{P}_C$ (or any other such symmetric bet).

8.3 Given that $\alpha \geq \beta > \gamma > 0$,

(a) The utility function is strictly increasing since

$$U'(W_1) = \begin{cases} \alpha e^{-\alpha W_1} + \beta e^{\beta(W_1 - W_0)} > 0, W_1 < W_0 \\ \alpha e^{-\alpha W_1} + \gamma e^{\gamma(W_0 - W_1)} > 0, W_1 > W_0 \end{cases}.$$

(b) The left- and right-hand derivatives at $W_1 = W_0$ are $\alpha e^{-\alpha W_0} + \beta$ and $\alpha e^{-\alpha W_0} + \gamma$, so with $\beta > \gamma$ there is a kink in U at W_0, the function falling off more rapidly as W_1 declines than it rises as W_1 increases. This would be consistent with risk aversion for symmetric bets. **(c)** The second derivatives are given by

$$U''(W_1) = \begin{cases} -\alpha^2 e^{-\alpha W_1} + \beta^2 e^{\beta(W_1 - W_0)}, W_1 < W_0 \\ -\alpha^2 e^{-\alpha W_1} - \gamma^2 e^{\gamma(W_0 - W_1)}, W_1 > W_0 \end{cases}.$$

This is uniformly negative for $W_1 > W_0$, so there would be risk aversion toward all prospects that offered gains but no losses. $U''(W_1)$ is also negative for $W_1 < \left[\beta W_0 + \ln\left(\alpha^2/\beta^2\right)\right] / (\beta + \alpha) < W_0$, so the individual would be willing to insure against large losses at actuarily fair rates. But since $U''(W_1) > 0$ for $\left[\beta W_0 + \ln\left(\alpha^2/\beta^2\right)\right]/(\beta+\alpha) < W_1 < W_0$, there would be risk-taking behavior for prospects with relatively small negative payoffs, consistent with the reflection effect in the experiments by Kahneman and Tversky.

9.1 From the moment-generating property of the MGF and the definition $\mathcal{L}(\zeta) \equiv \ln \mathcal{M}(\zeta)$ we have $\mathcal{M}^{(k)}(\zeta)\big|_{k=0} = EX^k, k \in \{0, 1, 2, ...\}$, so

(a)

$$\mathcal{L}'(\zeta) = \frac{\mathcal{M}'(\zeta)}{\mathcal{M}(\zeta)}$$

$$\mathcal{L}'(0) = \frac{EX}{1} \equiv \mu.$$

(b)

$$\mathcal{L}''(\zeta) = \frac{\mathcal{M}''(\zeta)}{\mathcal{M}(\zeta)} - \frac{\mathcal{M}'(\zeta)^2}{\mathcal{M}(\zeta)^2}$$

$$\mathcal{L}''(0) = EX^2 - \mu^2 \equiv \sigma^2.$$

(c)

$$\mathcal{L}'''(\zeta) = \frac{\mathcal{M}'''(\zeta)}{\mathcal{M}(\zeta)} - 3\mathcal{M}''(\zeta)\frac{\mathcal{M}'(\zeta)}{\mathcal{M}(\zeta)^2} + 2\frac{\mathcal{M}'(\zeta)^3}{\mathcal{M}(\zeta)^3}$$

$$\frac{\mathcal{L}'''(0)}{\mathcal{L}''(0)^{3/2}} = \frac{EX^3 - 3EX^2\mu + 2\mu^3}{\sigma^3}$$

$$= \frac{E(X-\mu)^3}{\sigma^3} = \kappa_3.$$

(d)

$$\mathcal{L}^{(4)}(\zeta) = \frac{\mathcal{M}^{(4)}(\zeta)}{\mathcal{M}(\zeta)} - 4\frac{\mathcal{M}'''(\zeta)\mathcal{M}'(\zeta)}{\mathcal{M}(\zeta)^2} - 3\frac{\mathcal{M}''(\zeta)^2}{\mathcal{M}(\zeta)^2}$$

$$+ 12\frac{\mathcal{M}''(\zeta)\mathcal{M}'(\zeta)^2}{\mathcal{M}(\zeta)^3} - 6\frac{\mathcal{M}'(\zeta)^4}{\mathcal{M}(\zeta)^4}$$

$$\frac{\mathcal{L}^{(4)}(0)}{\mathcal{L}''(0)^2} = \frac{EX^4 - 4\mu EX^3 - 3\left(EX^2\right)^2 + 12EX^2\mu^2 - 6\mu^4}{\sigma^4}$$

$$= \frac{E(X-\mu)^4 - 3\left(EX^2 - \mu^2\right)^2}{\sigma^4}$$

$$= \frac{E(X-\mu)^4}{\sigma^3} - 3 = \kappa_4 - 3.$$

9.3 The covariance is $\sigma_{12} = \sigma_1\sigma_2\rho_{12} = -.01$, so $\sigma_p^2 = .04p^2 - .02p\,(1-p) + .01\,(1-p)^2$ and $\mu_p = .01p + .005\,(1-p)$.

(a) Evaluating $d\sigma_p^2/dp$, equating to zero, and solving give $p_* = 2/7 \doteq .2857$ with $\sigma_{p_*}^2 \doteq .00429$ and $\mu_{p_*} \doteq .00643$.

(b) Since (R_1, R_2) are bivariate normal, any function of the form $a_0 + a_1R_1 + a_2R_2$ has a normal marginal distribution so long as at least one of a_1, a_2 is not zero. In particular, $R_{p_*} \sim N\,(.00643, .00429)$, so

$$\Pr\,(R_{p_*} > .10) = \Pr\left(\frac{R_{p_*} - .00643}{\sqrt{.00429}} > \frac{.10 - .00643}{\sqrt{.00429}}\right)$$
$$= \Pr\,(Z > 1.429) \doteq .0765,$$

where $Z \sim N(0, 1)$.

(c) We have $W_1/100 = 1 + R_p \sim N\,\left(1 + \mu_p, \sigma_p^2\right)$ and $EU\,(W_1) = 1 - Ee^{-1-R_p} = 1 - e^{-1}\exp\left(-\mu_p + \sigma_p^2/2\right)$. Maximizing this is equivalent to maximizing

$$\mu_p - \frac{\sigma_p^2}{2} = .01p + .005\,(1-p) - .02p^2 + .01p\,(1-p) - .005\,(1-p)^2,$$

and the maximum is attained at $p_0 = .025/.07 \doteq .3571$.

(d) We have $\mu_{p_0} \doteq .00679$, $\sigma_{p_0}^2 \doteq .004642$, so $\Pr\,(R_{p_0} > .10) \doteq \Pr(Z > 1.368) \doteq .0857$.

9.5 First, $\hat{E}_\tau R_{0,\tau} = \hat{\mu}\tau = .0004$ and $\tau = \frac{1}{252}$ requires $\hat{\mu} = .1008$. Next, $\hat{\kappa}_4 = 6 = 3 + 6/\left(\hat{k}\tau - 4\right)$ requires $\hat{k}\tau = 6$ and $\hat{k} = 1512$. Finally,

$$\hat{V}_\tau R_{0,\tau} = .0001 = \hat{\sigma}^2 \tau V T_{k\tau} = \hat{\sigma}^2 \tau\left(\frac{k\tau}{k\tau - 2}\right) = \hat{\sigma}^2\left(\frac{1}{252}\right)\left(\frac{6}{4}\right)$$

requires $\hat{\sigma} \doteq .1296$. With these parameters the mean, variance, and kurtosis of $R_{0,1}$ are $\hat{\mu} = .1008$, $\hat{\sigma}^2\,(1512/1510) \doteq .01682$, and $3 + 6/\,(1512 - 4) \doteq 3.0040$, respectively.

10.1 $\{X_n\}_{n=0}^{\infty}$ is clearly a martingale, so

(a) $E\,(X_{10} \mid \mathcal{F}_9) = X_9$

(b) $E\,(X_1 \mid \mathcal{F}_0) = X_0$

(c) $E(X_{10} \mid \mathcal{F}_0) = X_0$

(d) $E\,[E\,(X_{10} \mid \mathcal{F}_9) \mid \mathcal{F}_0] = E\,(X_9 \mid \mathcal{F}_0) = X_0.$

10.3 For $j = 1, 2, \ldots$ we have $W_j = W_{j-1}(1 + R_j)$ with $E(R_j \mid \mathcal{F}_{j-1}) = ER_j = \mu$, so

    (a) $E(W_{10}|\mathcal{F}_9) = W_9(1 + \mu)$

    (b) $E(W_{10}|\mathcal{F}_0) = W_0(1 + \mu)^{10}$

    (c) $E[E(W_{10}|\mathcal{F}_9)|\mathcal{F}_0] = W_0(1 + \mu)^{10}$

    (d) $b = (1 + \mu)^{-1}$.

10.5 For $t \in \{\tau, 2\tau, \ldots\}$ we have

$$
\begin{aligned}
E(P_t \mid \mathcal{F}_{t-\tau}) &= P_{t-\tau}e^{(\mu - \sigma^2/2)\tau}Ee^{\sigma\sqrt{\tau}Z_t} \\
&= P_{t-\tau}e^{(\mu - \sigma^2/2)\tau}e^{\sigma^2\tau/2} \\
&= P_{t-\tau}e^{\mu\tau}.
\end{aligned}
$$

Taking $\nu = \mu$ as the discount rate, the process $\{X_t \equiv e^{-\mu t}P_t\}$ is an $\{\mathcal{F}_t\}_{t\in\{0,\tau,2\tau,\ldots\}}$ martingale, since

$$
\begin{aligned}
E(X_t \mid \mathcal{F}_{t-\tau}) &= e^{-\mu t}E(P_t \mid \mathcal{F}_{t-\tau}) \\
&= e^{-\mu(t-\tau)}P_{t-\tau} \\
&= X_{t-\tau}
\end{aligned}
$$

and $E|X_t| = EX_t = X_0 = P_0 < \infty$.

10.7 For $t \geq 0$, W_t has the same distribution as $\sqrt{t}Z$, where $Z \sim N(0,1)$, so

    (a) $E|W_t| = \sqrt{t}E|Z| < \infty$ and $E(W_t \mid \mathcal{F}_s) = W_s + E(W_t - W_s \mid \mathcal{F}_s) = W_s$ for $0 \leq s \leq t$. Thus, $\{X_t \equiv W_t\}$ itself is an $\{\mathcal{F}_t\}_{t\geq0}$ martingale with $X_0 = 0$.

    (b) $E|\nu_t^{-1}e^{W_t}| = \nu_t^{-1}Ee^{\sqrt{t}Z} = \nu_t^{-1}e^{t/2} < \infty$ for any positive-valued deterministic process $\{\nu_t\}_{t\geq0}$, and $E(e^{W_t} \mid \mathcal{F}_s) = e^{W_s}E\left[e^{(W_t - W_s)} \mid \mathcal{F}_s\right] = e^{W_s+(t-s)/2}$ for $0 \leq s \leq t$. Thus, with $\nu_t = e^{t/2}$ and $Y_t = e^{-t/2+W_t}$ we have $E(Y_t \mid \mathcal{F}_s) = e^{-t/2+W_s+(t-s)/2} = Y_s$, so that $\{Y_t\}_{t\geq0}$ is a martingale with $Y_0 = 1$.

    (c) $E|W_t^2 - \beta_t| \leq EW_t^2 + \beta_t = t + |\beta_t| < \infty$ for any deterministic, finite process $\{\beta_t\}_{t\geq0}$. Also, for $0 \leq s \leq t$

$$
\begin{aligned}
E(W_t^2 \mid \mathcal{F}_s) &= E\left[W_s^2 + 2W_s(W_t - W_s) + (W_t - W_s)^2 \mid \mathcal{F}_s\right] \\
&= W_s^2 + 2W_s E\left[(W_t - W_s) \mid \mathcal{F}_s\right] + E\left[(W_t - W_s)^2 \mid \mathcal{F}_s\right] \\
&= W_s^2 + 0 + (t - s)EZ^2 \\
&= W_s^2 + (t - s),
\end{aligned}
$$

so taking $\beta_t = t$ makes $\{Q_t = W_t^2 - \beta_t\}_{t\geq0}$ an $\{\mathcal{F}_t\}_{t\geq0}$ martingale with $Q_0 = 0$.

11.1 Evaluating $\int_{[0,1]} (1+x) \cdot dF(x)$ and $\int_{(0,1]} (1+x) \cdot dF(x)$ with

(a) $dF(x) = dx$ gives $\int_{[0,1]} (1+x) \cdot dF(x) = \int_{(0,1]} (1+x) \cdot dF(x) = \int_0^1 (1+x) \cdot dx = \frac{3}{2}$.

(b) $dF(x) = e^{-x} \cdot dx$ gives $\int_{[0,1]} (1+x) \cdot dF(x) = \int_{(0,1]} (1+x) \cdot dF(x) = \int_0^1 (1+x) e^{-x} \cdot dx = 2 - 3e^{-1}$.

(c) $dF(x) = .25 \cdot 1_{\{0,1\}}(x) + .5 \cdot 1_{\{.5\}}(x)$ gives

$$\int_{[0,1]} (1+x) \cdot dF(x) = (1+0)(.25) + (1+.5)(.5) + (1+1)(.25) = 1.5$$

$$\int_{(0,1]} (1+x) \cdot dF(x) = (1+.5)(.5) + (1+1)(.25) = 1.25.$$

11.3 With $dX_t = g_t \cdot dt + h_t \cdot dW_t$ and $dM_t = M_t r \cdot dt$ we have

(a)

$$dX_t^2 = 2X_t \cdot dX_t + d\langle X \rangle_t$$
$$= \left(2g_t X_t + h_t^2 \right) \cdot dt + 2h_t X_t \cdot dW_t$$

(b)

$$de^{X_t^2} = \left[2X_t \cdot dX_t + \left(1 + 2X_t^2 \right) \cdot d\langle X \rangle_t \right] e^{X_t^2}$$
$$= \left[2g_t X_t + h_t^2 \left(1 + 2X_t^2 \right) \right] e^{X_t^2} \cdot dt + 2h_t X_t e^{X_t^2} \cdot dW_t$$

(c)

$$d\ln X_t = X_t^{-1} \cdot dX_t - X_t^{-2} \frac{d\langle X_t \rangle}{2}$$
$$= X_t^{-1} \left(g_t - \frac{h_t^2}{2} \right) + X_t^{-1} h_t \cdot dW_t$$

(d)

$$\frac{d\exp\left(\theta t + X_t^2\right)}{\exp\left(\theta t + X_t^2\right)} = \theta \cdot dt + 2X_t \cdot dX_t + \left(1 + 2X_t^2\right) \cdot d\langle X \rangle_t$$
$$= \left[\theta + 2g_t X_t + h_t^2 \left(1 + 2X_t^2 \right) \right] \cdot dt + 2h_t X_t \cdot dW_t$$

(e)

$$d\left(tX_t^{-1}\right) = X_t^{-1} \cdot dt - tX_t^{-2} \cdot dX_t + tX_t^{-3} d\langle X \rangle_t$$
$$= X_t^{-1} \left(1 - tg_t X_t^{-1} + th_t^2 X^{-2} \right) \cdot dt - th_t X_t^{-2} \cdot dW_t$$

(f)

$$d\left(M_t X_t\right) = M_t \cdot dX_t + X_t \cdot dM_t + d\left\langle X_t, M_t \right\rangle$$
$$= M_t \cdot dX_t + X_t \cdot dM_t + 0$$
$$= \left(M_t g_t + X_t r\right) \cdot dt + M_t h_t \cdot dW_t.$$

11.5 When $dX_t = g X_t \cdot dt + h X_t \cdot dW_t$ and X_0 is given,

(a) $X_t = X_0 \exp\left[(g - h^2/2)t + hW_t\right]$ solves the SDE, since

$$dX_t = X_t \left[\left(g - \frac{h^2}{2}\right) \cdot dt + h \cdot dW_t + \frac{h^2}{2} \cdot dt\right]$$
$$= g X_t \cdot dt + h X_t \cdot dW_t.$$

(b) X_t can be written as

$$X_t = X_0 \exp\left[gt - \frac{h^2 t}{2} + hW_s + h(W_t - W_s)\right]$$
$$= e^{g(t-s)} X_0 \exp\left[\left(g - \frac{h^2}{2}\right) s + hW_s - h^2 \frac{(t-s)}{2} + h(W_t - W_s)\right]$$
$$= e^{g(t-s)} X_s \exp\left[-h^2 \frac{(t-s)}{2} + h(W_t - W_s)\right],$$

so $E(X_t \mid \mathcal{F}_s) = e^{g(t-s)} X_s$.

(c) The above shows that $\{X_t\}_{t \geq 0}$ is not a martingale unless $g = 0$. Alternatively, note that unless $g = 0$ there is a nonzero drift term in the SDE.

11.7 We have

$$dX_t^* = \frac{1}{M_t} \cdot dX_t - \frac{X_t}{M_t^2} \cdot dM_t = X_t^* \cdot \frac{dX_t}{X_t} - X_t^* \cdot \frac{dM_t}{M_t}$$
$$= (\mu - r) X_t^* \cdot dt + \sigma X_t^* \cdot dW_t$$
$$dX_t^{**} = \frac{1}{Y_t} \cdot dX_t - \frac{X_t}{Y_t^2} \cdot dY_t - \frac{1}{Y_t^2} \cdot d\langle X, Y \rangle_t + \frac{X_t}{Y_t^3} \cdot d\langle Y \rangle_t$$
$$= \left(\mu - \nu - \rho\sigma\gamma + \gamma^2\right) X_t^{**} \cdot dt + \left(\sigma X_t^{**} \cdot dW_t - \gamma X_t^{**} \cdot dW_t'\right).$$

12.1 Since

$$\ln\left\{(W_1 - c_1)\left[1 + p_1 (R_2 - r_1)\right]\right\} = \ln(W_1 - c_1) + \ln\left[1 + p_1 (R_2 - r_1)\right],$$

the portfolio and consumption decisions at $t = 1$ are completely separate, with p_1^* as the solution to $\max_{p_1} E \ln\left[1 + p_1 (R_2 - r_1)\right]$ and $c_1^* = W_1/(1 + \delta)$ as solution to $\max_{c_1} \delta \ln c_1 + \delta^2 \ln(W_1 - c_1)$. At $t = 0$ the optimal portfolio solves $\max_{p_0} E \ln [1 + p_0 (R_1 - r_0)]$, so $p_0^* = p_1^*$ if R_1, R_2 are identically distributed and $r_0 = r_1$. (Of course, the precise value will depend on the common marginal distribution of R_1 and R_2 and the common value of r_0 and r_1, but p_0^* and p_1^* will be the same regardless of whether random variables R_1, R_2 are independent.) To find the optimal c_0, plug c_1^* and p_1^* into the objective function, ignore the portfolio terms, and solve

$$\max_{c_0} \ln c_0 + \delta \ln (W_0 - c_0) + \delta^2 \ln \left[(W_0 - c_0) \frac{\delta}{1 + \delta} \right]$$

$$= \max_{c_0} \ln c_0 + \left(\delta + \delta^2 \right) \ln (W_0 - c_0) + \delta^2 \ln \left(\frac{\delta}{1 + \delta} \right)$$

to obtain $c_0^* = W_0 / \left(1 + \delta + \delta^2 \right)$.

12.3 Our problem is $\max_{c_0, p_0} E_0 c_0 c_1 c_2$. At $t = 1$ we solve

$$\max_{c_1, p_1} c_0 c_1 (W_1 - c_1) E_1 [1 + r + p_1 (R_2 - r)]$$

subject to $0 \leq p_1 \leq 1$ and $0 \leq c_1 \leq W_1$. Independently of p_1 we find $c_1^* = W_1/2$, which satisfies the constraint. Since $E_1 R_2 = E R_2 = \mu > r$, $E_1 [1 + r + p_1 (R_2 - r)]$ increases linearly up to $p_1 = 1$, so $p_1^* = 1$. Letting $\theta_1 = (1 + \mu)$, we solve at $t = 0$

$$\max_{c_0, p_1} c_0 (W_0 - c_0)^2 E_0 [1 + r + p_0 (R_1 - r)] \frac{\theta_1}{4}.$$

Disregarding the irrelevant constant $\theta_1/4$, we find $c_0^* = W_0/3$ and $p_0^* = 1$.

13.1 We have $V_{j+1} = V_j [(1 - f_j) + f_j X]$, where $\Pr (X = 2) = 0.6 = 1 - \Pr (X = 0)$. Thus

$$E \ln V_{j+1} = \ln V_j + 0.6 \ln [(1 - f_j) + 2f_j] + 0.4 \ln (1 - f_j)$$
$$= \ln V_j + 0.6 \ln (1 + f_j) + 0.4 \ln (1 - f_j).$$

Differentiating yields

$$\frac{dE \ln V_{j+1}}{df_j} = \frac{.6}{1 + f_j} - \frac{.4}{1 - f_j}.$$

Equating to zero and solving for f_j give $f_j = .2$ for each j, and verifying that $d^2 E \ln V_{j+1}/df_j^2 \big|_{f_j = .2} < 0$ shows this to be the maximizing value.

13.3 We have

$$
\begin{aligned}
\frac{dV_t}{V_t} &= \frac{s_{1t}S_{1t}}{V_t}\frac{dS_{1t}}{S_{1t}} + \frac{s_{2t}S_{2t}}{V_t}\frac{dS_{2t}}{S_{2t}} + \frac{m_t M_t}{V_t}\frac{dM_t}{M_t} \\
&= p_{1t}\frac{dS_{1t}}{S_{1t}} + p_{2t}\frac{dS_{2t}}{S_{2t}} + (1 - p_{1t} - p_{2t})\frac{dM_t}{M_t} \\
&= p_{1t}(\mu_1 \cdot dt + \sigma_1 \cdot dW_{1t}) + p_{2t}(\mu_2 \cdot dt + \sigma_2 \cdot dW_{2t}) \\
&\quad + (1 - p_{1t} - p_{2t})r \cdot dt \\
&= [r + p_{1t}(\mu_1 - r) + p_{2t}(\mu_2 - r)] \cdot dt \\
&\quad + p_{1t}\sigma_1 \cdot dW_{1t} + p_{2t}\sigma_2 \cdot dW_{2t} \\
&= [r + \mathbf{p}'_t(\mu - r\mathbf{1})] \cdot dt + (p_{1t}\sigma_1, p_{2t}\sigma_2) \cdot d\mathbf{W}_t,
\end{aligned}
$$

where p_{1t} and p_{2t} are the proportional investments in the two stocks, $\mathbf{p}_t = (p_{1t}, p_{2t})'$, and $\mathbf{W}_t = (W_{1t}, W_{2t})'$ The quadratic variation is

$$
\langle V \rangle_t = \int_0^t V_s^2 \left(p_{1s}^2 \sigma_1^2 + 2p_{1s}p_{2s}\rho\sigma_1\sigma_2 + p_{2s}^2\sigma_2^2 \right) \cdot ds,
$$

so

$$
\begin{aligned}
d\langle V \rangle_t &= V_t^2 \left(p_{1t}^2 \sigma_1^2 + 2p_{1t}p_{2t}\rho\sigma_1\sigma_2 + p_{2t}^2\sigma_2^2 \right) \cdot dt \\
&= V_t^2 (p_{1t}, p_{2t}) \begin{pmatrix} \sigma_1^2 & \rho\sigma_1\sigma_2 \\ \rho\sigma_1\sigma_2 & \sigma_2^2 \end{pmatrix} \begin{pmatrix} p_{1t} \\ p_{2t} \end{pmatrix} \cdot dt \\
&= V_t^2 \mathbf{p}'_t \Sigma \mathbf{p}_t \cdot dt.
\end{aligned}
$$

Then from Itô's formula we obtain

$$
\begin{aligned}
d\ln V_t &= \frac{dV_t}{V_t} - \frac{d\langle V \rangle_t}{2V_t^2} \\
&= \left[r + \mathbf{p}'_t(\mu - r\mathbf{1}) - \frac{\mathbf{p}'_t \Sigma \mathbf{p}_t}{2} \right] \cdot dt + (p_{1t}\sigma_1, p_{2t}\sigma_2) \cdot d\mathbf{W}_t.
\end{aligned}
$$

The growth-optimal proportions maximize the expression in brackets. Provided that Σ is nonsingular ($|\rho| < 1$), the optimal investment shares are

$$
\begin{aligned}
\mathbf{p}'_t &= (\mu - r\mathbf{1})' \Sigma^{-1} \\
&= \left(\frac{\theta_1 - \rho\theta_2}{\sigma_1(1 - \rho^2)}, \frac{\theta_2 - \rho\theta_1}{\sigma_2(1 - \rho^2)} \right),
\end{aligned}
$$

where $\theta_j \equiv (\mu_j - r)/\sigma_j$ for $j \in \{1, 2\}$. The numbers of units held are $s_{1t} = p_{1t}V_t/S_{1t}$, $s_{2t} = p_{2t}V_t/S_{2t}$, and $m_t = V_t(1 - p_{1t} - p_{2t})/M_t$. When $\rho = \pm 1$ the assets are perfect substitutes and there is no unique solution for p_{1t}, p_{2t}.

13.5 The mean and variance of the instantaneous rate of return of optimal portfolio $\mathbf{p}_t = \boldsymbol{\Sigma}^{-1} (\mu - r\mathbf{1})$ are

$$ER_{\mathbf{p}_t} = r + (\mu - r\mathbf{1})' \, \boldsymbol{\Sigma}^{-1} (\mu - r\mathbf{1}) = r + \theta^2$$
$$VR_{\mathbf{p}_t} = (\mu - r\mathbf{1})' \, \boldsymbol{\Sigma}^{-1} (\mu - r\mathbf{1}) = \theta^2,$$

so its Sharpe ratio is $(ER_{\mathbf{p}_t} - r) / \sqrt{VR_{\mathbf{p}_t}} = \theta$.

14.1 $E_t S_{t+1}/\theta^{t+1} = E_t q D_{t+1}/\theta^{t+1} = q D_t/\theta^t = S_t/\theta^t$, so the fair-game property holds. The $\{\varepsilon_t\}$ being positive with probability one, we also have, assuming $0 \le D_0 < \infty$, $E \, |S_t \theta^{-t}| \equiv E \, (|S_t \theta^{-t}| \mid \mathcal{F}_0) = E \, (S_t \theta^{-t} \mid \mathcal{F}_0) = S_0 < \infty$, confirming that $\{S_t/\theta^t\}$ is a martingale adapted to $\{\mathcal{F}_t\}_{t=0,1,2,\dots}$.

14.3 Since $U_t' \, (D_t) = D_t^{-(\gamma+1)}$, the stochastic discount factor is

$$\frac{U_{t+1}'(D_{t+1})}{U_t' \, (D_t)} = \left(\frac{D_{t+1}}{D_t} \right)^{-(\gamma+1)},$$

and so the time t price of a unit discount bond that matures at $t + 1$ is

$$B(t, t+1) = E_t \left(\frac{D_{t+1}}{D_t} \right)^{-(\gamma+1)} B \, (t+1, t+1)$$
$$= E_t e^{-(\gamma+1)[\mu + \sigma(W_{t+1} - W_t)]} \cdot 1$$
$$= \exp \left[-(\gamma + 1) \mu + (\gamma + 1)^2 \, \frac{\sigma^2}{2} \right].$$

Thus

$$r \, (t, t+1) = -\ln B \, (t, t+1) = (\gamma + 1) \left[\mu - (\gamma + 1) \, \frac{\sigma^2}{2} \right],$$

and so we require $\mu > (\gamma + 1) \sigma^2 / 2$ to get $r \, (t, t+1) > 0$.

14.5 Letting $\hat{E}_t \, (\cdot) = \hat{E} \, (\cdot \mid \mathcal{F}_t)$ represent conditional expectation under $\hat{\mathbb{P}}$, we have

$$\hat{E}_t \left(\frac{S_{t+1}}{S_t} \right) = \hat{E}_t \left(\frac{D_{t+1}}{D_t} \right) = \hat{E}_t e^{\mu + \sigma(W_{t+1} - W_t)}$$
$$= \hat{E}_t e^{\mu - \nu + \sigma[W_{t+1} + \nu(t+1) - W_t - \nu t]}$$
$$= \hat{E}_t e^{\mu - \nu + \sigma(\hat{W}_{t+1} - \hat{W}_t)}$$
$$= e^{\mu - \nu + \sigma^2/2}.$$

This equals $B \, (t, t+1)^{-1} = e^{r(t,t+1)}$, the one-period total return on the discount bond, if and only if $\nu = \mu - r \, (t, t+1) + \sigma^2/2$.

15.1 With $U_t(c_t) = a_t + b_t c_t$ we have $\delta_{t,t+1} = U'_{t+1}(c_{t+1})/U'_t(c_t) = b_{t+1}/b_t$ and $E_t R_{t+1} = b_t/b_{t+1} - 1 = \mu_{t+1}$, where $\{\mu_t\}_{t=0,1,2,...}$ is $\mathcal{F}_0$-measurable. Then

$$
\begin{aligned}
E\left[X_t(R_{t+1} - \mu_{t+1})\right] &= E\left[E_t X_t(R_{t+1} - \mu_{t+1})\right]\\
&= E\left[X_t E_t(R_{t+1} - \mu_{t+1})\right]\\
&= 0.
\end{aligned}
$$

15.3 Applying the tower property of conditional expectation, we obtain

$$
E S_t \varepsilon_t = E S_t(S_t^* - S_t) = E\left[E_t S_t(S_t^* - S_t)\right] = E\left[S_t E_t(S_t^* - S_t)\right] = 0.
$$

Then

$$
\begin{aligned}
V S_t^* &= E\left(S_t^*\right)^2 - \left(E S_t^*\right)^2 = E\left(S_t^2 + 2S_t\varepsilon_t + \varepsilon_t^2\right) - \left[E\left(E_t S_t^*\right)\right]^2\\
&= E S_t^2 + 2E\left[E_t(S_t\varepsilon_t)\right] + E\varepsilon_t^2 - \left(E S_t\right)^2\\
&= E S_t^2 - \left(E S_t\right)^2 + E\varepsilon_t^2 = V S_t + E\varepsilon_t^2 \geq V S_t.
\end{aligned}
$$

16.1 From the first rows of μ and $\mathbf{B}$ we see that $r = .030$. The expected returns were constructed as $\mu_j = .030 + \lambda_1 \beta_{j1} + \lambda_2 \beta_{j2}$ with $\lambda_1 = .020$ and $\lambda_2 = .030$. These can be found approximately by solving any 2 of the 10 equations with nonzero β's.

16.3 Arbitrage profits can be made by buying bond 13 and selling appropriate portfolios of bonds (a) 1,4,7,11; (b) 2,4,7,11; (c) 3,4 7 11; (d) 1,5,7,11; (e) 1,6,7 11; (f) 1,4,8,11; (g) 1,4,9,11; (h) 1,4,10,11; also by selling bonds 1 and 4 and buying bond 6 in appropriate proportions.

16.5 The T-maturing discount bond is now our commodity; its time t spot price is $B(t,T)$; and it earns no cash flow. Thus, $B(t,T)$ in (4.7) corresponds to $S_t e^{-\delta(T-t)}$ in (16.7), while $B(t,t')$ in (4.7) and $B(t,T)$ in (16.7) each represent the cost at t of acquiring a one-dollar cash receipt on the delivery date.

16.7 Start out at $t = 0$ with $B(0,2)/B(0,1) \equiv B_0(1,2) < 1$ futures contracts, each pertaining to one unit of the commodity. Note that $B_0(1,2)$ represents the $t = 0$ forward price for one-day loans commencing at $t = 1$. At the end of day 1, marking to market will change the account balance by $(F_1 - F_0)B_0(1,2) = (f_1 - f_0)B_0(1,2)$, assuming that $F_0 = f_0$. Now, if this amount is positive, invest it by buying one-day bonds costing $B(1,2)$ each. If the amount is negative, finance it by issuing one-day bonds costing $B(1,2)$. Either way, this will contribute by day 2 a positive or negative amount of cash equal to $(f_1 - f_0)B_0(1,2)/B(1,2)$. But since $B(1,2)$ was, by assumption, known at $t = 0$, it must coincide with forward price $B_0(1,2)$, and so the cash increment is really $f_1 - f_0$. Continuing the replication process, on day 1 increase the futures position to one full contract at futures price $F_1 = f_1$. Then at expiration on day 2 the final mark to market will add $S_2 - f_1$ to the account. The total increment from this process is thus $(S_2 - f_1) + (f_1 - f_0) = S_2 - f_0$, precisely as if one had initiated a forward contract as of $t = 0$. Had F_0 not equaled f_0 to begin with, an arbitrage would have been possible by taking opposite positions in forwards and futures.

17.1 For $0 \le s \le t \le T$ we have

$$E\left(Q_t \mid \mathcal{F}_s\right) = E\left[E\left(Y_T \mid \mathcal{F}_t\right) \mid \mathcal{F}_s\right] = E\left(Y_T \mid \mathcal{F}_s\right) = Q_s$$

by the tower property of conditional expectation.

17.3 We have

$$
\begin{aligned}
dS_t^+ &= e^{-\delta T}\left(e^{\delta t} \cdot dS_t + \delta S_t e^{\delta t} \cdot dt\right) \\
&= S_t^+ \cdot \frac{dS_t}{S_t} + \delta S_t^+ \cdot dt \\
&= (\mu + \delta)\, S_t^+ \cdot dt + \sigma S_t^+ \cdot dW_t
\end{aligned}
$$

and

$$
\begin{aligned}
dV_t^+ &= p_t \cdot dS_t^+ + q_t \cdot dM_t \\
&= p_t S_t^+\left[(\mu + \delta) \cdot dt + \sigma \cdot dW_t\right] + \left(D - p_t S_t^+\right) r \cdot dt \\
&= \left[Dr + p_t S_t^+ (\mu + \delta - r)\right] \cdot dt + p_t S_t^+ \sigma \cdot dW_t.
\end{aligned}
$$

Equating the stochastic part to that of

$$dD\left(S_t, T - t\right) = \left(-D_{T-t} + D_S S_t \mu + \frac{D_{SS}}{2} S_t^2 \sigma^2\right) \cdot dt + D_S S_t \sigma \cdot dW_t$$

gives $p_t = D_S S_t / S_t^+ = e^{\delta(T-t)} D_S$. Then equating the deterministic parts of dV_t^+ and dD gives

$$0 = -D_{T-t} + D_S S_t(r - \delta) + \frac{D_{SS}}{2} S_t^2 \sigma^2 - Dr.$$

17.5 Using the specific symbol $\mathfrak{F}$ for our derivative, the fundamental PDE becomes

$$0 = -\mathfrak{F}_{T-t} + \mathfrak{F}_S S_t(r - \delta) + \frac{\mathfrak{F}_{SS}}{2} S_t^2 \sigma^2 - \mathfrak{F}r.$$

From $\mathfrak{F} \equiv \mathfrak{F}\left(S_t, T - t\right) = S_t e^{-\delta(T-t)} - f_0 e^{-r(T-t)}$ we find

$$
\begin{aligned}
-\mathfrak{F}_{T-t} &= \delta S_t e^{-\delta(T-t)} - r f_0 e^{-r(T-t)} \\
\mathfrak{F}_S S_t(r - \delta) &= (r - \delta) S_t e^{-\delta(T-t)} \\
\frac{\mathfrak{F}_{SS}}{2} S_t^2 \sigma^2 &= 0.
\end{aligned}
$$

Adding these components gives $r S_t e^{-\delta(T-t)} - r f_0 e^{-r(T-t)} = \mathfrak{F}r$, confirming that $\mathfrak{F}$ satisfies the PDE.

18.1 If $P(S_t, T - t) > B(t, T) X \equiv BX$, sell the put, receiving something more than BX in cash, and use the proceeds to buy $P(S_t, T - t)/B > X$ units of T-maturing discount bonds. If $S_T \geq X$, the put will expire worthless, and you will come away with $P(S_t, T - t)/B$ in cash. If $S_T < X$ the put will be exercised. In that case use X of the cash from your bonds to buy the share of stock, which has nonnegative value, and come away with $P(S_t, T - t)/B - X + S_T > 0$. For the lower bound, if $P(S_t, T - t) < 0$, you can buy puts for a negative sum and, if nothing better comes to mind, throw them away and pocket cash up front. If $0 \leq P(S_t, T - t) < X - S_t e^{-\delta(T-t)}$, then sell X discount, T-maturing, unit bonds and buy both the put and $e^{-\delta(T-t)}$ shares of stock. This will generate $BX - S_t e^{-\delta(T-t)} - P(S_t, T - t) > 0$ in cash up front. Reinvest the stock dividends as they come in, so that your stock holding will be worth S_T at T. You will also then have to pay X to the purchasers of the bonds. If the put is in the money ($S_T < X$), you can exercise it by selling the stock for X and then repay your debt to the bondholders, for zero net cash flow. If the put is out of the money ($S_T \geq X$), let it expire worthless, sell the stock, repay the debt, and come away with $S_T - X \geq 0$.

18.3 If $S_t e^{-\delta(t-t)} - BX > 0$, then $f_t \equiv S_t e^{-\delta(t-t)}/B > X$, $\ln(f_t/X) > 0$, and both $q^+(f_t/X, \tau)$ and $q^-(f_t/X, \tau)$ (the two arguments of Φ in formula (18.4)) approach $+\infty$ as $\sigma \downarrow 0$, in which case $C(S_t, \tau) \to Bf_t - BX = S_t e^{-\delta(t-t)} - BX$. If $S_t e^{-\delta(t-t)} - BX \leq 0$ then $\ln(f_t/X) \leq 0$, $q^+(f_t/X, \tau)$ and $q^-(f_t/X, \tau)$ approach $-\infty$, and $C(S_t, \tau) \to 0$, establishing (18.7). As $\sigma \uparrow \infty$, $q^+(f_t/X, \tau) \to +\infty$ and $q^-(f_t/X, \tau) \to -\infty$, regardless of the relation between f_t and X, and so $\Phi[q^+(f_t/X, \tau)] \to 1$ and $\Phi[q^-(f_t/X, \tau)] \to 0$, establishing (18.9). Verifications of (18.8) and (18.10) come similarly.

18.5 The replicating portfolio for the call is worth $V_t = p_t S_t + q_t M_t$ at time t, where $p_t = C_S$ and $q_t = C_X X/M_t$. We must show that $dV_t = C_S \cdot dS_t + (C_X X/M_t) \cdot dM_t$. Since $V_t = C(S_t, \tau) = C_S S_t + C_X X$ and $dM_t/M_t = r \cdot dt$, this is equivalent to

$$dC(S_t, \tau) = C_S \cdot dS_t + [C(S_t, \tau) - C_S S_t] r \cdot dt.$$

Applying Itô to $C(S_t, \tau)$ gives $dC(S_t, \tau) = \left[-C_\tau + C_{SS}\sigma^2 S_t^2/2\right] \cdot dt + C_S \cdot dS_t$. But fundamental PDE (17.3) implies that $-C_\tau + C_{SS}\sigma^2 S_t^2/2 = -[C(S_t, \tau) - C_S S_t] r$, from which the result follows.

19.1 We are given that $dS_t = .1 S_t \cdot dt + .2 S_t \cdot dW_t$ with $S_0 = 1$ and that $dM_t = .05 M_t \cdot dt$.

   (a) Clearly, we don't need martingale methods here, but it is nevertheless instructive to see that they lead to the answer we get from the present-value formula. Now the "process" $\{1\}_{0 \leq t \leq 1}$ is a martingale under natural measure $\mathbb{P}$ and under any equivalent measure. However, holding sterile cash is always dominated by holding the money fund, so we have to find a *normalized* process that is a martingale in some measure. Picking $\{M_t = M_0 e^{rt}\}$ as numeraire, we want the measure $\hat{\mathbb{P}}$ in which $\{M_t^{-1} D(S_t, T - t)\}$ is a martingale. But the process is deterministic, so $M_0^{-1} D(S_0, 1) = M_1^{-1} \hat{E} D(S_1, 0) = M_1^{-1} \cdot 1$, and $D(S_0, 1) = (M_0/M_1) = e^{-.05} = B(0, 1)$, the present value of a sure dollar receipt one year hence. Note that if short rate process $\{r_t\}_{0 \le t \le 1}$ had been stochastic, then $M_1 = M_0 \exp\left(\int_0^1 r_t \cdot dt\right)$ would have been uncertain, and to work under $\hat{\mathbb{P}}$ would have required a model for $\{r_t\}$. To avoid that, we could work under forward EMM $\mathbb{P}^1$, wherein bond process $\{B(t, 1)\}_{0 \le t \le 1}$ serves as numeraire. In this case we would have $B(0, 1)^{-1} D(S_0, 1) = E^1 B(1, 1)^{-1} D(S_1, 0)$. But since $B(1, 1) = 1$ and $D(S_1, 0) = 1$ are deterministic, they are the same in $\mathbb{P}^1$ as in $\mathbb{P}$, and so we again have $D(S_0, 1) = B(0, 1)$.

(b) $D(S_1, 0)$ is now stochastic, but since short rate process $\{r_t = r = .05\}$ is deterministic, we can use either $\{M_t\}$ or $\{B(t, 1)\}$ as numeraires with risk-neutral measure $\hat{\mathbb{P}}$. Thus

$$D(S_0, 1) = B(0, 1) \hat{E}\left[1 - \mathbf{1}_{(0,1)}(S_1)\right]$$
$$= e^{-.05}\left[1 - \hat{\mathbb{P}}\{0 < S_1 < 1\}\right]$$
$$= e^{-.05}\left[1 - \hat{\mathbb{P}}\{-\infty < \ln S_1 < 0\}\right]$$
$$= e^{-.05}\hat{\mathbb{P}}\{\ln S_1 \ge 0\}.$$

Since

$$S_1 = S_0 \exp\left(r - \frac{\sigma^2}{2} + \sigma \hat{W}_1\right)$$
$$= \exp\left(.03 + .2\hat{W}_1\right) \sim \exp(.03 + .2Z),$$

this gives $D(S_0, 1) = e^{-.05} \Pr(Z > -.15) \doteq .5323$.

(c) $D(S_0, 1) = B(0, 1) \hat{E}_0 (S_1 - 1)^+ = C(1, 1)$, where $C(1, 1)$ is found from formula (17.4) with $X = 1, \tau = 1, B = e^{-.05}, S_0 = 1, \sigma = .2$ as

$$C(1, 1) = \Phi\left(\frac{.05 + .04/2}{.2}\right) - e^{-.05} \Phi\left(\frac{.05 - .04/2}{.2}\right) \doteq .1045.$$

(d) $D(S_0, 1) = C(1, 1) + P(1, 1) \doteq .1045 + .0557$ from (17.4) and (17.5).

(e) $D(S_0, 1) = B(0, 1) \hat{E}_0 (S_1^2 - 2S_1 + 1) = B(0, 1) \hat{E}_0 S_1^2 - 2S_0 + B(0, 1)$. For the first term we find

$$B(0, 1) \hat{E}_0 S_1^2 = e^{-r} \hat{E}_0 S_0^2 e^{2r - \sigma^2 + 2\sigma \hat{W}_1} = S_0^2 e^{r + \sigma^2} = e^{.09},$$

so $D(S_0, 1) \doteq .0454$.

19.3 Using $\beta'(t) = e^{-bt} = 1 - b\beta(t)$ we have

$$dr_t = \left\{ -be^{-bt}r_0 + a\left[1 - b\beta(t)\right] \right\} \cdot dt + \sigma \cdot d\hat{W}_t'$$

$$- \left[\sigma \int_0^t e^{-b(t-s)} \cdot d\hat{W}_s' \right] b \cdot dt$$

$$= a \cdot dt - b \left[e^{-bt}r_0 + a\beta(t) + \sigma \int_0^t e^{-b(t-s)} \cdot d\hat{W}_s' \right] \cdot dt + \sigma \cdot d\hat{W}_t'$$

$$= (a - br_t) \cdot dt + \sigma \cdot d\hat{W}_t'.$$

19.5 Letting $B \equiv B(t,T)$, we have

$$dM_t^T = M_t^T \left(\frac{dM_t}{M_t} - \frac{dB}{B} + \frac{d\langle B \rangle_t}{B^2} \right),$$

where B is as given in (19.16). Itô's formula gives

$$\frac{dB}{B} = \left[a\beta(T-t) - \frac{\sigma^2}{2}\beta(T-t)^2 + e^{-b(T-t)}r_t + \frac{\sigma^2}{2}\beta(T-t)^2 \right] \cdot dt$$

$$- \beta(T-t) \cdot dr_t + \frac{1}{2}\beta(T-t)^2 \cdot d\langle r \rangle_t$$

$$= \left[a\beta(T-t) - \frac{\sigma^2}{2}\beta(T-t)^2 + e^{-b(T-t)}r_t + \frac{\sigma^2}{2}\beta(T-t)^2 \right] \cdot dt$$

$$- \beta(T-t) \left[(a - br_t) \cdot dt + \sigma d\hat{W}_t \right] + \frac{\sigma^2}{2}\beta(T-t)^2 \cdot dt$$

$$= r_t \cdot dt - \sigma\beta(T-t) \cdot d\hat{W}_t$$

$$\frac{d\langle B \rangle_t}{B^2} = \sigma^2\beta(T-t)^2 \cdot dt.$$

Combining and simplifying yield

$$\frac{dM_t^T}{M_t^T} = r_t \cdot dt - \left[r_t \cdot dt - \sigma\beta(T-t) \cdot d\hat{W}_t \right] + \sigma^2\beta(T-t)^2 \cdot dt$$

$$= \sigma\beta(T-t) \left[\sigma\beta(T-t) \cdot dt + d\hat{W}_t \right].$$

Putting $\alpha_t \equiv -\sigma \int_0^t \beta(T-s) \cdot ds$ and $W_t^T = \hat{W}_t - \alpha_t$ gives $dM_t^T/M_t^T = \sigma\beta(T-t) \cdot dW_t^T$. Thus, a measure $\mathbb{P}^T$ that makes $\{W_t - \alpha_t\}_{t\geq0}$ a standard Brownian motion makes $\{M_t^T\}_{t\geq0}$ a martingale.

19.7 We are given that $dF_t = \alpha F_t \cdot dt + \beta F_t \cdot dW_t$ under $\mathbb{P}$.

(a) Putting $\bar{S}_t(\gamma) \equiv \int_{-\infty}^t e^{-\gamma(t-s)} \cdot dS_s$, we have from (15.3) and (15.4), respectively, the following equations:

$$F_t = (1-c)^{-1}\left[S_t - c\bar{S}_t(\gamma) \right]$$

$$dS_t = -\frac{\gamma c}{1-c}\bar{S}_t(\gamma) \cdot dt + dF_t.$$

Applying the model for dF_t and simplifying give

$$dS_t = \left[\frac{\alpha - (\gamma + \alpha)c\bar{S}_t\,(\gamma)\,/S_t}{1 - c}\right]S_t \cdot dt + \left[\frac{\beta\left(1 - c\bar{S}_t\,(\gamma)\,/S_t\right)}{1 - c}\right]S_t \cdot dW_t.$$

The two expressions in brackets correspond to and define μ_t and σ_t, which clearly are not deterministic.

(b) With $dM_t = r_t M_t \cdot dt$ we have $dS_t^*/S_t^* = (\mu_t - r_t) \cdot dt + \sigma_t \cdot dW_t$, so $\hat{\mathbb{P}}$ would have to make $\left\{\hat{W}_t = W_t - \int_0^t (\mu_s - r_s)\,/\sigma_s \cdot ds\right\}$ a BM in order for $\{S_t^*\}_{0 \le t \le T}$ to be a martingale.

(c) Under $\hat{\mathbb{P}}$ the solution to SDE $dS_t^* = \sigma_t S_t^* \cdot d\hat{W}_t$ subject to the initial condition would be $S_t^* = S_0^* \exp\left(-\frac{1}{2}\int_0^t \sigma_s^2 \cdot ds + \int_0^t \sigma_s \cdot \hat{W}_s\right)$, and so we would have

$$S_T = S_t \exp\left[\int_t^T \left(r_s - \frac{\sigma_s^2}{2}\right) \cdot ds + \int_t^T \sigma_s \cdot d\hat{W}_s\right].$$

With this expression for S_T and with $\{r_t\}$ deterministic, our risk-neutral pricing formula would give $D(S_t, T - t) = B(t,T)\hat{E}_t D(S_T, 0)$ as usual. However, S_T is no longer conditionally lognormal, given that volatility σ_t is proportional to $\left[1 - c\bar{S}_t\,(\gamma)\,/S_t\right]$ and so depends on the entire history of the price process up to t. While the expected value could be approximated by simulation, there would be no simple formula corresponding to the Black–Scholes formulas for options.

20.1 Write $\sigma_t^2 = \delta + \alpha\sigma_{t-1}^2 + \gamma u_{t-1}^2$ as $\sigma_t^2(1 - \alpha L) = \delta + \gamma u_{t-1}^2$. When $|\alpha| < 1$, operator $1 - \alpha L$ is invertible, so

$$\sigma_t^2 = (1 - \alpha L)^{-1}\left(\delta + \gamma u_{t-1}^2\right)$$
$$= (1 - \alpha)^{-1}\delta + \gamma\sum_{j=0}^{\infty} \alpha^j L^j u_{t-1}^2$$
$$= (1 - \alpha)^{-1}\delta + \gamma\sum_{j=0}^{\infty} \alpha^j u_{t-j-1}^2.$$

20.3 Put $v_t \equiv \sigma_t^2$ and write (20.4) as $dv_t = (\alpha - \beta v_t) \cdot dt + \gamma\sqrt{v_t} \cdot dW_t'$. Then, with $\sigma_t = f(v_t) = v_t^{1/2}$, $f'(v_t) = \frac{1}{2}v_t^{-1/2}$, and $f''(v_t) = -\frac{1}{4}v_t^{-3/2}$ Itô's formula gives

$$dσ_t = f'(v_t) \cdot dv_t + \frac{1}{2}f''(v_t) \cdot d\langle v \rangle_t$$

$$= \frac{1}{2}v_t^{-1/2} \cdot dv_t - \frac{1}{8}v_t^{-3/2} \cdot d\langle v \rangle_t$$

$$= \frac{1}{2}v_t^{-1/2}\left[(\alpha - \beta v_t) \cdot dt + \gamma\sqrt{v_t} \cdot dW_t'\right] - \frac{1}{8}v_t^{-3/2}\gamma^2 v_t \cdot dt$$

$$= \frac{1}{2σ_t}\left(\alpha - \frac{\gamma^2}{4} - \beta σ_t^2\right) \cdot dt + \frac{\gamma}{2} \cdot dW_t'.$$

20.5 The call's arbitrage-free value at $t = 0$ is found via martingale pricing as

$$C(S_0, T) = B(0, T)\hat{E}(S_T - X)^+ \equiv B(0, T)\hat{E}\left[(S_T - X)^+ \mid \mathcal{F}_0\right].$$

Now, the BMs $\left\{\hat{W}_t\right\}_{0 \leq t \leq T}$ and $\left\{\hat{W}_t'\right\}_{0 \leq t \leq T}$ that drive increments to price and volatility are independent, and so the integrated squared volatility $\int_0^T σ_u^2 \cdot du$ is independent of $\left\{\hat{W}_t\right\}_{0 \leq t \leq T}$. We can, therefore, condition on $\int_0^T σ_u^2 \cdot du$, apply the tower property of conditional expectation, and write

$$C(S_0, T) = B(0, T)\hat{E}\left\{\hat{E}\left[(S_T - X)^+ \mid \int_0^T σ_u^2 \cdot du\right]\right\}.$$

But the inner expectation corresponds precisely to the result that pertains when volatility is deterministic, and so

$$C(S_0, T) = B(0, T)\hat{E}C\left(S_0, T; \{σ_u^2\}_{0 \leq u \leq T}\right).$$

21.1 With $\Pr(\Delta N_t = n) = (\lambda\Delta t)^n e^{-\lambda\Delta t}/n!$ for $n \in \{0, 1, 2, ...\}$ we have

(a) $\Pr(\Delta N_t = 0) = (\lambda\Delta t)^0 e^{-\lambda\Delta t}/0! = e^{-\lambda\Delta t}$. Taylor's theorem gives $e^{-\lambda\Delta t} = 1 - \lambda\Delta t + e^{-\lambda\theta\Delta t}(\Delta t)^2/2$ for some $\theta \in (0, 1)$. Thus

$$\lim_{\Delta t \to 0} \frac{\left|e^{-\lambda\Delta t} - 1 + \lambda\Delta t\right|}{\Delta t} = \lim_{\Delta t \to 0} \frac{1}{2}e^{-\lambda\theta\Delta t}\Delta t = 0,$$

so $e^{-\lambda\Delta t} = 1 - \lambda\Delta t + o(\Delta t)$.

(b)

$$\Pr(\Delta N_t = 1) = \frac{\lambda\Delta t e^{-\lambda\Delta t}}{1!} = \lambda(\Delta t)e^{-\lambda\Delta t}$$

$$= \lambda\Delta t \cdot (1 - \lambda\Delta t + o(\Delta t)) = \lambda\Delta t + o(\Delta t).$$

(c) For $n > 1$ we have

$$0 < \Pr\left(\Delta N_t = n\right) < \left(\lambda \Delta t\right)^{n-1} \Pr\left(\Delta N_t = 1\right) = o\left(\Delta t\right).$$

21.3 We have

$$
\begin{aligned}
E \ln S_t &= E\left[E\left(\ln S_t \mid N_t\right)\right] \\
&= \ln S_0 + \left(\mu - \frac{\sigma^2}{2}\right) t \\
&\quad + E\left\{\sigma E\left(W_t \mid N_t\right) + \sum_{j=0}^{N_t} E\left[\ln\left(1 + U_j\right) \mid N_t\right]\right\} \\
&= \ln S_0 + \left(\mu - \frac{\sigma^2}{2}\right) t + E\left(0 + N_t \theta\right) \\
&= \ln S_0 + \left(\mu - \frac{\sigma^2}{2} + \theta \lambda\right) t.
\end{aligned}
$$

$$
\begin{aligned}
V \ln S_t &= V\left[\sigma W_t + \sum_{j=0}^{N_t} \ln\left(1 + U_j\right)\right] \\
&= \sigma^2 t + V \sum_{j=0}^{N_t} \ln\left(1 + U_j\right),
\end{aligned}
$$

where the last step follows from the mutual independence of $\{W_t\}$, $\{N_t\}$, and the $\{U_j\}$. Expressing the variance of the sum as the expectation of the conditional variance plus the variance of the conditional mean and using the fact that $EN_t = VN_t = \lambda t$, we have

$$
\begin{aligned}
V \ln S_t &= \sigma^2 t + E\left\{V\left[\sum_{j=0}^{N_t} \ln\left(1 + U_j\right) \mid N_t\right]\right\} \\
&\quad + V\left\{E\left[\sum_{j=0}^{N_t} \ln\left(1 + U_j\right) \mid N_t\right]\right\} \\
&= \sigma^2 t + E\left(N_t \xi^2\right) + V\left(N_t \theta\right) \\
&= \left[\sigma^2 + \left(\xi^2 + \theta^2\right) \lambda\right] t.
\end{aligned}
$$

21.5 Using the tower property of conditional expectation and the relation of variance to conditional variance and conditional mean, we have

$$
\begin{aligned}
E \ln S_t &= \ln S_0 + \mu t + \gamma E \mathfrak{T}_t + \sigma E\left[E\left(W_{\mathfrak{T}_t} \mid \mathfrak{T}_t\right)\right] \\
&= \ln S_0 + \left(\mu + \gamma\right) t + 0. \\
V \ln S_t &= E\left[V\left(\gamma \mathfrak{T}_t + \sigma W_{\mathfrak{T}_t} \mid \mathfrak{T}_t\right)\right] + V\left[E\left(\gamma \mathfrak{T}_t + \sigma W_{\mathfrak{T}_t} \mid \mathfrak{T}_t\right)\right] \\
&= \sigma^2 E \mathfrak{T}_t + V\left(\gamma \mathfrak{T}_t\right) \\
&= \left(\sigma^2 + \gamma^2 \nu\right) t.
\end{aligned}
$$

21.7 We have $ES_{N_t}^* = S_0^* e^{\lambda t (\mu - 1 - r/\lambda)}$ under $\mathbb{P}$, where $\mu = EY = k\theta$. Now $\{S_{N_t}^*\}_{t \geq 0}$ is a martingale under some $\hat{\mathbb{P}}$ if and only if $\mu \equiv \hat{E}Y = 1 + r/\lambda$. However, in changing the mean as we move from $\mathbb{P}$ to $\hat{\mathbb{P}}$, we cannot change the support, since $\hat{\mathbb{P}}$ is equivalent to $\mathbb{P}$ only if $\hat{\mathbb{P}}(Y = y) > 0$ for $y \in \{0, 1, ..., k\}$ and $\sum_{y=0}^{k} \hat{\mathbb{P}}(Y = y) = 1$. One possibility is to preserve the binomial form and the value of k but set $\theta = (r + 1)/k$; however, any such equivalent measure with $\mu = 1 + r/\lambda$ would work.

22.1 If $1 + U_j$ is lognormal with parameters $\ln(1 + \eta) - \xi^2/2$ and ξ^2, then $\ln(1 + U_j)$ is distributed as normal with those same parameters. Thus, if $Z \sim N(0, 1)$, the distributions of $\ln(1 + U_j)$ and $\ln(1 + \eta) - \xi^2/2 + \xi Z$ are exactly the same, and so

$$
\begin{aligned}
EU_j &= Ee^{\ln(1+U_j)} - 1 \\
&= Ee^{\ln(1+\eta) - \xi^2/2 + \xi Z} - 1 \\
&= e^{\ln(1+\eta)} - 1 = \eta.
\end{aligned}
$$

22.3 Putting $Z = (Y - \mu)/\sigma$, we have

$$
\begin{aligned}
\phi_Z(\zeta) &= E \exp\left[i\zeta\left(\frac{Y - \mu}{\sigma}\right)\right] \\
&= e^{-i\zeta\mu/\sigma} Ee^{i\zeta Y/\sigma} = e^{-i\zeta\mu/\sigma}\phi_Y(\zeta/\sigma).
\end{aligned}
$$

22.5 The short answer is that when the variance of the increment to log price is infinite, there is no equivalent measure in which the normalized price process is a martingale. This is because the expected value of the future price itself is then infinite. To see it, let $Y \equiv \ln(1 + U)$ be symmetric stable (η, ξ, α) with $1 < \alpha < 2$. Then, assuming independence of $\{W_t\}$, $\{N_t\}$, and $\{Y_j\}_{j=1}^{\infty}$ as usual, normalized price under the jump–diffusion model is

$$
\begin{aligned}
ES_t^* &= S_0^* e^{(\mu - r - \sigma^2/2)t} E \exp\left(\sigma W_t + \sum_{j=0}^{N_t} Y_j\right) \\
&= S_0^* e^{(\mu - r)t} E\left\{E\left[\exp\left(\sum_{j=0}^{N_t} Y_j\right) \mid N_t\right]\right\} \\
&= S_0^* e^{(\mu - r)t} E\left(\prod_{j=0}^{N_t} Ee^{Y_j}\right).
\end{aligned}
$$

Now for any symmetric stable variate Y with median η (also the mean when $\alpha > 1$) and PDF f (symmetric about η) we have

$$
\begin{aligned}
Ee^Y &= e^{\eta} Ee^{Y - \eta} \\
&= e^{\eta}\left[\int_{-\infty}^{\eta} e^{y - \eta} f(y; \eta, \xi, \alpha) \cdot dy + \int_{\eta}^{\infty} e^{y - \eta} f(y; \eta, \xi, \alpha) \cdot dy\right] \\
&> e^{\eta} \int_{\eta}^{\infty} e^{y - \eta} f(y; \eta, \xi, \alpha) \cdot dy \\
&= e^{\eta} \int_{\eta}^{\infty}\left[1 + (y - \eta) + \frac{(y - \eta)^2}{2} e^{\theta(y - \eta)}\right] f(y; \eta, \xi, \alpha) \cdot dy,
\end{aligned}
$$

where (from Taylor's theorem) θ lies somewhere on $(0, 1)$. Thus

$$Ee^Y > \frac{e^\eta}{2} \int_\eta^\infty (y - \eta)^2 f(y; \eta, \xi, \alpha) \cdot dy$$

$$= \frac{e^\eta}{4} \int_{-\infty}^\infty (y - \eta)^2 f(y; \eta, \xi, \alpha) \cdot dy$$

$$= +\infty.$$

22.7 From the definition of θ we have that $\ln \hat{\phi}_{sT}(\zeta)$ equals $i\zeta(s_0 + rT)$ plus

$$\frac{T}{\nu} \left\{ i\zeta \ln \left[1 - \nu \left(\gamma + \frac{\sigma^2}{2} \right) \right] - \ln \left[1 - \left(i\zeta\gamma - \zeta^2 \frac{\sigma^2}{2} \right) \nu \right] \right\}.$$

Since $\ln(1 + x) = x + o(x)$ as $x \to 0$, the limit of this quantity is

$$\lim_{\nu \to 0} \frac{T}{\nu} \left[-i\zeta\nu \left(\gamma + \frac{\sigma^2}{2} \right) + \left(i\zeta\gamma - \zeta^2 \frac{\sigma^2}{2} \right) \nu + o(\nu) \right]$$

$$= \left[-i\zeta \left(\gamma + \frac{\sigma^2}{2} \right) + \left(i\zeta\gamma - \zeta^2 \frac{\sigma^2}{2} \right) \right] T$$

$$= - \left(i\zeta + \zeta^2 \right) \frac{\sigma^2 T}{2}.$$

22.9 Given that Y is supported on $\{0, 1, 2, ...\}$,

(a) $\Pi_Y(\varphi) = \sum_{y=0}^\infty \varphi^y \mathbb{P}(Y = y) = \varphi^0 \mathbb{P}(Y = 0) + \sum_{y=1}^\infty \varphi^y \mathbb{P}(Y = y)$,
so $\Pi_Y(0) = \mathbb{P}(Y = 0)$ and $\Pi_Y(1) = \sum_{y=0}^\infty \mathbb{P}(Y = y) = 1$, while

$$\Pi_Y'(\varphi)|_{\varphi=1} = \sum_{y=1}^\infty y\varphi^{y-1} \mathbb{P}(Y = y)\Big|_{\varphi=1} = \sum_{y=0}^\infty y\mathbb{P}(Y = y) = EY.$$

(b) This follows at once from (22.13) and the result of part (a).

(c) Refer to the figure on the following page, which plots two PGFs $\Pi_1(\varphi)$ and $\Pi_2(\varphi)$ on $[0, 1.1]$ for which $\mathbb{P}(Y_1 = 0) > \mathbb{P}(Y_2 = 0) > 0$. From the definition of the PGF, it should be clear that all the derivatives are positive on $[0, 1]$ unless $\mathbb{P}(Y = 0) = 1$. From here on we exclude that trivial case, and so all the PGFs we consider are upward sloping, strictly convex on $[0, 1]$, and satisfying $\Pi_Y(1) = 1$. Now, if $\mathbb{P}(Y = 0) = 0$, then $\Pi_Y(0) = 0$, and since $\Pi_Y(\varphi)$ is convex and $\Pi_Y(1) = 1$, the function lies entirely below the 45° line. On the other hand, if $\mathbb{P}(Y = 0) = \Pi_Y(0) > 0$, the function may either stay above the 45° line for $\varphi < 1$, as does Π_1, or it may intersect at one point before $\varphi = 1$, as does Π_2. How would the iterates behave for a function such as Π_1 that remains above the line? Since $\Pi_1(\varphi) > \varphi$ for $\varphi \in [0, 1)$, it is clear that $\Pi_1^{[2]}(0) \equiv \Pi_1[\Pi_1(0)] > \Pi_1(0)$, that $\Pi_1^{[3]}(0) \equiv \Pi_1\left[\Pi_1^{[2]}(0)\right] > \Pi_1^{[2]}(0)$, and so

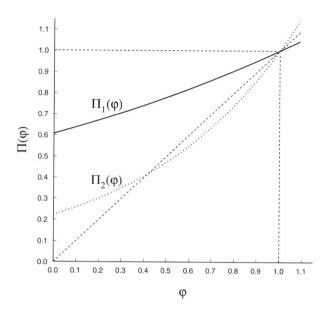

on, so that sequence $\left\{\Pi_1^{[n]}(0)\right\}_{n=1}^{\infty}$ is strictly increasing. Moreover, since $\Pi_1\left[\Pi_1^{[n-1]}(0)\right] > \Pi_1^{[n-1]}(0)$ so long as $\Pi_1^{[n-1]}(0) < 1$, the sequence increases right up to the upper bound of unity. Thus, $\lim_{n\to\infty}\Pi_1^{[n]}(0) = 1$. Since $\mathbb{P}(S_n = 0) = \Pi_1^{[n]}(0)^{S_0}$, it is also true that $\mathbb{P}(S_n = 0) \to 1$ as $n \to \infty$ whenever $\Pi_Y(\varphi)$ lies above the 45° line for $\varphi \in [0,1)$. Now under what condition does the PGF lie above the line? Since the function is increasing and strictly convex, it follows that first derivative $\Pi_Y'(\varphi)$ attains a maximum on $[0,1]$ at $\varphi = 1$. Thus, given that $\Pi_Y(0) > 0$, the function lies above the line if and only if $\Pi_Y'(1) = EY \leq 1$. (The event $S_n = 0$ corresponds to "extinction" of a population or, for a stock, to terminal bankruptcy. The condition $EY \leq 1$ holds if and only if the equation $\Pi_Y(\varphi) = \varphi$ has no root between 0 and 1. In that case, ultimate extinction is certain. When there is such a root, as for function Π_{Y_2} in the figure, the population or firm has a positive chance of surviving indefinitely.)

(d) Since for any $B > 0$ we have $\mathbb{P}(N_T > B) \to 1$ as $T \to \infty$, it follows that $N_T \to \infty$ with probability one when $\{N_t\}$ is a Poisson process. Thus, if $EY \leq 1$, we have $\lim_{T\to\infty}\mathbb{P}(S_{N_T} = 0) = 1$.

(e) If $\hat{\mathbb{P}}$ is an EMM, then it has the same null sets as $\mathbb{P}$. Thus, the result $\lim_{T\to\infty}\mathbb{P}(S_{N_T} \neq 0) = 0$ when $EY \leq 1$ implies $\lim_{T\to\infty}\hat{\mathbb{P}}(S_{N_T} \neq 0) = 0$ and therefore that $\hat{\mathbb{P}}(S_{N_T} = 0) = 1$ as well.

23.1 When $\alpha = \beta = \gamma = 0$, (23.8) and (23.9) reduce to $dg(\tau;\zeta) = i\zeta r \cdot d\tau$ and $dh(\tau;\zeta) = -\frac{1}{2}\left(i\zeta + \zeta^2\right) \cdot d\tau$, with solutions $g(\tau;\zeta) = i\zeta r\tau$ and $h(\tau;\zeta) =$

$-\frac{1}{2}\left(i\zeta+\zeta^2\right)\tau$. With $v_t=\sigma^2$ expression (23.7) then becomes

$$\hat{\phi}\left(\zeta;s_t,\sigma^2,\tau\right)=\exp\left[i\zeta r\tau-\frac{1}{2}\left(i\zeta+\zeta^2\right)\sigma^2\tau+i\zeta s_t\right]$$

$$=\exp\left[i\zeta(r-\frac{\sigma^2}{2})\tau-\zeta^2\frac{\sigma^2\tau}{2}+i\zeta s_t\right].$$

23.2 When $\gamma=0$ but α and β are positive,

(a) Expression(23.8) becomes $dh\left(\tau;\zeta\right)/d\tau=A-\beta h\left(\tau;\zeta\right)$, where $A\equiv-\left(i\zeta+\zeta^2\right)/2$. The solution subject to $h\left(0;\zeta\right)=0$ is $h\left(\tau;\zeta\right)=A\left(1-e^{-\beta\tau}\right)/\beta$. Then (23.7) is

$$\frac{dg\left(\tau;\zeta\right)}{d\tau}=i\zeta r+\alpha h(\tau;\zeta)$$

$$=\left(i\zeta r+\frac{\alpha A}{\beta}\right)-\frac{\alpha A}{\beta}e^{-\beta\tau}.$$

Integrating and imposing $g\left(0;\zeta\right)=0$ give

$$g(\tau;\zeta)=\int_0^\tau\left(i\zeta r+\alpha\frac{A}{\beta}-\frac{\alpha A}{\beta}e^{-\beta s}\right)\cdot ds$$

$$=\left(i\zeta r+\frac{\alpha A}{\beta}\right)\tau-\frac{\alpha A}{\beta^2}\left(1-e^{-\beta\tau}\right).$$

(b) When $\gamma=0$ (23.4) becomes $dv_t=(\alpha-\beta v_t)\cdot dt$. Given that $\beta>0$, the solution subject to $v_t|_{t=0}=v_0$ is

$$v_t=\frac{\alpha}{\beta}\left(1-e^{-\beta t}\right)+v_0e^{-\beta t}=\frac{\alpha}{\beta}+\left(v_0-\frac{\alpha}{\beta}\right)e^{-\beta t}.$$

Thus, as t increases the influence of v_0 diminishes and $v_t\to\alpha/\beta$.

23.3 From the solution in Exercise 23.2(b) we find the average squared volatility from t to $T\equiv t+\tau$ to be

$$\bar{\sigma}^2_{[t,T]}=\tau^{-1}\int_t^T v_u\cdot du$$

$$=\tau^{-1}\int_t^T\left[\frac{\alpha}{\beta}+\left(v_0-\frac{\alpha}{\beta}\right)e^{-\beta u}\right]\cdot du$$

$$=\frac{\alpha}{\beta}+\left(v_0-\frac{\alpha}{\beta}\right)\frac{e^{-\beta t}-e^{-\beta T}}{\beta\tau}$$

$$=\frac{\alpha}{\beta}+\left(v_0-\frac{\alpha}{\beta}\right)e^{-\beta t}\frac{1-e^{-\beta\tau}}{\beta\tau}$$

$$=\frac{\alpha}{\beta}+\left(v_t-\frac{\alpha}{\beta}\right)\frac{1-e^{-\beta\tau}}{\beta\tau}.$$

The solutions for $g(\tau;\zeta)$ and $h(\tau;\zeta)$ in 23.2(a) give

$$\ln \hat{\phi}(\zeta; s_t, v_t, \tau) = \left(i\zeta r + \frac{\alpha A}{\beta}\right)\tau - \frac{\alpha A}{\beta^2}\left(1 - e^{-\beta\tau}\right)$$
$$+ \frac{A}{\beta}\left(1 - e^{-\beta\tau}\right)v_t + i\zeta s_t$$
$$= i\zeta(s_t + r\tau) + A\tau\left[\frac{\alpha}{\beta} + \left(v_t - \frac{\alpha}{\beta}\right)\frac{1 - e^{-\beta\tau}}{\beta\tau}\right]$$
$$= i\zeta(s_t + r\tau) + A\bar{\sigma}_{[t,T]}^2\tau$$
$$= i\zeta\left(s_t + r\tau - \frac{\bar{\sigma}_{[t,T]}^2\tau}{2}\right) - \zeta^2\frac{\bar{\sigma}_{[t,T]}^2\tau}{2}.$$

Thus, with deterministic volatility the conditional distribution of $\ln S_T$ under $\hat{\mathbb{P}}$ is $N\left(\ln S_t + r\tau - \bar{\sigma}_{[t,T]}^2\tau/2, \bar{\sigma}_{[t,T]}^2\tau\right)$. In particular, when $t = 0$, we have $\ln S_T - \ln S_0 \sim \left(r - \bar{\sigma}_{[0,T]}^2/2\right)T + \bar{\sigma}_{[0,T]}\sqrt{T}Z$ as per (23.3).

References

1. Allais, M. (1953) "Le comportement de l'homme rationnel devant le risque," *Econometrica* 503–546.

2. Bachelier, L. (1900) "Theory of speculation," in Cootner (1964).

3. Barberis, N.; Huang, M.; and Santos, T. (2001) "Prospect theory and asset prices," *Quart. J. Econ.* 1–53.

4. Bates, D. S. (1996) "Jumps and stochastic volatility: Exchange rate processes implicit in Deutschemark options," *Rev. Finan. Stud.* 69–108.

5. Billingsley, P. (1995) *Probability and Measure,* 3d ed., Wiley: New York.

6. Black, F. (1972) "Capital market equilibrium with restricted borrowing," *J. Business* 444–454.

7. Black, F. and Scholes, M. (1973) "The pricing of options and corporate liabilities," *J. Polit. Econ.* 637–654.

8. Blattberg, R. C. and Gonedes, N. (1974) "A comparison of the stable and Student distributions as statistical models for stock prices," *J. Business* 244–280.

9. Brigo, D. and Mercurio, F. (2001) *Interest Rate Models: Theory and Practice*, Springer: Berlin.

10. Campbell, J. Y. (2003) "Two puzzles of asset pricing and their implications for investors," *Am. Economist* 48–74.

11. Campbell, J. Y. and Cochrane, J. (1999) "By force of habit: A consumption-based explanation of aggregate stock-market behavior," *J. Polit. Econ.* 205–251.

12. Campbell, J. Y.; Lo, A. W.; and MacKinlay, A. C. (1997) *The Econometrics of Financial Markets*, Princeton Univ. Press: Princeton, N.J.

13. Carr, P. and Wu, L. (2003) "The finite moment log stable process and option pricing," *J. Finance* 753–777.

14. Christie, A. A. (1982) "The stochastic behavior of common stock variance: Value, leverage, and interest rate effects," *J. Finan. Econ.* 407–432.

15. Cochrane, J. H. (2001) *Asset Pricing*, Princeton Univ. Press: Princeton, N.J.

16. Cootner, P. H. (1964) *The Random Character of Stock Market Prices*, MIT Press: Cambridge, MA.

17. Cox, J.; Ingersoll, J.; and Ross, S. (1985) "A theory of the term structure of interest rates," *Econometrica* 385–408.

18. Cox, J. and Ross, S. (1976) "The valuation of options for alternative stochastic processes," *J. Finan. Econ.* 145–166.

19. Duffie, D.; Pan, J.; and K. Singleton (2000) "Transform analysis and asset pricing for affine jump–diffusions," *Econometrica* 1343–1376.

20. Eberlein, E. and Keller, U. (1995) "Hyperbolic distributions in finance," *Bernoulli* 281–299.

21. Eberlein, E.; Keller, U.; and K. Prause (1998) "New insights into smile, mispricing, and value at risk: The hyperbolic model," *J. Business* 371–407.

22. Edwards, C. (2005) "Derivative pricing models with regime switching: A general approach," *J. Derivatives* 41–47.

23. Ellsberg, D. (1961) "Risk, ambiguity, and the Savage Axioms," *Quart. J. Econ.* 643–669.

24. Epps, T. W. (1979) "Comovements in stock prices in the very short run," *J. Am. Stat. Assoc.* 291–298.

25. Epps, T. W. (1996) "Stock prices as branching processes," *Commun. Stat.–Stochastic Models* 529–558.

26. Epps, T. W. (2007) *Pricing Derivative Securities,* 2d ed., World Scientific: Singapore.

27. Eraker, B.; Johannes, M.; and N. Polson (2003) "The impact of jumps in volatility and returns," *J. Finance* 1269–1300.

28. Eraker, B. (2004) "Do stock prices and volatility jump? Reconciling evidence from spot and option prices," *J. Finance* 1367–1403.

29. Fama, E. (1976) *Foundations of Finance,* Basic Books: New York.

30. Fama, E. and MacBeth, J. D. (1973) "Risk, return, and equilibrium: Empirical tests," *J.Polit. Econ.* 607–636.

31. Fama, E. and French, K. (1997) "Size and book-to-market factors in earnings and returns," *J. Finance* 131–155.

32. Fama, E. and French, K. (2004) "The CAPM: Theory and evidence," *J. Econ. Perspect.* (3), 25–46.

33. Fama, E.; Fisher, L.; Jensen, M.; and Roll, R. (1969) "The adjustment of stock prices to new information," *Internatl. Econ. Rev.* 1–21.

34. Fang, H. (2002) *Jumps with a Stochastic Jump Rate,* PhD dissertation, Univ. Virginia.

35. Feynman, R. (1948) "Space-time approach to nonrelativistic quantum mechanics," *Rev. Modern Phys.* 367–387.

36. Friedman, M. (1953) *Essays in Positive Economics*, Univ. Chicago Press: Chicago, IL.

37. George, T. and Hwang, C. (2004) "The 52-week high and momentum investing," *J. Finance* 2145–2176.

38. Grether, D. M. and Plot, C. R. (1979) "Economic theory of choice and the preference reversal phenomenon," *Am. Econ. Rev.* 623–638.

39. Hadar, J. and Russell, W. (1969) "Rules for ordering uncertain prospects," *Am. Econ. Rev.* 25–34.

40. Hadar, J. and Russell, W. (1971) "Stochastic dominance and diversification," *J. Econ. Theory* 288–305.

41. Hanoch, G. and Levy, H. (1969) "The efficiency analysis of choices involving risk," *Rev. Econ. Stud.* 335–346.

42. Hansen, L. and Jagannathan, R. (1991) "Implications of security market data for models of dynamic economies," *J. Polit. Econ.* 225–262.

43. Harris, C. and Laibson, D. (2001) "Dynamic choices of hyperbolic consumers," *Econometrica* 935–957.

44. Harris, T. E. (1989) *The Theory of Branching Processes*, Dover: Mineola, NY.

45. Heston, S. L. (1993) "A closed-form solution for options with stochastic volatility with applications to bond and currency options," *Rev. Finan. Stud.* 327–343.

46. Hobson, D. G. and Rogers, L. C. G. (1998) "Complete models with stochastic volatility with applications to bond and currency options," *Math. Finance* 27–48.

47. Ingersoll, J. E. (1987) *Theory of Financial Decision Making.* Rowman-Littlefield: Totowa, NJ.

48. Jegadeesh, N. and Titman, S. (1993) "Returns to buying winners and selling losers," *J. Finance* 65–91.

49. Joshi, M. (2003) *The Concepts and Practice of Mathematical Finance*, Cambridge Press: Cambridge, UK.

50. Kac, M. (1949) "On distributions of certain Wiener functionals," *Trans. Am. Math. Soc.* 1–13.

51. Keynes, J. M. (1935) *The General Theory of Employment, Interest, and Money*, Harcourt, Brace & World: New York.

52. Kuhn, T. S. (1970) *The Structure of Scientific Revolutions*, 2d ed., Univ. Chicago Press: Chicago, IL.

53. Kahneman, D. and Tversky, A. (1979) "Prospect theory: An analysis of decision under risk," *Econometrica* 61–72.

54. Kimball, M. S. (1990) "Precautionary saving in the small and in the large," *Econometrica* 53–73.

55. Laibson, D. (1997) "Golden eggs and hyperbolic discounting," *Quart. J. Econ.* 443–477.

56. Leroy, S. and Porter, R. (1981) "The present value relation: Tests based on variance bounds," *Econometrica* 555–577.

57. Lichtenstein, S. and Slovic, P. (1973) "Response-induced reversals of preference in gambling: An extended replication in Las Vegas," *J. Exper. Psychol.* 16–20.

58. Lintner, J. (1965) "The valuation of risk assets and the selection of risky investments in stock portfolios and capital budgets," *Rev. Econ. Stat.* 13–37.

59. Lucas, R. (1978) "Asset prices in an exchange economy," *Econometrica* 1426–1446.

60. Madan, D. B. and Seneta, E. (1990) "The variance gamma (v.g.) model for share market returns," *J. Business* 511–524.

61. Malkiel, B. (2007) *A Random Walk Down Wall Street*, 9th ed., W. W. Norton: New York.

62. Mandelbrot, B. (1963) "The variation of certain speculative prices," *J. Business* 394–419.

63. Mandelbrot, B. and Hudson, R. (2004) *The (Mis)Behavior of Markets,* Basic Books: New York.

64. Markowitz, H. M. (1952) "Portfolio selection," *J. Finance* 77–91.

65. Markowitz, H. M. (1959) *Portfolio Selection,* Wiley: New York.

66. Merton, R. C. (1973) "The theory of rational option pricing," *Bell J. Econ. Manage. Sci.* 141–183.

67. Merton, R. C. (1976) "Option pricing when underlying stock returns are discontinuous," *J. Finan. Econ.* 125–144.

68. Musiela, M. and Rutkowski, M. (2005) *Martingale Methods in Financial Modeling*, 2d ed., Springer: Berlin.

69. Ohlson, J. A. and Penman, S. H. (1985) "Volatility increases subsequent to stock splits: An empirical aberration," *J. Finan. Econ.* 251–266.

70. Perold, A. F. (2004) "The capital asset pricing model," *J. Econ. Perspect.* (3), 3–24.

71. Praetz, P. D. (1972) "The distribution of share price changes," *J. Business* 49–55.

72. Pratt, J. (1964) "Risk aversion in the small and in the large," *Econometrica* 122–136.

73. Press, J. (1967) "A compound events model for security prices," *J. Business* 317–335.

74. Pulley, L. B. (1981) "A general mean–variance approximation to expected utility for short holding periods," *J. Finan. Quant. Anal.* 361–373.

75. Quirk, J. P. and Saposnik, R. (1962) "Admissibility and measurable utility functions," *Rev. Econ. Stud.* 140–146.

76. Rabin, M. (2000) "Risk aversion and expected-utility theory: A calibration theorem," *Econometrica* 1281–1292.

77. Rabin, M. and Thaler, R. M. (2001) "Anomalies: Risk aversion," *J. Econ. Perspect.* (1), 219–232.

78. Roll, R. (1977) "A critique of the asset pricing theory's tests," *J. Finan. Econ.* 129–176.

79. Ross, S. (1976) "The arbitrage pricing theory of capital asset pricing," *J. Econ. Theory* 341–360.

80. Rubinstein, M. (1985) "Nonparametric tests of alternative option pricing models...," *J. Finance* 455–480.

81. Samuelson, P. (1965) "Proof that properly anticipated prices fluctuate randomly," *Indust. Manage. Rev.* 41–49.

82. Samuelson, P. (1970) "The fundamental approximation theorem of portfolio analysis in terms of means, variances, and higher moments," *Rev. Econ. Stud.* 537–542.

83. Savage, L. J. (1954) *The Foundations of Statistics*, Wiley: New York.

84. Sharpe, W. F. (1963) "A simplified model for portfolio analysis," *Manage. Sci.* 277–293.

85. Sharpe, W. F. (1964) "Capital asset prices: A theory of market equilibrium under conditions of risk," *J. Finance* 425–442.

86. Shiller, R. (1981) "Do stock prices move too much to be justified by subsequent changes in dividends?," *Am. Econ. Rev.* 421–436.

87. Shiller, R. (1982) "Consumption, asset markets, and macroeconomic fluctuations," *Carnegie-Rochester Conf. Public Policy* 203–238.

88. Shiller, R. (2003) "From efficient markets theory to behavioral finance," *J. Econ. Perspect.* (1), 83–104.

89. Shreve, S. (2004) *Stochastic Calculus for Finance II*, Springer: New York.

90. Vasicek, O. (1977) "An equilibrium characterization of the term structure," *J. Finan. Econ.* 177–188.

91. Verchenko, O. (2008) *Empirical Tests of Option Pricing Models*, PhD dissertation, Univ. Virginia.

92. von Neumann, J. and Morgenstern, O. (1944) *Theory of Games and Economic Behavior*, Princeton Univ. Press: Princeton, NJ.

INDEX